AF740831

Mae'r eitem yma ar gael i chi hyd y dyddiad neu'r amser a nodir isod, oni bai fod darllenydd arall yn ei alw'n ôl ynghynt.

This item is on loan to you until the date or time shown below, unless it is recalled sooner by another borrower.

08 11 2003

06 12 2003

~~18/12/03~~

24/10/04

04 11 2004

5. 12

~~23/3/06~~

26/5/06

BREEDING FOR DISEASE RESISTANCE IN FARM ANIMALS, 2ND EDITION

Breeding for Disease Resistance in Farm Animals, 2nd Edition

Edited by

R.F.E. Axford

School of Agricultural and Forest Sciences
University of Wales, Bangor
UK

S.C. Bishop

Roslin Institute (Edinburgh)
Roslin
UK

F.W. Nicholas

Department of Animal Science
University of Sydney, Sydney
Australia

J.B. Owen

School of Agricultural and Forest Sciences
University of Wales, Bangor
UK

CABI *Publishing*

U.W.B LIBRARY
P.C.B LLYFRGELL

UWB Deiniol Library
SF105 .B696 1999
30110005954061

CABI *Publishing* is a division of CAB *International*

CABI Publishing
CAB International
Wallingford
Oxon OX10 8DE
UK

Tel: +44 (0)1491 832111
Fax: +44 (0)1491 833508
Email: cabi@cabi.org

CABI Publishing
10 E 40th Street
Suite 3203
New York, NY 10016
USA

Tel: +1 212 481 7018
Fax: +1 212 686 7993
Email: cabi-nao@cabi.org

© CAB *International* 2000. All rights reserved. No part of this publication may be reproduced in any form or by any means, electronically, mechanically, by photocopying, recording or otherwise, without the prior permission of the copyright owners.

A catalogue record for this book is available from the British Library, London, UK.

Library of Congress Cataloging-in-Publication Data
Breeding for disease resistance in farm animals / edited by R.F.E.
Axford ... [et al.]. -- 2nd ed.
p. cm.
Includes bibliographical references.
ISBN 0-85199-325-7 (alk. paper)
1. Livestock--Breeding. 2. Livestock--Genetics. 3. Veterinary immunogenetics. I. Axford, R. F. E.
SF105.B696 1999
636.089'6079--dc21 99-27340
CIP

ISBN 0 85199 325 7

Typeset by Solidus (Bristol) Limited.
Printed and bound in the UK by Biddles Ltd, Guildford and King's Lynn.

Contents

Contributors

E. Authié, *Programme Santé Animale, CIRAD-EMVT, BP5035, 34032 Montpellier Cedex 1, France*
R.F.E. Axford, *School of Agricultural and Forest Sciences, University of Wales, Bangor, Gwynedd LL57 2UW, UK*
S.C. Bishop, *Roslin Institute (Edinburgh), Roslin, Midlothian EH25 9PS, UK*
R.A. Bowen, *Department of Physiology, Colorado State University, Fort Collins, CO 80523, USA*
J.O. Carlson, *Department of Microbiology, Colorado State University, Fort Collins, CO 80523, USA*
A.M. Crawford, *AgResearch Molecular Biology Unit and Centre for Gene Research, University of Otago, PO Box 56, Dunedin, New Zealand*
C.J. Davies, *Department of Microbiology and Immunology, College of Veterinary Medicine, Cornell University, Ithaca, NY 14853, USA*
J.J. de Castro, *FAO Sub Regional Office, PO Box 3730, Harare, Zimbabwe*
A. de la Concha-Bermejillo, *Department of Pathobiology, Texas A&M University, Texas Agricultural Experiment Station, San Angelo, TX 76901, USA*
J.C. DeMartini, *Department of Pathology, Colorado State University, Fort Collins, CO 80523, USA*
O. Distl, *Department of Animal Breeding and Genetics, School of Veterinary Medicine Hannover, PO Box 711180, 30545 Hannover, Germany*
K.G. Dodds, *AgResearch Invermay Agricultural Centre, Private Bag 50034, Mosgiel, New Zealand*
M.J. Doenhoff, *School of Biological Sciences, University of Wales, Bangor, Gwynedd LL57 2UW, UK*
I. Edfors-Lilja, *Department of Technology and Natural Sciences, University of Växjö, S-351 95, Växjö, Sweden*
L.C. Gasbarre, *USDA-ARS, LPSI, Immunology and Disease Resistance Laboratory, Beltsville Agricultural Research Center, Beltsville, MD 20705, USA*
G. Gettinby, *Department of Statistics and Modelling Science, University of*

Strathclyde, Livingstone Tower, 26 Richmond Street, Glasgow G1 1XH, UK
N. Hunter, *Institute for Animal Health, Neuropathogenesis Unit, West Mains Road, Edinburgh EH9 3JF, UK*
G.D.M. d'Ieteren, *International Livestock Research Institute (ILRI), PO Box 30709, Nairobi, Kenya*
S.J. Lamont, *Department of Animal Science, Iowa State University, Ames, IA 50011, USA*
E. Luiting, *Roslin Institute (Edinburgh), Roslin, Midlothian EH25 9PS, UK*
J.C. McEwan, *AgResearch Invermay Agricultural Centre, Private Bag 50034, Mosgiel, New Zealand*
B.A. Mallard, *Department of Pathobiology, University of Guelph, Guelph, Ontario, Canada N1G 2W1*
J.E. Miller, *Department of Epidemiology and Community Health, School of Veterinary Medicine and Department of Animal Science, Louisiana Agricultural Experiment Station, Louisiana State University, Baton Rouge, LA 70803, USA*
B. Minjauw, *International Livestock Research Institute (ILRI), PO Box 30709, Nairobi, Kenya*
C.A. Morris, *Ruakura Agricultural Research Centre, Private Bag 3123, Hamilton, New Zealand*
M. Murray, *Department of Veterinary Clinical Studies, University of Glasgow Veterinary School, Bearsden, Glasgow G61 1QH, UK*
F.W. Nicholas, *Department of Animal Science, University of Sydney, Sydney, NSW 2006, Australia*
J.B. Owen, *School of Agricultural and Forest Sciences, University of Wales, Bangor, Gwynedd LL57 2UW, UK*
H.W. Raadsma, *Centre for Sheep Research and Extension, The University of Sydney, PMB 3, Camden, New South Wales, Australia*
M.F. Rothschild, *Department of Animal Science, Iowa State University, Ames, IA 50011, USA*
K.A. Schat, *Department of Microbiology and Immunology, College of Veterinary Medicine, Cornell University, Ithaca, NY 14853, USA*
L. Skow, *Department of Veterinary Anatomy and Public Health, Texas A&M University, College Station, TX 77843, USA*
B.H. Thorp, *Ross Breeders Ltd, Newbridge, Midlothian EH28 8SZ, UK*
D. Wakelin, *School of Biological Sciences, University of Nottingham, University Park, Nottingham NG7 2RD, UK*
P. Wallgren, *National Veterinary Institute, S-750 07, Uppsala, Sweden*
B.N. Wilkie, *Department of Pathobiology, University of Guelph, Guelph, Ontario, Canada N1G 2W1*
N. Wissocq, *International Livestock Research Institute (ILRI), PO Box 30709, Nairobi, Kenya*

Introductory Editorial: Breeding for Disease Resistance in its Evolutionary Context

Present-day species of farm livestock have inherited a complex genome from their wild progenitors. Despite the bewildering proliferation of phenotypic variation in breeds within species, genetic structure studies indicate that genetic variation is much less than this would indicate and that genetic difference between extant breeds and their wild relatives is rather small (Rogdakis *et al.*, 1995).

This conserved genome from pre-domestication evolution owes much to the co-evolution of livestock host and its parasitic pathogens. This strategy (Khibnik and Kondrashov, 1997) is a continual battle to achieve an ecological symbiotic equilibrium, optimal for the joint survival of the two species.

Breeding for disease resistance, the subject of this book, has much to do with trying to lessen the impact of ecological perturbations involved in modern pharmaceutical intervention. It is relevant to note that the fossil record suggests that environmental perturbations of various magnitude, such as the major event that may have resulted in the demise of the dinosaurs 65 million years ago, had significant consequences for the evolution of species. Many species have become extinct and others have been newly created by branching speciation. Sometimes explosive multiplication of hitherto rare species occurred, as in the Cambrian explosion (Morris, 1998).

At domestication, another major perturbation in a species' evolution, many animal progenitors would be expected to be near an optimum equilibrium with their pathogens under the prevailing environmental conditions (including stocking rate). In spite of many changes in the environment and in selection objectives following domestication, the wild animal can still be regarded as an informative template for modern livestock, particularly in the fitness and behaviour traits, including disease resistance.

After Domestication

Fossil and other archaeological evidence, particularly inferences from the changes in skeletal features, suggest that domestication was initially a traumatic event for livestock species (Zeuner, 1963). Deterioration due to overstocking, resulting in poorer nutrition and greater parasite infection, probably accounts for the decreased skeletal size and changed conformation observed after domestication.

The new domesticates would have been subjected suddenly to novel conditions of severity and type of pathogen challenge, at a time when poorer nutrition would have reduced their acquired immunity.

However, domestication into nomadic/transhumant systems would have ameliorated some of the changes in conditions for newly domesticated pastoral species, particularly since the change in flock/herd movements might well have been minimally different from the seasonal movements of animals in the wild (Zeuner, 1963). As time went on there was the possibility, in the richer, more fertile areas, for domesticated farm livestock to enjoy improving conditions and optimal stocking rates. Indeed, some of the livestock kept by the present-day Bedouin nomads of the Middle East are amongst the best cared for animals anywhere in the world. Another instance of optimal care for stock seems to have been that under the Norfolk four-course rotation, with its sustainable fertility and its special provision of winter fodder as swedes, etc. This period of the late 19th and early 20th century in the UK, might well have been the golden age for domesticated livestock in terms of welfare and disease control.

Latter Half of the 20th Century

The advent of artificial nitrogen fertilizer in the 1930s and the subsequent development of intensive mechanized agriculture, caused a sea change in animal husbandry from the disease viewpoint, in developed Western countries. Highly intensive pig and poultry units developed, followed by an incipient trend towards a heavy stocking rate of land and buildings for the hitherto extensively managed pastoral ruminant species.

This development coincided with the discovery of penicillin, other antibiotics and the array of veterinary drugs which appeared to be the simple final answer, albeit expensive, to the disease problems of such intensive agriculture.

Evolutionary biology clearly indicated the inevitable outcome of such a strategy – the increase in frequency of genes for resistance, which is seen as the emergence of resistant strains of pathogens and parasites.

Figure I.1 illustrates the potential power of intense selection to target rare parasite genotypes, resistant to antiparasitic drugs. This is a simulation of a population of 5000 with a basal polygenic mean value 10 and SD 2.5 with 25 individuals carrying a single gene of values of 0.5, 1, 2 or 3 times the basal SD. It indicates that at high selection intensity and major gene value 3 × basal SD, the resistance gene becomes highly concentrated in the selected group.

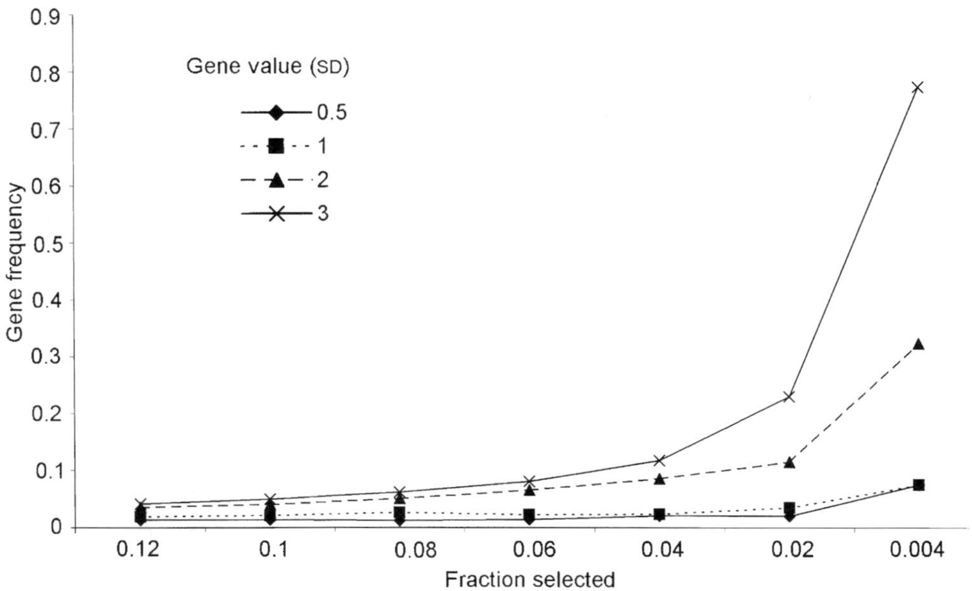

Fig. I.1. Effect of gene value (in SDs) and selection fraction on gene frequency.

A stochastic simulation of responses to selection in a multifactorial trait, such as disease resistance, was conducted by Owen *et al.* (1998), to examine the effect of perturbation, akin to the macro events before domestication, on pathogen ecology. Results from 15 generations of selection show that, under intense selection, nearly 100% of the total genetic response at some stages of selection can be accounted for by the increased frequency of a relevant major gene at a single locus, as opposed to the basal polygenic component which accounts for most of the response at more lax selection pressure (Fig. I.2).

It is interesting to see that the indications from this simulation are borne out by a study of the genetic control of domestication effects in the common bean (Koinange *et al.*, 1996).

Thus 'manufacturers' recommended dose rates for antiparasitic drugs, while clearly efficacious in killing the parasites in the short term, seem bound to accelerate the development of parasite resistance in the longer term, through the intense selection pressure exerted.

The initial forms of the resistant pathogens would be expected to be less potent than the 'wild' forms because of the selection emphasis on drug resistance. However, Bjorkman *et al.* (1998) have demonstrated that potency is quickly recovered and there are several instances of the subsequent emergence of super-pathogens that are more lethal and difficult to control than their predecessors.

Epistatic interaction is likely to set up a network of interactive changes in the genome arising from the emergence of a single gene for resistance. One of the resulting wonders of the genome is the wealth of dormant major genes (whether recessives or genes inhibited by the epistatic action of restrictor alleles at other loci) that lie in wait for an environmental change that fosters

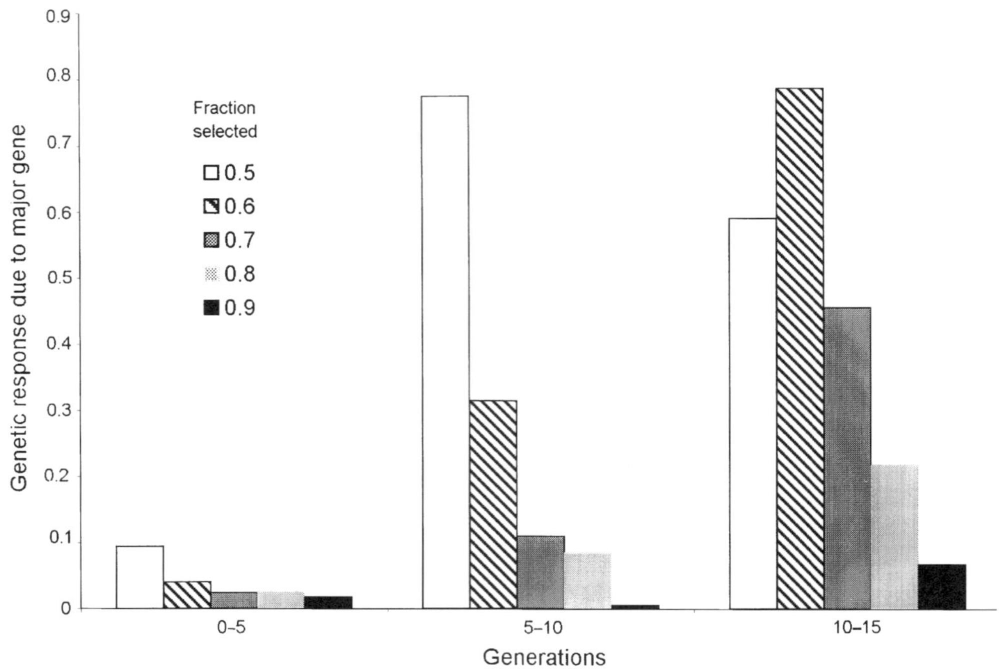

Fig. I.2. Effect of fraction selected on the proportion of total genetic response due to major gene frequency changes in generations 0–5, 5–10, 10–15 with basal $h^2 = 0.4$.

their overt emergence. A similar biological phenomenon (albeit in a different context) has been seen in the 'emergence' of major genes affecting production traits in farm animals, where selection for particular traits has resulted in high frequencies of particular alleles at the halothane locus in pigs, the double-muscling locus in cattle and the callipyge locus in sheep.

Future Strategy

Emphasis on breeding for disease resistance stems partly from awareness of the development of pathogen resistance to the therapeutic agent, which has led to calls for reduction in the use of antibiotics and other drugs throughout human and animal medicine. This emphasis is doubly important because of the possibility of between-species transfer of such resistance, particularly from livestock pathogens to human pathogens. However, in the rich democracies this is easier said than done. Medical GPs are struggling with the insistent right of mothers for their children to be given antibiotics even for mild viral colds. The large multinational pharmaceutical companies have evolved to meet just such a consumer demand, which may unwittingly turn out to have an uncontrollable suicidal will of its own.

Various options for future action need urgent consideration.

1. Therapeutic intervention. In view of the primacy of evolutionary short-termism over the predictions and avowed planning intentions of individuals and organizations hitherto, it may be considered prudent not to dismiss the possibility that the free rein of scientific ingenuity and commercial pressures may come up with similar novel solutions that abound in nature as a result of the similar simple genetic algorithm of evolution.

One such approach to therapy, which has had considerable attention in the field of pest control, especially in crop plants, is that of biological control. The difficulty of predicting its consequences in the complex ecological context of its projected use has led to unforseen difficulties in the past and considerably proscribes its usefulness in many applications. While, as yet, it may not have direct relevance to breeding for disease resistance, a recent development in this field, which holds out considerable promise to combat the difficulties alluded to with antibiotic use, is the use of bacteriophages. Barrow *et al.* (1998) have indicated the possibility of such an approach for the therapy of *Escherichia coli* septicaemia and meningitis in chickens and calves. There are also reports of its successful application to supplant antibiotics in human medicine (Morris, 1998). A merit of this approach is that it is strictly targeted to its quarry bacterium, rather than catholic in its appetite, as are the antibiotics. It also seems at present to be more difficult for bacteria to resist bacteriophages by their existing mechanisms.

2. Avoidance or reduction in the need for therapy. In animal disease control a moderate alternative is possible. This involves reduction of challenge by using lower stocking rates, thus allowing less frequent strategic drug intervention to control parasitic pathogens. In its extreme this could approximate the 'organic' approach.

3. Any of these and other viable future strategies may be replaced or considerably aided by various facets of breeding for disease resistance in our farm animals, as discussed in this book. This encompasses selection pressure within conventional breeding schemes on general immunological competence as well as on targeted resistance tests for specific major pathogens. This approach may result in a polygenic response as well as an increased frequency of favourable major genes.

In this respect an exciting potential exists for marker-assisted selection (MAS) acting on relevant candidate functional genes, or on DNA markers closely linked to quantitative trait loci (QTL). The immense investment in human genome mapping is likely to accelerate progress in the farm livestock mapping effort over the next few decades. The denser the maps available, the more likely it is that useful practical developments can occur at farm level, similar to that of the control of Marek's disease in poultry.

To what extent these approaches can reset the co-evolutionary oscillation between pests and pathogens at a lower level and enable the human population to be adequately fed on traditional plant and animal organisms remains to be seen.

This book examines these issues and the current state of play, in order to stimulate further thinking and cooperation in trying to solve one of mankind's difficult challenges. Since the first edition (1991) there has been a fruitful decade of research in breeding for disease resistance, and the present book aims to encapture and systemize these advances as a convenience to a widening spectrum of workers and students with interest in the subject. Some of the chapters remain with virtually the same titles and mainly update the reader on current developments. Other subjects, because of changes in emphasis over the decade, have been reclassified and do not have identical counterparts in the previous book. Still other chapter areas have been omitted because other fields have advanced relatively faster over the decade and therefore have been given priority.

References

Barrow, P., Lovell, M. and Berchieri, A. (1998) Use of lytic bacteriophage for control of experimental *Escherichia coli* septicaemia and meningitis in chickens and calves. *Clinical and Diagnostic Laboratory Immunology* 5, 294–298.

Bjorkman, J., Hughes, D. and Andersson, D.I. (1998) Virulence of antibiotic-resistant *Salmonella typhimurium*. *Proceedings of the National Academy of Science USA* 95, 3949–3953.

Khibnik, A.I. and Kondrashov, A.S. (1997) Three mechanisms of Red Queen dynamics. *Proceedings of The Royal Society of London Series B – Biological Sciences* 264 (1384), 1049–1056.

Koinange, E.M.K., Singh, S.P. and Gepts, P. (1996) Genetic control of the domestication syndrome in common bean. *Crop Science* 36, 1037–1045.

Morris, M. (1998) The virus that cures. *Biologist* 45(5), 196 (letter).

Owen, J.B., Ap Dewi, I. and Roberts, D. (1998) The effect of intensity of selection on the response to selection in the presence of a major gene and an underlying additive polygenic component. *Proceedings of the 6th World Congress on Genetics Applied to Livestock Production. Armidale, Australia. January 1998* 26, 21–24.

Morris, S.C. (1998) The evolution of diversity in ancient ecosystems: a review. *Philosophical Transactions of The Royal Society of London Series B – Biological Sciences* 353 (1366), 327–345.

Rogdakis, E., Kutsuli, P., Surdis, I. and Panopulu, E. (1995) Genetic-structure of Greek sheep breeds. *Journal of Animal Breeding and Genetics – Zeitschrift fur Tierzuchtung und Zuchtungsbiologie* 112(4), 255–266.

Zeuner, F.G. (1963) *A History of Domesticated Animals*. Hutchinson, London.

I Principles and Methods

1

DNA Markers, Genetic Maps and the Identification of QTL: General Principles

A.M. Crawford[1], K.G. Dodds[2] and J.C. McEwan[2]
[1]*AgResearch Molecular Biology Unit and Centre for Gene Research, University of Otago, Dunedin, New Zealand;*
[2]*AgResearch Invermay Agricultural Centre, Mosgiel, New Zealand*

Summary

Efficient quantitative trait loci (QTL) detection and identification methods in farm animals rely on the availability of numerous, cheap mapped polymorphic DNA markers, which preferably can also be linked comparatively to the better-mapped human and mouse genomes. There is still some way to go before this situation can be achieved in most farmed species, but sufficient markers exist to commence genomic scanning experiments for disease traits. We envision that QTL detection and identification will use a wide variety of techniques, each appropriate to its specific situation. Where possible, preliminary screens based on complex segregation analysis will identify segregating QTL, which will then be identified to a specific genomic region using some form of appropriate genomic screening. This will be followed by independent validation and finer mapping using a combination of techniques. Two logical end-points then exist, identification of the actual locus or marker-assisted selection (Meuwissen and Goddard, 1997). The latter is probably more cost effective in the near future and is particularly suited for difficult-to-measure traits such as disease resistance.

Introduction

Primary linkage maps covering all chromosomes are now available for all the major domestic livestock species and most research today is not concerned with map generation *per se* but the use of markers on these maps to search for regions of the genome containing loci affecting traits of economic importance. This process has become known as a search for quantitative trait loci (QTL) or, more recently, a genome scan. This chapter reviews the methodologies that led to the development of genetic linkage maps and genome scans. First, various commonly used DNA marker technologies are described, with

© CAB *International* 2000. *Breeding for Disease Resistance in Farm Animals*
(eds R.F.E. Axford, S.C. Bishop, F.W. Nicholas and J.B. Owen)

particular emphasis placed on describing their relative strengths and weaknesses. Then DNA mapping techniques are examined, followed by discussion on QTL identification and localization methods relevant to the study of disease traits in farmed livestock.

DNA Markers

From the time of Mendel until the 1980s the only single-locus genetic markers available to geneticists were either simple phenotype markers, such as eye colour in *Drosophila*, or protein polymorphisms, such as the haemoglobin blood groups. Using these markers, some quite detailed genetic linkage maps were developed for model species such as mice and *Drosophila*, but there were great limitations for linkage map construction in outbred livestock species. The advent of recombinant DNA technologies, in particular the polymerase chain reaction (PCR) technique, suddenly removed the obstacle of marker availability so that linkage mapping projects for any species could be planned and implemented. This section details the numerous DNA marker types that have been developed over the past decade and describes their relative advantages and disadvantages for QTL searches, comparative linkage mapping and the measurement of genetic diversity.

Multilocus markers

Minisatellite or VNTR (variable number tandem repeat) markers

The minisatellites were the first tandem repeat markers with multiple alleles to be developed. Discovered by Jeffreys *et al.* (1985), they were the first markers to be sufficiently informative to reveal a unique genotype in each individual. With the advent of microsatellites they have fallen out of favour for linkage studies because they are technically demanding and require a large quantity of DNA for analysis. They were mostly dominant markers with only one identifiable allele. Minisatellites were also located mainly in telomeric regions so that good coverage of the genome using these markers could not be achieved.

A few highly informative single-loci minisatellites were identified in livestock (Georges *et al.*, 1990) and these have been found useful. The first marker linked to the callipyge gene of sheep was a single-locus minisatellite (Cockett *et al.*, 1994).

RAPD (random amplified polymorphic DNA fragments) markers

These were one of the first PCR-based markers to be used (Williams *et al.*, 1990). Small primers (8–10 bp) were used to amplify a mixture of random fragments of DNA from the genome. The size of the primers was set such that about 20 bands were amplified by each PCR reaction. Some of these bands would be polymorphic and could be used as genetic markers. These markers have the great advantage of being very easily generated and requiring only small quantities of DNA. For this reason many linkage maps, especially in plants, have used RAPD markers. Since heterozygous and homozygous individuals cannot be differentiated, the markers are dominant. The appearance (and disappearance) of bands is very sensitive to slight changes in PCR conditions, so that RAPD markers are not easily reproducible. The most serious disadvantage, however, is that a new map must be generated for each new pedigree being examined, as there is no locus specificity in the primers used. The bands generated by a particular primer in one pedigree may not bear any relation to bands generated by the same primer in a second pedigree.

AFLP (amplified fragment length polymorphisms) markers

These are now the multilocus markers of choice (Vos *et al.*, 1995). Genomic DNA is cut with restriction endonucleases and linkers are ligated to each end of the restriction fragments. Selective PCR primers are then used to amplify subsets of fragments from the mixture of genomic restriction fragments. The selective primers encompass the linkers added on to the end of the restriction fragments and include additional bases to the 3′ end of the primer to give additional specificity. The amplified fragments are then separated according to size. Those bands that are present in some individuals but absent in others can be used as genetic markers.

These markers have the same advantage as RAPD, being easily generated, but they are much less reliant on PCR conditions staying exactly the same in order to obtain reproducible amplification products. To obtain a new set of markers, small changes in the 3′ bases of the amplifying primers are all that is required, so this technology can provide a limitless supply of new markers. The markers have the disadvantages of being dominant, requiring the generation of a new map for each new set of pedigrees being studied.

Single-locus markers

Restriction fragment length polymorphisms (RFLPs)

These, the first DNA markers developed, predate the development of PCR methods. The RFLP method utilizes restriction endonuclease digestion of genomic DNA, its separation by size using agarose gel electrophoresis, and detection and analysis of the DNA sequence by Southern blotting. The RFLP marker detects the presence or absence of a restriction site. A very full description is found in Sambrook *et al.* (1989).

RFLP markers are codominant. The example shown in Fig. 1.1 is the simplest case. Here a large fragment of DNA (allele A) has a point mutation in a restriction site such that it will sometimes be cut into two pieces by the enzyme *Eco*RI. Other times it will stay intact. In this particular case, only one of the two fragments (allele B) produced by restricting the A allele can be detected. The other fragment will probably have run off the gel. The enzymes most likely to give an RFLP marker in ruminants are *Msp*I and *Taq*I (Montgomery *et al.*, 1995). The two major disadvantages of these markers are that they are technically demanding and require large quantities of DNA (approximately 5 µg per lane, compared with 1 ng for any PCR-based test).

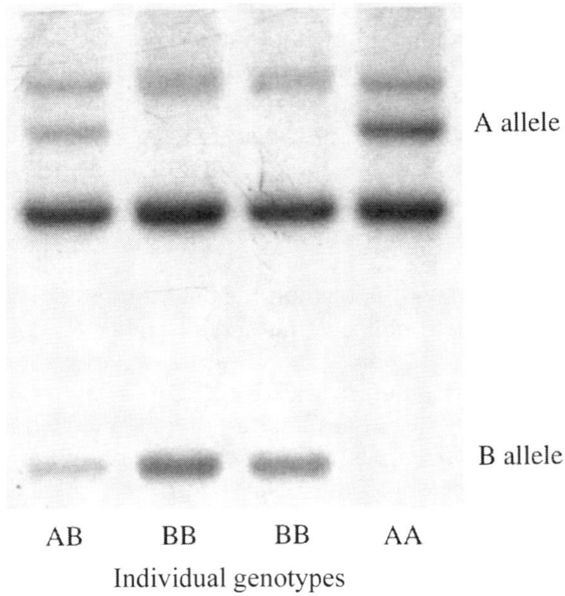

Fig. 1.1. The result of an RFLP analysis of sheep genomic DNA from four animals, digested with *Eco*RI. (Kindly supplied by Dr S.H. Phua.)

Microsatellites

Microsatellites are the markers responsible for the recent expansion in genetic linkage maps of farm animals (see p. 13 and Table 1.1). Like the minisatellites, they are multiallelic tandem repeats. However, they are single-locus, codominant, spread throughout the genome, require only small amounts of template DNA and are relatively easy to find and characterize.

At the heart of any microsatellite is a simple sequence, either a mono-, di-, tri- or tetranucleotide, that is repeated between 10 and 50 times. Virtually all of the microsatellites that have been found for livestock have the sequence (AC/GT) as the repeat unit. The reason for this is not that the other types do not exist but that this type is the most abundant within the livestock genomes and hence is much easier to find and characterize. The variation between alleles of the microsatellites is due to variation in the number of simple sequence repeats. Microsatellites are typed by designing primers to the unique DNA sequences on either side of the repeat and, using PCR, amplifying the region containing the repeat. The size of the PCR product is then measured, usually by electrophoresis on a DNA sequencing gel. Alleles differ in size according to the difference in number of simple-sequence repeats (Fig. 1.2).

Single nucleotide polymorphisms (SNPs)

The remaining variation found within the genome of animals occurs as changes in single nucleotides. The RFLP, the first DNA marker developed for eukaryotes, detects the variations that occur within a restriction endonuclease

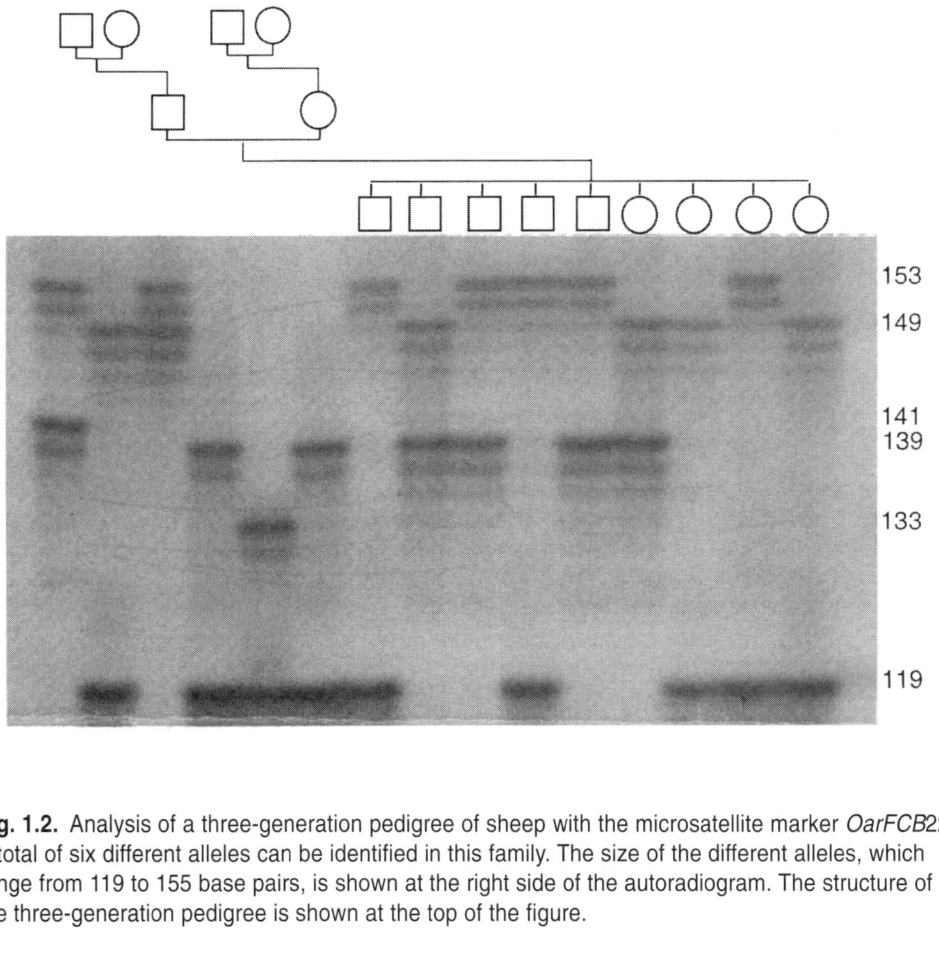

Fig. 1.2. Analysis of a three-generation pedigree of sheep with the microsatellite marker *OarFCB226*. A total of six different alleles can be identified in this family. The size of the different alleles, which range from 119 to 155 base pairs, is shown at the right side of the autoradiogram. The structure of the three-generation pedigree is shown at the top of the figure.

cut site. This, however, is only a small subset of all the polymorphic nucleotides in the genome. Very few of the nucleotide changes have any deleterious effect, so the preferred name is single nucleotide polymorphisms (SNPs) rather than point mutations. It has been estimated that an SNP occurs about once every kilobase of unique sequence in humans (Cooper *et al.*, 1985). In other mammals, such as livestock, the figure is likely to be similar, so at least two to three million SNPs remain to be identified and characterized in any livestock species. SNPs represent by far the richest source of genetic variation available for research purposes.

With the advent of PCR any unique region of DNA could be amplified and scanned for variation in its DNA sequence. These SNPs can be detected in numerous ways. Once detected and characterized a wide variety of systems is now available for typing SNPs.

If SNPs occur at sufficiently high density, such as the hypervariable exon 2 region of *DRB* and *DQB* genes, direct sequencing of the DNA can be a cost-effective method of typing. More often, however, SNPs are hundreds of base pairs from their nearest neighbour, which means that other methods of detection are employed.

A base change in a short (approximately 100 bp) amplified fragment of DNA will often change the secondary structure of the DNA when it is melted and allowed to reanneal as single strands (Orita *et al.*, 1989). The differences in secondary structure can be detected by electrophoresis on non-denaturing polyacrylamide gels, where one structure travels at a different speed through the gel. Ion gradients, gel density gradients and strictly controlled gel temperature have all been found useful in resolving these differences in secondary structure. The different methods of band resolution have led to many different acronyms, but the most commonly used name for this methodology is single-stranded conformational polymorphism (SSCP).

If the DNA sequence around the SNP is known, a number of typing methods become available. The most simple is that of Zhu and Clarke (1996). Four primers are designed. Two primers amplify the region containing the point mutation. Two further primers are designed such that their 3′ ends correspond to one of the bases at the SNP. Depending on which primer was present in the reaction and which mutation(s) were in the template DNA, a smaller band would or would not be amplified. In this way, two PCR reactions can be used to type the SNP.

The invention of the 'Taqman' system of typing (Holland *et al.*, 1991) allows typing of SNP markers without needing an electrophoresis step. Instead, the different alleles are detected by a change in the colour of the PCR reaction. Using a fluorogenic probe complementary for the target DNA sequences being amplified, the system detects and quantifies cycle by cycle increases in the level of PCR products. The probe consists of an oligonucleotide with a reporter and quencher dye attached. Uncoupling of the two dyes occurs when the probe, bound to the internal sequence of the PCR product, is cleaved by the nucleolytic activity of the *Taq* polymerase, which results in an increase in the fluorescence intensity of the reporter dye.

For SNP analysis, competition between oligonucleotides differing only at the point mutation is used. The two allele-specific oligonucleotides, which differ only by one base-pair change, are labelled with different dyes. Binding of the 'correct' oligonucleotide is detected by increased fluorescence dye used to label the 'correct' oligonucleotide. With a heterozygote, a mixture of the two dyes is released and detected. This method should reduce both the cost and time required to type SNPs.

The recent development of 'DNA chips' containing high-density arrays of DNA (Chee *et al.*, 1996) may speed up the analysis process even further. Perhaps now is the time to begin searching for more SNPs in livestock, so that a large battery of SNPs is available when these new technologies become available.

Genetic Mapping

Genetic maps are sets of loci arranged in order and separated by distances (in units depending on the type of map), with each set corresponding to a chromosome pair. Locating genes of interest by scanning the genome in a linkage study is best conducted using a set of markers evenly spaced throughout the genome. These may be chosen from a marker map. Once linkage has been established, these and other types of maps help refine the location. Maps of genes or expressed sequences may provide clues to an effect's cause. Maps are also useful in selecting markers for other uses, such as population studies, tracking the inheritance of DNA segments using flanking markers, and parentage identification.

Linkage maps

This section discusses the construction of linkage maps. Ott's book (1991) on linkage analysis in humans is largely applicable to linkage studies in farmed animals, although there are some aspects particular to farmed animals.

Genetic linkage

Under Mendelian laws of segregation, loci on different chromosomes segregate independently. Loci on the same chromosome may show evidence of co-segregation, i.e. alleles passed to progeny are often of the same grandparental origin. The proportion of times that alleles are not of the same grandparental origin is the recombination fraction (rf). This statistical measure of the distance between a pair of loci ranges from 0 for tightly linked loci to 0.5 for loci that are far apart or on different chromosomes. The knowledge of which alleles of a parent came from which grandparent is termed the 'phase'.

Populations useful for developing linkage maps

A mapping population needs to provide information on the transmission of alleles from grandparents to progeny. Grandparental genotypes are not necessary for families of at least two in size, as some phase information can be inferred. The construction of linkage maps can be made more efficient by using appropriately designed populations. Optimal designs for livestock maps have been considered by Hetzel (1991), van der Beek and van Arendonk (1993) and Elsen *et al.* (1994). For a fixed total size, increasing family size will decrease the proportion of parents that need to be genotyped. There is a risk, particularly with a single family, that a locus may not be heterozygous (and so potentially informative for mapping) but could have been useful for applications of the map. However, the increased efficiency and precision of map distances more than compensates for the loss of such information, which can also be offset by the development of new markers. Full-sib families are useful for obtaining linkage information for both male and female meioses, further decreasing the proportion of parents genotyped per meiosis, and allowing mapping of the homogametic sex chromosome. Although genotype information on grandparents results in less reliance on inferring phase, additional progeny also helps establish phase.

In some farmed animal species, such as sheep and cattle, large paternal half-sib pedigrees are common. These can be used for mapping studies, rather than expensive, large, full-sib families. In these cases, however, there is no information on female meioses, including X-linkage. There is usually little information to be gained from genotyping the dams (Dodds *et al.*, 1993; Elsen *et al.*, 1994).

The efficiency of mapping is improved by the use of highly polymorphic markers. Another strategy to increase the informativeness of families is to use breed or subspecies crosses, since the most prevalent alleles may be different in different breeds. Breed crosses are therefore more likely to be heterozygous. Using a cross with grandparents of at least three different breeds (but with each parent being a breed cross) would maximize the chance of informative meioses under this scenario. This strategy has been particularly successful in deer and chickens.

The construction of rudimentary maps does not require large populations. Linkage at a true rf of 0.2 can be established with 90% power when there are 60–70 coinformative meioses in a few families. Using highly polymorphic markers results in up to 50% of meioses being coinformative. In practice a few half-sib families with 150–200 total progeny should be adequate. In this size of population there is a 90% chance of observing a recombination between two loci if the true rf between them is 0.05. Ordering loci on a finer scale will therefore require larger mapping populations.

Linkage between pairs of markers

When the phase is known, the rf can be established simply as the proportion of recombinants to total informative meioses. When phase is not known (as is usually the case), likelihood methods are used to estimate the rf. Geneticists have traditionally presented such results as a lod score, the $\log_{10}$ of the ratio of the likelihood maximized with respect to the estimate of rf, to that with the rf set to 0.5. A general algorithm for calculating likelihoods on pedigrees was published by Elston and Stewart (1971) and has since been extended to more complicated pedigree types. This algorithm forms the basis of linkage programs for general pedigrees such as LINKAGE (Lathrop *et al.*, 1984) and MENDEL (Lange *et al.*, 1988), and also for ANIMAP (Nielsen *et al.*, 1995) which is designed for half-sib pedigrees. The algorithm involves calculating the likelihood for every possible combination of genotypes. The number of combinations gets large quickly if there are many individuals with unknown genotypes. CRI-MAP (Lander and Green, 1987) uses a simpler algorithm in that it fills in unknown genotypes only where they can be deduced. MAPMAKER (Lander *et al.*, 1987) has been designed to handle crosses between inbred lines rather than general pedigrees. For some pedigree structures it may be possible to code the data in such a way that an analysis similar to that given by CRI-MAP can be obtained.

If a founder has a missing genotype, then the Elston–Stewart algorithm uses allele frequencies to assign probabilities to the various possible genotypes for that individual. This requires suitable estimates of the allele frequencies relevant to the ungenotyped founders. Allele frequencies can be estimated from pedigree data, or simultaneously estimated during linkage analysis (Boehnke, 1991; Dodds *et al.*, 1993). For half-sib families, estimates based on counts of unambiguous alleles passed by the dams can give misleading results, but maximum likelihood estimates can be obtained by ignoring the major genotype in equations (3) and (4) of Dodds *et al.* (1993). The usefulness of incorporating information from allele frequencies decreases as the locus increases in polymorphism, but the computing requirements increase.

The rf may vary according to specific factors, such as the sex of the parent in which a meiosis is recorded, or the family in the experiment. Allowing a single value gives estimates of the sex-averaged rf (which will depend on the number of informative male and female meioses in the data). In some cases it is possible to allow for separate male and female rates by including both parameters in the likelihood model. For human autosomes, the female map is about 90% longer than the male map, but this relationship does not appear to be universal, with an estimated increase of 30% in pigs (Archibald *et al.*, 1995) and less than 5% in cattle (Barendse *et al.*, 1997; Kappes *et al.*, 1997). A special case of allowing sex-specific rates is that of sex-linked loci.

Since mapping involves comparing a particular locus with many others, a stringent threshold must be set. A commonly used threshold, developed in the context of human linkage but suitable for other species with similar genome sizes, is a lod score of 3 (Ott, 1991). This ensures a less than 5% chance of a marker showing significant linkage to a marker on a different chromosome. If an analysis of autosomal loci uses sex-specific recombination rates, the threshold should be increased by 0.3 to accommodate the extra degree of freedom.

Assigning loci to linkage groups

Assignment to linkage groups uses the linkage results for each pair of markers (two-point analyses) using a strategy such as the following:

1. Placing the first locus into a linkage group.
2. Adding any loci to the group that are linked to the first locus.
3. Adding any other loci to the group that are linked to any other locus in that group.
4. Taking the set of unassigned loci and repeating 1 to 3 until no unassigned loci remain.

Usually a more stringent threshold (e.g. lod = 4 and rf < 0.3) than that used to declare two-point linkage is used to prevent false linkage erroneously joining linkage groups from different chromosomes. Providing such errors are not too common, they will usually become evident when trying to order the loci.

Ordering sets of loci

For a group of m loci, a comparison of likelihoods will indicate which order is the most likely, and how well supported that order is. Many linkage packages can calculate likelihoods, maximized over interloci distances, for a given order, by extending the two-point methodology. CRI-MAP is often chosen for such analyses when highly polymorphic markers are used, because it has lower computing requirements and disregards only a small amount of the information in the data.

The number of possible orders, $m!/2$, increases rapidly with m, so that a comparison of all possible orders is generally not feasible. Algorithms have been devised to construct maps without evaluating all possible orders. Although these are not guaranteed to find the most likely order, they tend to work well in practice.

One such algorithm starts with a pair of loci (ideally highly informative and with rf ~ 0.2), and then tries adding one locus at a time in such a way that the best supported order for the placement of the new locus is at least 1000 times more likely (support for the order of lod 3) than any other. The criterion for support for the order may need to be relaxed after initially adding as many loci as possible, to allow all other loci to be added to the map.

Another strategy is to obtain an approximate order using a simple method, and then comparing likelihoods of this order with perturbed orders (e.g. by transposing the order of each pair of adjacent loci, or by looking at all possible orders with permutations of each set of *n* adjacent loci). Any order which has a higher likelihood than the current working order becomes the new working order. Several methods are available for obtaining an initial order – minimize the sum of adjacent rfs; maximize the sum of adjacent lod scores; and multidimensional scaling methods which collapse a set of distances in $m - 1$ dimensions to a set in one dimension.

Over short regions it is reasonable to assume that multiple recombinants do not occur. For dense maps, loci may be ordered by requiring no close multiple recombinants. If two alleles passed by an individual come from different grandparents, there has been a recombination event; all those from one grandparent are at one end of the group, those from the other grandparent are at the other end. Continuing with each meiosis will allow these groups to be subdivided further. If there has been recombination in each interval, all the loci can be ordered by this method.

Estimating distances between loci

Distance estimates arise from maximizing the likelihood with respect to the distances between the loci. An increased complexity from the two-point case is the need to consider how distances between adjacent pairs of loci combine to give a distance between the outer pair of loci. This is done with the aid of mapping functions which transform rfs to a scale that is additive. Linkage packages commonly use either the Haldane or Kosambi map functions. The Haldane map function assumes that there is no interference (i.e. recombination occurs independently in adjacent intervals). The Kosambi map function models interference as high over short distances, but decreasing with distance. The unit for genetic distance, expressed by a map function, is a Morgan; distances are often specified in centimorgans (cM) giving the distances multiplied by 100.

Data checking

Genotyping and other errors may be present in linkage datasets. Such errors can have a large effect on the results (e.g. a 1% error between tightly linked loci will increase the estimated distance between them by ~1 cM). These errors often reveal themselves through inconsistencies or unlikely events in maps, such as: (i) apparent pedigree inconsistency; (ii) unusual segregation ratios; (iii) multiple recombination over short distances; (iv) family differences in recombination rates; and (v) inflation of overall map length by insertion of a new locus. Mapping projects need to make these checks to reduce map errors.

Presentation of maps

Linkage maps are generally presented as scaled linear structures with positions of loci denoted. In some cases, different fonts are used to denote different types of markers. Sometimes a comprehensive map (including all loci) is given. In other cases a framework map, in which orders are supported by a lod > 3, is given, with the positions of the additional loci denoted alongside, either as a best estimate, or as the intervals that attain a threshold likelihood relative to the best position estimate (e.g. lod 3 support).

Maps of farm animals

Some recent maps of farm animals are listed in Table 1.1. The pig and cattle maps are estimated to cover 99% of the genome, while those for sheep and chicken cover 90–95% of the genome. These estimates provide only a rough guide of the coverage, since estimates of total genome length are not very reliable, and estimated map lengths tend to be biased upward by genotyping errors. These genetic linkage maps consist primarily of microsatellite markers with a smaller number of RFLPs based on known genes. Microsatellite-based maps of man and mice each contain between 5000 and 10,000 microsatellites. The densest livestock map is that for cattle with 1300 microsatellites, but most current livestock species have at least 200 markers. Maps for pig, chicken, deer, sheep, and cattle can be found at http://www.ri.bbsrc.ac.uk./genome_mapping.html.

Other mapping methods

Other methods can assign genetic markers to chromosomes or chromosomal locations. These generally do not require the target region to be polymorphic, as is required in linkage mapping.

Somatic cell hybrids

Hybrid cell lines containing chromosomes or chromosomal fragments of the species of interest fused to the genome of another species (usually hamster) can be probed with a gene of interest. Selection of these hybrids can ensure a panel which will enable the gene to be located to a particular chromosome, by observation of which member(s) of the panel it binds to.

In situ hybridization

A gene of interest can be radioactively or fluorescently labelled and allowed to hybridize to metaphase chromosome spreads. The gene can then be assigned to its chromosome, and its position relative to the chromosome banding.

Table 1.1. Summary information on genetic maps of farmed animal species.

Species	Reference	No. loci[a]	Spacing[b] (cm)	No. progeny	No. full-sib families	No. Half-sib families	URL
Cattle	Barendse *et al.* (1997)	703	5.5	328	29	15	//spinal.tag.csiro.au/
Cattle	Kappes *et al.* (1997)	1250	2.5	180	22	4	//sol.marc.usda.gov/genome/cattle/cattle.html
Pig	Rohrer *et al.* (1996)	1042	2.2	94	8	2	//sol.marc.usda.gov/genome/swine/swine.html
Pig	Archibald *et al.* (1995)	239	8.3	118	13		//www.ri.bbsrc.ac.uk/pigmap/
Sheep	De Gortari *et al.* (1998)	519	6.4	345	53	8	//zaphod.agresearch.cri.nz/
Chicken	Cheng (1997)	589	4.9	400	18	2	//poultry.mph.msu.edu/
Deer	Tate (1998)	197	7.8	352		7	//www.ri.bbsrc.ac.uk/cgi-bin/arkdb/browsers/ browser.sh?species=deer
Goat	Vaiman *et al.* (1996)	223	11.9			12	//locus.jouy.inra.fr

[a] Number of loci in linkage groups.
[b] Average distance between markers.

Artificial chromosomes

Genomic DNA can be cut and fragments cloned into a vector such as yeast (to create yeast artificial chromosomes, YACs) or bacteria (to create BACs). A fragment containing a particular genetic marker can be isolated, and this can then be used to isolate fragments that overlap with it. Continuing this process allows contiguous segments to be isolated. These can be used for a variety of purposes, such as developing new markers near an existing marker or searching for gene-coding sequences near a particular marker.

Radiation hybrids

DNA can be broken into segments by exposing it to radiation. These segments can then be probed with a pair of markers. The proportion of times a fragment contains both markers is then a measure of the distance between the two markers. This proportion also depends on the strength of the radiation, and may make it difficult to compare measures across studies, but has the advantage that different radiation strengths can be applied to obtain different mapping resolutions.

Comparative mapping

Genetic mapping of a species rarely advances in isolation of maps being developed in other species. There is much conservation of DNA sequences (more so for coding than non-coding sequences) and relative locus position in related species. Some knowledge of the relationships between the genetic maps of different species can aid in the development of their maps.

Comparative mapping data can be obtained in two ways. The first is the comparison of a set of common loci across two or more species. A panel of 410 loci to help in this task has been developed (Lyons *et al.*, 1997). The second is by chromosomal painting – where DNA derived from a chromosome from one species is fluorescently labelled and allowed to hybridize to a chromosome spread of the other species. This identifies the regions corresponding to the target chromosome, providing information at a coarser but more complete level.

Comparative mapping can be used to isolate a gene contributing to phenotypic differences. For example, a comparison of corresponding regions of the human chromosome to that in pigs which contained the locus responsible for the porcine stress syndrome has led to the isolation of the gene responsible (Fujii *et al.*, 1991).

Information from a well-mapped species can direct the mapping effort in a related species. For example, the development of the sheep map has relied heavily on using its relationship to the cattle map to target areas of poor marker density (De Gortari *et al.*, 1998).

Considerable effort may be required to develop polymorphic markers for a locus to be comparatively mapped. One alternative is to use an intermediary species in which it is easier to develop such markers. For example, comparative mapping between ruminants and other species may be most efficient by using deer, in which there is an interspecies hybrid mapping population, to establish the relationships with other species (Tate *et al.*, 1995). The relationships within ruminants can then proceed with less well conserved, and usually more polymorphic, regions of DNA.

QTL Detection

There are many methods to detect QTL affecting multigenic traits such as disease resistance, and these methods have been comprehensively reviewed recently (Lynch and Walsh, 1997). All methods except segregation analysis rely on linkage disequilibrium between the genetic markers described in the first section and the loci affecting the disease trait. While this basic theoretical principle has been known for many years, the explosion of new genetic markers has resulted in rapid development of data analysis techniques to detect QTL. These allow a much wider range of pedigree structures to be utilized. The actual method chosen is determined by several factors, including available knowledge about host factors influencing disease severity, available financial and animal resources, and comparative information of QTL affecting the trait from other species. Often there is no clear distinction between many of the methods and, in practice, several techniques will be required if the ultimate objective is to identify the actual mutation and its effects. The following gives a brief summary of approaches currently being used in farmed animals and describes several recent developments.

Phenotype measurement

A crucial factor that is often overlooked in any study to identify QTL is the measurement of the disease trait. By their very nature, disease traits are extremely difficult to measure. At the extreme, disease measurement is a binary measurement (affected/not affected), with this classification only crudely reflecting an underlying distribution of susceptibility. Additional complications arise because responses are often affected by the level of challenge, health status of the host and previous exposure to the disease. These problems are also faced by quantitative genetics studies and fortunately, in many cases these researchers have already developed suitable challenge protocols and measurement techniques.

The disease-resistance measurement should ideally be on a continuous scale and highly correlated with actual field resistance. An example of this is the γ-glutamyl transferase (GGT) levels used to measure facial eczema resistance in sheep (Towers and Stratton, 1978) and faecal egg counts and parasite resistance in sheep (Woolaston *et al.*, 1990). Experimental measurement error should be reduced where possible and additional traits thought to be strongly correlated with the resistance trait should be measured. For example, Crawford *et al.* (1997a) reported a QTL experiment investigating host resistance to internal parasites where they measured strongyle faecal egg counts in three separate samples at the end of each challenge in order to reduce measurement error. Often the cost of genotyping animals is so high that it is cheaper to make multiple phenotypic measurements and so reduce the animal numbers and genotyping required.

Measurement of correlated traits has the benefit that in some cases a QTL may be detected with the correlated trait, but fails to reach significance with the trait of interest. An example of this situation is a study by Comuzzie *et al.* (1997) where a QTL affecting leptin levels, part of an obesity control feedback loop, was localized to chromosome 2. In contrast, the primary interest of the study, fat content, while showing the same region to be important, failed to achieve significance. A trait may be affected by several biological control pathways. Multiple measurements of key parameters of these control mechanisms can give an insight into the specific control mechanism for a disease QTL. In the case of disease resistance mediated by the host immune system, measurement of cytokines responsible for Th1 or Th2 immune responses are obvious candidates.

Segregation analysis

Segregation analyses, designed to detect genetic segregation in the trait of interest in the absence of genotype marker information, have developed rapidly in recent years. Based originally on mixture models and complex segregation analysis (Elston and Stewart, 1971), their current invocation is mixed model inheritance programs FINDGENE (Kinghorn *et al.*, 1993) and Maggic (Janns *et al.*, 1995). The advantage of these new computer algorithms is they can handle large and arbitrary pedigree structures commonly encountered in animal breeding situations. They can screen existing data sets rapidly to identify if large QTL are segregating, their likely mode of inheritance, size of effect and frequency. Their detection limit is a QTL contributing approximately 8% of the phenotypic variation, which is not as powerful as the linkage disequilibrium techniques discussed later. There are two potential problems for many disease traits: multigenerational pedigrees of disease susceptibility are often not available, and disease-trait measurements are often grossly non-normal. These methods are sensitive to non-normality and it is often not clear whether the observed non-normality of the disease phenotype is a consequence of segregation or due to extraneous factors.

Despite the problems described above, segregation analyses have already been used to good effect in disease-resistance studies to identify a major gene for tick resistance in cattle (Kerr *et al.*, 1994) and host resistance to internal nematodes in sheep (McEwan and Kerr, 1998). Similar programs have also been used in humans to identify major genes segregating for host resistance to leprosy (Abel and Demenais, 1988), *Schistosoma mansoni* (Abel *et al.*, 1991) and malaria (Abel *et al.*, 1992).

Genome scan

A genome scan consists of genotyping animals, segregating for a QTL affecting the trait of interest, with sufficient markers to detect the QTL. In practice, most experiments limit the search to the autosomes. For outbred species, such as farmed livestock, the magnitude of the resources required for such experiments, often exceeding US$500,000, has prompted a large number of publications investigating their optimal design and analysis. Traditionally, genome scans have consisted of rather simple mating designs, in part a result of the difficulties of analysing arbitrary pedigree structures. Currently, three common designs are utilized: outcross, backcross and F_2. Backcross and outcross designs are common in sheep and beef cattle experiments because they have fewer resource constraints, but they achieve this at a small expense in power. A variation on the outcross design is the granddaughter design (Weller *et al.*, 1990), where the granddaughters of a large half-sib male progeny group are phenotyped but only the sires and their parents are genotyped. Obviously this structure mimics that of existing dairy evaluation schemes and this is its primary objective.

Typically, for the outcross or backcross design divergent breeds or selection lines for the trait of interest are crossed, and the resulting progeny are mated to a large number of unrelated animals or back to one or both of the parental lines. Generally, the aim is to produce as many progeny per sire as possible. In practice, a minimum of 150 or more progeny are required if the QTL experiment is to have even moderate power. Several half-sib families are generated in order to account for the possibilities that the parental lines also may not be fixed at the QTL and individual sires may not be informative at the markers for a particular region of interest.

The progeny are challenged with the disease and susceptibility is measured. At a minimum, only the sires and the progeny are genotyped, but commonly the grandparents are also genotyped. Dams are usually not genotyped, except for the F_2 design, for reasons identical to those discussed in the mapping section. Optimum marker spacing for the initial scan has been investigated by several authors and depends to a small extent on the informativeness of the markers used and their availability at specific chromosome locations. For highly polymorphic, evenly spaced markers, a 20–30 cM spacing is appropriate (Davarsi and Soller, 1994a). Where markers are less polymorphic (e.g. SNPs) or animal numbers are constrained, then a closer spacing may be warranted. An alternative is to use the same method, but in another species more amenable to study. A genomic region conferring tolerance to trypanosomiasis has been identified in mice (Kemp *et al.*, 1996) which, it is hoped, will be useful in finding similar regions in cattle.

Extreme tails

The large cost of genotyping, currently close to US$3 per microsatellite genotype, means that many methods have been examined to reduce the number required while still obtaining most of the information. One such method is genotyping only the extreme phenotypic tails of the progeny. This technique is particularly suitable for disease traits, because only one trait is of direct interest in the population under study. The exact proportion of progeny to genotype depends on the relative costs of genotyping and phenotyping individuals, but in general somewhere between 15 and 25% of progeny in each tail is a good compromise (Lander and Botsein, 1989; Darvasi and Soller, 1992). This method was used by Crawford *et al.* (1997a,b) and Phua *et al.* (1998) to examine disease traits in farmed livestock. An important implication of this technique is that the magnitude of detected QTL will be overestimated using standard analytical methods.

DNA pooling

For initial screening, another technique to reduce genotyping costs is to pool the DNA samples from the extreme tails. The sire allele frequencies in the pools are then estimated by densitometry after adjustment using the sire allele densities. Approximate likelihood of the allele frequency differences can be calculated for the particular experiment, and thresholds can be set. This results in a further ten- to 100-fold reduction in the number of genotypes required, albeit at the additional expense of the pool creation and measurement. At a minimum it would only require markers spaced at approximately 20 cM intervals by three samples per sire, or a total of 450 genotypes per half-sib family for a genome length of 3000 cM. In practice, however, some replication is desirable. DNA pooling could enable a genomic scan involving four or five sires to be completed by one person in 1 or 2 months. The power of this technique was described theoretically by Darvarsi and Soller (1994b) and Taylor and Phillips (1996) have used the technique to identify QTL in inbred mouse lines using microsatellite markers.

Data analysis

In recent years there has been an explosion of methods proposed to analyse the data from the experiments described above. The techniques can be divided broadly into two classes, single-marker and multiple-marker methods. In practice, multiple-marker methods have more power than single-marker techniques and all the multiple-marker methods are similar in their detection power, so often other factors will determine the techniques actually used (Bovenhuis *et al.*, 1997; Lynch and Walsh, 1997).

The simplest single-marker technique for outbred species with half-sib outcross or backcross designs is to use regression, nesting the inherited sire allele within the sire. The advantage is that the method can use standard statistical software and is incremental in nature, encouraging its use after each new marker has been completed.

For the designs described previously, two multiple-marker analytical methods are commonly used. The first involves maximum likelihood techniques, and the program ANIMAP (Nielsen *et al.*, 1995) is often utilized for farmed livestock. The second involves the use of regression on the conditional probabilities of the parental phase inherited. This technique was developed independently by Haley and Knott (1992) and Martinez and Curnow (1992).

No matter what technique is used, an extremely important factor in these analyses is the calculation of appropriate error thresholds due to the large number of comparisons being made. Lander and Kruglyak (1995) provide a series of graded criteria ranging from suggestive linkage, through to confirmed linkage. The calculation of the appropriate thresholds for a particular experiment can also be undertaken independently using permutation tests where the analysis is replicated many times accompanied by random shuffling of the original data (Churchill and Doerge, 1994).

Fine mapping and candidate genes

If a QTL has been detected by a method described above, usually the locus is defined only to a 10–30 cM region. Alternative approaches are required to fine map the QTL efficiently. Often the most difficult facet of this technique is to find appropriate polymorphic markers in the region of interest in the putatively segregating sires. Various methods have been used to detect QTL once these markers have been developed.

Candidate-gene techniques differ from genomic scans in that no attempt is made to exclude chromosomal regions using anonymous markers; rather, specific regions are examined using prior knowledge about the disease. Previously, this has largely involved knowledge that a particular host gene was involved in the disease process. However, in the future it will be more common to involve a gene that has been shown previously to be important in other species, or to be located in an important region identified by a prior genomic scan. This exemplifies the key feature of the candidate-gene approach: it is a 'fine mapping' technique and it cannot be used to infer that no QTL exist where no significant results are obtained.

Multiple generations

Where several intermediate generations of inter-se matings have passed before progeny are evaluated, the linkage disequilibrium between the markers and the linked QTL reduces exponentially in proportion to the map distance separating them. This relationship can be exploited to 'fine map' any QTL identified in an initial scan (Keightley *et al.*, 1996).

This technique is dependent on the animals remaining capable of breeding after evaluation. This will not be possible in many cases involving disease resistance, unless gametes are preserved prior to evaluation. It will also involve excessive time and resources for the majority of farmed livestock if no existing populations are available. One alternative, which uses markedly fewer resources and only one additional generation, is to identify male progeny of a sire segregating for the QTL in which the paternally inherited chromosome has recombined within this interval. If sufficient markers with known map locations are available, animals whose paternal chromosome have recombined at equi-spaced intervals can be selected and progeny tested using an appropriate sized half-sib family. This should be able to restrict the QTL location to a 2–5 cM region, which is sufficiently small to make physical mapping of the region a practical possibility.

Association

The most commonly used technique is to examine the association between the candidate gene marker allele and the animal's phenotype. Using these techniques, Gulland *et al.* (1993) reported an association between host nematode faecal egg count and the ADA locus, located on chromosome 13 of sheep. In a separate study, Stear *et al.* (1996) found a significant association between host nematode faecal egg count and alleles in the major histocompatibility complex (MHC) region located on chromosome 20. Great care has to be used in these analyses to remove spurious associations, particularly sire effects. Obviously, the designs described for genome scans are ideal for these studies and Phua *et al.* (1998) have reported results from five candidate loci for facial eczema resistance using such a resource.

Divergent selection lines

For many farmed livestock, quantitative genetics studies have already produced divergent selection lines from common foundation animals. These flocks are analogous to the multiple-generation fine-mapping resource described previously. As such they are ideal for testing candidate genes, particularly when a region has already been identified from a genome scan experiment using the same experimental population. The basic technique is to measure the selection line allele frequencies of the polymorphism associated with the candidate gene. Appropriate statistical tests are required which take account of founder effects, subsequent genetic drift and sampling effects. Fortunately, accurate pedigree records are also available for many of these flocks, and this allows calculation of the exact probability of the allele frequency difference using simulation (Dodds and McEwan, 1997).

Haplotype analysis

Once one or several of the above methods has restricted the chromosomal location to a small region, say less than 1 cM or even to within a known gene, alternative methods need to be applied. One method is to develop several markers in the region and consider the various haplotypes (the relevant combination of alleles at all these loci) present in the animals evaluated. Essentially this technique relies on ancestral linkage disequilibrium caused by the QTL mutation occurring on a specific background haplotype. The drawback is that there can be a very large number of haplotypes with correspondingly small numbers in each subgroup. Templeton *et al.* (1987) suggest that this problem can be overcome by incorporating information of the inferred evolutionary relationships between the sampled haplotypes and then performing nested ANOVAs. Obviously, this method could also be combined with the divergent selection line method detailed in the previous section.

References

Abel, L. and Demenais, F. (1988) Detection of major genes for susceptibility to leprosy and its subtypes in a Caribbean Island: Desirade Island. *American Journal of Human Genetics* 42, 256–266.

Abel, L., Demenais, F., Prata, A., Souza, A.E. and Dessein, A. (1991) Evidence for the segregation of a major gene in human suceptibility/resistance to infection by *Schistosoma mansoni*. *American Journal of Human Genetics* 48, 959–970.

Abel, L., Cot, M., Mulder, L., Carnevale, P. and Feingold, J. (1992) Segregation analysis detects a major gene controlling blood infection levels in human malaria. *American Journal of Human Genetics* 50, 1308–1317.

Archibald, A.L., Haley, C.S., Brown, J.F., Couperwhite, S., McQueen, H.A., Nicholson, D., Coppieters, W., Van de Weghe, A., Stratil, A., Wintero, A.K., Fredholm, M., Larsen, N.J., Nielsen, V.H., Milan, D., Woloszyn, N., Robic, A., Dalens, M., Riquet, J., Gellin, J., Caritez, J.C., Burgaud, G., Ollivier, L., Bidanel, J.P., Vaiman, M., Renard, C., Geldermann, H., Davoli, R., Ruyter, D., Verstege, E.J.M., Groenen, M.A.M., Davies, W., Høyheim, B., Keiserud, A., Andersson, L., Ellegren, H., Johansson, M., Marklund, L., Miller, J.R., Anderson Dear, D.V., Signer, E., Jeffreys, A.J., Moran, C., Le Tissier, P., Muladno, Rothschild, M.F., Tuggle, C.K., Vaske, D., Helm, J., Liu, H.C., Rahman, A., Yu, T.P., Larson, R.G. and Schmitz, C.B. (1995) The PiGMaP consortium linkage map of the pig (*Sus scrofa*). *Mammalian Genome* 6, 157–175.

Barendse, W., Vaiman, D., Kemp, S.J., Sugimoto, Y., Armitage, S.M., Williams, J.L., Sun, H.S., Eggen, A., Agaba, M., Aleyasin, S.A., Band, M., Bishop, M.D., Buitkamp, J., Byrne, K., Collins, F., Cooper, L., Coppettiers, W., Denys, B., Drinkwater, R.D., Easterday, K., Elduque, C., Ennis, S., Erhardt, G., Ferretti, L., Flavin, N., Gao, Q., Georges, M., Gurung, R., Harlizius, B., Hawkins, G., Hetzel, J., Hirano, T., Hulme, D., Jorgensen, C., Kessler, M., Kirkpatrick, B.W., Konfortov, B., Kostia, S., Kuhn, C., Lenstra, J.A., Leveziel, H., Lewin, H.A., Leyhe, B., Lil, L., Martin Burrie, I., McGraw, R.A., Miller, J.R., Moody, D.E., Moore, S.S., Nakane, S., Nijman, I.J., Olsaker, I., Pomp, D., Rando, A., Ron, M., Shalom, A., Teale, A.J., Thieven, U., Urquhart, B.G.D., Vage, D.-I., Van de Weghe, A., Varvia, S., Velmala, R., Vilkki, J., Weikard, R., Woodside, C., Womack, J.E., Zanotti, M. and Zaragoza, P. (1997) A medium-density genetic linkage map of the bovine genome. *Mammalian Genome* 8, 21–28.

Boehnke, M. (1991) Allele frequency estimation from data on relatives. *American Journal of Human Genetics* 48, 22–25.

Bovenhuis, H., Vanarendonk, J.A.M., Davis, G., Elsen, J.M., Haley, C.S., Hill, W.G., Baret, P.V., Hetzel, D.J.S. and Nicholas, F.W. (1997) Detection and mapping of quantitative trait loci in farm animals. *Livestock Production Science* 52, 135–144.

Chee, M., Yang, R., Hubbell, E., Berno, A., Huang, X.C., Stern, D., Winkler, J., Lockhart, D.J., Morris, M.S. and Fodor, S.P. (1996) Accessing genetic information with high-density DNA arrays. *Science* 274, 610–614.

Cheng, H.H. (1997) Mapping the chicken genome. *Poultry Science* 76, 1101–1107.

Churchill, G.A. and Doerge, R.W. (1994) Empirical threshold values for quantitative trait mapping. *Genetics* 138, 963–971.

Cockett, N.E., Jackson, S.P., Shay, T.L., Nielsen, D., Moore, S.S., Steele, M.R., Barendse, W., Green, R.D. and Georges, M. (1994) Chromosomal localization of the callipyge gene in sheep (*Ovis aries*) using bovine DNA markers. *Proceedings of the National Academy of Sciences USA* 91, 3019–3023.

Comuzzie, A.G., Hixson, J.E., Almasy, L., Mitchell, B.D., Mahaney, M.C., Dyer, T.D., Stern, M.P., MacCluer, J.W. and Blangero, J. (1997) A major quantitative trait locus determining serum leptin levels and fat mass is located on human chromosome 2. *Nature Genetics* 15, 273–276.

Cooper, D.N., Smith, B.A., Cooke, H.J., Niemann, S. and Schmidtke, J. (1985) An estimate of unique DNA sequence heterozygosity in the human genome. *Human Genetics* 69, 201–205.

Crawford, A.M., McEwan, J.C., Dodds, K.G., Bisset, S.A., Macdonald, P.A., Knowler, K.J., Greer, G.J., Green, R.S., Cuthbertson, R.P., Wright, C.S., Vlassoff, A., Squire, D.R., West, C.J., Paterson, K.A. and Phua, S.H. (1997a) Parasite resistance: a genome scan approach to finding markers and genes. *Proceedings of the New Zealand Society of Animal Production* 57, 297–300.

Crawford, A.M., McEwan, J.C., Dodds, K.G., Wright, C.S., Bisset, S.A., Macdonald, P.A., Knowler, K.J., Greer, G.J., Green, R.S., Shaw, R.J., Paterson, K.A., Cuthbertson, R.P., Vlassoff, A., Squire, D.R., West, C.J. and Phua, S.H. (1997b) Resistance to nematode parasites in sheep: how important are the MHC genes? *Proceedings of the Association for the Advancement of Animal Breeding and Genetics* 12, 58–62.

Darvasi, A. and Soller, M. (1992) Selective genotyping for determination of linkage between a marker locus and a quantitative trait locus. *Theoretical and Applied Genetics* 85, 353–359.

Darvasi, A. and Soller, M. (1994a) Optimum spacing of genetic markers for determining linkage between marker loci and quantitative trait loci. *Theoretical and Applied Genetics* 89, 351–357.

Darvasi, A. and Soller, M. (1994b) Selective DNA pooling for determination of linkage between a molecular marker and a quantitative trait locus. *Genetics* 138, 1365–1373.

De Gortari, M.J., Freking, B.A., Cuthbertson, R.P., Kappes, S.M., Keele, J.W., Stone, R.T., Leymaster, K.A., Dodds, K.G., Crawford, A.M. and Beattie, C.W. (1998) A second-generation linkage map of the sheep genome. *Mammalian Genome* 9, 204–209.

Dodds, K.G. and McEwan, J.C. (1997) Calculating exact probabilities of allele frequency differences in divergent selection lines. *Proceedings of the Association for the Advancement of Animal Breeding and Genetics* 12, 556–560.

Dodds, K.G., Montgomery, G.W. and Tate, M.L. (1993) Testing for linkage between a marker locus and a major gene locus in half-sib families. *Journal of Heredity* 84, 43–48.

Elsen, J.M., Mangin, B., Goffinet, B. and Chevalet, C. (1994) Optimal structure of protocol designs for building genetic linkage maps in livestock. *Theoretical and Applied Genetics* 88, 129–134.

Elston, R.C. and Stewart, J. (1971) A general model for the analysis of pedigree data. *Human Heredity* 21, 523–542.

Fujii, J., Otsu, K., Zorzato, F., de Leon, S., Khanna, V.K., Weiler, J.E., O'Brien, P.J. and MacLennan, D.H. (1991) Identification of a mutation in porcine ryanodine receptor associated with malignant hyperthermia. *Science* 253, 448–451.

Georges, M., Lathrop, M., Hilbert, P., Marcotte, A., Schwers, A., Swillens, S., Vassart, G. and Hanset, R. (1990) On the use of DNA fingerprints for linkage studies in cattle. *Genomics* 6, 461–474.

Gulland, F.M.D., Albon, S.D., Pemberton, J.M., Moorcroft, P.R. and Clutton-Brock, T.H. (1993) Parasite-associated polymorphism in a cyclic ungulate population. *Proceedings of the Royal Society of London* 254, 7–13.

Haley, C.S. and Knott, S.A. (1992) A simple regression method for mapping quantitative trait loci in line crosses using flanking markers. *Heredity* 69, 315–324.

Hetzel, D.J.S. (1991) The use of reference families for genome mapping in domestic livestock. In: Schook, L.B., Lewin, H.A. and McLaren, D.G. (eds) *Gene-mapping Techniques and Applications.* Marcel Dekker, New York, pp. 51–64.

Holland, P.M., Abrahamson, R.D., Watson, R. and Gelfand, D.H. (1991) Detection of specific polymerase chain reaction products by utilising the 5′ to 3′ exonuclease activity of *Thermus aquaticus* DNA polymerase. *Proceedings of the National Academy of Sciences USA* 88, 7276–7280.

Janss, L.L.G., Thompson, R. and Van Arendonk, J.A.M. (1995) Application of Gibbs sampling for inference in a mixed major gene–polygenic inheritance model in animal populations. *Theoretical and Applied Genetics* 91, 1137–1147.

Jeffreys, A.J., Wilson, V. and Thien, S.L. (1985) Hypervariable minisatellite regions in human DNA. *Nature* 314, 67–73

Kappes, S.M., Keele, J.W., Stone, R.T., McGraw, R.A., Sonstegard, T.S., Smith, T.P.L., Lopez-Corrales, N.L. and Beattie, C.W. (1997) A second-generation linkage map of the bovine genome. *Genome Research* 7, 235–249.

Keightley, P.D., Hardge, T., May, L. and Bulfield, G. (1996) A genetic map of quantitative trait loci for body weight in the mouse. *Genetics* 142, 227–235.

Kemp, S.J., Darvasi, A., Soller, M., Teale, A.J., Shewen, P.E., Lunney, J.K. and Gershwin, L.J. (1996) Genetic control of resistance to trypanosomiasis. *Veterinary Immunology and Immunopathology* 54, 239–243.

Kerr, R.J., Frisch, J.E. and Kinghorn, B.P. (1994) Evidence for a major gene for tick resistance in cattle. *Proceedings 5th World Congress on Genetics Applied to Livestock Production* 20, 265–268.

Kinghorn, B.P., Kennedy, B.W. and Smith, C. (1993) A method of screening for genes of major effect. *Genetics* 134, 351–360.

Lander, E.S. and Botstein, D. (1989) Mapping Mendelian factors underlying quantitative traits using RFLP linkage maps. *Genetics* 121, 185–199.

Lander, E.S. and Green, P. (1987) Construction of multilocus genetic linkage maps in humans. *Proceedings of the National Academy of Sciences USA* 84, 2363–2367.

Lander, E. and Kruglyak, L. (1995) Genetic dissection of complex traits: guidelines for interpreting and reporting linkage results. *Nature Genetics* 11, 241–247.

Lander, E.S., Green, P., Abrahamson, J., Barlow, A., Daly, M.J., Lincoln, S.E. and Newburg, L. (1987) MAPMAKER: an interactive computer package for constructing primary genetic linkage maps of experimental and natural populations. *Genomics* 1, 174–181.

Lange, K., Weeks, D. and Boehnke, M.. (1988) Programs for pedigree analysis: MENDEL, FISHER and dGENE. *Genetic Epidemiology* 5, 471–472.

Lathrop, G.M., Lalouel, J.M., Julier, C. and Ott, J. (1984) Strategies for multilocus analysis in humans. *Proceedings of the National Academy of Sciences USA* 81, 3443–3446.

Lynch, M. and Walsh, B. (1997) *Genetics and Analysis of Quantitative Traits.* Sinauer Associates, Massachusetts.

Lyons, L.A., Laughlin, T.F., Copeland, N.G., Jenkins, N.A., Womack, J.E. and O'Brien, S.J. (1997) Comparative anchor tagged sequences (CATS) for integrative mapping of mammalian genomes. *Nature Genetics* 15, 47–56.

McEwan, J.C. and Kerr, R.J. (1998) Further evidence that major genes affect host resistance to nematode parasites in Coopworth sheep. *Proceedings of the 6th World Congress of Genetics Applied to Livestock Production* 27, 335–338.

Martinez, O. and Curnow R.N. (1992) Estimating the locations and the sizes of the effects of quantitative trait loci using flanking markers. *Theoretical and Applied Genetics* 85, 480–488.

Meuwissen, T.H.E. and Goddard, M.E. (1996) The use of marker haplotypes in animal breeding schemes. *Genetics, Selection, Evolution* 28, 161–176.

Montgomery, G.W., Penty, J.M., Henry, H.M., Sise, J.A., Lord, E.A., Dodds, K.G. and Hill D.F. (1995) Sheep linkage mapping: RFLP markers for comparative mapping studies. *Animal Genetics* 26, 249–259

Nielsen, D., Cockett, N.E. and Georges, M. (1995) Mapping markers and quantitative traits in large half-sib pedigrees. *Proceedings of the Western Section of the American Society of Animal Science* 46, 205–208.

Orita, M., Iwahana, H., Kanazawa, H., Hayashi, K. and Sekiya T. (1989) Detection of polymorphisms of human DNA by gel electrophoresis as single-strand conformation polymorphisms. *Proceedings of the National Academy of Sciences USA* 86, 2766–2770.

Ott, J. (1991) *Analysis of Human Genetic Linkage.* Johns Hopkins University Press, Baltimore.

Phua, S.H., Dodds, K.G., Morris, C.A., Towers, N.R. and Crawford, A.M. (1998) Antioxidant enzymes as candidate genes for disease resistance in sheep facial eczema. *Proceedings of the 6th World Congress of Genetics Applied to Livestock Production* 27, 273–276.

Rohrer, G.A., Alexander, L.J., Hu, Z.L., Smith, T.P.L., Keele, J.W. and Beattie, C.W. (1996) A comprehensive map of the porcine genome. *Genome Research* 6, 371–391.

Sambrook, J., Fritsch, E.F. and Maniatis, T. (1989) *Molecular Cloning: a Laboratory Manual.* 2nd Edn. Cold Spring Harbor Laboratory Press, Cold Spring Harbor, New York.

Stear, M.J., Bairden, K., Bishop, S.C., Buitkamp, J., Epplen, J.T., Gostomski, D., McKellar, Q.A., Schwaiger, F.W. and Wallace, D.S. (1996) An ovine lymphocyte antigen is associated with reduced faecal egg counts in four-month-old lambs following natural, predominantly *Ostertagia circumcincta* infection. *International Journal for Parasitology* 26, 423–428.

Tate, M.L. (1998) Evolution of ruminant chromosomes. PhD thesis, University of Otago, Dunedin, New Zealand.

Tate, M.L., Mathias, H.C., Fennessy, P.F., Dodds, K.G., Penty, J.M. and Hill, D.F. (1995) A new gene mapping resource: interspecies hybrids between Père David's deer (*Elaphurus davidianus*) and red deer (*Cervus elaphus*). *Genetics* 139, 1383–1391.

Taylor, B.A. and Phillips, S.J. (1996) Detection of obesity QTLs on mouse chromosomes 1 and 7 by selective DNA pooling. *Genomics* 34, 389–398.

Templeton, A.R., Boerwinkle, E. and Sing, C.F. (1987) A cladistic analysis of phenotypic associations with haplotypes inferred from restriction endonuclease mapping. I. Basic theory and an analysis of alcohol dehydrogenase activity in *Drosophila. Genetics* 117, 343–351.

Towers, N.R. and Stratton, G.C. (1978) Serum gamma-glutamyltransferase as a measure of sporidesmin-induced liver damage in sheep. *New Zealand Veterinary Journal* 26, 109–112.

Vaiman, D., Schibler, L., Bourgeois, F., Oustry, A., Amigues, Y. and Cribiu, E. (1996) A genetic linkage map of the male goat genome. *Genetics* 144, 279–305.

van der Beek, S. and van Arendonk, J.A.M. (1993) Criteria to optimize designs for detection and estimation of linkage between marker loci from segregating populations containing several families. *Theoretical and Applied Genetics* 86, 269–280.

Vos, P., Hogers, R., Bleeker, M., Reijans, M., van de Lee, T., Hornes, M., Frijters, A., Pot, J., Peleman, J., Kuiper, M. and Zabeau, M. (1995) AFLP: a new technique for DNA fingerprinting. *Nucleic Acids Research* 23, 4407–4414.

Weller, J.I., Kashi, Y. and Soller, M. (1990) Power of daughter and granddaughter designs for determining linkage between marker loci and quantitative trait loci in dairy cattle. *Journal of Dairy Science* 73, 2525–2537.

Williams, J.G., Kubelik, A.R., Livak, K.J., Rafalski, J.A. and Tingey, S.V. (1990) DNA polymorphisms amplified by arbitrary primers are useful as genetic markers. *Nucleic Acids Research* 18, 6531–6535.

Woolaston, R.R., Barger, I.A. and Piper, L.R. (1990) Response to Helminth infection of sheep selected for resistance to *Haemonchus contortus*. *International Journal for Parasitology* 20, 1015–1018.

Zhu, K.Y. and Clarke, J.M. (1996) Addition of a competitive primer can dramatically improve the specificity of PCR amplification of specific alleles. *Biotechniques* 21, 586–587.

2

Modelling Farm Animal Diseases

S.C. Bishop[1] and G. Gettinby[2]
[1]Roslin Institute (Edinburgh), Roslin, UK; [2]Department of Statistics and Modelling Science, University of Strathclyde, Glasgow, UK

Summary

The modelling of diseases in domestic animal populations is often required to understand the complex interactions between the pathogen and the host animal, and to help to devise strategies to control or minimize the impact of the disease. A wide variety of approaches are taken to model diseases, including mathematical models derived using differential equations, simulation and transition-type matrices. These models may be formulated using both stochastic and deterministic settings. Diseases which have been modelled include a variety of microparasitic (e.g. viral and bacterial) and macroparasitic (e.g. helminth and arthropod) infections in cattle, sheep and pigs.

Few disease epidemiology models have been published which include the effect of host genotype. This is despite the fact that from an animal breeding perspective, incorporation of host genotype for resistance into these models is of particular interest, as the average host genotype of the population will mediate the disease epidemiology. Using simulation, it has been shown that responses to selection for resistance to nematode infections in sheep are considerably greater than predicted by quantitative genetic theory, as the observed responses are a function of changes in both host genotype and the disease epidemiology. Thus, estimation of the true benefits of selecting for disease resistance requires the use of epidemiological models.

The current state of knowledge regarding the modelling of host genotype by disease epidemiology interactions is considered in detail in this chapter. General mathematical frameworks are seen as a requirement before such modelling techniques become widespread. It is suggested that the use of the basic reproductive ratio (R_0), defined as a trait of the host population, may be a means of linking host genotype and disease epidemiology in general terms. Finally, the developments required for epidemiological models to become widely used to support decision making by animal breeders are considered.

© CAB *International* 2000. *Breeding for Disease Resistance in Farm Animals*
(eds R.F.E. Axford, S.C. Bishop, F.W. Nicholas and J.B. Owen)

Introduction

This chapter describes the modelling of diseases in domestic livestock populations and the importance of such epidemiological modelling when attempting to quantify the benefits of selection for resistance to an infectious disease. Why is it necessary to take an epidemiological modelling approach to address selection for disease resistance? A fundamental difference exists between the genetic improvement of productivity traits and traits describing resistance to infectious diseases. For production traits, animals express their genetic merit independently of each other, and benefits from selection can simply be summed across the population. However, for disease resistance there is an interaction between animals in the expression of the trait. In simple terms, animals infect each other and the expression of the disease-resistance trait is dependent on the epidemiology of the actual disease itself. Thus, an interaction exists between the average host genotype and the disease epidemiology. If this interaction is not accounted for, incorrect inferences may be made regarding the benefits of selecting for resistance.

The structure of this chapter is as follows. It reviews briefly the role of epidemiological modelling in understanding disease processes, the modelling approaches available, and provides examples of different disease models. It then considers, in more detail, cases where host genotype for resistance has been considered as a factor in the model and shows how genotype × epidemiology interactions may be quantified. Next, it considers the incorporation of the genetics of the pathogen into the model. Finally, it considers developments required to improve modelling of the host genotype effects on disease epidemiology and the use of such models in animal breeding.

The Attractions of Modelling

Quantifying and describing the processes and impacts of infectious diseases in domestic farmed animals presents a considerable challenge. Complex interactions occur between the host animal population and the pathogen, and these interactions will vary temporally, spatially and with the host genotype. Moreover, the infection and subsequent host immune-response processes are multifactorial. In order to predict the effect of any factor, including treatment strategies or genetic selection of resistant hosts, on an infectious disease at the population level, an epidemiological model is required. This model should capture and quantify the dynamics of the disease and allow the impact of various factors affecting the disease severity and epidemiology to be predicted.

Major uses of epidemiological models have traditionally included: (i) evaluating the effects of various control strategies, e.g. vaccination or management options, on the disease severity; (ii) evaluating the impact of local weather conditions on the disease severity; and (iii) evaluating the impact of various chemotherapy strategies on likely evolution of drug resistance in the pathogen. Examples of each of these uses are given below.

The chapter therefore considers in detail the use of disease epidemiology models for predicting the effects of genetic change in the host on the epidemiology of the disease. This is perhaps an area which has been largely neglected in the past, but is likely to become of increasing importance within the context of farm animal production.

Modelling Approaches and Definitions

It is necessary when considering the modelling of diseases, and the epidemiological consequences of given infections, to consider both the modelling approach and the type of pathogen being modelled. Models are generally formulated either deterministically or stochastically, although there is overlap between these categories. Likewise, the modelling approach may be based on differential equations, simulation or transition-type matrices. Finally, the infection being modelled may be classified into those caused by microparasites, e.g. bacteria or viruses, or macroparasites, e.g. ticks or helminths. These breakdowns may be considered cross-classified, e.g. one can use either deterministic or stochastic formulations when modelling either macro- or microparasitic infections. A brief summary of these concepts is given in this section.

Model formulation

The distinction between deterministic and stochastic models is familiar to animal breeders. Deterministic models generally treat the values of input parameters as fixed, and produce point estimates of an outcome. Stochastic models incorporate random variation in processes or parameters, and produce probability distributions of outcomes. In disease modelling, stochastic models are particularly appropriate for modelling the random contacts between susceptible and infected animals, e.g. as a time-dependent binomial distribution, hence producing a probability distribution of numbers of infected animals.

Modelling approach

Differential equations

The disease epidemic may be described by rates of change in, for example, the numbers of susceptible or infected animals. Hence the essence of the process can often be captured in a few simple differential equations. For example, the rate of change in the number of adult (macro)parasites in a single host at time t, $\mathrm{d}M(t)/\mathrm{d}t$, when the host is constantly exposed to infection at a rate δ and the adult parasite dies at constant per capita rate μ, may be described as follows:

$$\mathrm{d}M(t)/\mathrm{d}t = \delta - \mu M(t)$$

with solution: $M(t) = M^* [1 - \exp(-\mu t)]$

where M^* is an equilibrium population.

Extending such an approach to describing the parasite life cycle is conceptually simple. For example, as a first approximation, a ruminant helminth infection can be described by two differential equations, the first describing the change in host–parasite burden with time and the second describing the change in pasture larval contamination with time. The equations may include terms describing the uptake of larvae by the host, the establishment probabilities of the parasite, worm burdens in the host, worm fecundity, host immunity, parasite survival probabilities, the density of the hosts on the pasture and the density of the larvae on the pasture. Roberts and Grenfell (1991) include a third equation describing changes in host immunity. Although this approach is deterministic in structure, it can nevertheless contain probabilistic elements to describe the distribution of parasite numbers per host, in the case of macroparasites. An equivalent, simple, approach for microparasitic infections might consist of two equations, one describing the rate of change of numbers of susceptible animals, the other describing the rate of change of immune animals.

This approach, using differential equations to capture the essence of the epidemic, is widespread in epidemiological modelling, due to its flexibility and relative simplicity. The procedure is usually a simplified case of the more complete approach using fully stochastic models. However, such stochastic models are often somewhat intractable to analytical investigation, owing to the many non-linearities that are inherent in biological problems.

Simulation

Simulation models take many forms, with the main aim being to describe the disease patterns in relation to events that may change either deterministically or stochastically. Inputs into the simulation models may be explanatory variables describing the infection process, developed in the same way as the inputs into the mathematical differential-equations approach, or simply empirical relationships between observed phenomena – sometimes with no strong theoretical justification.

Classic applications of models using empirical relationships are the weather-based models describing helminth infection in cattle (Gettinby and Gardiner, 1980) and sheep (Barnes *et al.*, 1988). In these models, the parasite epidemiology is calculated as a function of the prevailing weather conditions in any particular season, and inputs to the model include a database of meteorological records over many years.

Many simulation models include a combination of empirical and explanatory relationships, couched in both deterministic and stochastic terms. Such an example is the model of Bishop and Stear (1997), investigating the inheritance of resistance to nematode parasites in sheep. In their model, the life cycle of the parasite is defined in explanatory deterministic terms, empirical relationships are used to describe the development of host immunity, and stochastic between-animal variation is used to describe the variability in the host–parasite interactions at the population level.

Finally, fully stochastic models may be formulated using Markov-chain processes, in the cases where analytical solutions to the deterministic or stochastic models are intractable. Not only will this technique supply probability densities for various outcomes, it may even supply answers which are qualitatively different from simpler deterministic models. For example, as well as describing 'major epidemics' where most of the population is infected, such simulation models will also give rise to 'minor epidemics' where there is a probability of the disease arising in a population, then dying out through chance events. An elegant exposition of such models is given by Renshaw (1991; Chapter 10).

Transition-type matrices

Transition-type matrices are often useful in describing the flow of the disease in the population, especially when dealing with discrete time events. This use of matrices became established when Leslie (1945) published his so-called Leslie matrix. The host population may be divided into *n* discrete age classes, and the transition rates are defined by probabilities of infective animals of age class *j* infecting susceptible animals of age class *i*, at time *t*. This is sometimes known as the WAIFW ('who acquires infection from whom') matrix (Anderson and May, 1992).

The power of such an approach arises when there is heterogeneity in the host population, e.g. different husbandry classes, geographical differences or genetic differences, and the explicit relationships required for a simulation or mathematical approach are difficult to specify. An excellent example is that given by De Jong *et al.* (1994), describing the dynamics of a viral infection in a pig farm. A transition-type matrix, the *next generation operator* (M), is specified, describing the probabilities of the infection process throughout the population. It accounts for the age structure of animals in the host population, the susceptibility and infectivity of animals of each class, contact rates amongst animals, possible paths of infection through the population, etc. A similar approach is given by Roberts and Heesterbeek (1995) for nematode infections in ruminants where there are multiple host types, e.g. adult females and juveniles. However, in this case the elements of the matrix are the explicit solutions from differential equations describing the epidemic.

Pathogen type

Microparasites

Most viral and bacterial parasites, and many protozoan and fungal parasites, may be classified as microparasites. Microparasites may be thought of as those that have direct reproduction, usually at very high rates, within the host (Anderson and May, 1979). They tend to be characterized by small size, a short generation time (often just a fraction of that of their host) and a short duration of infection. Hosts that recover from infection usually acquire immunity against reinfection for some time, and often for life.

Unlike macroparasites such as nematodes, microparasites seldom have a closed life cycle which can be, or even requires to be, modelled. For such pathogens, the host population is often simply divided into a small number of classes of individuals, e.g. susceptible, infected, recovered-and-immune. An operational definition of a microparasite is an organism whose population biology can be sensibly described by such compartmental models (Anderson and May, 1992). Hence, little or no account is generally taken of the degree of severity of infection, e.g. the abundance of the parasite within the host, due to the rapid proliferation of the parasite once it is established. Animals are considered to be either infected or not. The reality of infected animals with differing nutritional, environmental or genetic status is replaced by simple concepts of infected, immune, latent but not yet infectious, etc.

Macroparasites

Most helminths and arthropods may be classified as macroparasites. Macroparasites may be thought of as those having no direct reproduction within the host (Anderson and May, 1979). Typically they are larger and have longer generation intervals than microparasites, with the generation interval often being an appreciable proportion of the host's life span. Macroparasitic infections are typically of a persistent nature, with hosts being continually reinfected. The classic examples in domestic livestock are nematode infections of cattle, sheep and goats.

For macroparasite infections, various factors describing the interaction between the host and the pathogen must be described when attempting to model the infection, e.g. egg output per parasitic female, pathogenic effects on the host, immune response, parasitic death rates, etc. Thus, the simple compartmental models which are adequate for microparasite infections must be replaced by more sophisticated models which take account of the distribution of parasites among the host. An operational definition of a macroparasite is an organism whose population biology requires a full description of the distribution of parasites among hosts (Anderson and May, 1992).

Macroparasites are rarely distributed in a random way among their hosts. Typically they show an aggregated or clumped distribution, with the minority of hosts harbouring the majority of the pathogens. The distribution of parasites among hosts often conforms to a negative binomial distribution. The 80:20 rule is a useful approximation to such distributions: often 80%, or more, of the macroparasites will be contained in 20%, or fewer, of the hosts (Anderson and May, 1992). Pareto (1906) observed that the same rule-of-thumb applies to the distribution of wealth in human populations.

Basic reproduction ratio – R_0

Fundamental to describing the epidemiology of the disease in many situations, especially for microparasites, is the concept of the basic reproductive ratio, R_0. R_0 may be defined as the expected number of secondary cases produced in a completely susceptible population, by a typical infected individual during its entire period of infectiousness (Diekmann *et al.*, 1990). A disease can invade and maintain itself in a host population only if $R_0 > 1$. If $R_0 < 1$ the disease cannot maintain itself and, under most circumstances, will become extinct. This is the threshold theorem of Kermack and McKendrick (1927), and its use in determining long-term microparasitic disease dynamics, especially when considering selection for resistance, is readily apparent.

For macroparasitic infections, the definition of R_0 is less straightforward. The number of parasites constituting the infection is important and R_0 must be formulated in terms of the dynamics of the parasitic population. For macroparasites, a definition of R_0 may be 'the average number of offspring (female offspring in a dioecious species) produced throughout the reproductive life span of a mature parasite that themselves survive to reproductive maturity in the absence of density-dependent constraints on population growth' (Anderson and May, 1992). This is directly equivalent to Fisher's definition of net reproductive rate for free-living species (Fisher, 1930). A formal definition of a parameter with threshold properties analogous to R_0, for macroparasitic infections, is given by Heesterbeek and Roberts (1994) who define a term called the basic reproductive quotient, Q_0. Q_0 has the same properties as R_0 in a microparasitic infection, i.e. $Q_0 > 1$ implies that the parasite can invade the population, and $Q_0 < 1$ implies that the parasite population is not expected to persist.

Many formulations are present in the literature for the calculation of R_0. In the case of domestic farm animals, where one deals with heterogeneous populations, perhaps the most useful and flexible means of estimating R_0 is the algorithm of De Jong *et al.* (1994). In such circumstances, R_0 may be calculated as the dominant eigenvalue of the *next generation operator* (M) matrix, described above (Diekmann *et al.*, 1990). Equivalently, for nematodes (macroparasites), Roberts and Heesterbeek (1995) demonstrate that Q_0 is also the dominant eigenvalue of their transition-type matrix. Implicitly, such an approach can be used to investigate the impact of animals in the population with differing genotypes for susceptibility.

Epidemiological Models – Overview

This section gives a brief and selective overview of recent epidemiological models of common disease problems in domestic livestock. It attempts to do no more than give a flavour of available models and what they achieve – a full summary is beyond the scope of this chapter. The majority of published models describe disease problems in cattle, with fewer models for pig and sheep diseases and a glaring lack of recent models describing poultry diseases.

Cattle

Published epidemiological models for cattle diseases are wide ranging, covering both macro- and microparasitic infections, and use a wide range of modelling techniques. At the extreme of macroparasitic infections, considerable effort has been put into the modelling of tick and tick-borne diseases in cattle. Notable examples include the simulation models of host–parasite interactions for cattle fever transmitted by *Boophilus* cattle ticks (Haile *et al.*,

1992; Teel *et al.*, 1996) and East Coast fever transmitted by *Rhipicephalus appendiculatus* ticks (Byrom and Gettinby, 1992). Gettinby *et al.* (1988) have also modelled, by simulation, the evolution of genetic resistance to acaricides in *R. appendiculatus* populations.

For gastrointestinal parasite infections, although simulation models have been published (e.g. Gettinby *et al.*, 1979), recent models tend to be deterministic mathematical models. Comprehensive mathematical models for *Ostertagia ostertagi* infections include those of Grenfell *et al.* (1987) and Smith and Guerrero (1993), whereas Roberts and Grenfell (1991) developed a mathematical model with seasonal effects generally applicable to nematode infections in ruminants.

An important class of disease problems is that where there is a wildlife reservoir for the disease, e.g. bovine tuberculosis in badger or possum populations. Elegant simulation models, using both deterministic and stochastic formulations, are given for possums harbouring bovine tuberculosis in New Zealand by Barlow (1991, 1993) and Barlow *et al.* (1997). A broad summary of such wildlife models is given by Barlow (1995).

Viral diseases have also been modelled extensively in cattle. Stochastic simulations of bovine viral diarrhoea are presented by Sørenson and Enevoldsen (1994) and Innocent *et al.* (1997). A mathematical model of rinderpest infection is described by Tillé *et al.* (1991) using both deterministic and stochastic approaches – importantly they found that the variability due to stochastic factors was important and the deterministic model alone was not an adequate description. Other recently simulated viral disease problems include foot-and-mouth disease (Hutber and Kitching, 1996), bovine leukaemia virus (Casal *et al.*, 1990), bluetongue virus (Ward and Carpenter, 1996) and bovine syncytial virus (De Jong *et al.*, 1996).

Finally, considerable attention has turned recently to transmissible spongiform encephalopathies, especially BSE, the cattle form. A comprehensive mathematical model describing the BSE epidemic in Great Britain is described by Anderson *et al.* (1996), Donnelly *et al.* (1997a) and Ferguson *et al.* (1997a). This model was used subsequently to enable interpretations to be made of the results of experimental data on BSE transmission (Donnelly *et al.*, 1997b; Ferguson *et al.*, 1997b) – something that would not have been possible without recourse to such models. For example, analyses using the model show that maternal transmission generally explains disease transmission within the experimental data better than genetic susceptibility, and only for certain epidemiological assumptions does the model support both maternal transmission and genetic susceptibility as predisposing factors to BSE.

Sheep

Sheep epidemiological models reflect the fact that sheep tend to be extensively farmed and the tractable disease problems are mainly macroparasitic. The major emphasis in published models is on nematode parasite infections. An early fully stochastic framework for studying parasite populations in sheep, given by Tallis and Leyton (1966), served to show the true complexity of the processes. Mathematical models describing the nematode population dynamics are outlined by Barnes and Dobson (1990a), Roberts and Heesterbeek (1995) and Grenfell *et al.* (1995). Simulation models abound, including those of Gettinby *et al.* (1989), Barnes and Dobson (1990b), Leathwick *et al.* (1992), Beecham *et al.* (1994) and Bishop and Stear (1997). The majority of these simulation models are weather-based, i.e. they take as input parameters either expected weather (temperature and rainfall) patterns or a database of actual weather records. A major use of such models has been to study the predicted rates of evolution of anthelmintic resistance amongst the nematode parasite populations (e.g. Dobson *et al.*, 1987; Gettinby *et al.*, 1989; Barnes and Dobson, 1990b; Echevarria *et al.*, 1993; Leathwick *et al.*, 1995).

Sheep-disease models are not restricted to nematode worm problems, however. Other sheep-disease problems that have been modelled include weather-based simulations of fascioliasis (Gettinby and Byrom, 1991), mathematical models of cestode infections (Roberts, 1994) – for cestode infections there are often 'wild' animal (e.g. dog, fox or vole) reservoirs of the disease – and mathematical models of scrapie (Stringer *et al.*, 1998; Woolhouse *et al.*, 1998).

Pigs

The pig epidemiological models reflect the fact that pigs are generally intensively housed and farmed. Hence, the disease problems tend to be acute microparasitic infections, with macroparasitic problems being of less concern. Aujesky's virus (pseudorabies) is perhaps the disease that has received the most attention in modelling terms, with simulation-model investigations reported by Miller *et al.* (1994), Buijtels *et al.* (1997) and Casal *et al.* (1997), looking at virus transmission and vaccination strategies. The latter paper also addresses virus transmission for foot-and-mouth disease. Hone (1994) reports a mathematical model for porcine transmissible gastroenteritis. Finally, a discrete-time Markov-chain approach to modelling atrophic rhinitis in pigs is described by Turner *et al.* (1993). In an interesting development, the same authors (Turner *et al.*, 1997) use fuzzy logic to incorporate environmental effects into their model. Fuzzy logic is a departure from 'crisp' Boolean logic as it deals with degrees of truth and allows incorporation of expert opinion. Thus, it attempts to reflect real-life decision-making processes.

Epidemiological Models – Incorporating Host Genetics

The models described in the previous section cover a wide range of disease scenarios and are used for a wide variety of purposes. However, very few of the models explicitly include host genetics. Yet, it is perhaps the incorporation of host genetics that will lead to the richest development of such models – both as an approach for making breeding decisions and towards better understanding the disease dynamics.

One of the first authors to acknowledge the influence of host genotype for resistance and to attempt to quantify the effects of host genotype on disease epidemiology was Barger (1989). Using the model for nematode infection in sheep subsequently published by Barnes and Dobson (1990b), he demonstrated that the assumed genetic resistance of the sheep population does indeed regulate the disease epidemiology. In simple terms, more resistant sheep lead to lower pasture larval contamination and these sheep require fewer anthelmintic treatments to keep parasite burdens below a pre-defined threshold. Therefore, a positive feedback loop is created – resistant sheep put fewer parasites back onto the pasture, which then lowers the subsequent challenge to the sheep, resulting in even lower parasite contamination on the pasture, and so on. That this is a true epidemiological change in the disease dynamics is readily seen from the fact that all sheep subsequently grazing the pasture will benefit from the reduced larval challenge, not only selected 'resistant' sheep but also unselected 'susceptible' sheep.

These concepts were developed further by Bishop and Stear (1997) who modelled the epidemiological consequences of selecting for reduced faecal egg count in sheep (faecal egg count is a convenient indicator of resistance to nematode parasite infection). Their model defined heritable between-animal variation for pasture intake (hence larval intake) and for each major interaction between the parasite and the host. This created heritable variation in faecal egg count which mimicked field data and allowed the effects of selection to be investigated. Selection responses predicted by this model were always greater than those predicted by quantitative genetic theory. Although the actual responses depended on the grazing management assumptions, they were, in general, up to two times greater than predicted. The implication is that responses to selection, in chronic macroparasitic infections, combine factors due to changes in both the host genetics and the disease epidemiology. This observation alters the relative emphasis that breeders may wish to place on disease-resistance traits relative to production traits.

The genotype × epidemiology interaction becomes even more important when productivity traits are incorporated into the epidemiological model. Bishop and Stear (1999) developed their nematode parasite model to include growth rate of the host sheep and growth rate losses due to infection. This highlighted the fact that not only are there epidemiological benefits in selecting for resistance, but these epidemiological benefits also result in improved productivity. Figure 2.1 shows responses in liveweight gain from selection for reduced faecal egg for two situations: (i) where the sheep graze the same pasture each year, and hence fully exploit the epidemiological effects; and (ii) where they graze different but equally contaminated fields. Fully exploiting the epidemiological effects can lead to very large productivity benefits – in this case the correlated improvements in productivity were sometimes greater than the direct responses from selection for productivity. Even after the first year of selection, the correlated responses in liveweight were more than double that predicted by quantitative genetic theory. Crucially, these correlated epidemiological benefits are not symmetric – selection for increased productivity will not lead to correspondingly large improvements

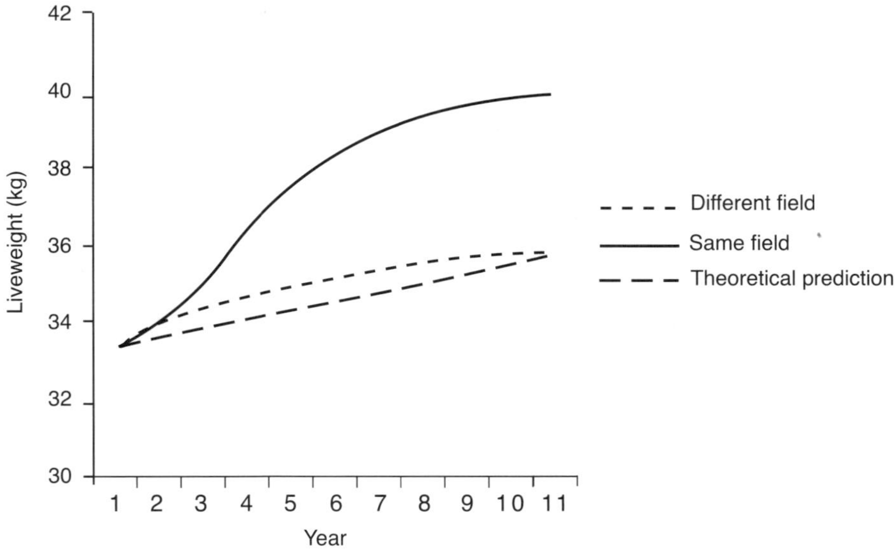

Fig. 2.1. Predicted responses in liveweight resulting from selection for increased resistance to nematode infection in sheep, as assessed by faecal egg count. The 'same field' scenario assumes that epidemiological effects are fully exploited. 'Different field' assumes that different but equally contaminated fields or pastures are grazed each year. The 'theoretical prediction' is the prediction obtained from quantitative genetic theory, ignoring epidemiological effects. The difference between the theoretical prediction and the other lines represents the epidemiological contribution to the apparent selection response.

in disease resistance. Therefore, correctly weighting disease resistance and productivity traits in breeding programmes requires the use of such modelling techniques.

Can the principles of genotype × epidemiology interactions for macroparasitic diseases be incorporated into a general mathematical framework? Although the answer is yes, relatively little work appears to have been done towards this end. An attempt was made by Roberts and Heesterbeek (1995), who defined a mathematical model for nematode infection in sheep which incorporated host heterogeneity in resistance, although it only allowed discrete categories of resistance. Their technique defined proportions of animals in the flock which must be 'resistant' in order for the parasite population to no longer be viable, i.e. Q_0 (the R_0 analogue) less than 1.0. The scrapie model of Stringer *et al.* (1998) is the most complete mathematical model to date explicitly incorporating host genotype for resistance. In this model, specific alleles with Mendelian inheritance are defined to denote resistance or susceptibility, reflecting PrP genotype associations with scrapie resistance/susceptibility. The model then predicts epidemics of several decades' duration which, with no intervention, die out through natural selection of resistant genotypes. Interestingly, the epidemic ends before the resistance allele reaches fixation. Woolhouse *et al.* (1998) then used this model to compare the dynamics of breeding for scrapie resistance with other control strategies – demonstrating the power of the modelling approach.

In considering how host genetics might be incorporated into future epidemiological models, it is the use of R_0 (the basic reproductive ratio) that provides the most likely means of achieving this aim for a wide range of diseases. R_0 is often thought of as a property of the disease. However, it is equally well a function of the mean host genotype of the infected population. The algorithm of De Jong *et al.* (1994) provides a means of calculating R_0 for discrete-time epidemic models, where between-animal differences exist in both susceptibility and infectivity. This method of quantifying host–genotype effects on the disease epidemiology is explored for the case of viral diseases in pigs, by MacKenzie and Bishop (1999). For acute microparasitic diseases, the question that must be framed is slightly different from that for chronic macroparasitic infections, where the feedback loop is easily defined. The phenomenon to be quantified is the likely spread of a disease in the population, should the infection strike. MacKenzie and Bishop (1999) showed that the disease dynamics does indeed depend on the host genotype and that the utility of selecting for resistance to a particular disease depends critically upon the initial R_0 value. For population and diseases where R_0 is high (e.g. > 5), selection is unlikely to be feasible, as even after many years of successful selection R_0 would remain sufficiently high that an epidemic would still sweep quickly through the population, as described below. For cases where the initial R_0 is only just above 1.0 or where a major gene for resistance exists, then selection may well be a feasible means of addressing the disease problem. In these circumstances R_0 would quickly be reduced to values where the population would not support an epidemic.

Woolhouse *et al.* (1998) also used the concept of R_0 to summarize and compare their scrapie control strategies. R_0 is related not only to the frequency of the resistance allele in the population but also to the level of inbreeding. R_0 will tend to rise with inbreeding (Woolhouse, 1998), making epidemics more likely.

Detailed consideration of R_0 reveals the true non-linearities in the relationships between host genotype and disease epidemiology. Two functions of R_0 are the probability of an animal being infected during the course of the epidemic ($I = 1 - \exp(-R_0 I)$) and the maximum proportion of animals infected at any one time ($Y_{max} = 1 - (1 + \ln R_0)/R_0$). These two parameters plotted against R_0 are shown in Fig. 2.2. It can be seen that for highly infectious diseases, successful selection which reduces mean susceptibility to the disease, hence R_0, may have no apparent effect on the disease dynamics or the probability of an animal becoming infected. Conversely, for populations and diseases where R_0 is less than 3, for example, selection may have quick and dramatic effects.

A phenomenon displayed by all models which have considered discrete categories of resistance is that the risk of an epidemic becomes negligible even when susceptible animals remain in the population (Roberts and Heesterbeek, 1995; MacKenzie and Bishop, 1999; Stringer *et al.*, 1998; Woolhouse *et al.*, 1998). This result is of critical importance when determining breeding strategies incorporating a disease-resistance gene, as it implies that it is not necessary to take the beneficial allele to fixation. This differs from the production trait situation, where it is desirable to increase the frequency of a desirable

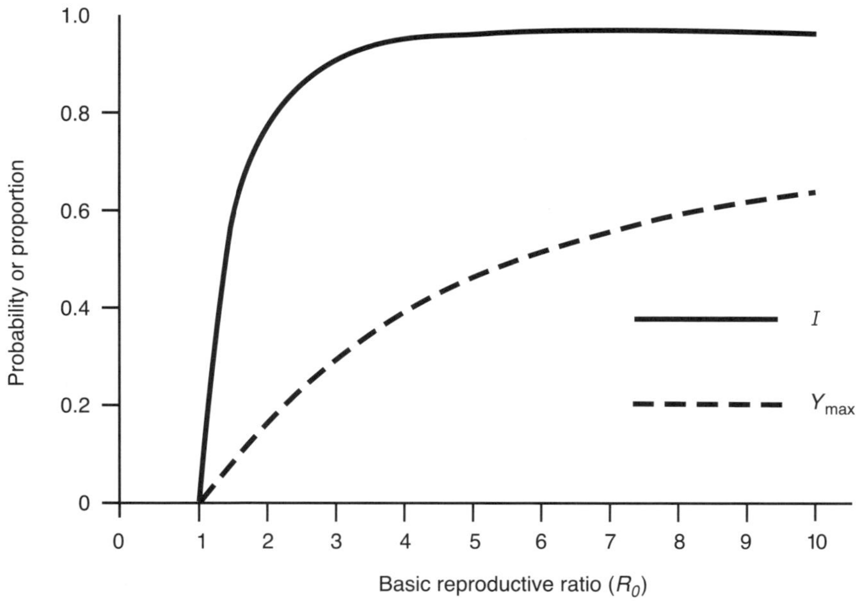

Fig. 2.2. The relationship of the probability of an animal being infected during an epidemic (I) and the maximum proportion of animals infected at any given time (Y_{max}) with R_0.

allele as far as is feasible. The required proportion of resistant animals in the population is a function of the initial R_0, and an example is shown in Fig. 2.3 for the case of introgressing a disease resistance gene into a pig population (MacKenzie and Bishop, 1999). This is a result of importance to breeders that would not be obtainable without recourse to epidemiological models.

Epidemiological Models – Incorporating Parasite Genetics

More effort has been put into modelling the parasite genotype than the host genotype. As mentioned above, one of the major uses of disease epidemiology models has been to investigate the evolution of drug resistance in the parasite, especially for anthelmintic resistance in nematodes (e.g. Dobson *et al.*, 1987, 1996; Gettinby *et al.*, 1989; Barnes and Dobson, 1990b; Echevarria *et al.*, 1993; Leathwick *et al.*, 1995). These models generally assume that resistance of the nematode to the drug is controlled by a single locus, with alleles *R* for resistance and *S* for susceptible. The mode of inheritance (i.e. recessive or dominant) of the *R* and *S* alleles is known to differ for different families of anthelmintics, and it may even be dose-dependent (Dobson *et al.*, 1996). Differential fitness for the different parasite genotypes can also be incorporated (Leathwick *et al.*, 1995). These models are very useful for devising strategies to extend the life of available anthelmintics. Importantly, they demonstrate that dosing frequency *per se* is a poor indicator of selection pressure for resistance (Leathwick *et al.*, 1995), and that comprehensive modelling approaches are

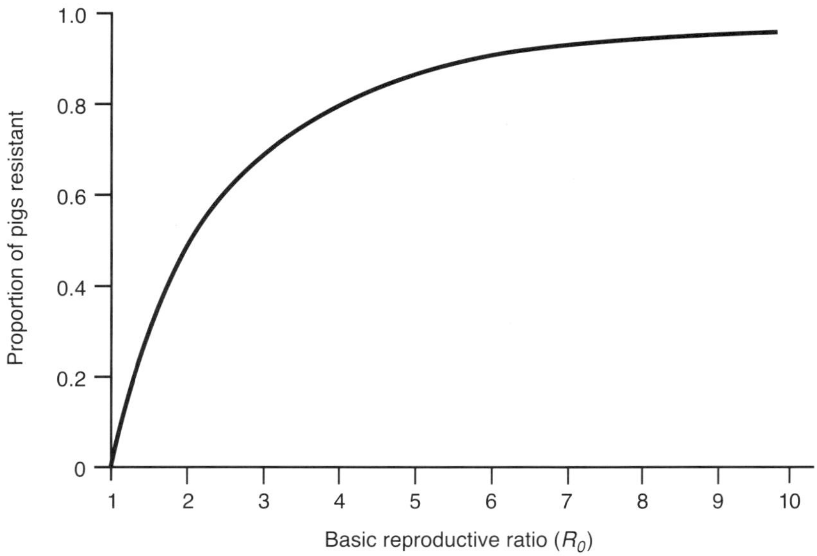

Fig. 2.3. The required proportion of genetically resistant pigs in a population to avoid risk of an epidemic from a specified disease, as a function of the R_0 for that disease in the unselected population.

required to devise dosing strategies to minimize the evolution of anthelmintic resistance.

A major longer-term concern with the selection of domestic animal populations for resistance to specific diseases is the co-evolution of the pathogen in response to the genetic changes in the host. This issue has been investigated in detail in plants but to a lesser extent in animals. Traditionally, a simple gene-for-gene model has been used, where it is hypothesized that for each gene determining resistance in the host, there is a corresponding gene for virulence in the parasite (Schmid-Hempel and Koella, 1994). The first author to investigate the more realistic situation of continuous genetic variation in both the host and parasite populations was Frank (1994). Theoretically this is a complex area, with Anderson and May (1992, p. 653) concluding that 'both theory and empirical evidence indicate that the coevolution between parasites and hosts can follow many paths, depending on the relationships between virulence and transmissibility of the parasite, and the cost to the host of evolving resistance'. For directional selection of the host, the phenomenon may be more tractable.

In domestic livestock populations, co-evolution of the parasite in response to selection for host resistance has yet to be modelled. One may expect large differences in parasite co-evolution rates between genetic selection of the host and control procedures (e.g. anthelmintic treatment) which kill the majority of the parasite population. The former process is gradual, with weak selection pressure being placed on the parasite, whereas the latter places intense selection pressure on the parasite population. Moreover, one may expect large differences between macro- and microparasites, merely as a consequence of the difference in generation intervals. For many macroparasites, the generation interval of the parasite is often an appreciable proportion of the generation interval of the host. A third important factor is the genetic heterogeneity of host resistance. Resistance due to a single gene arguably poses less of a challenge to the parasite than the multifactorial resistance, such as is seen for resistance to nematodes. These factors will all affect likely co-evolution rates.

The Way Forward

While there has been considerable effort in modelling diseases in domestic animals populations, the use of these techniques for investigating host genetic effects and selection for disease resistance is in its infancy. Moreover, the realization amongst animal breeders that epidemiological effects may contribute to apparent selection responses is only just beginning. At the time of writing, there are no examples where epidemiological models have been fully utilized for both quantifying the benefits of selection and for deriving appropriate economic weights for the disease-resistance traits. Epidemiological models will become especially valuable when they can be used as a tool for animal breeders as well as a tool for understanding the disease, itself.

Considerable work needs to be done before epidemiological models are used routinely by animal breeders. First, the generality of the results reported by Bishop and Stear (1997) needs to be investigated – by exploring other environmental situations and also other disease problems. Models incorporating host genetics then need to be fully integrated with other models which describe the disease epidemiology in greater detail. Secondly, a more general mathematical framework combining quantitative genetics and epidemiology theory is required. The concept of defining R_0 as a trait of the host population (MacKenzie and Bishop, 1999) provides a means of achieving this.

Having developed flexible models combining genetics and epidemiology, the next challenge is to use these models in decision-making processes. Two obvious situations are: (i) deciding upon the feasibility of selection in the first instance; and (ii) using the models to help to derive relative economic weights for production and disease-resistance traits. Again, theoretical development is required, but situations are easily envisaged where the importance placed on disease resistance is a function of the prevalence of the disease in that environment.

Finally, fully integrated models which include parasite genetics need to be developed. Co-evolution to both the chemical treatment and the change in the host genotype may be considered. In the longer term, with greater emphasis on biological and vaccine control, models need to be developed which describe co-evolution to these control methods as well. Developing these fully integrated models will help in the devising of both treatment strategies and selection strategies in the longer term.

In summary, it can be seen that developing disease models which describe the host genotype for resistance is an exciting new area of research. It will have practical benefits to the animal breeder and it will also provide additional insight into the dynamics of disease problems in domestic animal populations.

Acknowledgement

The contribution by SCB was funded by BBSRC CSG and PIC.

References

Anderson, R.M. and May, R.M. (1979) Population biology of infectious diseases: Part 1. *Nature* 280, 361–367.

Anderson, R.M. and May, R.M. (1992) *Infectious Diseases of Humans. Dynamics and Control.* Oxford University Press, Oxford.

Anderson, R.M., Donnelly, C.A., Ferguson, N.M., Woolhouse, M.E.J., Watt, C.J., Udy, H.J., Mawhinney, S., Dunstan, S.P., Southwood, T.R.E., Wilesmith, J.W., Ryan, J.B.M., Hoinville, L.J., Hillerton, J.E., Austin, A.R. and Wells, G.A.H. (1996) Transmission dynamics and epidemiology of BSE in British cattle. *Nature* 382, 779–788.

Barger, I.A. (1989) Genetic resistance of hosts and its influence on epidemiology. *Veterinary Parasitology* 32, 21–35.

Barlow, N.D. (1991) A spatially aggregated disease/host model for bovine Tb in New Zealand possum populations. *Journal of Applied Ecology* 28, 777–793.

Barlow, N.D. (1993) A model for the spread of bovine Tb in New Zealand possum populations. *Journal of Applied Ecology* 30, 156–164.

Barlow, N.D. (1995) Critical evaluation of wildlife disease models. In: Grenfell, B.T. and Dobson, A.P. (eds) *Ecology of Infectious Diseases in Natural Populations.* Cambridge University Press, Cambridge, pp. 230–259.

Barlow, N.D., Kean, J.M., Hickling, G., Livingstone, P.G. and Robson, A.B. (1997) A simulation model for the spread of bovine tuberculosis within New Zealand cattle herds. *Preventative Veterinary Medicine* 32, 57–75.

Barnes, E.H. and Dobson, R.J. (1990a) Population dynamics of *Trichostrongylous colubriformis* in sheep: mathematical model of worm fecundity. *International Journal for Parasitology* 20, 375–380.

Barnes, E.H. and Dobson, R.J. (1990b) Population dynamics of *Trichostrongylous colubriformis* in sheep: computer model to simulate grazing systems and the evolution of anthelmintic resistance. *International Journal for Parasitology* 20, 823–831.

Barnes, E.H., Dobson, R.J., Donald, A.D. and Waller, P.J. (1988) Predicting populations of *Trichostrongylous colubriformis* infective larvae on pasture from meteorological data. *International Journal for Parasitology* 18, 767–744.

Beecham, J.A., Wright, I.A. and Gettinby, G. (1994) An integrated model of helminth epidemiology in ryegrass pastures grazed by sheep. *Proceedings of the 45th Meeting of the European Association of Animal Production*, paper S1.5.

Bishop, S.C. and Stear, M.J. (1997) Modelling responses to selection for resistance to gastrointestinal parasites in sheep. *Animal Science* 64, 469–478.

Bishop, S.C. and Stear, M.J. (1999) Genetic and epidemiological relationships between productivity and disease resistance: gastrointestinal parasite infection in growing lambs. *Animal Science*, in press.

Buijtels, J., Huirne, R., Dijkhuizen, A., De Jong, M.C.M. and van Nes, A. (1997) Computer simulation to support policy making in the control of pseudorabies. *Veterinary Microbiology* 55, 181–185.

Byrom, W. and Gettinby, G. (1992) Using the computer model ECFXPERT to study ticks and East Coast fever. *Insect Science and Applications* 13, 527–535.

Casal, J., Learte, P. and Torre, E. (1990) A path model of factors influencing bovine leukemia virus transmission between cattle herds. *Preventative Veterinary Medicine* 10, 47–61.

Casal, J., Moreso, J.M., Planas-Cuchí, E. and Casal, J. (1997) Simulated airborne spread of Aujesky's disease and foot-and-mouth disease. *The Veterinary Record* 140, 672–676.

De Jong, M.C.M., Diekmann, O. and Heesterbeek, J.A.P. (1994) The computation of Ro for discrete-time epidemic models with dynamic heterogeneity. *Mathematical Biosciences* 119, 97–114.

De Jong, M.C.M., van der Poel, W.H.M., Kramps, J.A., Brand, A. and van Oirschot, J.T. (1996) Quantitative investigation of population persistance and recurrent outbreaks of bovine respiratory syncytial virus on dairy farms. *American Journal of Veterinary Research* 57, 628–633.

Diekmann, O., Heesterbeek, J.A.P. and Metz, J.A.J. (1990) On the definition and the computation of the basic reproduction ratio R_0 in models for infectious diseases in heterogeneous populations. *Journal of Mathematical Biology* 28, 365–382.

Dobson, R.J., Griffiths, D.A., Donald, A.D. and Waller, P.J. (1987) A genetic model describing the evolution of Levamisole resistance in *Trichostrongylus colubriformis*, a nematode parasite of sheep. *IMA Journal of Mathematics Applied in Medicine and Biology* 4, 279–293.

Dobson, R.J., Le Jambre, L. and Gill, J.H. (1996) Management of anthelmintic resistance: inheritance of resistance and selection with persistent drugs. *International Journal for Parasitology* 26, 993–1000.

Donnelly, C.A., Ferguson, N.M., Ghani, A.C., Woolhouse, M.E.J., Watt, C.J. and Anderson, R.M. (1997a) The epidemiology of BSE in cattle herds in Great Britain. 1. Epidemiological processes, demography of cattle and approaches to control by culling. *Philosophical Transactions of the Royal Society of London, Series B – Biological Sciences* 352, 781–801.

Donnelly, C.A., Ghani, A.C., Ferguson, N.M., Wilesmith, J.W. and Anderson, R.M. (1997b) Analysis of the bovine spongiform encephalopathy maternal cohort study: evidence for direct maternal transmission. *Applied Statistics* 46, 321–344.

Echevarria, F.A.M., Gettinby, G. and Hazelwood, S. (1993) Model predictions for anthelmintic resistance amongst *Haemonchus contortus* populations in southern Brazil. *Veterinary Parasitology* 47, 315–325.

Ferguson, N.M., Donnelly, C.A., Woolhouse, M.E.J. and Anderson, R.M. (1997a) The epidemiology of BSE in cattle herds in Great Britain. 2. Model construction and analysis of transmission dynamics. *Philosophical Transactions of the Royal Society of London, Series B – Biological Sciences* 352, 803–838.

Ferguson, N.M., Donnelly, C.A., Woolhouse, M.E.J. and Anderson, R.M. (1997b) Genetic interpretation of heightened risk of BSE in offspring of affected dams. *Proceedings of the Royal Society of London, Series B – Biological Sciences* 264, 1445–1455.

Fisher, R.A. (1930) *The Genetical Theory of Natural Selection.* Clarendon, Oxford.

Frank, S. (1994) Coevolutionary genetics of hosts and parasites with quantitative inheritance. *Evolutionary Ecology* 8, 74–94.

Gettinby, G. and Byrom, W. (1991) Weather-based computer experiments on parasites. *Preventative Veterinary Medicine* 11, 293–308.

Gettinby, G. and Gardiner, W.P. (1980) Disease incidence forecasts by means of climatic data. *Biometeorology* 7, 87–103.

Gettinby, G., Bairden, K., Armour, J. and Benitez-Usher, C. (1979) A prediction model for bovine ostertagiasis. *Veterinary Record* 105, 57–59.

Gettinby, G., Newson, R.M., Calpin, M.M.J. and Paton, G. (1988) A simulation model for genetic resistance to acaricides in the African brown ear tick, *Rhipicephalus appendiculatus* (Acarina: Ixodidae). *Preventative Veterinary Medicine* 6, 183–197.

Gettinby, G., Soutar, A., Armour, J. and Evans, P. (1989) Anthelmintic resistance and the control of ovine ostertagiasis: a drug action model for genetic selection. *International Journal for Parasitology* 19, 369–376.

Grenfell, B.T., Smith, G. and Anderson, R.M. (1987) A mathematical model of the population biology of *Ostertagia ostertagi* in calves and yearlings. *Parasitology* 95, 389–406.

Grenfell, B.T., Wilson, K., Isham, V.S., Boyd, H.E.G. and Dietz, K. (1995). Modelling patterns of parasite aggregation in natural populations: trichostrongylid nematode–ruminant interactions as a case study. *Parasitology* 111, S135–S151.

Haile, D.G., Mount, G.A. and Cooksey, L.M. (1992) Computer simulation of *Babesia bovis* (Babes) and *B. bigemina* (Smith & Kilborne) transmission by *Boophilus* cattle ticks (Acari: Ixodidae). *Journal of Medical Entomology* 29, 246–258.

Heesterbeek, J.A.P. and Roberts, M.G. (1994) Threshold properties for helminth infections. *Journal of Mathematical Biology* 33, 415–434.

Hone, J. (1994). A mathematical model of detection and dynamics of porcine transmissible gastroenteritis. *Epidemiology and Infection* 113, 187–197.

Hutber, A.M. and Kitching, R.P. (1996) The use of vector transition in the modelling of intraherd foot-and-mouth disease. *Environmental and Ecological Statistics* 3, 245–255.

Innocent, G., Morrison, I., Brownlie, J. and Gettinby, G. (1997) A computer simulation of the transmission dynamics and the effects of duration of immunity and survival of persistently infected animals on the spread of bovine viral diarrhoea virus in dairy cattle. *Epidemiology and Infection* 119, 91–100.

Kermack, W.O. and McKendrick, A.G. (1927) Contributions to the mathematical theory of epidemics – 1. *Proceedings of the Royal Society* 115A, 700–721. Reprinted as (1991) *Bulletin of Mathematical Biology* 53, 33–55.

Leathwick, D.M., Vlassof, A. and Barlow, N.D. (1992) A model of nematodiasis in New Zealand lambs. *International Journal for Parasitology* 22, 789–799.

Leathwick, D.M., Vlassof, A. and Barlow, N.D. (1995) A model of nematodiasis in New Zealand lambs: the effect of drenching regime and grazing management on the development of anthelmintic resistance. *International Journal for Parasitology* 25, 1–12.

Leslie, P. (1945) On the use of matrices in certain population mathematics. *Biometrika* 35, 183–212.

MacKenzie, K. and Bishop, S.C. (1999) A discrete-time epidemiological model to quantify selection for disease resistance. *Animal Science*, in press.

Miller, G.Y., Forster, D.L., Tsai, J. and Bech Nielsen, S. (1994) Predicting the number of herds infected with pseudorabies virus in the United States. *American Journal of Veterinary Research* 55, 628–635.

Pareto, V.F.D. (1906) *Manuale d'Economica Politica*. Societa Editrice Libraria, Milan.

Renshaw, E. (1991) *Modelling Biological Populations in Time and Space*. Cambridge University Press, Cambridge.

Roberts, M.G. (1994) Modelling of parasitic populations: cestodes. *Veterinary Parasitology* 54, 145–160.

Roberts, M.G. and Grenfell, B.T. (1991) The population dynamics of nematode infections of ruminants: periodic perturbations as a model for management. *IMA Journal of Mathematics Applied in Medicine and Biology* 8, 83–93.

Roberts, M.G. and Heesterbeek, J.A.P. (1995) The dynamics of nematode infections of farmed ruminants. *Parasitology* 110, 492–502.

Schmid-Hempel, P. and Koella, J.C. (1994) Variability and its implications for host–

parasite interactions. *Parasitology Today* 10, 98–102.

Smith, G. and Guerrero, J. (1993) Mathematical models for the population biology of *Ostertagia ostertagi* and the significance of aggregated parasite distributions. *Veterinary Parasitology* 46, 243–257.

Sørenson, J.T. and Enevoldsen, C. (1994) Dynamic stochastic simulation as a tool for studying bovine virus diarrhoea virus infections at the herd level. *Veterinary Research* 25, 317–321.

Stringer, S.M., Hunter, N. and Woolhouse, M.E.J. (1998) A mathematical model of the dynamics of scrapie in a sheep flock. *Mathematical Biosciences* 153, 79–98.

Tallis, G.M. and Leyton, M. (1966) A stochastic approach to the study of parasite populations. *Journal of Theoretical Biology* 13, 251–260.

Teel, P.D., Martin, S.L. and Grant, W.E. (1996) Simulation of host–parasite–landscape interactions: influence of season and habitat on cattle fever tick (*Boophilus* sp.) population dynamics. *Ecological Modelling* 84, 19–30.

Tillé, A., Lefèvre, C.L., Pastoret, P.-P. and Thiry, E. (1991) A mathematical model of rinderpest infection in cattle populations. *Epidemiology and Infection* 107, 441–452.

Turner, L.W., Wathes, C.M. and Audsley, E. (1993) Dynamic, probabilistic modeling of respiratory disease in swine, including production and economic effects. In: Collins, E. and Boon, C. (eds) *Proceedings of the Fourth International Livestock Environment Symposium, University of Warwick, Coventry, England. ASAE, St Joseph, Michigan, USA*, 6–9 July, 1993, pp. 89–97.

Turner, L.W., Gates, R.S., Wathes, C.M. and Audsley, E. (1997) Using fuzzy logic for assessment of dust and ammonia effects on swine respiratory disease. In: *Proceedings of the Fifth International Livestock Environment Symposium, Bloomington, Minnesota, USA*, 29–31 May, 1997, pp. 638–646.

Ward, M.P. and Carpenter, T.E. (1996) Simulation modeling of the effect of climatic factors on bluetongue virus infection in Australian cattle herds. I. Model formulation, verification and validation. *Preventative Veterinary Medicine* 27, 1–12.

Woolhouse, M.E.J. (1998) Scrapie: epidemiology. 1998 Wellcome Trust Research Workshop: 'Using Genetics to Control Sheep Diseases'. Pentlands Science Park, Midlothian, UK, 29 September, 1998.

Woolhouse, M.E.J., Stringer, S.M., Matthews, L., Hunter, N. and Anderson, R.M. (1998) Epidemiology and control of scrapie within a sheep flock. *Proceedings of the Royal Society of London, Series B* 265, 1205–1210.

The Immune System

3

M.J. Doenhoff
School of Biological Sciences, University of Wales, Bangor, UK

Summary

Three sets of genes affect resistance: those responsible for innate immunity, those that determine the specificity of the adaptive immune response, and those affecting the 'quality' of acquired immune reponses. In the small number of instances in which just a single gene/gene product is clearly responsible for a particular resistance/susceptibility profile, steady progress is being made in characterizing the gene and its product, and determining their mode of action, e.g. the macrophage protein 1 (*Nramp1*) gene effecting resistance against intracellular pathogens, and chemokine receptor genes, which render lymphocytes resistant to infection by human immunodeficiency virus 1 (HIV-1).

Adaptive immune responses are distinguished from innate immune mechanisms by a higher degree of specific reactivity for the inducing agent and recall memory. Adaptive responses can be cell-mediated and/or antibody-mediated, and involve three distinct cell lineages: antigen-presenting cells (APC), thymus-derived cells (T cells) and bursa- or bone marrow-derived cells (B cells). The specific activity of these cells against proteinaceous antigens is respectively determined by genes coding for three different types of antigen-recognition molecules: major histocompatibility complex (MHC) class I or class II proteins on APCs, T-cell antigen receptors (TCR) and immunoglobulins (Ig) on the surface of B cells. Surface CD1 molecules appear to be involved in recognition of lipid and glycolipid antigens by APCs.

There is only limited evidence that MHC, TCR and/or Ig genes which confer enhanced resistance against infection can be selected for. This contrasts with the 'quality' of the immune state that results from the complex interactions between APCs, T cells and B cells. T-helper (Th1) cells, interleukin-12 (IL-12) and γ-interferon (γ-IFN) are associated with cell-mediated inflammatory immune responsiveness, while Th2 cells, IL-10 and IL-4 drive antibody, particularly IgE, production. However, the recombinant DNA technologies involving gene 'knockout', 'transfection' and 'cloning' are more likely to have an important economic impact as an aid to tackling resistance to infection in farm animals.

© CAB *International* 2000. *Breeding for Disease Resistance in Farm Animals*
(eds R.F.E. Axford, S.C. Bishop, F.W. Nicholas and J.B. Owen)

Introduction

Individuals vary markedly in their capacity to resist, control and/or reject infections, and in their susceptibility to diseases in general. That such differences in susceptibility are to a considerable extent genetically determined allows for the possibility that 'useful' genetic traits can be selected. So far, however, in only relatively few instances have particular disease resistance or susceptibility traits been attributed significantly to single genes; in the majority of cases disease profiles appear to be multifactorial. Perhaps this is not surprising because, for example, 16 different parameters of immunity to *Trichinella spiralis* are known to vary between strains of mice (Wakelin, 1989), and five distinct regions of the mouse genome have been identified carrying genes affecting susceptibility to infection by species of the genus *Leishmania* (Blackwell, 1996). In humans at least 12 genes are suspected to contribute to resistance to malaria (Hill, 1996b).

Host defences against infection are classically differentiated into mechanisms of 'innate' immunity and the 'adaptive' immune response. In turn, genes that affect a host's response to a pathogen may be distinguished according to whether they: (i) control susceptibility or resistance to the acquisition of infection, i.e. are concerned with innate immunity; (ii) govern the specificity of acquired adaptive immune responses, the products of these genes being histocompatibility molecules and the antigen receptors on T cells (the T-cell receptor, TCR) and B lymphocytes (immunoglobulin, Ig); or (iii) affect the 'quality' of specific immune responses; for example specifying whether the response will be primarily cell-mediated and result in inflammation, or in the production of humoral antibody.

In the first edition of this book, note was taken of the considerable inherent potential for variation in the antigenic receptors possessed by cells involved in the adaptive immune response. There therefore seemed then to be limited scope for enhancing the specificity of acquired immune responses by selective breeding, and there is little new evidence from recent work to indicate that this conclusion is no longer valid. However, progress has been made, first, in defining the characteristics of some innate defence mechanisms at the molecular level, and secondly, in elucidating how the 'quality' of adaptive immune responses is regulated, particularly by subpopulations of lymphocytes and their cytokine products.

A Problem of Terminology

Much of what is written below will be concerned with how genes influence 'resistance' to infection and other diseases. Unfortunately when used in the literature this key word can have different connotations depending, for example, on whether the setting is experimental or clinical. Experimentally, host phenotypes defined as 'resistant' can range from those that are completely impervious to entry, let alone growth, of pathogens, through to those that allow pathogen survival and replication, but in which the infection is eventually controlled. Often, more virulent isolates of a pathogen will overcome the defences of hosts that have been typed as resistant to isolates of lesser virulence. There is sometimes discrepancy between different parameters that are used to estimate resistance, for example the apparent lack of association between measurements of cumulative host mortality and the number of brain cysts following peroral infection of different recombinant inbred strains of mice with *Toxoplasma gondii* (McLeod *et al.*, 1989). In veterinary practice it may therefore be necessary to measure several different parameters in order to obtain an accurate estimate of resistance, as quantitative phenotyping of N'Dama cattle for trypanotolerance has indicated (Trail *et al.*, 1994).

Innate Immunity

Factors traditionally listed as contributing to innate resistance include barriers to penetration such as tails, hair, skin, mucus and acidic secretions, ciliary activity in the respiratory tract, lysozyme in sweat and tears, and expropriation of habitable niches by normally harmless bacteria in the intestine and skin. At the molecular level, numerous factors contributing to innate immunity are now known. These include carbohydrate- and lipopolysaccharide-binding lectins, complement-activating and acute-phase proteins, the interferons and other cytokines, and the lipid-derived prostaglandins and leukotrienes (reviewed in Fearon and Locksley, 1996; Medzhitov and Janeway, 1997; Parish and O'Neill, 1997). Innate immunity is either expressed constitutively or its expression can be induced rapidly. It has a longer evolutionary history than the adaptive immune response, as illustrated by the recent finding that humans have a genetic homologue of the *Drosophila* Toll protein (Medzhitov *et al.*, 1997). In the invertebrate this is involved in antifungal protection (Lemaitre *et al.*, 1996). The human homologue of Toll has a cytoplasmic domain that is similar to that of the human interleukin-1 (IL-1) receptor (Medzhitov *et al.*, 1997).

Many innate factors suspected of conferring resistance to an infection appear to have no obvious relationship with, or effect on, the acquired immune response: the haemoglobinopathies and other red blood cell polymorphisms which protect humans against malaria (Alison, 1954; Weatherall *et al.*, 1988; Weatherall, 1996) are examples. However, elsewhere innate mechanisms of resistance do have an impact on the adaptive immune response, and in a way that is now being subjected to intensive scrutiny (see below).

The high degree of interspecies variation in susceptibility to particular diseases is perhaps a reflection of the importance of innate resistance mechanisms. There is marked variation between species of laboratory and domestic animals in susceptibility to agents such as anthrax, botulism and tetanus (Rumyantsev, 1992). This is probably mainly attributable to differences in innate rather than adaptive mechanisms of resistance, though the genetic or other factors responsible for such interspecies differences are largely unknown, and opportunities to exploit them are thus probably some way off.

The genetic and molecular factors responsible for differences in resistance are likely to be more readily identifiable by comparative studies between individuals in one host species, or between species that are closely related, because of the greater degree of homogeneity of genetic background thus provided. Progress in identification of protective factors may still be slow, however. Thus, even for a disease as economically important as that caused by the African trypanosomes, it is still not clear whether the distinction between the trypanotolerant host *Bos taurus* and the more susceptible *B. indicus* is due to innate or adaptive immunity (Vickerman *et al.*, 1993; Authie, 1994), although progress in identifying genetic regions determining susceptibility to trypanosomiasis is being made by comparative studies in mice and cattle (Kemp and Teale, 1998).

The problem is due, first, to the complexity of the genetic background of many innate mechanisms: for example, in cattle there are over 30 genes coding for type 1 interferons that have an effect on the severity of herpes-1 virus infections (Ryan and Womack, 1993, 1997), such high degrees of polymorphism appearing even in some defence mechanisms of invertebrates (Clark and Wang, 1997). Secondly, there is heritable variability in control of gene expression or the metabolism of gene products, as in the example of concentrations of the plasma protein, conglutinin. This is an inherited characteristic in cattle, and low levels of the protein predispose to respiratory infections (Holmskov *et al.*, 1998). It should also be remembered that many of the cellular and molecular elements that are involved in the complexity of an inflammatory response evolved initially as innate mechanisms, although they are now subject to considerable modulation by effectors of the adaptive immune response.

In only relatively few examples has a difference in the resistance or susceptibility to a disease been ascribed to mutations in one gene. One example that is currently the object of considerable interest is the gene for the chemokine receptor, CCR-5, that acts as a co-receptor with the CD4 molecule for infection of cells by the human immunodeficiency virus HIV-1 (Y. Feng *et al.*, 1996). Homozygosity for deletion of a 32 nucleotide base pair sequence in the *CCR5* gene confers resistance to infection by some non-syncytium-inducing, macrophage-tropic HIV-1 isolates in Caucasian populations (Liu *et al.*, 1996; Paxton *et al.*, 1998), and heterozygosity delays disease progression (Bratt *et al.*, 1998). HIV infection may have a slower rate of spread in populatons with high frequencies of the gene (Voevodin *et al.*, 1998). Another chemokine receptor, CXCR-4, appears to act as the receptor for T-cell tropic, syncytia-inducing HIV strains (Hesselgesser *et al.*, 1998).

An intensely studied determinant of innate resistance is the gene *Nramp* (natural resistance-associated macrophage protein) which confers resistance against some intracellular pathogens. It is worth recounting the *Nramp1* gene story in a little detail for the lessons it conveys, particularly with respect to the degree of specificity of the resistance conferred by single genes, and the difficulties inherent in trying to extrapolate results from one host species to another.

Nramp1, a gene controlling resistance to intracellular pathogens

Over 20 years ago Bradley found that inbred mouse strains segregated into two discrete groups with respect to the rate at which *Leishmania donovani* grew in the liver during the first 2 weeks after administration of the infection (Bradley, 1974, 1977). Hybridization and backcross studies indicated that this difference was under the control of a single autosomal gene. The gene was mapped to mouse chromosome 1 (Bradley *et al.*, 1979). It was noted that the distribution of *L. donovani*-resistant and susceptible mouse strains mimicked that which had been observed for infections of *Salmonella typhimurium* (Plant and Glynn, 1974), and subsequently of *Mycobacterium bovis* in the form of bacille Calmette–Guérin (BCG) (Skamene *et al.*, 1982). Further comparative and gene mapping studies confirmed complete concordance in patterns of resistance to the three intracellular pathogens (Blackwell, 1989a, b): hence the original designation of this gene *Bcg/Lsh/Ity*. A linkage group of five other loci congenic to *Lsh/Ity/Bcg* were found to be conserved in the same order on the distal region of human chromosome 2 (Schurr *et al.*, 1990).

One of two paralogous genes in mammals, *Nramp1* encodes an integral membrane phosphoglycoprotein that is structurally homologous with prokaryotic and eukaryotic ion transport or permeation systems (Vidal *et al.*, 1993; Govoni *et al.*, 1996). Its expression is restricted to 'professional' phagocytic cells, such as macrophages (Cellier and Gross, 1997). Although its specific function in this setting is still unclear, susceptibility to intracellular pathogen infection is associated with a glycine to aspartic acid substitution at amino acid residue 169 of the protein (Govoni *et al.*, 1996; Vidal *et al.*, 1996). There is evidence that *Nramp1* is involved in regulation of major histocompatibility complex (MHC) class II genes, expression of the latter being upregulated in the 'resistant' phenotype (Barrera *et al.*, 1997).

Mouse strains that are resistant to BCG are also unsusceptible to infection by other mycobacteria species, including *Mycobacterium lepraemurium*, *M. intracellulare* and *M. smegmatis* (Schurr *et al.*, 1990), but not to infection with *M. tuberculosis*, the pathogen principally responsible for human tuberculosis (Medina and North, 1996). The resistance to BCG could be overridden by administration of substrains of high virulence (Orme *et al.*, 1985). In humans there is also no support for the notion that the homologue of *Nramp1* controls resistance to tuberculosis (Blackwell, 1998), although it appears to be linked to resistance to leprosy, another important human mycobacterial disease (Abel *et al.*, 1998). The failure of *Nramp1* in mice to confer resistance against all species of mycobacteria applies also to *Leishmania* parasites. Thus, in inbred mouse strains the *L. donovani* resistance/susceptibility profile shows little concordance with that for cutaneous or viscerotropic infections of *L. major*, or for *L. mexicana* (Mock *et al.*, 1985; Blackwell, 1989b).

Homologues of the *Nramp1* gene have been found in the genomes of a range of farm and other animals, including those of bovines (J.W. Feng *et al.*, 1996), ovines (Pitel *et al.*, 1995), pigs (Tuggle *et al.*, 1997), chickens (Hu *et al.*, 1996; Girardsantosuosso *et al.*, 1997) and deer (Mathews and Crawford, 1998). An association between *Nramp1* and resistance to an intracellular pathogen (salmonellosis) has been found in chickens (Hu *et al*., 1997), and it seems likely that similar evidence will accrue for other species of intracellular pathogen and host.

In passing it may be noted that the *Nramp2* gene appears to be involved in iron transport, with mutations resulting in microcytic anaemia in mice (Fleming *et al*., 1997) and rats (Fleming *et al*., 1998).

The relationship between innate and acquired immunity

It was for long accepted that, with some exceptions (e.g. the complement system), there was little connection between innate mechanisms of resistance and the adaptive immune response. This view is changing, with hypotheses now being tabled suggesting that primary adaptive immune responses are initiated only after innate mechanisms have operated (Fearon and Locksley, 1996; Medzhitov and Janeway, 1997, 1998; Janeway, 1998). These authors argue that a set of molecules that are innate 'pattern recognition receptors' (PRR) have evolved and been selected for because they react with invariant molecular structures on pathogens (the latter being invariant because they are essential to the pathogenicity or survival of the invading organism). It is argued that the adaptive response can only become active after interaction between a pathogen and a PRR in the host. This results in production of one or more signals that mediate inflammatory responsiveness or are involved in co-stimulation and regulation of cells of the adaptive response (Medzhitov and Janeway, 1997).

As described below, mechanisms of innate and adaptive immunity show further signs of integration, with the products of cells of the adaptive response in turn influencing the activities of the innate system.

The Natural Killer (NK) Cell, an Intermediary Between Innate and Adaptive Immunity

NK cells are lymphoid cells that can kill a variety of virus-infected and tumour cells *in vitro* without the cell donor having been previously sensitized by the virus or tumour cell antigens. Unlike lymphocytes of the adaptive response, NK cells do not exhibit target-specific memory. They are larger and have more granular cytoplasm than T and B lymphocytes, properties that allow them to be isolated for study. In contrast to cytotoxic T cells, they are found in animals deficient in adaptive immune responsiveness. They do not exhibit the rearrangement of genetic elements typical of T-cell antigen receptors, and are not subject to the same degree of MHC restriction as cytotoxic T cells. MHC class I molecules do, however, regulate the function of NK cells (Lanier and Phillips, 1996; Rolstad and Seaman, 1998). Studies to define the nature of the receptors that NK cells use for recognition of target cells are still under way, but they may include lectin-like molecules, receptors for the immunoglobulin Fc, and other membrane-bound signal transducing molecules (Hofer *et al*., 1992; Yokoyama and Seaman, 1993).

As well as being important in the control of viruses (Delano and Brownstein, 1995; Bruunsgaard *et al.*, 1997) and bacteria (Way *et al.*, 1998), NK cells have been implicated in inhibition of protozoan infections (Laskay *et al.*, 1995; Sharton-Kerston and Sher, 1997). Their role seems to be particularly important in restricting the replication of pathogens during the very early stages of infection. However, NK cells may continue to be important even after immune responses are under way; for example, in patients who had low NK cell counts, HIV infections were found to progress more rapidly to AIDS (Bruunsgard *et al.*, 1997).

The activity of NK cells and adaptive immune responsiveness are coordinated by cytokines (Kos, 1998). In particular, NK cell metabolism, cytotoxicity and replication are enhanced by the cytokine interleukin-12 (IL-12). This is a cytokine that is produced by phagocytic cells that have been activated by exposure to pathogens or their products and it plays a crucial role in cell-mediated immune responses. In turn, activated NK cells produce γ-interferon (γ-IFN) that drives these adaptive responses (Sharton-Kerston and Scott, 1995; Trinchieri, 1995; Stern *et al.*, 1996; Unanue, 1997).

The Adaptive Immune Response

The characteristics that distinguish adaptive immune responses are 'specificity' of interaction with antigen, their ability to distinguish 'self' from 'non-self' and 'their possession of memory'. 'Specificity' is also shown by the above-mentioned PRRs of innate immunity, and 'self' becomes the target of the adaptive response in autoimmune diseases. This leaves 'memory' (i.e. the production of faster and quantitatively greater and qualitatively improved responses against a second and subsequent challenges of the same pathogen or antigen than against the first), as one of the distinctive characteristics of adaptive immunity. Another important distinction is the mechanism of rearrangement of DNA elements involved in genes expressing antigen receptors on lymphocytes.

Three cell types are involved in generation of adaptive immune responses: antigen-presenting cells (APC), thymus-derived lymphocytes (T cells) and bursal-equivalent or bone marrow-derived lymphocytes (B cells). The effector mechanisms generated by immune responses can be differentiated into those that are cell-mediated, and independent of B cells and antibody, or humoral, in which interactions between all three cell types culminate in the production of specific antibody. The specificity of an adaptive immune response is governed by antigen handling and recognition molecules on each of the three cell types: major histocompatibility antigens (MHC) on antigen-presenting cells (APC), T-cell receptors (TCR) on T cells and immunoglobulin (Ig) molecules on B cells.

Antigen-presenting Cells and MHC Molecules

APCs 'present antigen' to T cells at two different times in the lifetime of the latter. The first is during the differentiation and maturation of the T cells in the thymus, prior to these cells being seeded into the periphery in an immunocompetent state. The exact mechanisms are unclear, but it is accepted that the interactions between precursor T cells and APCs, mediated through respective cell-surface-borne TCR and MHC molecules (the latter carrying 'self'-derived peptides) are important in ensuring correct selection of the mature T-cell repertoire. Current hypotheses invoke the avidity of interaction between MHC and the newly expressed TCR on a developing T cell to explain T-cell selection (Ashton-Rickardt and Tonegawa, 1994). Thus, failure of the TCR to recognize 'self' MHC in the thymus results in death of the T cell by apoptosis, predicating that a mature T cell will only be activated if its TCR reacts with antigen that is presented by 'self' MHC. A cell with a TCR which, in the thymus, reacts avidly with MHC bearing self-derived peptides will also be eliminated, i.e. 'negative selection', which accounts for 'self-tolerance' in adaptive immunity. There is thus considerable cell attrition in the thymus (Nossal, 1994). The only developing T cells that survive are the minority which express TCRs that interact with peptide-bearing MHC molecules in the thymus with an intermediate avidity ('positive selection'). These emerge into the mature peripheral T-cell pool, where they can detect 'non-self' antigenic peptides presented by 'self' MHC. Early observations showed that 'histocompatibility' was required for activation of T cells by APC (Rosenthal and Shevach, 1973). Antigen-specific interaction between helper T cells and B cells (Kindred and Schreffler, 1972) and the killing of target cells by cytotoxic T cells (Zinkernagel and Doherty, 1997) were thus accounted for.

MHC molecules are found throughout the vertebrates. Mammalian *Mhc*, coding for 'classical' MHC class I and class II antigen-presenting molecules, is the most important set of genetic loci responsible for tissue allograft rejection. However, many other proteins (class III), including those involved in antigen peptide processing, complement components, heat-shock proteins and tumour necrosis factors (TNF) are also encoded within the *Mhc* complex.

Classical class I and II MHC molecules have four characteristics by which their function is defined: a high degree of polymorphism, high-level expression in particular cells, and the ability to bind small polypeptide molecules and present them to T cells (Kaufman *et al.*, 1994). The peptides constituting the two types of molecule are distinctive members of the Ig superfamily of proteins. Class I molecules are composed of a heavy chain (40–45 kDa) comprised of three Ig-like domains, two of which (1 and 2) form the peptide-binding groove lying above the 3 domain and a non-covalently associated and lighter (12 kDa) β_2-microglobulin chain. The structure of a class I molecule was originally derived by X-ray crystallography by Bjorkman *et al.* (1987a, b), and it provides a good illustration of concordance between the structure and the antigen (peptide)-binding function of the molecule. Class II molecules are also heterodimers, but in contrast to class I molecules, are each comprised of an α and a β peptide chain of approximately 35 kDa and 28 kDa, respectively. Nevertheless, the deduced tertiary configurations of class II molecules (Brown *et al.*, 1993) are nearly superimposable, thus giving class II molecules an antigen peptide-binding groove similar to that possessed by class I molecules. Not unexpectedly, the high degree of genetic polymorphism in MHC is localized mainly in, or close to, the peptide-binding groove areas of the class I and class II molecules.

Class I and class II MHC molecules are different in some respects, such differences being important with regard to their respective roles in antigen presentation. Class I molecules are expressed on the surfaces of virtually all nucleated cells. The antigenic peptides that associate with class I MHC are in general eight or nine amino acids long (Rammensee *et al.*, 1993), and derive mostly from endogenous proteins in the cytosol or nucleus of the cell carrying the class I molecules. Class I molecules present antigenic peptides mainly to the TCRs of $CD8^+$ cytotoxic T cells, the principal immune function of which is considered to be the killing of virus-infected cells and tumour cells. The distribution of class II molecules is, however, restricted to those cells deemed to have antigen uptake, processing and presentation functions (e.g. macrophages, dendritic cells and B cells). Peptides carried by class II molecules are thus derived from exogenous proteins that have been ingested and processed for presentation to the TCR of $CD4^+$ helper T cells. These are generally longer than those presented by class I molecules (Chicz *et al.*, 1993).

MHC molecules and disease

The discovery that MHC molecules were involved in antigen recognition led to the suspicion that the high rate of polymorphism of MHC molecules is an adaptation that provides protection against pathogens, either in terms of resisting or reducing the severity of infection. The advantages could be in terms of the heterozygous state itself allowing qualitatively better immune response to be elicited than homozygosity. It could also be due to the persistence of rare useful alleles in the population.

In West Africa, class I antigen HLA-B53 and class II antigen HLA-DRB1*1302 were found significantly less frequently in children with severe falciparum malaria infections (Hill *et al.*, 1991). Analysis of the HLA-B53 association indicated that the mechanism responsible was the generation of cytotoxic T cells by a particular parasite epitope presented by this MHC molecule (Hill *et al.*, 1992).

In other animals, one of the clearest examples of an association between MHC and disease resistance is that of the B21 haplotype and Marek's disease, caused by a herpesvirus, in chickens. In birds the *Mhc* gene complex seems to be much smaller and less complex than in mammals (Kaufman *et al.*,1995). More recently *Rfp-Y* haplotypes have also been shown to have an influence on Marek's disease independent of that of the *B* gene complex (Wakenell *et al.*, 1996).

Resistance possibly associated with the MHC complex has been noted in respect of viral hepatitis B (Thursz, 1997), tuberculoid leprosy (van Eden *et al.*, 1985) and chronic Lyme disease (Dwyer and Winchester, 1993) in humans; susceptibility to lymphocytic choriomeningitis (Oldstone *et al.*, 1973) in mice; and susceptibility to lymphocytosis caused by bovine leukaemia virus (Lewin and Bernoco, 1986) and infection-induced high somatic cell counts in milk

(Dietz *et al*., 1997) in cattle. Direct evidence that MHC alleles which increase disease severity have been negatively selected may be difficult to find, but long-term studies on the outcome of an apparent association between DR5 and HIV infection (Cruse *et al*., 1991) could prove interesting in this respect.

There is contention over whether MHC gene polymorphism evolved primarily as a defence against infectious agents. Studies of animal models of infectious diseases have often indicated that the cumulative influence of other 'background' genes is of greater significance than MHC in defining resistance to infection (Wakelin and Blackwell, 1988). There may thus be other reasons why MHC gene polymorphism evolved, including mate selection (Apanius *et al*., 1997), and that once it had evolved, it was retained as a useful adaptation for control of infectious diseases.

The evidence for associations between *Mhc* and autoimmune disease are less equivocal. That between human class I *HLA-B27* and the human rheumatic disease ankylosing spondylitis was one of the earliest to be recognized (Brewerton *et al*., 1973). Other human autoimmune diseases that are associated with particular HLA class II haplotypes associations include rheumatoid arthritis, insulin-dependent diabetes and systemic lupus erythematosus. There are several hypotheses to explain these associations, including the presentation of pathogenic peptides by MHC molecules, inappropriate selection of the T-cell repertoire by MHC/self peptide combinations in the thymus, and cross-reactivity between peptides derived from microorganisms and self-proteins.

CD1 molecules and lipid antigen presentation

The 'classical' mode of processing and presenting antigens, performed by molecules of the MHC complex, is generally considered to be restricted to proteins and their peptide derivatives. It has recently been found that molecules of the CD1 family of proteins also have a role in antigen presentation to T cells, particularly with respect to antigenic lipid and glycolipid molecules (Porcelli *et al*., 1998). The importance of this new pathway of antigen presentation in disease control has yet to be determined.

T Cells and T-cell Receptors

Two main lineages of T cell can be distinguished by the nature of the T-cell receptor (TCR) heterodimer they express on their surface membranes, i.e. either $\alpha\beta$ or $\gamma\delta$ receptors. Those expressing the $\alpha\beta$ heterodimer account for nearly all known antigen-specific, T-cell-dependent regulatory and effector functions of humoral and cellular immune responses (Hedrick and Eidelman, 1993). The α and β peptides of the heterodimer are covalently associated by disulphide bonding. They are glycoproteins of approximately 42–45 kDa, with glycosylation accounting for approximately 10 kDa.

The primary structure of the TCR heterodimer is determined by a process of gene rearrangement that occurs during the maturation of T cells in the thymus. Up to four genetic elements are randomly incorporated into the mature gene: these are the element for the carboxy-terminal constant domain and elements that code for variable (V), joining (J) and, in the case of β chains, diversity (D) regions of the variable amino-terminal antigen-recognition portion of the receptor molecule. Random addition of nucleotides at V–J, V–D and D–J junctions during gene rearrangement (i.e. N-region addition) creates further structural diversity in the TCR repertoire (Lewis and Gellert, 1989; Hedrick and Eidelman, 1993). The rearrangement process for TCR genes allows V-region and J-region elements to combine randomly, such that, for example, in construction of the mouse α chain one of 100 V elements can combine with one of 50 J elements to give up to 5000 different sequences. Similarly, about 500 different β chain genes can be made up from 20 V genes, 2 D regions and 12 β J genes. It is considered unlikely that all the different gene elements have equal likelihood of being used, but theoretically random combinations of α and β chains can thus give rise to 2.5×10^6 heterodimers. When all mechanisms for creating diversity in TCR structure are taken into account it is calculated that up to 10^{15} different αβ receptors are possible (Davis and Bjorkman, 1988).

When in its membrane position, the heterodimer is associated with several other peptides, forming the CD3 complex. Two subpopulations of T cells emerge from the thymus: $CD4^+$ helper T cells and $CD8^+$ cytotoxic T cells. The CD4 and CD8 accessory molecules determine that the reactivity of the TCRs on these cells is directed, respectively, towards class II and class I MHC molecules on antigen-presenting cells (see above). T cells generally will react to antigenic peptides only if the latter are presented to the TCR in the context of 'self-MHC'. However, some products of microbial and protozoan pathogens, the so-called 'superantigens', can short-circuit the relative specificity of the interaction between TCR and MHC/antigen peptide, and cause polyclonal T-cell activation (Herman *et al.*, 1991).

T-helper cells can be divided into two subpopulations, Th1 and Th2 cells, depending on the respective pattern of cytokines produced by the cells during the course of immune activation (see below). The thymus also produces another population of mature T cells bearing a second type of receptor, the heterodimeric TCR that is also expressed in association with CD3. T cells with γδ receptors do not carry the CD4 or CD8 molecules. Diversity in the γδ peptides is generated in a manner similar to that described for the αβ receptor, with an estimated 10^{18} different constructs possible. T cells with αβ TCRs tend to concentrate in epithelial tissues such as skin (Alaibac *et al.*, 1997) and intestine (Kagnoff, 1998). Their function is not yet well defined, but there is some evidence for their involvement in resistance to microbial infections (Ladel *et al.*, 1995; Takahashi *et al.*, 1996).

During the course of an infection, T cells with particular TCRs may be selected (Wang *et al.*, 1993). However, as the foregoing makes evident, the potential for diversity of T-cell receptors is considerably larger than the number of T cells in mammalian bodies. It is therefore perhaps not surprising that gaps in the T-cell repertoire have not so far been held to account for a general failure in immune responsiveness to infectious agents. SJL mice and related strains, and NZW mice are known to have a deletion on the β-chain gene region (Behlke *et al.*, 1986). They do not, however, appear to have an increased propensity to infection, although an increase in prevalence of autoimmune disorders is associated with the NZW phenotype (Noonan *et al.*, 1986; Kotzin and Palmer, 1987).

It is commonly held that activation of T cells via the TCR requires a high degree of specificity of interaction between the receptor and the antigenic peptide presented to it 'in the context of self MHC'. However, Mason (1998) has recently argued that an essential feature of a T cell is its ability to cross-react productively with a very wide range of MHC-associated peptide epitopes.

B Cells and Immunoglobulins

B lymphocytes mature in the bursa of Fabricius of birds (hence the abbreviation; Roitt *et al.*, 1969) or the bone marrow of mammals, and are specialized for the synthesis and secretion of specific antibodies in the form of immunoglobulins (Ig).

The genes for coding for Ig proteins are generated by gene segment rearrangement processes that occur during B-cell maturation in the primary lymphoid organ source, and that are similar to those described for TCR genes, but elucidated earlier (Tonegawa, 1983). Thus, Ig heavy-chain diversity results from recombination of V, D and J gene segments, with N region additions, while light-chain diversity stems from recombination of V and J regions (Max, 1993). Theoretical calculations indicate a potential for up to 10^{11} different immunoglobulin variable domain sequences for the total antibody repertoire (Davis and Bjorkman, 1988). As with T cells, the B-cell antigen-receptor repertoire is subject to education in respect of self and non-self discrimination (tolerance), although there is still debate as to how this is achieved (de St Groth, 1998).

Somatic mutation is a further mechanism for generation of additional B-cell antigen-receptor diversity, and it contributes to increasing affinity of antibody produced during the course of an immune response (French *et al.*, 1989; Wabl and Steinberg, 1996). Somatic mutation or its equivalent has not been found in TCR genes.

During the course of an immune response the genetic elements coding for the variable domains of the Ig molecule (Fab) can recombine successively with different constant-region genes coding for the various Ig isotypes (i.e. class or isotype switching). The constant domains, constituting the Fc portion of the molecule, define biological properties of the antibody molecule, giving the different isotypes particular properties; for example, activation of complement or the ability to interact with Fc receptors on the surfaces of granulocytes and other cells.

V(D)J recombination in developing T cells and B cells is controlled by two 'recombinase activation genes' *RAG-1* and *RAG-2* (Alt *et al.*, 1992). Loss of *RAG* gene function in mice results in severe combined immunodeficiency, due to the absence of mature T and B lymphocytes in central or peripheral lymphoid organs and tissues (Mombaerts *et al.*, 1992; Shinkai *et al.*, 1992). Mutation in a different gene, *scid*, which also gives rise to severe combined immunodeficiency due to inability of lymphocytes to undergo normal V(D)J recombination, was discovered in mice by Bosma *et al.* (1983). The putative function of the product of this gene is general repair of breaks in double-stranded DNA (Bosma and Carroll, 1991). The genes and gene products that affect the DNA recombination events resulting in isotype (Ig class) switching have not been identified, although T-cell-derived cytokines are influential in this process (Coffman *et al.*, 1993; Snapper and Mond, 1993).

As with T-cell subpopulations, there is evidence for the existence of distinct lineages of B cells, with $CD5^+$ (B1) cells being ontogenetically the first subset of B cells to be produced by the bone marrow. Compared with other B cells, B1 cells have minimal rearrangement of germ-line genes and express mainly IgM low avidity and autoreactive antibodies (Lydyard and Youinou, 1994). The role of this subpopulation in the adaptive immune response has yet to be fully clarified.

In conclusion of this section, it may be noted that selection for the capacity to produce antibody responses of high titre *per se* does not necessarily result in good protective immunity. Thus, 'Biozzi' mice selected for high antibody responsiveness were not consistently more resistant to infection than those selected for low responsiveness (Biozzi *et al.*, 1985). Studies in sheep have also shown that: (i) there was no relationship between either immunoglobulin λ chain constant region or either TCR or chain polymorphisms and resistance to *Haemonchus contortus* (Blattman and Beh, 1994); and (ii) resistance to blowfly strike caused by *Lucilia caprina* was found to be independent of the titre of antibody reactive against the insect's antigens (O'Meara *et al.*, 1997). The ability of B cells to recognize certain epitopes as immunodominant may be more critical. For example, it is suggested that trypanotolerance in N'Dama cattle is attributable to their ability to respond to cryptic antigens of the variable surface glycoprotein of *Trypanosoma congolense* with higher antibody titres than infection-susceptible cattle (Williams *et al.*, 1996).

Thus, in summary, T cells and B cells mature and differentiate in the thymus and bone marrow (or bursa) respectively, these being central or primary lymphoid organs. The mature cells are immunologically reactive by virtue of possessing receptors for recognition of antigen, and selection of the T-cell repertoire is governed by antigen-presenting cells in the thymus presenting 'self' peptides. The mature cells are disseminated throughout the body, and for the most of their lifetimes remain inactive as small lymphocytes circulating throughout the body via blood and lymph. They are found in high densities in certain areas (i.e. T-dependent or B-dependent areas) of peripheral or secondary lymphoid organs, including the spleen, lymph nodes and Peyer's patches. Such tissues provide an appropriate environment for T cells and B cells to respond to an antigen for the first time in a primary immune response. The elicitation of T-cell responsiveness requires that non-self, foreign antigenic peptides be presented to the TCR, again by histocompatible antigen-presenting cells. After participation in an immune response, cells may leave the lymphoid organ in which the reaction occurred, some B cells completing their differentiation into antibody-secreting cells elsewhere. Other T and B cells dedifferentiate into small lymphocyte form again and resume circulation and recirculation throughout the body.

As a result of a primary immune response, immunological memory, the principal factor that distinguishes adaptive from innate immunity, is established. It is due to this memory that re-exposure to the same antigen induces an immune response that, when compared with the primary response, begins earlier, is quantitatively larger (e.g. higher antibody titres) and qualitatively better (e.g. antibody with higher affinity for the antigen). Both T cells (Dutton *et al.*, 1998) and B cells (Lane, 1996) contribute to immunological memory. B cells that have responded to antigen previously can be relatively easily recognized through possession of Ig genes that have undergone somatic mutation. In contrast, the memory component in T cells may be more easily detectable as a function rather than as cells with a distinct phenotype, although memory T cells may have distinct surface markers (Dutton *et al.*, 1998).

Regulation of the Immune Response

Classic experiments indicated that the properties of the antigen influenced the nature of the resulting immune response; for example, whether the response would be mainly cell-mediated inflammation or antibody production (Parish, 1996). There are, however, very few adaptive immune responses that are independent of mediation by T cells. Thus, one or more subpopulations of T cells are involved in all 'cell-mediated' responsiveness, and in most antibody responses. The exceptions are antibody responses to antigens such as polysaccharides and some proteins and glycoproteins with repeated and/or polymeric epitopes. They are known as T-cell-independent antigens type 2 (not to be confused with 'Type-2' immune responses discussed below), and can stimulate antibody production in the absence of MHC class II-restricted T-cell help (Mond *et al.*, 1995).

It has also been clear for some time that subpopulations of T cells have distinct immunological roles, with $CD8^+$ Tc cells being cytotoxic and responding to endogenously produced antigenic peptides presented by MHC class I. These properties suit Tc cells for a defensive role against viral and other intracellular pathogens. In contrast, $CD4^+$ T-helper (Th) cells respond to antigenic peptides presented by MHC class II molecules that occur on a more restricted range of antigen-presenting cells. Th cells regulate the activities of B cells with which they cooperate in antibody responses to thymus-dependent antigens, and of other cell types such as macrophages and $CD8^+$ Tc cells during the course of cell-mediated immune responses.

Type-1 and Type-2 Responses and the Cytokine Network

Current research in immunology is being profoundly influenced by the observations of a decade or so ago which showed that, at least in the mouse, there are two subpopulations of $CD4^+$ Th cells. The respective populations exercise their pivotal role in adaptive immune responses through a differential pattern of cytokine secretion (Mosmann and Coffman, 1987, 1989; Bottomly, 1988). Mice that had different resistance properties with respect to cutaneous *Leishmania major* infections were instrumental in setting the scene for this revolution, with resistant mice displaying Th1-type responsiveness and susceptible mice Th2-type responsiveness (Heinzel *et al.*, 1989; Reiner and Locksley, 1995). While γ-IFN and IL-4 secretion typify the responsiveness of Th1 and Th2 cells respectively, it is now clear that overall patterns of cytokine production are more complex, and that reactive Th cells can, and do, secrete a range of molecular messages that regulates the activities of other cell types. Thus, in addition to γ-IFN, Th1 cells secrete IL-2 and lymphotoxin(s), and Th2 cells produce IL-5, IL-6, IL-9, IL-10 and IL-13 in addition to IL-4 (Abbas *et al.*, 1996). An immune response can be dominated by one Th cell subset as a result of the cytokines it produces having a down-modulatory effect on the other subset. Th1-type (Type-1) responses are often categorized as cell-mediated or inflammatory, while Th2 cell-mediated (Type-2) responses are more associated with humoral immunity. These distinctions are not exclusive, however, and production of antibodies of some isotypes, particularly IgG1, IgG2a and IgG2b, can occur during Type-1 reactions in mice.

Differentiation of progenitor T cells into Th1 cells is promoted by IL-12 that, as noted above, is produced by phagocytic cells early in the response to an infection. Mice, which in the wild-type state express Type-1 resistance to infection with *L. major*, become susceptible when their genes for IL-12 expression are disrupted (Mattner *et al.*, 1996). Administration of this cytokine simultaneously with antigen can be used to solicit preferentially Type-1 responses. For example, it has been used in mice to increase the degree of resistance induced against *Klebsiella pneumoniae* (Greenberger *et al.*, 1996), *Mycobacterium tuberculosis* (Flynn *et al.*, 1995), *Cryptosporidium parvum* (Urban *et al.*, 1996), *Cryptococcus neoformans* (Kawakami *et al.*, 1996) and *Schistosoma mansoni* (Wynn *et al.*, 1996). It is likely that in many of these situations the enhancing activity of IL-12 activity is actually mediated through $CD4^+$ cells and γ-IFN (Schijns *et al.*, 1996; Sharton-Kerston *et al.*, 1996, 1998; Mohan *et al.*, 1997; Romani *et al.*, 1997). The extent to which IL-12 can be exploited to enhance particular immune responses could, however, be limited by its potential for inducing damage due to its potent pro-inflammatory activity (Alexander *et al.*, 1997).

It is clear that mouse immunology is being revolutionized by the Th1/Th2 paradigm. Evidence indicating that immune responses of other animal species are similarly regulated is accumulating more slowly, however, and in any case there are reasons for treating the seemingly emphatic message from mice with caution. The classification of an immune response as Type-1 or Type-2 does not always correspond with intuition: for example, the granulomatous response around tissue-bound *S. mansoni* eggs is clearly cell-mediated inflammation, but evidence from cytokine expression patterns indicates a Type-2 profile (Brunet *et al.*, 1998). Many of the characteristics of immune responses to helminth infections (IgE, eosinophilia, mast-cell involvement) are Type-2 (Finkelman *et al.*, 1997), yet it is not always clear to what extent those responses contribute to host protection or parasite survival (Finkelman *et al.*, 1991; Wilson, 1993). For example, in *S. mansoni* infections in mice, depletion of the ability of the host to synthesize IgE antibodies gave no advantage to the parasite (Amiri *et al.*, 1994; El Ridi *et al.*, 1998). Immune responses against infectious agents are complex affairs that cannot always be packaged neatly into compartments labelled Type-1 or Type-2 (Allen and Maizels, 1997).

Some Remaining Problems . . .

Parasites are subject to the 'Red Queen' predicament, and to survive have had to evolve and adapt themselves to the increasingly sophisticated defence mechanisms that have developed in their hosts. Mechanisms such as immunosuppression, antigenic variation and drift, antigenic mimicry, and adoption of intracellular or other immunologically privileged sites, have long been seen as strategies by which pathogenic organisms evade host immune defences. However, the new information about how immune responses are regulated is, in turn, revealing novel mechanisms of parasite and pathogen survival: these include interference with antigen processing and presentation (Hill, 1996a; Garcia *et al.*, 1997; Plebanski *et al.*, 1997), modulation of cell-signalling pathways (Schofield and Tachado, 1996), inhibition of apoptotic cytolysis (Cuff and Ruby, 1996) and modulation of the cytokine network (Riffkin *et al.*, 1996; Donelson *et al.*, 1998). That many parasites, both micro- and macro-, are well adapted and integrated into the lives of their hosts is indicated not only by their ability to evade immune and other defences of the host, but by their ability also to actively exploit them (Doenhoff, 1997; Doenhoff and Chappell, 1997).

Even if a high degree of host resistance can be selectively bred or otherwise induced (see below), the inherent genotypic polymorphism that exists in many species of pathogen may ensure that new infective isolates can evolve.

Another perceived problem is that when resistance occurs, or is selected for, it is effective against only a limited number, often only one, species of pathogen. This has been noted above with respect to species of *Leishmania* and mycobacteria. Another example is the ability of some breeds of sheep to resist infections of *Fasciola gigantica*, but not *F. hepatica* (Spithill *et al.*, 1997).

. . . and Future Solutions . . .

The science of immunology is at an exciting crossroad. Discoveries made in the 1960s and 1970s about the function of the thymus and cell cooperation in immune responses laid the basis for the rapid progress that has been made in the past two decades in understanding how adaptive immune responses are regulated, and integrated with innate defences. Accumulation of knowledge has been accelerated by recombinant DNA methodologies that have generated genetically modified experimental animals more rapidly than by the selection and breeding of desired traits.

The new methods of gene 'transfection' and 'knockout' are likely to be exploited for the generation of farm animals with desirable characteristics (Muller and Brem, 1996; Osterrieder and Wolf, 1998). Genes that give improved responses to vaccination and chemotherapy, as well as those that confer resistance to disease *per se*, are likely to be included in future work programmes. In view of recent arguments that some 'germiness' in our lives is beneficial (Rook and Stanford, 1998), it may be appropriate also to apply the new technologies to the design of pathogens that have reduced pathogenicity while retaining whatever useful features they may have.

Ultimately, selective breeding and other means of eliciting genetically based resistance are likely to prove most effective when they are part of an integrated approach to management of disease. Many other aspects of good animal husbandry, including nutrition, biological control, prophylaxis and therapy, are important parts of the whole picture (Uilenberg, 1996; Fox, 1997).

References

Abbas, A.K., Murphy, K.M. and Sher, A. (1996) Functional diversity of helper T lymphocytes. *Nature* 383, 787–793.

Abel, L., Sanchez, F.O., Oberti, J., Thuc, N.V., Van Hoa, L., Lap, V.D., Skamene, E., Lagrange, P.H. and Schurr, E. (1998) Susceptibility to leprosy is linked to the human *NRAMP1* gene. *Journal of Infectious Diseases* 177, 133–145.

Alaibac, M., Morris, J. and Chu, A.C. (1997) Gamma delta T-cells in human cutaneous immunity. *International Journal of Clinical and Laboratory Research* 27, 158–164.

Alexander, J., Sharton-Kerston, T.M., Yap, G., Roberts, C.W., Liew, F.Y. and Sher, A. (1997) Mechanisms of innate resistance to *Toxoplasma gondii* infection. *Philosophical Transactions of the Royal Society of London Series B – Biological Sciences* 352, 1355–1359.

Alison, A.C. (1954) Protection afforded by sickle cell trait against subtertian malarial infection. *British Medical Journal* 1, 290–294.

Allen, J.E. and Maizels, R.M. (1997) Th1–Th2: reliable paradigm or dangerous dogma. *Immunology Today* 18, 387–392.

Alt, F.W., Oltz, E.M., Young, F., Gorman, J., Taccioli, G. and Chen, J. (1992) *VDJ* recombination. *Immunology Today* 13, 306–314.

Amiri, P., Haakfrendschi, M., Robbins, K., McKerrow, J.H., Stewart, T. and Jardieu, P. (1994) Antiimmunoglobulin-E treatment decreases worm burden and egg-production in *Schistosoma mansoni*-infected normal and interferon-gamma knockout mice. *Journal of Experimental Medicine* 180, 43–51.

Apanius, V., Penn, D., Slev, P.R., Ruff, L.R. and Potts, W.K. (1997) The nature of selection on the major histocompatibility complex. *Critical Reviews in Immunology* 17, 179–224.

Ashton-Rickardt, P.G. and Tonegawa, S. (1994) A differential avidity model for T-cell selection. *Immunology Today* 15, 362–366.

Authie, E. (1994) Trypanosomiasis and trypanotolerance in cattle: a role for congopain. *Parasitology Today* 10, 360–364.

Barrera, L.F., Kramnik, I., Skamene, E. and Radzioch, D. (1997) I-a-beta gene expression regulation in macrophages derived from mice susceptible or resistant to infection with m-bovis bcg. *Molecular Immunology* 34, 343–355.

Behlke, M.A., Chou, A.S., Huppi, K. and Loh, D.Y. (1986) Murine T-cell receptor mutants with deletions of beta-chain variable regions. *Proceedings of the National Academy of Sciences USA* 83, 767–771.

Biozzi, G., Mouton, D., Siqueira, M. and Stiffel, C. (1985) Effect of genetic modification of immune responsiveness on anti-infection and anti-tumour resistance. *Progress in Leukocyte Biology* 3, 3–18.

Bjorkman, P.J., Saper, M.A., Samraoui, B., Bennett, W.S., Strominger, J.L. and Wiley, D.C. (1987a) Structure of the human class I histocompatibility antigen, HLA-A2. *Nature* 329, 506–512

Bjorkman, P.J., Saper, M.A., Samraoui, B., Bennett, W.S., Strominger, J.L. and Wiley, D.C. (1987b) The foreign antigen binding site and T-cell recognition of class I histocompatibility antigens. *Nature* 329, 512–518.

Blackwell, J.M. (1989a) Bacterial infections. In: Wakelin, D.M. and Blackwell, J.M. (eds) *Genetics of Resistance to Bacterial and Parasitic Infection.* Taylor and Francis, London, pp. 63–101.

Blackwell, J.M. (1989b) Protozoan infections. In: Wakelin, D.M. and Blackwell, J.M. (eds) *Genetics of Resistance to Bacterial and Parasitic Infection.* Taylor and Francis, London, pp. 103–151.

Blackwell, J.M. (1996) Genetic susceptibility to leishmanial infections: studies in mice and man. *Parasitology* 112 (Suppl.), S67–S74.

Blackwell, J.M. (1998) Genetics of host resistance and susceptibility to intramacrophage pathogens: a study of multicase families of tuberculosis, leprosy and leishmaniasis in north-eastern Brazil. *International Journal for Parasitology* 28, 21–28.

Blattman, A.N. and Beh, K.J. (1994) T-cell receptor and immunoglobulin polymorphisms and resistance to *Haemonchus contortus* in sheep. *Journal of Animal Breeding and Genetics* 111, 65–74.

Bosma, G.C., Custer, R.P. and Bosma, M.J. (1983) A severe combined immunodeficiency mutation in the mouse. *Nature* 301, 527–530.

Bosma, M.J. and Carroll, A.M. (1991) The SCID mouse mutant: definition, characterization and potential uses. *Annual Review of Immunology* 9, 323–350.

Bottomly, K. (1988) A functional dichotomy in $CD4^+$ lymphocytes-T. *Immunology Today* 9, 268–274.

Bradley, D.J. (1974) Genetic control of natural resistance to *Leishmania donovani. Nature* 250, 353–354.

Bradley, D.J. (1977) Regulation of *Leishmania* populations within the host. II. Genetic control of acute susceptibility of mice to *Leishmania donovani* infection. *Clinical and Experimental Immunology* 30, 130–140.

Bradley, D.J., Taylor, B.A., Blackwell, J.M., Evans, E.P. and Freeman, J. (1979) Regulation of *Leishmania* populations within the host. III. Mapping of the locus controlling susceptibility to visceral leishmaniasis in the mouse. *Clinical and Experimental Immunology* 37, 7–14.

Bratt, G., Leandersson, A.S., Albert, J., Sandstrom, E. and Wahren, B. (1998) MT-2 tropism and CCR-5 genotype strongly influence disease progression in HIV-1-infected individuals. *AIDS* 12, 729–736.

Brewerton, D.A., Caffery, M., Hart, F.D., James, D.C.O., Nichols, A. and Sturrock, R.D. (1973) Ankylosing spondylitis and HL-A27. *Lancet* i, 904–907.

Brown, J.H., Jardetzky, T.S., Gorga, J.C., Stern, L.J., Urban, R.G., Strominger, J.L. and Wiley, D.C. (1993) Three-dimensional structure of the human class II histocompatibility antigen HLA-DR1. *Nature* 364, 33–39.

Brunet, L.R., Dunne, D.W. and Pearce, E.J. (1998) Cytokine interaction and immune responses during *Schistosoma mansoni* infection. *Parasitology Today* 14, 422–427.

Bruunsgaard, H., Pedersen, C., Skinhol, P. and Pedersen, B.K. (1997) Clinical progression of HIV infection: role of NK cells. *Scandinavian Journal of Immunology* 46, 91–95.

Cellier, M. and Gros, P. (1997) The NRAMP1 gene: resistance to intracellular infections and antimicrobial activity of phagocytes. *M S-Medecine Sciences* 13, 501–508.

Chicz, R.M., Urban, R.G., Gorga, J.C., Vignali, D.A.A., Lane, W.S. and Strominger, J.L. (1993) Specificity and promiscuity among naturally processed self peptides bound to HLA-DR alleles. *Journal of Experimental Medicine* 178, 27–47.

Clark, A.G. and Wang, L. (1997) Molecular population genetics of *Drosophila* immune systems. *Genetics* 147, 713–724.

Coffman, R.L., Lebman, D.A. and Rothman, P. (1993) Mechanism and regulation of immunoglobulin isotype switching. *Advances in Immunology* 54, 229–270.

Cruse, J.M., Brackin, M.N., Lewis, R.E., Meeks, W., Nolan, R. and Brackin, B. (1991) HLA disease association and protection in HIV-infection among African–Americans and Caucasians. *Pathobiology* 59, 324–328.

Cuff, S. and Ruby, J. (1996) Evasion of apoptosis by DNA viruses. *Immunology and Cell Biology* 74, 527–537.

Davis, M.M. and Bjorkman, P.J. (1988) T-cell antigen receptor genes and T-cell recognition. *Nature* 334, 395–402.

Delano, M.L. and Brownstein, D.G. (1995) Innate resistance to lethal mousepox is genetically linked to the NK-gene complex on chromosome 6 and correlates with early restriction of virus-replication by cells with an NK phenotype. *Journal of Virology* 69, 5875–5877.

de St Groth, B.F. (1998) Nature versus nurture: contributions of developmental programming and the microenvironment to B cell tolerance. *Immunology and Cell Biology* 76, 369–372.

Dietz, A.B., Cohen, N.D., Timms, L. and Kehrli, M.E. (1997) Bovine lymphocyte antigen class II alleles as risk factors for high somatic cell counts in milk of lactating dairy cows. *Journal of Dairy Science* 80, 406–412.

Doenhoff, M.J. (1997) A role for granulomatous inflammation in the transmission of infectious disease: schistosomiasis and tuberculosis. *Parasitology* 117 (Suppl.), S113–S125.

Doenhoff, M.J. and Chappell, L.H. (eds) (1997) Survival of parasites, microbes and tumours: strategies for evasion, manipulation and exploitation of the immune response. *Parasitology* 115 (Suppl.).

Donelson, J.E., Hill, K.L. and El Sayed, N.M.A. (1998) Multiple mechanisms of immune evasion by African trypanosomes. *Molecular and Biochemical Parasitology* 91, 51–66.

Dutton, R.W., Bradley, L.M. and Swain, S.L. (1998) T cell memory. *Annual Review of Immunology* 16, 201–223.

Dwyer, E. and Winchester, R. (1993) Genetic basis of chronic Lyme disease. In: Coyle, P. (ed.) *Immunogenetics of Lyme Disease.* Mosby Year Book, St Louis, Missouri.

El Ridi, R., Ozaki, T. and Kamiya, H. (1998) *Schistosoma mansoni* infection in IgE-producing and IgE-deficient mice. *Journal of Parasitology* 84, 171–174.

Fearon, D.T. and Locksley, R.M. (1996) The instructive role of innate immunity in the acquired immune response. *Science* 272, 50–54.

Feng, J.W., Li, Y.J., Hashad, M., Schurr, E., Gros, P., Adams, L.G. and Templeton, J.W.

(1996) Bovine natural resistance associated macrophage protein 1 (NRAMP1) gene. *Genome Research* 6, 956–964.

Feng, Y., Broder, C.C., Kennedy, P.E. and Berger, E.A. (1996) HIV-1 entry cofactor: Functional cDNA cloning of a seven-transmembrane, G protein-coupled receptor. *Science* 272, 872–877.

Finkelman, F.D., Pearce, E.J., Urban, J.F. and Sher, A. (1991) Regulation and biological function of helminth-induced cytokine responses. *Immunoparasitology Today: Combined Issue of Immunology Today and Parasitology Today* 3, A62–A66.

Finkelman, F.D., Shea-Donohue, T., Goldhill, J., Sullivan, C.A., Morris, S.C., Madden, K.B., Gauser, W.C. and Urban, J.F. (1997) Cytokine regulation of host defense against parasitic gastrointestinal nematodes. *Annual Review of Immunology* 15, 505–533.

Fleming, M.D., Trenor, C.C., Su, M.A., Foernzler, D., Beier, D.R., Dietrich, W.F. and Andrews, N.C. (1997) Microcytic anaemia mice have a mutation in NRAMP2, a candidate iron transporter gene. *Nature Genetics* 16, 383–386.

Fleming, M.D., Romano, M.A., Su, M.A., Garrick, L.M., Garrick, M.D. and Andrews, N.C. (1998) NRAMP2 is mutated in the anaemic belgrade (b) rat — evidence for a role for NRAMP2 in endosomal iron transport. *Proceedings of the National Academy of Sciences USA* 95, 1148–1153

Flynn, J.L., Goldstein, M.M., Triebold, K.J., Sypek, J., Wolf, S. and Bloom, B.R. (1995) IL-12 increases resistance of BALB/C mice to *Mycobacterium tuberculosis* infection. *Journal of Immunology* 155, 2515–2524.

Fox, M.T. (1997) Pathophysiology of infection with gastrointestinal nematodes in domestic ruminants: recent developments. *Veterinary Parasitology* 72, 285–297.

French, D.L., Laskov, R. and Scharff, M.D. (1989) The role of somatic hypermutation in the generation of antibody diversity. *Science* 244, 1152—1157.

Garcia, M.R., Graham, S., Harris, R.A., Beverley, S.M. and Kaye, P.M. (1997) Epitope cleavage by *Leishmania* endopeptidase(s) limits the efficiency of the exogenous pathway of major histocompatibility complex class I-associated antigen presentation. *European Journal of Immunology* 27, 1005–1013.

Girardsantosuosso, O., Bumstead, N., Lantier, I., Protais, J., Colin, P., Guillort, J.F., Beaumont, C., Malo, D. and Lantier, F. (1997) Partial conservation of the mammalian NRAMP1 syntenic group on chicken chromosome 7. *Mammalian Genome* 8, 614–616.

Govoni, G., Vidal, S., Gauthier, S., Skamene, E., Malo, D. and Gros, P. (1996) The Bcg/Ity/Lsh locus: genetic transfer of resistance to infections in C57BL/6J mice transgenic for the Nramp1(Gly169) allele. *Infection and Immunity* 64, 2923–2929.

Greenberger, M.J., Kunkel, S.L., Strieter, R.M., Lukacs, N.W., Bramson, J., Gauldie, J., Graham, F.L., Hitt, M., Danforth, J.M. and Standiford, T.J. (1996) IL-12 protects mice in lethal *Klebsiella* pneumonia. *Journal of Immunology* 157, 3006–3012.

Hedrick, S.M. and Eidelman, F.J. (1993) T lymphocyte antigen receptors. In: Paul, W.E. (ed.) *Fundamental Immunology*. Raven Press, New York, pp. 383–420.

Heinzel, F.P., Sadick, M.D., Holaday, B.J., Coffman, R.L. and Locksley, R.M. (1989) Reciprocal expression of interferon-gamma or interleukin-4 during the resolution or progression of murine leishmaniasis: evidence for expansion of distinct helper T-cell subsets. *Journal of Experimental Medicine* 169, 59–72.

Herman, A., Kappler, J.W., Marrack, P. and Pullen, M. (1991) Superantigens – mechanism of T-cell stimulation and role in immune responses. *Annual Review of Immunology* 9, 745–772.

Hesselgesser, J., Liang, M., Hoxie, J., Greenberg, M., Brass, L.F., Orsini, M.J., Taub, D. and Horuk, R. (1998) Indentification and characterization of the CXCR4 chemokine receptor in human T cell lines: ligand binding, biological activity, and HIV-1

infectivity. *Journal of Immunology* 160, 877–883.
Hill, A.B. (1996a) Mechanisms of interference with MHC class I-restricted pathway of antigen presentation by herpesviruses. *Immunology and Cell Biology* 74, 523–526.
Hill, A.V.S. (1996b) Genetic susceptibility to malaria and other infectious diseases: from the MHC to the whole genome. *Parasitology* 112 (Suppl.), S75–S84.
Hill, A.V.S., Allsopp, C.E.M., Kwiatkowski, D., Anstey, N.M., Twumasi, P., Rowe, P.A., Bennett, S., Brewster, D., McMichael, A.J. and Greenwood, B.M. (1991) Common west African HLA antigens are associated with protection from severe malaria. *Nature* 352, 595–600.
Hill, A.V.S., Elvin, J., Williks, A.C., Aidoo, M., Allsopp, C.E.M., Gotch, F.M., Gaco, X.M., Takiguchi, M., Greenwood, B.M., Townsend, A.R., McMichael, A.J. and Whittle, H.C. (1992) Molecular analysis of the association of HLA-B53 and resistance to severe malaria. *Nature* 360, 434–439.
Hofer, E., Duchler, M., Fuad, S.A., Houchins, J.P., Yabe, T. and Bach, F.H. (1992) Candidate natural-killer-cell receptors. *Immunology Today* 13, 429–430.
Holmskov, U., Jensenius, J.C., Tornoe, I. and Lovendahl, P. (1998) The plasma levels of conglutinin are heritable in cattle and low levels predispose to infection. *Immunology* 93, 431–436.
Hu, J.X., Bumstead, N., Skamene, E., Gros, P. and Malo, D. (1996) Structural organization, sequence, and expression of the chicken NRAMP1 gene encoding the natural resistance-associated macrophage protein 1. *DNA and Cell Biology* 15, 113–123.
Hu, J.X., Bumstead, N., Barrow, P., Sebastiani, G., Olien, L., Morgan, K. and Malo, D. (1997) Resistance to salmonellosis in the chicken is linked to NRAMP1 and tnc. *Genome Research* 7, 693–704.
Janeway, C.A. (1998) The road less traveled by: the role of innate immunity in the adaptive immune response. *Journal of Immunology* 161, 539–544.
Jones, T.M. (1997) Quantitative aspects of the relationship between the sickle cell gene and malaria. *Parasitology Today* 13, 107–111.
Kagnoff, M.F. (1998) Current concepts in mucosal immunity – III. Ontogeny and function of gamma delta T cells in the intestine. *American Journal of Physiology – Gastrointestinal and Liver Physiology* 37, G445–G458.
Kaufman, J., Salomonsen, J. and Flajnik, M. (1994) Evolutionary conservation of MHC class I and class II molecules: different yet the same. *Seminars in Immunology* 6, 411–422.
Kauffman, J., Volk, H. and Wallny, H.-J. (1995) A 'minimal essential MHC' and an 'unrecognized MHC': two extremes in selection for polymorphism. *Immunological Reviews* 143, 63–88.
Kawakami, K., Tohyama, M., Xie, Q. and Saito, A. (1996) IL-12 protects mice against pulmonary and disseminated infection caused by *Cryptococcus neoformans*. *Clinical and Experimental Immunology* 104, 208–214.
Kemp, S.J. and Teale, A.J. (1998) Genetic basis of trypanotolerance in cattle and mice. *Parasitology Today* 14, 450–454.
Kindred, B. and Schreffler, D.C. (1972) H-2 dependence of co-operation between T and B cells *in vivo*. *Journal of Immunology* 109, 940–943.
Kos, F.J. (1998) Regulation of adaptive immunity by natural killer cells. *Immunologic Research* 17, 303–312.
Kotzin, B.L. and Palmer, E. (1987) The contribution of NZW genes to lupus-like disease in (NZB × NZQ) F_1 mice. *Journal of Experimental Medicine* 165, 1237–1251.
Ladel, C.H., Blum, C., Dreher, A., Reifenberg, K. and Kaufmann, S.H.E. (1995) Protective role of gamma/delta T-cells and alpha/beta T-cells in tuberculosis. *European Journal of Immunology* 25, 2877–2881.

Lane, P. (1996) Development of B-cell memory and effector function. *Current Opinion in Immunology* 8, 331–335.
Lanier, L.L. and Phillips, J.H. (1996) Inhibitory MHC class I receptors on NK cells and T cells. *Immunology Today* 17, 86–91.
Laskay, T., Diefenbach, A., Rollinhoff, M. and Solbach, W. (1995) Early parasite containment is decisive for resistance to *Leishmania major* infection. *European Journal of Immunology* 25, 2220–2227.
Lemaitre, B., Nicolas, E., Michaut, L., Reichart, J.M. and Hoffmann, J.A. (1996) The dorsoventral regulatory gene cassette spatzle/Toll/cactus controls the potent antifungal response in *Drosophila* adults. *Cell* 86, 973–983.
Lewin, H.A. and Bernoco, D. (1986) Evidence for BoLA-linked resistance and susceptibility to subclinical progression of bovine leukemia virus infection. *Animal Genetics* 17, 197–207.
Lewis, S. and Gellert, M. (1989) The mechanism of antigen receptor gene assembly. *Cell* 59, 585–588.
Liu, R., Paxton, W.A., Choe, S., Ceradini, D., Martin, S.R., Horuk, R., MacDonald, M.E., Stuhlmann, H., Koup, R.A. and Landau, N.R. (1996) Homozygous defect in HIV-1 coreceptor accounts for resistance of some multiply-exposed individuals to HIV-1 infection. *Cell* 86, 367–377.
Lydyard, P.M. and Youinou, P.Y. (1994) Human CD5+ cells in health and disease. *Fundamental and Clinical Immunology* 2, 9–25.
McLeod, R., Eisenhauer, P., Mack, D., Brown, C., Filice, G. and Spitalny, G. (1989) Immune-responses associated with early survival after peroral infection with *Toxoplasma gondii*. *Journal of Immunology* 143, 3031–3034.
Mason, D. (1998) A very high level of crossreactivity is an essential feature of the T-cell receptor. *Immunology Today* 19, 395–404.
Mathews, G.D. and Crawford, A.M. (1998) Cloning, sequencing and linkage mapping of the NRAMP1 gene of sheep and deer. (1998) *Animal Genetics* 29, 1–6.
Mattner, F., Magram, J., Ferrante, J., Launois, P., Di Padova, K., Behin, R., Gately, M.K., Louis, J.A. and Alber, G. (1996) Genetically resistant mice lacking interleukin-12 are susceptible to infection with *Leishmania major* and mount a polarized Th2 cell response. *European Journal of Immunology* 26, 1553–1559.
Max, E.E. (1993) Immunoglobulins: molecular genetics. In: Paul, W.E. (ed.) *Fundamental Immunology*. Raven Press, New York, pp. 315–382.
Medina, E. and North, R.J. (1996) Evidence inconsistent with a role for the Bcg gene (NRAMP1) in resistance of mice to infection with virulent *Mycobacterium tuberculosis*. *Journal of Experimental Medicine* 183, 1045–1051.
Medzhitov, R. and Janeway, C.A. (1997) Innate immunity: impact on the adaptive immune response. *Current Opinion in Immunology* 9, 4–9.
Medzhitov, R. and Janeway, C.A. (1998) Innate immune recognition and control of adaptive immune responses. *Seminars in Immunology* 10, 351–353.
Medzhitov, R., Preston-Hurlburt, P. and Janeway, C.A. (1997) A human homologue of the *Drosophila* Toll protein signals activation of adaptive immunity. *Nature* 388, 394–397.
Mock, B.A., Fortier, A.H., Potter, M., Blackwell, J. and Nacy, C.A. (1985) Genetic control of systemic *Leishmania major* infection: identification of subline differences for susceptibility to disease. *Current Topics in Microbiology and Immunology* 122, 115–121.
Mohan, K., Moulin, P. and Stevenson, M.M. (1997) Natural killer cell cytokine production, not cytotoxicity, contributes to resistance against blood-stage *Plasmodium chabaudi* infection. *Journal of Immunology* 159, 4990–4998.
Mombaerts, P., Iacomini, J., Johnson, R.S., Herrup, K., Tonegawa, S. and Papaioannou,

V.E. (1992) RAG-1 deficient mice have no mature lymphocytes-B and lymphocytes-T. *Cell* 68, 869–877.

Mond, A.J., Lees, A. and Snapper, C.M. (1995) T cell-independent antigens Type 2. *Annual Review of Immunology* 13, 655–692.

Mosmann, T.R. and Coffman, R.L. (1987) 2 types of mouse helper T-cell clone: implications for immune regulation. *Immunology Today* 8, 223–227.

Mosmann, T.R. and Coffman, R.L. (1989) Th1 and Th2 cell – different patterns of lymphokine secretion lead to different functional properties. *Annual Review of Immunology* 7, 145–173.

Muller, M. and Brem, G. (1996) Intracellular, genetic or congenital immunization: transgenic approaches to increase disease resistance of farm animals. *Journal of Biotechnology* 44, 233–242.

Noonan, D.J., Kofler, R., Singer, P.A., Cardenas, G., Dixon, F.J. and Theofilopolous, N.A. (1986) Delineation of a defect in T cell receptor genes of NZW mice predisposed to autoimmunity. *Journal of Experimental Medicine* 163, 644–653.

Nossal, G.V. (1994) Negative selection of lymphocytes. *Cell* 76, 229–239.

Oldstone, M.A., Dixon, F.J., Mitchell, G. and McDevitt, H. (1973) Histocompatibility linked genetic control of disease susceptibility. Murine lymphocytic choriomeningitis virus infection. *Journal of Experimental Medicine* 137, 1201–1212.

O'Meara, T.J., Nesa, M. and Sandeman, R.M. (1997) Antibody responses to *Lucilia cuprina* in sheep selected for resistance or susceptibility to *L. cuprina*. *Parasite Immunology* 19, 535–543.

Orme, I.M., Stokes, R.W. and Collins, F.M. (1985) Only two out fifteen BCG strains follow the *Bcg* pattern. *Progress in Leukocyte Biology* 3, 285–289.

Osterrieder, N. and Wolf, E. (1998) Lessons from knockouts. *Revue Scientifique et Technique de L'Office International des Epizooties* 17, 351–364.

Parish, C.R. (1996) Immune deviation: a historical perspective. *Immunology and Cell Biology* 74, 449–456.

Parish, C.R. and O'Neill, E.R. (1997) Dependence of the adaptive immune response on innate immunity: some questions answered but new paradoxes emerge. *Immunology and Cell Biology* 75, 523–527.

Paxton, W.A., Kang, S. and Koup, R.A. (1998) The HIV type 1 coreceptor CCR5 and its role in viral transmission and disease progression. *AIDS Research and Human Retroviruses* 14, S89–S92.

Pitel, F., Cribiu, E.P., Yerle, M., Lahbibmansais, Y., Lenneluc, I., Lanier, F. and Gellin, J. (1995) Regional localization of the ovine NRAMP gene to chromosome 2q41->q42 by *in situ* hybridization. *Cytogenetics and Cell Genetics* 70, 116–118.

Plant, J. and Glynn, A.A. (1974) Natural resistance to *Salmonella* infection, delayed hypersensitivity and Ir genes in different strains of mice. *Nature* 248, 345–347.

Plebanski, M., Lee, E.A.M. and Hill, A.V.S. (1997) Immune evasion in malaria: altered peptide ligands of the circumsporozoite protein. *Parasitology* 115 (Suppl.), S55–S66.

Porcelli, S.A., Segelke, B.W., Sugita, M., Wilson, I.A. and Brenner, M.B. (1998) The CD1 family of lipid antigen-presenting molecules. *Immunology Today* 19, 362–368.

Rammensee, H.G., Falk, K. and Rotzschke, O. (1993) Peptides naturally presented by MHC class I molecules. *Annual Review of Immunology* 11, 213–244.

Reiner, S.L. and Locksley, R.M. (1995) The regulation of immunity to *Leishmania major*. *Annual Review of Immunology* 13, 151–177.

Ridley, M. (1994) *The Red Queen*. Penguin Books, London.

Riffkin, M., Seow, H.F., Jackson, S., Brown, L. and Wood, P. (1996) Defence agains the immune barrage: helminth survival strategies. *Immunology and Cell Biology* 74, 564–574.

Roitt, I.M., Torrigiani, G., Greaves, F.M., Brostoff, J. and Playfair, J.H. (1969) The cellular basis of immunological responses. *Lancet* ii, 367–371.

Rolstad, B. and Seaman, W.E. (1998) Natural killer cells and recognition of MHC class I molecules: new perspectives and challenges in immunology. *Scandinavian Journal of Immunology* 47, 412–425.

Romani, L., Puccetti, P. and Bistoni, F. (1997) Interleukin-12 in infectious diseases. *Clinical Microbiology Reviews* 10, 611–638.

Rook, G.A.W. and Stanford, J.L. (1998) Give us this day our daily germs. *Immunology Today* 19, 113–116.

Rosenthal, A.S. and Shevach, E.M. (1973) Function of macrophages in antigen recognition by guinea pig T lymphocytes. *Journal of Experimental Medicine* 138, 1194–1212.

Rumyantsev, S.N. (1992) Observations on constitutional resistance to infection. *Immunology Today* 13, 184–187.

Ryan, A.M. and Womack, J.E. (1993) Type-1 interferon genes in cattle: restriction-fragment-length-polymorphisms, gene numbers and physical organization on bovine chromosome-8. *Animal Genetics* 24, 9–16.

Ryan, A.M. and Womack, J.E. (1997) A molecular genetic approach to improved animal health: the effect of interferon genotype on the severity of experimental bovine herpesvirus-1 infection. *Veterinary Clinics of North America – Food Animal Practice* 13, 401–411.

Schijns, V.E.J.C., Wierda, C.M.H., van Hoeij, M. and Horzinek, M.C. (1996) Exacerbated viral hepatitis in IFN-gamma receptor-deficient mice is not suppressed by IL-12. *Journal of Immunology* 157, 815–821.

Schofield, L. and Tachado, S.D. (1996) Regulation of host cell function by glycophosphatidylinositols of the parasitic protozoa. *Immunology and Cell Biology* 74, 555–563.

Schurr, E., Buschman, E., Malo, D., Gros, P. and Skamene, E. (1990) Immunogenetics of mycobacterial infections: mouse–human homologies. *Journal of Infectious Diseases* 161, 634–639.

Sharton-Kerston, T., Nakajima, H., Yap, G., Sher, A. and Leonard, W.J. (1998) Infection of mice lacking the common cytokine receptor gamma-chain (gamma(c)) reveals an unexpected role for CD4(+) lymphocytes in early IFN-gamma-dependent resistance to *Toxoplasma gondii*. *Journal of Immunology* 160, 2565–2569.

Sharton-Kerston, T.M. and Scott, P. (1995) The role of the innate immune response in Th1cell-development following *Leishmania major* infection. *Journal of Leukocyte Biology* 57, 515–522.

Sharton-Kerston, T.M. and Sher, A. (1997) Role of natural killer cells in innate resistance to protozoan infections. *Current Opinion in Immunology* 9, 44–51.

Sharton-Kerston, T.M., Wynn, T.A., Denkers, E.Y., Bala, S., Grunvald, E., Hieny, S., Gazzinelli, R.T. and Sher, A. (1996) In the absence of endogenous IFN-gamma, mice develop unimpaired IL-12 responses to *Toxoplasma gondii* while failing to control acute infection. *Journal of Immunology* 157, 4045–4054.

Shinkai, Y., Rathbun, G., Lam, K.P., Oltz, E.M., Stewart, V., Mendelsohn, M., Charron, J., Datta, M., Young, F., Stall, A.M. and Alt, F.W. (1992) RAG-2-deficient mice lack mature lymphocytes owing to inability to initiate V(D)J rearrangement. *Cell* 68, 855–867.

Skamene. E., Gros, P., Forget, A., Kongshavn, P.A.L., St Charles, D. and Taylor, B.A. (1982) Genetic regulation of resistance to intracellular pathogens. *Nature* 297, 506–509.

Snapper, C.M. and Mond, J.J. (1993) Towards a comprehensive view of immunoglobulin class switching. *Immunology Today* 14, 15–17.

Spithill, T.W., Piedrafita, D. and Smooker, P.M. (1997) Immunological approaches for the control of fasciolosis. *International Journal for Parasitology* 27, 1221–1235.

Stern, A.S., Magram, J. and Presky, D.H. (1996) Interleukin-12 an integral cytokine in the immune response. *Life Sciences* 58, 639–654.

Takahashi, I., Nakagawa, I., Xu, L. and Hamada, S. (1996) Role of alpha beta T cells and gamma delta T cells in protective immunity to *Actinobacillus actinomycetemcomitans* in mutant mice by targeted gene disruption. *Journal of Periodontal Research* 31, 589–595.

Thursz, M.R. (1997) Host genetic factors influencing the outcome of hepatitis. *Journal of Viral Hepatitis* 4, 215–220.

Tonegawa, S. (1983) Somatic generation of antibody diversity. *Nature* 302, 575–581.

Trail, J.C.M., Wissocq, N., Dieteren, G.D.M., Kakiese, O. and Murray, M. (1994) Quantitative phenotyping of Ndama cattle for aspects of trypanotolerance under field tsetse challenge. *Veterinary Parasitology* 55, 185–195.

Trinchieri, G. (1995) Interleukin-12: a proinflammatory cytokine with immunoregulatory functions that bridge innate resistance and antigenic-specific adaptive immunity. *Annual Review of Immunology* 13, 251–276.

Tuggle, C.K., Schmitz, C.B. and Gingerichfeil, D. (1997) Cloning of a pig full-length natural resistance associated macrophage protein (NRAMP1) cDNA. *Journal of Animal Science* 75, 277.

Uilenberg, G. (1996) Integrated control of tropical animal parasitoses. *Tropical Animal Health and Production* 28, 257–265.

Unanue, E.R. (1997) Studies in listeriosis show the strong symbiosis between the innate cellular system and the T-cell response. *Immunological Reviews* 158, 11–25.

Urban, J.F., Fayer, R., Chen, S.J., Gause, W.C., Gately, M.K. and Finkelman, F.D. (1996) IL-12 protects immunocompetent and immunodeficient neonatal mice against infection with *Cryptosporidium parvum*. *Journal of Immunology* 156, 263–268.

van Eden, W., Gonzalez, N.M., de Vries, R.R.P., Conva, J. and van Rood, J.J. (1985) HLA-linked control of predisposition to lepromatous leprosy. *Journal of Infectious Diseases* 151, 9–14.

Vickerman, K., Myler, P.J. and Stuart, K.D. (1993) African trypanosomiasis. In: Warren, K.S. (ed.) *Immunology and Molecular Biology of Parasitic Infections*. Blackwell Scientific, Oxford, pp. 170–212.

Vidal, A.M., Pinner, E., Lepage, P., Gauthier, S. and Gros, P. (1996) Natural-resistance to intracellular infections – Nramp1 encodes a membrane phosphoglycoprotein absent in macrophages from susceptible (NRAMP1(D169)) mouse strains. *Journal of Immunology* 157, 3559–3568.

Vidal, S.M., Malo, D., Vogan, K., Skamene, E. and Gros, P. (1993) Natural resistance to infection with intracellular parasites: isolation of a candidate for BCG. *Cell* 73, 469–485.

Voevodin, A., Samilchuk, E. and Dashti, S. (1998) A survey for 32 nucleotide deletion in CCR-5 chemokine receptor gene (Delta ccr-5) conferring resistance to human immunodeficiency virus type 1 in different ethnic groups and in chimpanzees. *Journal of Medical Virology* 55, 147–151.

Wabl, M. and Steinberg, C. (1996) Somatic hypermutability. *Current Topics in Microbiology and Immunology* 217, 203–219.

Wakelin, D. (1989) Helminth infections. In: Wakelin, D.M. and Blackwell, J.M. (eds) *Genetics of Resistance to Bacterial and Parasitic Infection*. Taylor and Francis, London, pp. 153–224.

Wakelin, D.M. and Blackwell, J.M. (1988) *Genetics of Resistance to Bacterial and Parasitic Infection*. Taylor and Francis, London.

Wakenell, P.S., Miller, M.M., Goto, R.M., Gauderman, W.J. and Briles W.E. (1996) Association between the Rfp-Y haplotype and the incidence of Marek's disease in chickens. *Immunogenetics* 44, 242–245.

Wang, X.H., Ohmen, J.D., Uyemura, K., Rea, T.H., Kronenberg, M. and Modlin, R.L. (1993) Selection of lymphocytes-T bearing limited T-cell receptor beta chains in the response to a human pathogen. *Proceedings of the National Academy of Sciences USA* 90, 188–192.

Way, S.S., Borrzuk, A.C., Dominitz, R. and Goldberg, M.B. (1998) An essential role for gamma interferon in innate resistance to *Shigella flexneri* infection. *Infection and Immunity* 66, 1342–1348.

Weatherall, D.J. (1996) Host genetics and infectious disease. *Parasitology* 112 (Suppl.), S23–S29.

Weatherall, D.J., Bell, J.I., Clegg, J.B., Flint, F., Higgs, D.R., Hill, A.V.S., Pasvol, G. and Thein, S.L. (1988) Genetic factors as determinants of infectious disease transmission in human communities. *Philosophical Transactions of the Royal Society (London) Series B* 321, 1–22.

Williams, D.J.L., Taylor, K., Newson, J., Gichuki, B. and Naessens, J. (1996) The role of anti-variable surface glycoprotein antibody responses in bovine trypanotolerance. *Parasite Immunology* 18, 209–218.

Wilson, R.A. (1993) Immunity and immunoregulation in helminth infections. *Current Opinion in Immunology* 5, 538–547.

Wynn, T.A., Reynolds, A., James, S., Cheever, A.W., Caspar, P., Hieny, S., Jankovic, D., Strand, M. and Sher, A. (1996) IL-12 enhances vaccine-induced immunity to schistosomes by augmenting both humoral and cell-mediated immune responses against the parasite. *Journal of Immunology* 157, 4068–4078.

Yokoyama, W.M. and Seaman, W.E. (1993) The Ly-49 and NKR-P1 gene families encoding lectin-like receptors on natural-killer-cells: the NK gene complex. *Annual Review of Immunology* 11, 613–635.

Zinkernagel, R.M. and Doherty, P.C. (1997) The discovery of MHC restriction. *Immunology Today* 18, 14–17.

4

The Major Histocompatibility Complex and its Role in Disease Resistance and Immune Responsiveness

M.F. Rothschild[1], L. Skow[2] and S.J. Lamont[1]

[1]*Department of Animal Science, Iowa State University, Ames, USA;* [2]*Department of Veterinary Anatomy and Public Health, Texas A&M University, College Station, USA*

Summary

Central to an animal's defence system is the ability to differentiate self from non-self and, thus, protect itself from disease. This unique quality is determined, in large part, by the major histocompatibility complex (MHC) genes of each species. In farm animals, the knowledge of the MHC genes has been growing, especially since the rapid advances in molecular biology. The general structure of the MHC of the pig, cow, chicken and horse is presented in this chapter. Disease and immune response associations are also described. Although in many instances these data are limited, it is clear that the MHC is strongly associated with health traits in farm animals. Understanding these MHC–disease resistance and immune-responsiveness relationships will allow genetic improvement to be made by marker-assisted selection, rather than by direct challenge of populations with pathogens.

Introduction

The major histocompatibility complex (MHC) genes are a group of genes that encode molecules that are intimately involved in the control of immune response and disease resistance. First discovered in the mouse (Gorer, 1937), then the chicken (Schierman and Nordskog, 1961) and subsequently in humans (1963; see Klein, 1986), the MHC genes have more recently been characterized in other livestock species. These genes are organized into three classes (class I, class II and class III) in mammals and an additional class IV in avian species. The class I and class II genes encode glycoproteins that bind and present peptides to T cells of the immune system, which results in their stimulation. The class III genes include some of the complement components. The function of the class IV genes found in poultry is not yet completely clear. The MHC region is perhaps one of the best-studied gene regions in domestic

© CAB *International* 2000. *Breeding for Disease Resistance in Farm Animals* (eds R.F.E. Axford, S.C. Bishop, F.W. Nicholas and J.B. Owen)

animal species. These studies have revealed that the genes are extremely polymorphic and there is great similarity among the MHC genes of many species. Evidence to date suggests that the MHC has evolved in response to selection pressure on the immune system to recognize and rebuke the invasion of microorganisms and parasites (Andersson, 1996). Recent MHC research with several domestic species has been reviewed (Srivastara *et al.*, 1991; Schook and Lamont, 1996). The purpose of this chapter is to present an overview of the MHC of pigs, cattle, chickens and horses and the role these genes play in disease resistance and immune responsiveness.

In this chapter the commonly used MHC nomenclature will be cited. An alternative nomenclature proposed by Klein *et al.* (1990), using abbreviations of the scientific name of each species (i.e. *Bota* for domestic cattle, *Bos taurus*), has not been generally accepted by workers in animal immunogenetics.

The Porcine Major Histocompatibility Complex

Introduction

The MHC of the pig is called the swine leucocyte antigen (SLA) complex and was first discovered by Vaiman (Vaiman *et al.*, 1970) and Viza (Viza *et al.*, 1970). Knowledge of the SLA complex lags behind that of the mouse and human MHCs, but steady progress has been made. Initial research dealt with the role of the SLA complex in graft acceptance and rejection, and serological testing revealed that there was a considerable level of genetic diversity within the complex. More recent research has examined the molecular genetic structure of the SLA complex, and it is now the best understood chromosomal region in the pig. Several complete reviews have been published, including Warner and Rothschild (1991), Lunney (1994), Schook *et al.* (1996) and Lunney and Butler (1998).

Chromosomal map and location

The chromosomal location of the SLA complex has been mapped to the region that surrounds the centromere of chromosome 7 (Geffrotin *et al.*, 1984; Rabin *et al.*, 1985; Smith *et al.*, 1995). The SLA complex is composed of three subregions corresponding to the different classes. The order of these three regions is class I–class III–class II (Xu *et al.*, 1992). The class I and III regions are on the short arm of chromosome 7 while the class II region is on the long arm (Smith *et al.*, 1995). The SLA complex is approximately 1–1.5 cM and 2 Mb of DNA. This makes the SLA complex one of the smallest mammalian MHCs.

Genetic variability and organization

Initially, serological methods were used to determine the extent of genetic variability of the SLA complex and the association of SLA haplotypes with immune responsiveness, disease resistance and traits of economic importance. Analysis of over 500 SLA serologically informative families has revealed at least 68 haplotypes in numerous commercial breeds and lines. Use of molecular techniques has now given us a more complete understanding of the genomic structure of the SLA complex (Fig. 4.1).

The first class I gene was cloned and expressed by Singer *et al.* (1982). Additional research revealed that a total of seven class I genes existed in the National Institutes of Health (NIH) miniature pigs. Of these genes, it appears that at least three are expressed (Singer *et al.*, 1988). Subsequent use of restriction fragment length polymorphism (RFLP) analyses revealed that seven to ten class I genes existed in commercial swine (Chardon *et al.*, 1985; Vaiman *et al.*, 1998). A total of nine class I genes were identified by screening of a porcine cosmid library with human cDNA probes (Gaycken *et al.*, 1994). More recently, researchers found ten class I genes, of which four appear to be transcribed (Velten *et al.*, 1998). Many porcine class I sequences have been identified, and PCR–RFLP and PCR-sequencing methods are now available for genotyping individual animals.

The SLA class II serology has been less well developed, due to the difficulty of developing useful typing sera (Chardon *et al.*, 1981). Studies have revealed, however, that while there are many class II genes, only the SLA-DQ and SLA-DR molecules are expressed at the protein level. RFLP analyses suggest that there are at least three α-chain genes and from five to eight β-chain genes (Shia *et al.*, 1991, 1995). In addition, this region includes the genes encoding proteins that chaperone (DM) or transport (TAP) foreign antigens as they become associated with the class II antigens, for T-cell receptor presentation (Lunney and Butler, 1998). The class II region has been estimated to be about 450 kb (Vaiman *et al.*, 1998).

The class III region of the SLA is approximately 700 kb and contains 33 genes that have been characterized to date (Peelman *et al.*, 1996). The porcine class III region is very similar to that of the human class III region. It contains a considerable array of genes, including serum complement products (C2, C4, Bf), the steroid 21-hydroxylase enzyme (CYP21), the cytokine tumour necrosis factor (TNF) genes, heat-shock protein (Hsp70), lymphotoxins (LTA, LTB) and some non-immune factors. One exception to the similarity with the human class III region is that there appears to be only one CYP21 gene (Geffrotin *et al.*, 1991) and one C4 gene in the pig – additional copies are found in humans, mice and other species (Vaiman *et al.*, 1998).

Class I, class II and class III proteins and function

Similar to those of other species, class I SLA antigens consist of two chains, a heavy chain encoded in the SLA, and a light chain, β_2-microglobulin, that is encoded on chromosome 1 (Rogel-Gaillard *et al.*, 1998). Class I heavy chains are *c.* 45 kDa transmembrane glycoproteins. These heavy chains are in tight non-covalent association with β_2-microglobulin, an extracellular protein of

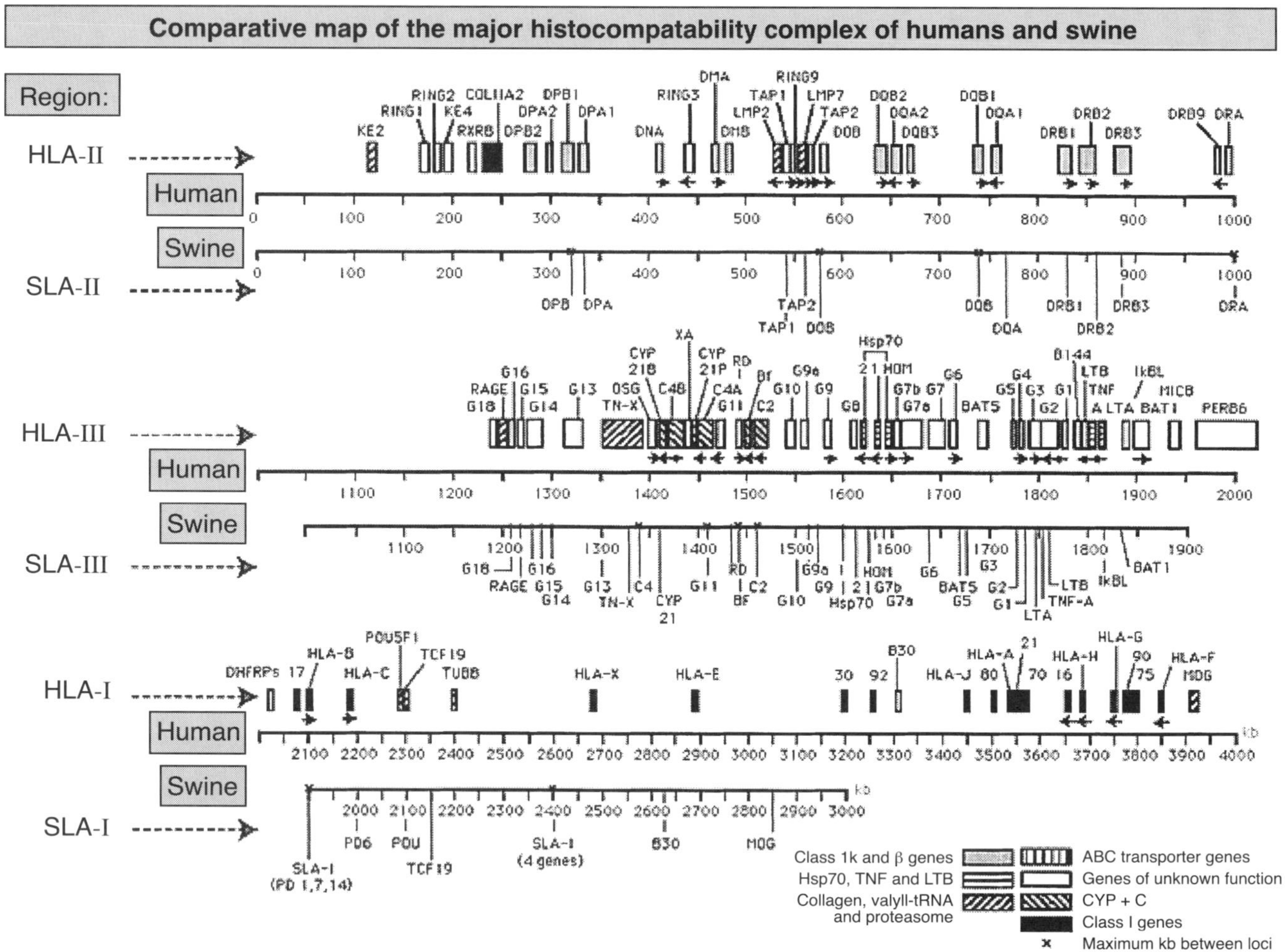

Fig. 4.1. Detailed map of the SLA complex compared with the human leucocyte antigen (HLA) complex. This SLA map was developed by Lunney (1994), based on data from colleagues worldwide, and compared to the HLA map published by Campbell and Trowsdale (1993, 1997). For additional details see Lunney and Butler (1998).

molecular mass 12 kDa. Class I antigens are expressed on all cells of the body in varying amounts.

Class II SLA antigens are also similar to those of other species. The molecules are heterodimers consisting of an α heavy chain with molecular mass of 34 kDa and a β light chain of approximately 29 kDa (Gustafsson *et al.*, 1990). The two non-covalently associated chains have a transmembrane region and a cytoplasmic tail and are encoded by genes within the class II region. Class II antigens of the SLA complex are found primarily on B cells and macrophage cells, and on approximately 60% of circulating T cells. Some other cells in the body also may express class II antigens.

The class III proteins are involved in several functions, including those directly and indirectly involved with immune-system function. Those that are a part of the complement cascade include C2, C4 and Bf, and are found in serum. Early studies showed that these complement components were controlled by genes within the SLA complex (Kirszenbaum *et al.*, 1985; Lie *et al.*, 1987). The large number of other genes in this region and the proteins they produce have not been fully characterized.

Association of the SLA complex with immune response and complement activity

Studies in humans and mice have revealed a considerable role of the MHC in immune response and complement activity. Evidence of such involvement in the pig is more limited but still convincing. This evidence has generally been obtained using inbred NIH miniature pigs bred for SLA homozygosity (Sachs *et al.*, 1976) and from outbred strains analysed using the considerable panel of SLA typing reagents developed primarily by French researchers (Renard *et al.*, 1988; Vaiman *et al.*, 1998).

One of the first reports was that of Vaiman *et al.* (1978b) on the role of the SLA complex on serum haemolytic complement levels. These researchers, using Large White pigs, found that CH50 levels were significantly associated with SLA complex differences. However, this result was not confirmed by Mallard *et al.* (1989b) using the NIH SLA miniature pigs.

The role of the SLA complex in controlling the immune response has been confirmed by several studies. The first significant association of the SLA complex with immune response was found using the same pigs as those of Vaiman *et al.* (1978a) and represents the first such report in livestock. In this study, pigs with SLA haplotypes H10 and H12 were challenged with hen eggwhite lysozyme (HEWL). The initial immune response to the HEWL challenge was higher for H10/H12 animals than for H10 or H12 homozygotes. Secondary challenge produced a significantly higher immune response in the homozygous H12 and heterozygous H12/H10 animals when compared to H10 homozygotes (Vaiman *et al.*, 1978a). A confirming report by Lunney *et al.* (1984), using the NIH minature pigs, demonstrated that the SLA complex was associated with immune response following challenge using HEWL. Using a synthetic immunogen, (T,G)-A--L, MHC influence on immune response has also been shown in NIH miniature pigs (Lunney *et al.*, 1984): SLA aa and cc animals were high responders while the dd and gg animals were not. These results confirmed that the SLA class II region was involved because the SLA gg pigs (the g haplotype is a recombinant between the c and d haplotypes in which the class I genes are from the c haplotype and the class II genes are from the d haplotype) responded in the same way as the SLA dd animals but not the cc animals. Mallard and co-workers (Mallard *et al.*, 1989a; Appleyard *et al.*, 1992) have confirmed these results using HEWL, sheep red blood cells and (T,G)-A--L.

Other experiments have examined the role of the swine MHC in controlling immune response. Rothschild and co-workers (Rothschild *et al.*, 1984a, b; Meeker *et al.*, 1987) searched for associations of the SLA complex with immune response following vaccination with commercial vaccinations for *Bordetella bronchiseptica* and pseudorabies virus. Only the IgA response to the *Bordetella bronchiseptica* vaccine appeared to be associated with SLA haplotype.

Cellular activity and its relationship to the SLA complex have also been investigated, using the NIH miniature pigs and the delayed hypersensitivity test to bacillus Calmette–Guérin. The challenge injection of purified tuberculin derivative 21 days later demonstrated considerable differences in reactions between some SLA genotypes. Hypersensitivity to dinitrochlorobenzene was also quite different among some SLA genotypes (Mallard *et al.*, 1989c).

Association of the SLA complex with disease resistance

Given the expense of conducting actual disease challenges, *in vitro* tests have been used as indicators of resistance. Lacey and colleagues (1989) examined phagocytic and bactericidal action of blood monocytes of 4- and 8-week-old pigs against *Salmonella typhimurium* and *Staphylococcus aureus* in NIH miniature pigs. SLA haplotype effects on uptake and killing of each bacterium were generally significant. Of particular interest was the observation that the better responding haplotypes to the bacterial challenges were not the same as those better responding haplotypes to challenges with sheep red blood cells, HEWL or (T,G)-A--L (Mallard *et al.*, 1989a). Additionally, serum-agglutinating antibody titre and *O*-polysaccharide-specific peripheral-lymphocyte blastogenesis were measured following two parenteral vaccinations with a mutant of *S. typhimurium* and an oral challenge of the bacteria. Some relationship of immune-response parameters was seen to be associated with SLA haplotype but most differences were related instead to parity differences (Lumsden *et al.*, 1993).

The SLA complex has been shown to be significantly associated with cutaneous malignant melanoma in Sinclair miniature swine (Hook *et al.*, 1979; Tissot *et al.*, 1987; Blangero *et al.*, 1996). The evidence suggests a two-locus model for the occurrence of melanocytic lesions in this line of pigs, with one of the loci being within the SLA complex. However, other reports of similar lesions in German miniature pigs have not been associated with differences in the SLA complex (Müller *et al.*, 1995).

Resistance against the nematode *Trichinella spiralis* has been examined in

NIH miniature pigs. Results (Lunney and Murrel, 1988) suggest that SLA cc animals had a lower burden of larvae in the muscle than did SLA aa or dd animals, and this was associated with serum antibody activity and a heightened cellular response to the *T. spiralis* antigens. When secondary infection was also investigated it appeared that pigs with the a haplotype had fewer encapsulated muscle larvae from primary infection when compared to pigs of the other SLA haplotypes (Madden *et al.*, 1990, 1993). However, other studies with the foodborne protozoan parasite *Toxoplasma gondii* have shown no SLA complex association (Dubey *et al.*, 1996).

Other associations have been reported. In her review of disease-resistance research in the pig, Lunney cites possible associations of skin diseases with the swine MHC (Lunney and Butler, 1998). Expression differences of class I and class II antigens were observed in the spleens of Yorkshire pigs inoculated with different African swine fever virus isolates (Gonzalez-Juarrero *et al.*, 1992). Differences in SLA molecule expression were observed. Further experiments with NIH miniature pigs showed some SLA haplotype differences (Martins *et al.*, 1988). There has also been some suggestion that resistance to porcine reproductive and respiratory syndrome virus (PRRSV) may be influenced by the SLA complex (Kristensen, 1995).

The SLA complex has been studied for its associations with pig production traits. While not directly related to disease association, these give some evidence of differences in indirect measures of resistance. The SLA complex has been shown to be associated with early and later growth, backfat and reproductive measures, as summarized by Vaiman *et al.* (1988, 1998) and Warner and Rothschild (1991).

The results summarized above point to some SLA complex involvement in immune response, disease resistance and production. As understanding of the molecular biology of the SLA complex improves and genes within the region are uncovered, it is likely that we will better understand the role of the SLA complex in disease resistance.

The Bovine Major Histocompatibility Complex

Introduction

Twenty years have passed since the discovery of the MHC in cattle (Amorena and Stone, 1978; Spooner *et al.*, 1978). The designation BoLA (bovine lymphocyte antigen) was adopted at the First International Bovine Lymphocyte Antigen (BoLA) Workshop (Spooner *et al.*, 1979).

Chromosomal map and location

The BoLA complex is located on chromosome Bta23, but in contrast to the single cluster of genes that comprise the typical mammalian MHC, the MHC genes in cattle are found in two distinctly separate regions of Bta23

(Andersson *et al.*, 1988; van Eijk *et al.*, 1993; Skow *et al.*, 1996). The larger gene cluster is located at Bta23 band 22 and apparently contains all of the bovine class I and class III sequences, and genes encoding both subunits of the classical class II proteins DQ and DR. The remaining BoLA class II loci are located in a cluster near the centromere at Bta23 band 12–13 (Fig. 4.2). This separated arrangement of MHC genes observed in cattle has also been reported in sheep, goats and deer, and probably represents an ancient evolutionary event in artiodactyl evolution. Recent determination of BoLA gene order by radiation and hybrid analysis suggests that the different arrangement of genes in BoLA, as compared to homologous gene regions in humans and mice, can be explained by a single large inversion, with breakpoints near the centromere and within the ancestral class II region (Band *et al.*, 1998).

Genetic variability and organization

Analysis of bovine class I restriction fragment patterns produced by hybridization to a human class I cDNA suggests that the BoLA complex contains up to 20 class I sequences (Lindberg and Andersson, 1988). Several studies have identified from two to six transcribed genes (Ennis *et al.*, 1988; Joosten *et al.*, 1988; Bensaid *et al.*, 1991; Ellis *et al.*, 1992; Garber *et al.*, 1994), indicating that at least three class I genes are expressed transcriptionally in bovine lymphocytes. Additionally, at least 15 other class I sequences have been identified in bovine genomic libraries (Hu, 1996). Nine of these sequences are transcribed at low levels and the remainder represent pseudogenes and gene fragments. Although transcripts for at least 12 class I genes have been identified, it

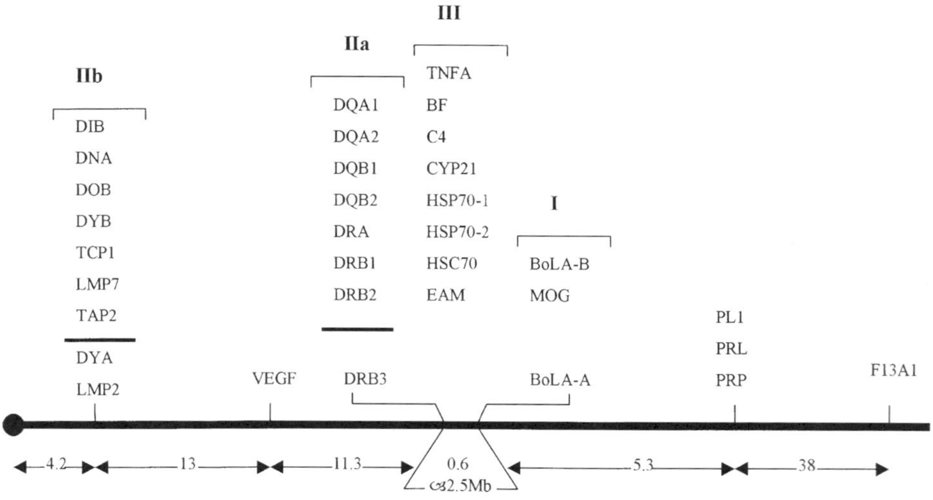

Fig. 4.2. A map of the BoLA regions on bovine chromosome 23, determined by recombination (see review by Lewin, 1996) and by physical mapping (Bank *et al.*, 1998; Gallagher *et al.*, 1998; Grosz *et al.*, 1998; Hess *et al.*, 1998). Distances, in centiMorgans, were estimated from the Illinois Reference/Resource familes (Lewin, 1994).

remains to be determined how many of these genes are functional. All of the class I genes so far identified map to the class I region of BoLA (DiPalma *et al.*, 1998; Gallagher *et al.*, 1998).

Eleven class II loci have been identified, but only products encoded at the DR and DQ loci (DRA, DRB3, DQA, DQB1, DQB2) have been identified on bovine leucocytes. A possible third class II locus similar to the HLA DP gene was reported by Ababon *et al.* (1994), but it is unclear whether this gene represents a new class II locus or is identical to a locus in the IIb region.

Disease associations with the bovine MHC

In 1993, Park *et al.* described a chronic progressive neurological disorder, posterior spinal paresis (PSP), in Holstein bulls possessing the A8 BoLA specificity. Two clinically similar conditions occur in humans, both of which are associated with HLA determinants. Ossification of posterior longitudinal ligament (OPLL) is a recessive autosomal disease in humans characterized by chronic spinal cord compression and progressive paralysis of the lower limbs. OPLL has been localized to a chromosomal segment that contains the candidate gene for collagen alpha 2 (COL II A2), about 45 kb centromeric from HLA-DP1 (Koga *et al.*, 1998). The second human disorder, ankylosing spondylitis (AS), is a dominant, incompletely penetrant autoimmune disease that occurs subsequent to bacterial infections (Rubin *et al.*, 1994). It is strongly associated with the B27 allele at the HLA B locus. As in the two human disorders, PSP is mediated by a cytotoxic T-cell response to a self-antigen found in tissues of the vertebral joint. Acute inflammation of the spine is thought to result from immunity induced by the cross-reacting bacteria (Benjamin and Parham, 1990). Linkage analysis of bovine PSP failed to establish a marker that was more specific for PSP susceptibility than the originally serologically defined A8. However, tight linkage (no recombinants) was detected between BoLA class I and PSP, leading Park *et al.* (1993) to conclude that bovine PSP is homologous to human AS.

Haemochromatosis is a condition produced by a defect in iron metabolism and has been described in three members of the Salers breed of cattle (House *et al.*, 1994). The disease phenotype is a consequence of severe iron overload that produces fibrotic lesions in the liver and pancreas, cardiac myopathy and hyperpigmentation of skin. Gene mapping in humans has localized the haemochromatosis gene to a non-classical class I locus, designated HLA H (Calandro *et al.*, 1995; Feder *et al.*, 1996). If haemochromatosis turns out to be a single-gene disorder in cattle, the gene is predicted to map to the class I region of the BoLA complex.

Resistance to infectious diseases

In cattle, BoLA haplotypes are often associated with phenotypic differences in generalized immune response (Lie *et al.*, 1986; Glass *et al.*, 1990; Weigel *et al.*,

1990; Newman *et al.*, 1996; Dietz *et al.*, 1997). In some instances, however, variation in immune response may not be caused by genetic differences in the inherent capacity to respond to antigens. Rather, it is a consequence of a biased response in animals of certain BoLA specificities to preferentially present 'immunodominant' peptides on the surface of infected cells (Taracha *et al.*, 1995).

Specific associations of BoLA haplotypes with resistance/susceptibility have been observed in cattle exposed to a variety of viral and bacterial infections, intracellular parasites and ectoparasites (Lewin, 1989; Østergard *et al.*, 1989; Lewin *et al.*, 1991; Andersson and Davies, 1994), but the causal genes responsible for disease resistance/susceptibility remain largely unidentified. For a few diseases, however, evidence is accumulating that implicates differences at the class II locus DRB3, specifically in exon 2 (DRB3.2) as the primary determinants of many examples of MHC-linked disease resistance. Genetic studies of mastitis and bovine leucosis constitute the largest body of experimental evidence for DRB3 involvement, and are discussed below. However, it is noteworthy that Maillard *et al.* (1998) localized genetic susceptibility to the bacterial skin infection, dermatophilosis, in Brahman cattle to a glutamic acid–serine motif at positions 20 and 22 in DRB3.2.

In dairy cattle, infections of bovine leukaemia virus progress to lymphosarcomas and persistent lymphocytosis in a small percentage of infected cattle, but result in substantial losses to the dairy industry through declining production (Da *et al.*, 1993). Genetic resistance to persistent lymphocytosis is inherited as a dominant trait associated with the BoLA A14 haplotype (Lewin, 1989, 1994; Ernst *et al.*, 1997), and has been localized to the DRB3 molecule (Xu *et al.*, 1993). At least three DRB3 alleles (*DRB3.*11*, **23*, **28*) have been associated with resistance to bovine leukaemia virus (Sulimova *et al.*, 1995). The resistance allele *A14-DRB3*11* is present at low to moderate frequency in most Holstein herds, leading to the recommendation that genetic resistance to persistent lymphocytosis can be selected for, based on the BoLA haplotype (Xu *et al.*, 1993). The feasibility of this approach is enhanced by independent studies which show the A14 haplotype is associated with significantly increased milk production traits and profitability (Weigel *et al.*, 1990; Batra *et al.*, 1996).

Clinical mastitis has a complex aetiology. At least four pathogens have been identified as causing the disease (Lewin *et al.*, 1991), and susceptibility to mastitis is strongly influenced by environmental conditions. Different BoLA haplotypes have been associated with resistance or susceptibility to mastitis in different cattle breeds (Mejdell *et al.*, 1994; Mallard *et al.*, 1995). The A11 and A16 haplotypes are associated with susceptibility in Norwegian Red and Swedish Red and White cattle (Lundén *et al.*, 1990; Väge *et al.*, 1992), but A11 is associated with resistance in Holstein cattle (Weigel *et al.*, 1990). It is not clear whether the association of the same haplotypes with resistance and susceptibility is attributable to different causative organisms or to variation in the genetic content of A11 haplotypes among different cattle breeds. More recent studies in Holsteins implicate the DRB3.2 alleles **11* and **8* with susceptibility to clinical mastitis (Kelm *et al.*, 1997). Therefore, resistance and susceptibility phenotypes may share a common genetic determinant with resistance/susceptibility to bovine leucosis. There is little evidence for association of BoLA haplotypes with resistance or susceptibility to subclinical mastitis (Aarestrup *et al.*, 1995).

The Chicken Major Histocompatibility Complex

Introduction

The chicken MHC (or *B* complex) contains several classes of highly polymorphic genes located on microchromosome 16 (Bloom and Bacon, 1985; Bloom *et al.*, 1987). The class I and II genes (*B-F* and *B-L*, respectively) resemble their mammalian counterparts. The class IV (*B-G*) genes encode the B blood group antigens, which can be identified by serological blood typing. The class III region seems to be split in chickens into MHC-linked and non-MHC-linked elements. The *Rfp-Y* system, a system that bears strong similarity to the MHC, maps to the same microchromosome as the MHC. The MHC genes show associations with response to pathogens as diverse as virally induced neoplasia, bacterial, parasitic and autoimmune diseases (Table 4.1).

Table 4.1. Associations of immune response, disease resistance and pathology of infectious diseases with the pig, cow, chicken and horse MHCs[a].

Pig MHC	**Chicken MHC**
Disease response	Neoplastic diseases
Melanoma response	Marek's disease neoplasia
Response to *Trichinella spiralis* infection	Marek's disease transient paralysis
Cellular responses	Rous sarcoma
DTH response	Lymphoid leucosis
Bacterial phagocytosis anti-*Salmonella*	Parasitic diseases
Bacteriocidal activity	*Eimeria tenella*
Parasite antigen proliferation	*Eimeria acervulina*
Antibody response	Bacterial diseases
IgG and IgM responses	*Staphylococcus aureus*
Several synthetic antigens	Fowl cholera (*Pasteurella multocida*)
Several vaccines	*Salmonella enteritidis*
Several haemolytic complement responses	
Cattle MHC	**Horse MHC**
Disease associations	Disease associations
Posterior spinal paresis	Equine sarcoid
Mastitis	Hypersensitivity to insect bites
Bovine leucosis	Antibody response
Viral and bacterial infections	Several measures
Intracellular parasites and ectoparasites	Some vaccines
Several synthetic antigens	

[a]Discussions are included in the chapter.

Genetic variability, organization and structure

The MHC of the chicken was first identified by Briles and colleagues as a highly polymorphic blood group system (Briles and McGibbon, 1950) and was named the *B* system. Gilmour (1959) independently reported the same blood-group locus. A role of the B blood group in fitness, and a selective advantage of *B* heterozygosity for survival was suggested by persistence of multiple alleles of the *B* system, even in small, closed and inbred populations (Briles and McGibbon, 1948; Gilmour, 1959). The discovery that the *B* blood-group system is the MHC established the biological mechanism by which the *B* system (or linked genes) exerted genetic control on health traits (Schierman and Nordskog, 1961).

By 1977, it was evident that there were at least three loci in the chicken MHC (Pink *et al.*, 1977). After the advent of molecular approaches to study the MHC, an unanticipated level of complexity of genomic organization was found (Fig. 4.3) (Guillemot and Auffray, 1989; Guillemot *et al.*, 1989a, b; Kroemer *et al.*, 1990; Kaufman *et al.*, 1990, 1991b; Miller, 1991). The microchromosome on which the MHC is located is estimated to be 8 Mbp, with 6 Mbp of nucleolar organizer region (NOR) encoding approximately 400 rRNA genes in a tandem cluster (Delaney *et al.*, 1991; Riegert *et al.*, 1996).

The genomic organization of the chicken MHC differs from that of most mammalian MHCs. The chicken MHC is very compact (approximately 30–100 kb) compared with the MHC of mammals. The gene number is relatively small (approximately two class II β, one class II α and one to two class I α genes) (Kaufman *et al.*, 1995; Kaufman, 1996). The classical class I loci are in the *B-F/B-L* region. Other genes are also located in this region. Non-classical

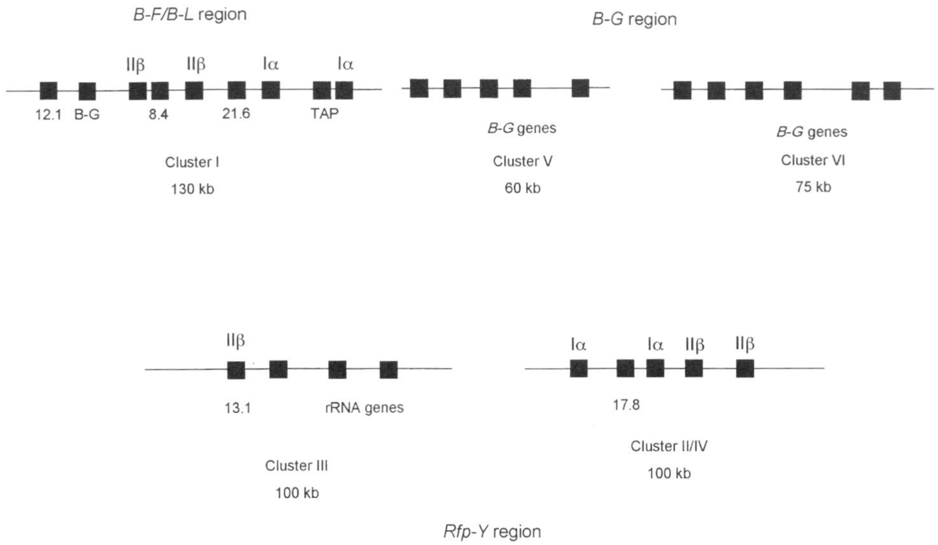

Fig. 4.3. General genomic organization of the chicken MHC (modified from Kaufman and Lamont, 1996).

(non-class I, non-class II) MHC genes mapping to the MHC region include ones that encode a G protein-like molecule, two TAP (transporter associated with antigen processing) molecules, and a member of the C-type animal lectin superfamily (Guillemot *et al.*, 1989b; Kaufman *et al.*, 1991b; Bernot *et al.*, 1994). As in mammals, the β_2-microglobulin gene is on a different microchromosome (Pickle *et al.*, 1990; Riegert *et al.*, 1996).

The class III genes in the chicken may be located in both MHC- and non-MHC-linked sites. Electrophoretic polymorphisms of glyoxylase 1 (GLO) and factor B, found in the MHC of mammals, do not segregate with the *B* complex (Rubinstein *et al.*, 1981; Koch, 1986). However, a DNA polymorphism detected by G9a (Milner and Campbell, 1993), a clone from the class III region of the human MHC, has been mapped to the chicken MHC (Spike and Lamont, 1995), demonstrating that some part of the class III region is associated with the chicken MHC.

Some class I α and class II β RFLP bands segregate independently of the serologically defined *B* complex (Briles *et al.*, 1993). These bands define the *Rfp-Y* complex. The MHC and the newly discovered *Rfp-Y* complexes were mapped to the same microchromosome by *in situ* hybridization (Fillon *et al.*, 1996). The *Rfp-Y* region is hypothesized to be involved in natural killer recognition and thus may play an important role in disease resistance (Miller *et al.*, 1990b).

Genes encoding the MHC class IV, or *B-G* erythrocyte alloantigens, are closely linked to the *B-F/B-L* region (Simonsen *et al.*, 1982; Crone and Simonsen, 1987; Goto *et al.*, 1988; Kaufman *et al.*, 1989, 1991b; Döhring *et al.*, 1993). The *B-G* genes form an extensive multigene family. Although generally identified as erythrocyte antigens, there is evidence that B-G molecules are located on other cell types: there are multiple B-G molecules present on erythrocytes, and very similar molecules are present on thrombocytes, lymphocytes, and stromal cells of the caecal tonsil, bursa, thymus and the epithelial cells of the small intestine (Salomonsen *et al.*, 1987, 1991a; Kline *et al.*, 1988; Kaufman *et al.*, 1990). These B-G chains vary significantly in size (35–55 kDa), mainly in the length of their cytoplasmic tails (Kaufman *et al.*, 1990).

Their function is not yet confirmed, but the structure of B-G molecules suggests typical adhesion molecules, with the extracellular region interacting with other cells and the environment, and the cytoplasmic tail being involved in signal transduction. The extensive polymorphism may help them respond over a wide variety of stimulating antigens. The B-G molecules exhibit some specific immune phenomena, including an 'adjuvant effect', initiating a rapid and strong antibody response, and the presence of 'natural antibodies' (Schierman and McBride, 1967; Longenecker *et al.*, 1979; Hala *et al.*, 1991; Kaufman *et al.*, 1991a; Salomonsen *et al.*, 1991b).

Immune response and the chicken MHC

Regulation of cellular communication in the immune response is a critical function of the chicken MHC (Dietert, 1987). The MHC cell-surface proteins distinguish 'self' from 'non-self'. This allows immune reaction to occur against foreign antigens of a virtually infinite variety while simultaneously preserving the self-integrity of the organism. The MHC cell-surface molecules interact with both the foreign antigen and with the complementary structure of other immune cells, generating an immune response that is specific for the inducing antigen. Signals are also transduced from the cell surface to the cell nucleus to change gene expression as a result of antigen stimulation.

To communicate effectively, cells of the immune system must share at least one MHC haplotype. This phenomenon, known as MHC restriction, may occur at the level of immune cell differentiation or of interactions between mature cells in the immune response (Vainio and Toivanen, 1987; Vainio *et al.*, 1983, 1988). The T–B cell cooperation needed for antibody production and for generation of germinal centres in the chicken spleen is MHC-restricted (Toivanen and Toivanen, 1997). Interaction of antigen-presenting cells and T cells, evaluated by the *in vitro* proliferation of T cells in response to specific antigen in the presence of antigen-presenting cells, is also MHC restricted. The *B-F/B-L* antigens are the restriction elements for all of these cellular interaction phenomena (Vainio *et al.*, 1984, 1988). Some T-cell cytotoxic reactions with virus-infected and/or transformed chicken cells have been shown to be MHC-restricted (MacCubbin and Schierman, 1986; Schat, 1994). An antigen-restricted class II-positive T cell with cytotoxic properties functions as the effector cell responsible for MHC-restricted cell-mediated cytotoxicity in the chicken against REV-infected cells (Weinstock and Schat, 1987).

The MHC influences a wide variety of immune responses in the chicken. Antibody production against a variety of antigens, such as immune response to synthetic polypeptides, is associated with the chicken MHC (Gunther *et al.*, 1974; Benedict *et al.*, 1975; Pevzner *et al.*, 1979). Antibody titres to several soluble antigens (e.g. bovine serum albumin, BSA), viral antigens (Bacon *et al.*, 1987) and to cellular antigens (e.g. *Salmonella pullorum* bacterin and sheep erythrocytes) (Pevzner *et al.*, 1975; Loudovaris *et al.*, 1990) are associated with allelic variation at the chicken MHC, as are total serum IgG levels (Rees and Nordskog, 1981). Genetic control of cell-mediated immunity is also associated with the chicken MHC. Polyclonal activators of T cells, such as phytohaemagglutinin (Taylor *et al.*, 1987), and specific inducers, such as *Staphylococcus aureus* (Cotter *et al.*, 1987), an important pathogen of poultry, show cell-mediated immune responses associated with the MHC. The levels of total serum haemolytic complement have also been associated with the MHC in chickens (Chanh *et al.*, 1976). Chemotactic activity of chicken-blood mononuclear leucocytes, and activity and recruitment to the peritoneal cavity of macrophages, differed among B-congenic lines (Qureshi *et al.*, 1986, 1988). Chicken B-congenic lines were also used to demonstrate MHC associations with cell-surface CD4 and CD8 lymphocyte percentages and ratios (Hala *et al.*, 1981).

The MHC has been studied in chicken lines selected long-term for traits of immunoresponsiveness. In long-term selection experiments for an antibody response to sheep red blood cells, correlated changes of MHC allelic frequencies with the antibody selection occurred (Dunnington *et al.*, 1984; Pinard and van der Zijpp, 1993). Although MHC genotype explains part of the variation in antibody levels, background genome also has a substantial effect (Dunnington *et al.*, 1989; Pinard *et al.*, 1993). After divergent selection for an early antibody response to *Escherichia coli* vaccination in meat-type chicken lines, differences in frequency of MHC class IV RFLP bands were present (Uni *et al.*, 1993). In an F_2 generation, polymorphisms of class I and TAP2 were associated with the antibody response (Yonash *et al.*, 1999). In Leghorn lines divergently selected for an index of immunocompetence traits (antibodies, cell-mediated response, and reticuloendothelial activity) (Cheng *et al.*, 1991), differences occurred in MHC genotype frequencies, identified both by serology (Kean *et al.*, 1994) and by DNA analysis (Lakshmanan and Lamont, unpublished; Weigend and Lamont, 1999). Therefore many diverse facets of the immune response are partly under MHC genetic control.

Disease resistance and the chicken MHC

Several recent reviews have summarized the association of the chicken MHC with resistance to disease (Bacon, 1987; Lamont, 1989, 1991, 1993, 1998; Gavora, 1990; Bumstead *et al.*, 1991; Stevens, 1991; Bumstead, 1996; Kaufman and Lamont, 1996).

The best-known association of the MHC with resistance to disease is that of Marek's disease (MD), a lymphoma induced by herpes virus. The recent emergence of highly virulent strains of MD reinforces the need for control measures that include genetic resistance approaches (Bumstead, 1996). Using MHC recombinants, MHC involvement in resistance to MD has been mapped to the *B-F/B-L* region (Briles *et al.*, 1983; Hepkema *et al.*, 1993). The MHC is associated with MD-related traits as diverse as incidence of tumour formation, mortality and transient paralysis. The B^{21} haplotype conveys MD resistance in many different genetic backgrounds. Variation in response to MD virus by sublines of birds identical at the *B* locus, however, illustrates the important role of non-MHC linked genes (Bacon, 1987). The role of the *Rfp-Y* region in resistance to MD is just beginning to be studied. Two studies found no association of *Rfp-Y* variation with resistance to MD (Vallejo *et al.*, 1997; Lakshmanan and Lamont, 1998), but another study (Wakenell *et al.*, 1996) reported that homozygous Y^3 genotype birds had double the risk of tumour development compared to all other genotypes. The differences between these studies most likely result from the specific *Rfp-Y* alleles and background genes, as well as the MD virus strain used for challenge and the specific disease traits studied. Genetic complementation of MHC alleles with other genes, such as *Ir-GAT* (Steadham *et al.*, 1987) and the background genome (Hartmann, 1997) also affects the susceptibility to MD.

Besides MD, the MHC also influences the response to other virally induced diseases, including Rous sarcoma virus tumours (Schierman and Collins, 1987), and avian leucosis (Yoo and Sheldon, 1992). Other categories of disease have also been found to be associated with the MHC (Gavora, 1990; Bacon and Witter, 1994). These include fowl cholera, a bacterial disease caused by *Pasteurella multocida* (Lamont *et al.*, 1987), spontaneous autoimmune thyroiditis in the genetic Obese strain (Rose, 1994) and coccidiosis, caused by several species of the parasite *Eimeria* (Lillehoj *et al.*, 1989; Bumstead *et al.*, 1991). MHC congenic lines have often been used to investigate MHC associations for diseases as diverse as MD (Bacon and Witter, 1992; Schat *et al.*, 1994), tumours induced by Rous sarcoma virus (White *et al.*, 1994) and v-*src*, (Taylor *et al.*, 1992), and *S. aureus* (Cotter *et al.*, 1992). The range and variety of diseases influenced by the chicken MHC is extensive (Table 4.1).

Studies of resistance to *Salmonella enteritidis* in several genetic lines have demonstrated variation in LD_{50}, organ contamination and bacterial burden (Bumstead and Barrow, 1988, 1993; Guillot *et al.*, 1995; Protais *et al.*, 1996), but no association with the MHC. However, recent studies utilizing MHC congenic lines showed an association of the MHC with *S. enteritidis* infection-induced morbidity and mortality (Cotter *et al.*, 1997), thereby illustrating the need for detailed studies of the MHC–*Salmonella* relationship.

No single haplotype performs optimally in all genetic backgrounds in response to all disease challenges. That is not surprising, given the complexity of the immune mechanisms involved in resistance to each specific disease, and the many genetic control points determined by the MHC. The B blood-group antigens are excellent candidates to be used as genetic markers, because they are highly polymorphic and can be typed inexpensively and rapidly from a small blood sample. Information on the MHC can be used to alter MHC allelic frequencies selectively for improvement of associated traits, such as disease resistance and vaccine response. The extensive polymorphism of the MHC class IV (dozens of alleles in Leghorns and in meat-type birds) allows many choices of specific B alleles in grandparent lines and therefore heterozygous combinations at the commercial crossbred level. However, the interaction of MHC alleles with the background genome is an important practical issue to be considered in any application of altering allele frequencies.

Because antigen-binding specificity is critical to effective poultry vaccine design, detailed knowledge of specific MHC types in a population will allow specific choices to be made to optimize the vaccine–host MHC combination or allow design of targeted recombinant vaccines (Pharr *et al.*, 1993; Bacon and Witter, 1994; Schat, 1994; Witter and Hunt, 1994). Efficacy of vaccination to MD in commercial genetic backgrounds is associated with MHC alleles (Bacon and Witter, 1994). The DNA can also be examined directly for allelic forms of the MHC genes of each class with the use of specific gene tests. Association with disease resistance of variation in structural genes and also in regulatory elements must be determined as allelic diversity for each individual class of MHC genes that is defined.

The Equine Major Histocompatibility Complex

Introduction

The equine MHC is commonly called ELA (equine leucocyte antigen). Compared to the MHCs of human, mouse and most domestic animal species, the organizational features and genetic content of ELA are poorly understood at present. Available evidence suggests that the genetic content and level of complexity within ELA is comparable to that observed in the MHCs of other species (Vaiman *et al.*, 1986; Alexander *et al.*, 1987; Bailey *et al.*, 1995). However, several intriguing peculiarities have also been identified in the ELA complex (see below).

Chromosomal map and location

ELA sequences have been located to chromosome *Eca* 20q14–22 by fluorescence *in situ* hybridization using heterologous class I cDNA probes (Ansari *et al.*, 1988; Mäkinen *et al.*, 1989). However, recent genetic mapping of a polymorphism in a class II DQA-like sequence to *Eqa* 5 (Fraser *et al.*, 1998) suggests that the MHC of horses may be disrupted even more than it is in cattle and chickens: if the presence of ELA loci on chromosome 5 is confirmed, this would be the first example in mammals where MHC sequences are located on different chromosomes.

Genetic variability and organization

Analysis of genomic class I sequences by Southern-blot hybridization indicates that the horse genome may contain as many as 30 class I loci, two or three times as many as observed in humans and in other domestic animals, and more similar to the mouse H-2 complex. As reported in other species, only a small subset of these genes is expressed on lymphocytes. Transcriptional analysis of class I sequences expressed on lymphocytes from a serologically heterozygous horse (ELA-A3/A7) identified seven unique sequences consistent with the presence of at least four expressed genes (Ellis *et al.*, 1995). It is not known how many of these gene products are translated into functional class I proteins. Sequence comparisons among coding regions of horse class I loci did not reveal firm phylogenetic relationships among the loci, perhaps indicating an unusual evolutionary history for ELA class I loci.

Sera from some, but not all, horses contain an 'equine soluble class I substance' (ESCI) encoded at a locus linked to ELA-A (Lew *et al.*, 1986a). ESCI was identified in other species of Perissodactyla but not in the wild Przewalski horse nor among species in the order Artiodactyla (Lew *et al.*, 1986b). Soluble class I protein has only been observed in other species of the rodent genus *Mus* where it is encoded at the H-2 non-classical Q10 locus (Lew *et al.*, 1986c). No function has been ascribed to ESCI or Q10, and the absence of this protein in sera from individuals that are otherwise phenotypically normal argues against the gene being functionally important.

Genes encoding the functional α and β subunits of the class II DQ and DR molecules have been described in horses (Albright *et al.*, 1991; Szalai *et al.*, 1993, 1994b; Gustafsson and Andersson, 1994), but the class II genes demonstrate several unusual properties. At least three alleles occur at the equine DR α locus (Hänni *et al.*, 1988; Bailey, 1994) in contrast to little or no variation at this locus in many other species. The presence of divergent sequences in two breed-specific alleles of DQA (Szalai *et al.*, 1994a), in the absence of additional allelic variation, and the apparent contradictory observations of little within-gene sequence diversity but high levels of RFLP, may reflect the analysis of two unlinked DQA sequences in the horse genome (Fraser *et al.*, 1998). The equine DQB also demonstrates several unusual properties, including the absence of shared amino acid motifs among alleles (Szalai *et al.*, 1993) and a mature polypeptide that is eight amino acids longer than commonly observed among mammals, and more similar to the mouse DQB homologue (Szalai *et al.*, 1994b).

The ELA complex and disease associations

As in other species, ELA has been associated with quantitative measures of the immune response (Lazary *et al.*, 1978; Bodo *et al.*, 1996). Generally, however, association of immune response with ELA haplotype was weak.

Only a few studies have searched for associations between susceptibility to specific diseases and ELA genotypes (Marti *et al.*, 1996) and only two statistically significant associations have been identified. Sarcoids are wart-like growths caused by papilloma virus, and are the most common form of skin tumours in horses. The class II DW13 antigen appears to be most predictive of susceptibility (Lazary *et al.*, 1985; Meredith *et al.*, 1986; Broström *et al.*, 1988). DW13 is commonly found with the class I antigens A3, A5 and A15, due to linkage disequilibrium, which likely explains the weak association of susceptibility with class I determinants. ELA class II-associated predisposition to sarcoids is even stronger when examined in half-sibling families from obligate heterozygous stallions (Gerber *et al.*, 1988) where nearly all of the affected offspring shared a common paternal ELA haplotype. The differences observed in ELA haplotype sarcoid susceptibility among different populations and with different class I or class II antigens most likely indicates that the causative gene lies within the ELA complex and that class I and class II genes serve as fortuitous genetic markers.

Inherited predisposition to allergic hypersensitivity to insect bites from blackflies (genus *Simulans*) and midges (genus *Culicoides*) (sweet itch or summer dermatitis) has been observed in Icelandic horses and in Swiss Warmbloods. In a population study of Icelandic horses, Halldorsdottir *et al.* (1991) observed that the class II antigen DW22 occurred more frequently in affected than unaffected animals, with a relative risk value of 2.53. Haplotyping of two half-sib families of Swiss Warmbloods demonstrated an even stronger association with ELA class II antigen DW23 (Marti *et al.*, 1992; Lazary *et al.*, 1994), but it is not clear whether this association is detectable at the population level in Warmblood breeds. Further study should reveal additional information on the role of ELA and disease susceptibility.

Overall Conclusions

Given the information obtained to date, the MHC of farm animals is an important region of the genome for immune responsiveness and disease resistance. Genetic selection for specific alleles within the MHC may be a desirable approach to improving health-related traits. However, despite the extensive research conducted in this area, the only case of the use of MHC alleles in marker-assisted selection for disease resistance is that using the *B* allele in chicken for selection to improve resistance to Marek's disease. Understanding more clearly the MHC–disease resistance relationships may enable such use of MHC alleles in the future.

Acknowledgements

The skilful editorial assistance of Gretchen Triplett in synthesis of the co-authors' contributions into the final draft is gratefully acknowledged. This is Journal Paper No. J-18285 of the Iowa Agriculture and Home Economics Experiment Station, Projects No. 3215 and 2237.

References

Aarestrup, F.M., Jensen, N.E. and Ostergard, H. (1995) Analysis of associations between major histocompatibility complex (BoLA) class I haplotypes and subclinical mastitis of dairy cows. *Journal of Dairy Science* 78, 1684–1692.

Ababon, A., Goyeneche, J., Davis, W.C. and Levy, D. (1994) Evidence of the expression of three different BoLA-class II molecules on the bovine BL-3 cell line: determination of a non-DR non-DQ gene product. *Journal of Leucocyte Biology* 56, 182–186.

Albright, D., Bailey, E. and Woodward, J.G. (1991) Nucleotide sequence of a cDNA clone of the horse (*Equus caballus*) DRA gene. *Immunogenetics* 34, 136–138.

Alexander, A.J., Bailey, E. and Woodward, J.G. (1987) Analysis of the equine lymphocyte antigen system by Southern blot hybridization. *Immunogenetics* 25, 47–54.

Amorena, B. and Stone, W.H. (1978) Serologically defined (SD) locus in cattle. *Science* 201, 159–160.

Andersson, L. (1996) Major histocompatibility complex evolution. In: Schook, L.B. and Lamont, S.J. (eds) *The Major Histocompatibility Complex in Domestic Animal Species*. CRC Press, Boca Raton, Florida, pp. 1–15.

Andersson, L. and Davies, C.J. (1994) The major histocompatibility complex. In: Goddeeris, B.M. and Morrison, W.I. (eds) *Cell-Mediated Immunity in Ruminants*. CRC Press, Boca Raton, Florida, pp. 37–57.

Andersson, L., Lundén, A., Sigurdardottir, S., Davies, C.J. and Rask, L. (1988) Linkage relationships in the bovine Mhc region. High recombination frequency between class II subregions. *Immunogenetics* 27, 273–280.

Ansari, H.A., Hediger, R., Fries, R. and Stranzinger, G. (1988) Chromosomal localization of the major histocompatibility complex of the horse (ELA) to chromosome 20 by *in situ* hybridization. *Hereditas* 110, 93–98.

Appleyard, G.D., Mallard, B.A., Kennedy, B.W. and Wilkie, B.N. (1992) Antibody avidity in

swine lymphocyte antigen-defined miniature pigs. *Canadian Journal of Veterinary Research* 56, 303–307.
Bacon, L.D. (1987) Influence of the MHC on disease resistance and productivity. *Poultry Science* 66, 802–811.
Bacon, L.D. and Witter, R.L. (1992) Influence of turkey herpesvirus vaccination on the B-haplotype effect on Marek's disease resistance in 15.B-congenic chickens. *Avian Disease* 36, 378–385.
Bacon, L.D. and Witter, R.L. (1994) B haplotype influence on the relative efficacy of Marek's disease vaccines in commercial chickens. *Poultry Science* 73, 481–487.
Bacon, L.D., Ismail, N. and Motta, J.V. (1987) Allograft and antibody responses of $15I_5$-B congenic chickens. In: Weber, W.T. and Ewert, D.L. (eds) *Avian Immunology II.* Alan R. Liss, New York, pp. 219–233.
Bailey, E. (1994) Variation within the antigen-binding site of the major histocompatibility complex DRA gene of domestic horses. In: Nakajima, H. and Plowright, W. (eds) *Equine Infectious Diseases VII.* R. & W. Publications, Newmarket, UK, pp. 123–126.
Bailey, E., Graves, K.T., Cothran, E.G., Reid, R., Lear, T.L. and Ennis, R.B. (1995) Synteny-mapping horse microsatellite markers using a heterohybridoma panel. *Animal Genetics* 26 (3), 177–180.
Band, M., Larson, J.H., Womack, J.E. and Lewin, H.A. (1998) A radiation hybrid map of BTA23: identification of a chromosomal rearrangement leading to separation of the cattle MHC class II subregions. *Genomics*, 53 (3), 269–275.
Batra, T.R., Stear, M.J. and Macdonald, P.A. (1996) Association of class I bovine lymphocyte antigens with profitability and lifetime yields in the Holstein breed. *Canadian Journal of Animal Science* 76, 145–148.
Benedict, A.A., Pollard, L.W., Morrow, P.R., Abplanalp, H.A., Mauer, P.H. and Briles, W.E. (1975) Genetic control of immune responses in chickens. I. Responses to a terpolymer of poly($glu^{60}ala^{30}tyr^{10}$) associated with the major histocompatibility complex. *Immunogenetics* 2, 313–324.
Benjamin, R. and Parham, P. (1990) Guilt by association: HLA-B27 and ankylosing spondylitis. *Immunology Today* 11, 137–142.
Bensaid, A., Kaushal, A., Baldwin, C.L., Clevers, H., Young, J.R., Kemp, S.J., MacHugh, N.D., Toye, P.G. and Teale, A.J. (1991) Identification of expressed bovine class I MHC genes at two loci and demonstration of physical linkage. *Immunogenetics* 33, 247–254.
Bernot, A., Zoorob, R. and Auffray, C. (1994) Linkage of a new member of the lectin supergene family to chicken Mhc genes. *Immunogenetics* 39, 221–229.
Blangero, J., Tissot, R.G., Beattie, C.W. and Amoss, M.S. Jr (1996) Genetic determinants of cutaneous malignant melanoma in Sinclair swine. *British Journal of Cancer* 73, 667–671.
Bloom, S. and Bacon, L. (1985) Linkage of the major histocompatibility (B) complex and the nucleolar organizer region in the chicken. *Journal of Heredity* 76, 146–154.
Bloom, S.E., Briles, W.E., Briles, R.W., Delaney, M.E. and Dietert, R.R. (1987) Chromosomal localization of the major histocompatibility (B) complex (MHC) and its expression in chickens aneuploid for the major histocompatibility complex/ribosomal deoxyribonucleic acid microchromosome. *Poultry Science* 66, 782–789.
Bodo, G., Marti, E., Gallard, C., Weiss, M., Bruckner, L., Gerber, H. and Lazary, S. (1996) Association of the immune response with the major histocompatibility complex in the horse. In: Nakajima, H. and Plowright, W. (eds) *Equine Infectious Diseases VII.* R. & W. Publications, Newmarket, UK, pp. 143–151.
Briles, W.E. and McGibbon, W.H. (1948) Heterozygosity of inbred lines of chickens at

two loci effecting cellular antigens. *Genetics* 33, 605 (abstract).

Briles, W.E., McGibbon, W.H. and Irwin, M.R. (1950) On multiple alleles affecting cellular antigens in the chicken. *Genetics* 35, 633–652.

Briles, W.E., Briles, R.W., Taffs, R.E. and Stone, H.A. (1983) Resistance to a malignant lymphoma in chickens is mapped to subregion of major histocompatibility (B) complex. *Science* 219, 977–979.

Briles, W.E., Goto, R.M., Auffray, C. and Miller, M.M. (1993) A polymorphic system related to but genetically independent of the chicken major histocompatibility complex. *Immunogenetics* 37, 408–414.

Broström, H., Fahlbrink, E., Dubath, M.-L. and Lazary, S. (1988) Association between equine leucocyte antigens (ELA) and equine sarcoid tumors in the populations of Swedish halfbreds and some of their families. *Veterinary Immunology and Immunopathology* 19, 215–223.

Bumstead, N. (1996) Breeding for disease resistance. In: Davison, T.F., Morris, T.R. and Payne, L.N. (eds) *Poultry Immunology.* Carfax Publishing Company, Abingdon, UK, pp. 405–415.

Bumstead, N. and Barrow, P. (1988) Genetics of resistance to *Salmonella typhimurium* in newly hatched chicks. *British Poultry Science* 29, 521–529.

Bumstead, N. and Barrow, P. (1993) Resistance to *Salmonella gallinarum*, *S. pullorum*, and *S. enteritidis* in inbred lines of chickens. *Avian Disease* 37, 189–193.

Bumstead, N., Millard, B.M., Barrow, P. and Cook, J.K.A. (1991) Genetic basis of disease resistance in chickens. In: Owen, J.B. and Axford, R.F.E. (eds) *Breeding for Disease Resistance in Farm Animals.* CAB International, Melksham, UK, pp. 10–23.

Calandro, L.M., Baer, D.M. and Sensabaugh, G.F. (1995) Characterization of a recombinant that locates the hereditary hemochromatosis gene telomeric to HLA-F. *Human Genetics* 96, 339.

Campbell, R.D. and Trowsdale, J. (1993) Map of human MHC. *Immunology Today* 14, 349–352.

Campbell, R.D. and Trowsdale, J. (1997) A map of the human major histocompatibility complex. *Immunology Today* 18, 43 (poster).

Chanh, T.C., Benedict, A.A. and Abplanalp, H. (1976) Association of serum hemolytic complement levels with the major histocompatibility complex in chickens. *Journal of Experimental Medicine* 144, 555–561.

Chardon, P., Renard, Ch. and Vaiman, M. (1981) Characterization of class II histocompatibility antigens in pigs. *Animal Blood Groups and Biochemical Genetics* 12, 59–65.

Chardon, P., Vaiman, M., Kirszenbaum, M., Geffrotin, C., Renard, C. and Cohen, D. (1985) Restriction fragment length polymorphism of the major histocompatibility complex of the pig. *Immunogenetics* 21, 161–171.

Cheng, S., Rothschild, M.F. and Lamont, S.J. (1991) Estimates of quantitative genetic parameters of immunological traits in the chicken. *Poultry Science* 780, 2023–2027.

Cotter, P.F., Taylor, R.L., Wing, T.L. and Briles, W.E. (1987) Major histocompatibility (B) complex-associated differences in the delayed wattle reaction to Staphylococcal antigen. *Poultry Science* 66, 203–208.

Cotter, P.F., Taylor, R.L. Jr and Abplanalp, H. (1992) Differential resistance to *Staphylococcus aureus* challenge in major histocompatibility (B) complex congenic lines. *Poultry Science* 7 (1), 1873–1878.

Cotter, P.F., Taylor, R.L. Jr and Abplanalp, H. (1997) B-complex (chicken MHC) associated immunity to *Salmonella enteritidis.* In: Colin, P., LeGoux, J.M. and Clement, G. (eds) *Salmonella and Salmonellosis Proceedings,* 20–22 May, Ploufragan, France, pp. 281–285.

Crone, M. and Simonsen, M. (1987) Avian major histocompatibility complex. In: Toivanen, A. and Toivanen, P. (eds) *Avian Immunology: Basis and Practice*, Vol. 2. CRC Press, Boca Raton, Florida, pp. 25–42.

Da, Y., Shanks, R.D., Stewart, J.A. and Lewin, H.A. (1993) Milk and fat yields decline in bovine leukemia virus-infected Holstein cattle with persistent lymphocytosis. *Proceedings of the National Academy of Sciences USA* 90, 6538–6541.

Delaney, M.E., Muscarella, D.E. and Bloom, S.E. (1991) Formation of nucleolar polymorphisms in trisomic chickens and subsequent microevolution of rRNA gene clusters in diploids. *Journal of Heredity* 82 (3), 213–215.

Dietert, R. (1987) Introduction: the major histocompatibility complex as a communication gene complex. *Poultry Science* 66, 774–775.

Dietz, A.B., Detilleux, J.C., Freeman, A.E., Kelley, D.H., Stabel, J.R. and Kehrli, M.E. (1997) Genetic association of bovine lymphocyte antigen DRB3 alleles with immunological traits of Holstein cattle. *Journal of Dairy Science* 80, 400–405.

DiPalma, F., Young, J.R. and Ellis, S.A. (1998) Analysis of the bovine MHC class I region using a BAC library constructed from an MHC homozygous animal. XXVI International Conference on Animal Genetics, Auckland, New Zealand. *Animal Genetics* 29 (*Suppl.* 1), p. 25.

Döhring, C., Riegert, P., Salomonsen, J., Skjødt, K. and Kaufman, J. (1993) The extracellular Ig V-like regions of the polymorphic B-G antigens of the chicken Mhc lack structural features expected for antibody variable regions. In: Coudert, F. (ed.) *Avian Immunology in Progress*. INRA, Paris, France, p. 145.

Dubey, J.P., Lunney, J.K., Shen, S.K., Kwok, O.C.H., Ashford, D.A. and Thulliez, P. (1996) Infectivity of low numbers of *Toxoplasma gondii* oocysts to pigs. *Journal of Parasitology* 82, 438–443.

Dunnington, E.A., Briles, R.E., Briles, W.E., Gross, W.B. and Siegel, P.B. (1984) Allelic frequencies in eight alloantigen systems of chickens selected for high and low antibody response to sheep red blood cells. *Poultry Science* 63, 1470–1472.

Dunnington, E.A., Martin, A., Briles, R.W., Briles, W.E., Gross, W.B. and Siegel, P.B. (1989) Antibody responses to sheep erythrocytes for White Leghorn chickens differing in haplotypes of the major histocompatibility complex (B). *Animal Genetics* 20, 213–216.

Ellis, S.A., Braem, K.A. and Morrison, W.I. (1992) Transmembrane and cytoplasmic domain sequences demonstrate at least two expressed bovine MHC class I loci. *Immunogenetics* 37, 49–56.

Ellis, S.A., Martin, A.J., Holmes, E.C. and Morrison, W.I. (1995) At least four MHC class I genes are transcribed in the horse: phylogenetic analysis suggests an unusual evolutionary history for the MHC in this species. *European Journal of Immunogenetics* 22 (3), 249–260.

Ennis, P.D., Jackson, A.P. and Parham, P. (1988) Molecular cloning of bovine class I MHC cDNA. *Journal of Immunology* 141, 642–651.

Ernst, L.K., Sulimova, G.E., Orlova, A.R., Udina, I.G. and Pavlenko, S.P. (1997) Peculiarities of distribution of BoLA-A antigens and alleles of the BoLA-DRB3 gene in black pied cattle in relation to association with leukemia. *Genetika* 33, 87–95.

Feder, J.N., Gnirke, A., Thomas, W., Tsuchihashi, Z., Ruddy, D.A., Basava, A., Dormishian, F., Domingo, R. Jr, Ellis, M.C., Fullan, A., Hinton, L.M., Jones, N.L., Kimmel, B.E., Kronmal, G.S., Lauer, P., Lee, V.K., Loeb, D.B., Mapa, F.A., McClelland, E., Meyer, N.C., Mintier, G.A., Moeller, N., Moore, T., Morikang, E., Prass, C.E., Quintana, L., Starnes, S.M., Schatzman, R.C., Brunke, K.J., Drayna, D.T., Risch, N.J., Bacon, B.R. and Wolff, R.K. (1996) A novel MHC class I-like gene is mutated in patients with hereditary hemochromatosis. *Nature Genetics* 13, 399–408.

Fillon, V., Zoorob, R., Yerle, M., Auffray, C. and Vigual, A. (1996) Mapping of the genetically independent chicken major histocompatibility complexes B and RFP-Y to the same microchromosome by two-color fluorescent *in situ* hybridization. *Cytogenetics and Cell Genetics* 75, 7–9.

Fraser, D.G., Bailey, E., Swinburne, J., Binns, M., Lear, T.L. and Skow, L.C. (1998) Two MHC class II DQA loci for the horse, XXVI International Conference on Animal Genetics. *Animal Genetics* 29 (*Suppl.* 1) p. 29.

Gallagher, D.S., McShane, R., Davis, S.K., Taylor, J.F. and Skow, L.D (1998) Ordering of BoLA genomic sequences by high resolution FISH. XXVI International Conference on Animal Genetics, Auckland, New Zealand. *Animal Genetics* 29 (*Suppl.* 1) p. 24.

Garber, T.L., Hughes, A.L., Watkins, D.I. and Templeton, J.W. (1994) Evidence for at least three transcribed *BoLA* class I loci. *Immunogenetics* 39, 257–265.

Gavora, J.S. (1990) Disease genetics. In: Crawford R.D. (ed.) *Poultry Breeding and Genetics*. Elsevier, Amsterdam, pp. 805–846.

Gavora, J.S., Simonsen, M., Spencer, J.L., Fairful, R.W. and Gowe, R.S. (1986) Changes in the frequency of major histocompatibility haplotypes in chickens under selection for both high egg production and resistance to Marek's disease. *Zeitschrift für Tierzüchtung und Züchtungsbiologie* 103, 218–226.

Gaycken, U., Shabhang, M., Meyer, J.N. and Glodek, P. (1994) Molecular characterization of the porcine MHC class I region. *Animal Genetics* 25, 357–363.

Geffrotin, C., Popescu, C.P., Cribiu, E.P., Boscher, J., Renard, Ch., Chardon, P. and Vaiman, M. (1984) Assignment of MHC in swine to chromosome 7 by *in situ* hybridization and serological typing. *Annales de Génétiques* 27, 213–219.

Geffrotin, C., Renard, Ch., Chardon, P. and Vaiman, M. (1991) Marked genomic polymorphism of the swine steroid 21-hydroxylase gene and its location between the SLA class I and class II regions. *Animal Genetics* 22, 311–322.

Gerber, H., Dubath, M.-L. and Lazary, S. (1988) Association between predisposition to equine sarcoid and MHC in multiple-case families. In: Powell, D.G. (ed.) *Equine Infectious Diseases V*. The University Press of Kentucky, pp. 272–277.

Gilmour, D.G. (1959) Segregation of genes determining red cell antigens at high levels of inbreeding in chickens. *Genetics* 44, 14–33.

Glass, E.J., Oliver, R.A. and Spooner, R.L. (1990) Variation in T cell responses to ovalbumin in cattle: evidence for Ir gene control. *Animal Genetics* 21, 15–28.

Gonzalez-Juarrero, M., Lunney, J.K., Sanchez-Vizcaino, J.M. and Mebus, C. (1992) Modulation of splenic macrophages, and swine leukocyte antigen (SLA) and viral antigen expression following African swine fever virus (ASFV) inoculation. *Archives of Virology* 123, 145–156.

Gorer, P.A. (1937) The genetic and antigenic basis of tumour transplantation. *Journal of Pathology of Bacteriology* 44, 691–697.

Goto, R., Miyada, C.G., Young, S., Wallace, R.B., Abplanalp, H., Bloom, S.E., Briles, W.E. and Miller, M.M. (1988) Isolation of a cDNA clone from the B-G subregion of the chicken histocompatibility (B) complex. *Immunogenetics* 27, 102–109.

Grosz, M.D., Davis, S.K., Taylor, J.F. and Skow, L.C. (1998) Mhc-linked HSP70 gene cluster architecture is conserved in cattle. XXVI International Conference on Animal Genetics, Auckland, New Zealand. *Animal Genetics* 29 (*Suppl.* 1) p. 34.

Guillemot, F. and Auffray, C. (1989) The molecular biology of the chicken major histocompatibility complex. *CRC Critical Review of Poultry Biology* 2, 255–275.

Guillemot, F., Billault, A. and Auffray, C. (1989a) Physical linkage of a guanine nucleotide-binding protein related gene to the chicken major histocompatibility complex. *Proceedings of the National Academy of Sciences USA* 86, 4594–4598.

Guillemot, F., Kaufman, J., Skjødt, K. and Auffray, C. (1989b) The major histocompati-

bility complex of the chicken. *Trends in Genetics* 57, 300–304.
Guillot, J.F., Beaumont, C., Bellatif, F., Mouline, C., Lantier, F., Colin, P. and Protais, J. (1995) Comparison of resistance of various poultry lines to infection by *Salmonella enteritidis*. *Veterinary Research* 26, 81–86.
Gunther, E., Balcarova, J., Hala, K., Rude, E. and Hraba, T. (1974) Evidence for an association between immune responseness of chicken to (T,G)-A--L and the major histocompatibility system. *European Journal of Immunology* 4, 548–553.
Gustafsson, K.T. and Andersson, L. (1994) Structure and polymorphism of equine MHC class II DRB genes: convergent evolution on the antigen binding site. *Immunogenetics* 39, 355–358.
Gustafsson, K., Germana, S., Hirsch, F., Pratt, K., LeGuern, C. and Sachs, D.H. (1990) Structure of miniature swine class II DRB genes: conservation of hypervariable amino acid residues between distantly related mammalian species. *Proceedings of the National Academy of Sciences USA* 87, 9798–9802.
Hala, K., Plachy, J. and Schulmannova, J. (1981) Role of the B-G-region antigen in the humoral immune response to the B-F-region antigen of chicken MHC. *Immunogenetics* 14, 393–401.
Hala, K., Vainio, O., Plachy, J. and Bock, G. (1991) Chicken major histocompatibility complex congenic lines differ in the percentages of lymphocytes bearding CD4 and CD8 antigens. *Animal Genetics* 22, 279–284.
Halldorsdottir, S., Lazary, S., Gunnarsson, E. and Larsen, H.J. (1991) Distribution of leucocyte antigens in Icelandic horses affected with summer eczema compared to non-affected horses. *Equine Veterinarian Journal* 23, 300–302.
Hänni, H., Hesford, F., Lazary, S. and Gerber, H. (1988) Restriction fragment length polymorphisms of horse class II MHC genes observed using various human alpha- and beta-chain cDNA probes. *Animal Genetics* 19, 395–408.
Hartmann, W. (1997) Evaluation of major genes affecting resistance to disease in poultry. *World's Poultry Science Journal* 53 (9), 231–252.
Hepkema, B.G., Blankert, J., Albers, G.A.A., Tilanus, M.G.J., Egberts, E., van der Zijpp, A.J. and Hensen, E.J. (1993) Mapping of susceptibility to Marek's disease within the major histocompatibility (B) complex by refined typing of White Leghorn chickens. *Animal Genetics* 24, 283–287.
Hess, M., Eggen, A., Gelhous, A., Goldammer, T., Schwerin, M., Bishop, M.D. and Horstrom, R.D. (1998) YAC contigs covering the bovine MHC. *Animal Genetics* 29 (*Suppl.* 1) p. 24.
Hook, R.R., Aultman, M.D., Adeltein, E.H., Oxenhandler, R.W., Millikan, L.E. and Middleton, C.C. (1979) Influence of selective breeding on the incidence of melanomas in sinclair miniature swine. *International Journal of Cancer* 24, 668–672.
House, J.K., Smith, B.P., Mass, J., Lane, V.M., Anderson, B.C., Graham, T.W. and Pino, M.V. (1994) Hemachromatosis in Salers Cattle. *Journal of Veterinary Internal Medicine* 8 (2), 105–111.
Hu, S. (1996) Studies on bovine MHC (BoLA) Class I gene family: identification and expression of BoLA class I genes. Doctoral Dissertation, Texas A&M University, College Station, Texas.
Joosten, I., Oliver, R.A., Spooner, R.L., Williams, J.L., Hepkema, B.G., Sanders, M.F. and Hensen, E.J. (1988) Characterization of the class I bovine lymphocyte antigens (BoLA) by one-dimensional isolectric focusing. *Animal Genetics* 19, 103–113.
Kaufman, J. (1996) Structure and function of the major histocompatibility complex of chickens. In: Davison, T.F., Morris, T.R. and Payne, L.N. (eds) *Poultry Immunology*. Carfax Publishing Company, Abingdon, UK, pp. 67–82.
Kaufman, J., Salomonsen, J. and Skjødt, K. (1989) B-G cDNA clones have multiple small

repeats and hydridize to both chicken MHC regions. *Immunogenetics* 30, 440–451.

Kaufman, J., Salomonsen, J., Skjødt, K. and Thorpe, D. (1990) Size polymorphism of the chicken MHC-encoded B-G molecules is due to length variation in the cytoplasmic heptad repeat region. *Proceedings of the National Academy of Sciences USA* 87, 8277–8281.

Kaufman, J., Salomonsen, J., Riegert, P. and Skjødt, K. (1991a) Using chicken class I sequences to understand how xenoantibodies crossreact with MHC-like molecules in nonmammalian vertebrates. *American Zoology* 31, 570–579.

Kaufman, J., Skjødt, K. and Salomonsen, J. (1991b) The B-G multigene family of the chicken major histocompatibility complex. *Critical Review of Immunology* 11, 113–143.

Kaufman, J., Völk, H. and Wallny, H.-J. (1995) A 'minimal essential MHC' and an 'unrecognized MHC': two extremes in selection for polymorphism. *Immunology Review* 143, 63–88.

Kaufman, J.F. and Lamont, S.J. (1996) The chicken major histocompatibility complex. In: Schook, L.B. and Lamont, S.J. (eds) *The Major Histocompatibility Complex in Domestic Animal Species.* CRC Press, Boca Raton, Florida, pp. 35–64.

Kean, R.P., Briles, W.E., Cahaner, A., Freeman, A.E. and Lamont, S.J. (1994) Differences in major histocompatibility complex frequencies after multitrait, divergent selection for immunocompetence. *Poultry Science* 73, 7–17.

Kelm, S.C., Detilleuh, J.C., Freeman, A.E., Kehrli, M.E., Dietz, A.B., Fox, L.K., Butler, J.E., Kasckovics, I. and Kelley, D.H. (1997) Genetic association between parameters of innate immunity and measures of mastitis in periparturient Holstein cattle. *Journal of Dairy Science* 80, 1767–1775.

Kirszenbaum, M., Renard, C., Geffrotin, C., Chardon, P. and Vaiman, M. (1985) Evidence for mapping pig C4 genes within the pig major histocompatibility complex (SLA). *Animal Blood Groups and Biochemical Genetics* 16, 65–72.

Klein, J. (1986) *Natural History of the Major Histocompatibility Complex.* John Wiley & Sons, New York.

Klein, J., Bontrop, R.E., Dawkins, R.L., Erlich, H.A., Gyllensten, U.B., Heise, E.R., Jones, P.P., Parham, P., Wakeland, E.K. and Watkins, D.I. (1990) Nomenclature for the major histocompatibility complexes of different species: a proposal. *Immunogenetics* 31, 217–219.

Kline, K., Briles, W.E., Bacon, L. and Sanders, B.G. (1988) Characterization of two distinct disulfide-linked B-G molecules in the chicken. *Journal of Heredity* 79, 249–256.

Koch, C. (1986) A genetic polymorphism of the complement component factor B in chickens is not linked to the major histocompatibility complex (MHC). *Immunogenetics* 23, 364–367.

Koga, H., Sakou, T., Taketomi, E., Hayashi, K., Numasawa, T., Harata, S., Yone, K., Matsunaga, S., Otterud, B., Inoue, I. and Leppert, M. (1998) Genetic mapping of ossification of the posterior longitudinal ligament of the spine. *American Journal of Human Genetics* 62 (6), 1460–1467.

Kristensen, B. (1995) Possible influences of parental MHC class I on survival of offspring from sows naturally infected with porcine reproduction and respiratory syndrome virus (PRRSV). *Proceedings of the International Veterinary Immunology Symposium.* Davis, California.

Kroemer, G., Bernot, A., Behar, G., Chause, A.-M., Gastinel, L.-N., Guillemot, F., Park, I., Thoraval, P., Zoorob, R. and Auffray, C. (1990) Molecular genetics of the chicken MHC: current status and evolutionary aspects. *Immunology Review* 3, 118–145.

Lacey, C., Wilkie, B.N., Kennedy, B.W. and Mallard, B.A. (1989) Genetic and other effects on bacterial phagocytosis and killing by cultures peripheral blood monocytes of

SLA-defined miniature pigs. *Animal Genetics* 20, 371–382.
Lakshmanan, N. and Lamont, S.J. (1998) Research Note: Rfp-Y region polymorphism and Marek's disease resistance in multitrait immunocompetence-selected chicken lines. *Poultry Science* 77, 538–541.
Lamont, S.J. (1989) The chicken major histocompatibility complex in disease resistance and poultry breeding. *Journal of Dairy Science* 72, 1328–1333.
Lamont, S. (1991) Immunogenetics and the major histocompatibility complex. *Veterinary Immunology and Immunopathology* 30, 121–127.
Lamont, S.J. (1993) The major histocompatibility complex. In: Etches, R.J. and Verrinder, A.M. (eds) *Manipulation of the Avian Genome.* CRC Press, Boca Raton, Florida, pp. 185–203.
Lamont, S.J. (1998) Impact of genetics on disease resistance. *Poultry Science* 77, 1111–1118.
Lamont, S.J., Bolin, C. and Cheville, N. (1987) Genetic resistance to fowl cholera is linked to the major histocompatibility complex. *Immunogenetics* 25, 284–289.
Lazary, S., Kuslys, A., Bullen, S., De Weck, A.L. and Gerber, H. (1978) Leukocyte antigens (ELA) in the antibody production to tetanus toxoid and toxin in the horse. In: Bryans, J.T. and Gerbert, H. (eds) *Equine Infectious Diseases IV. Journal of Equine Medicine and Surgery*, Supplement 1. Veterinary Publications, Princeton, New Jersey, pp. 237–242.
Lazary, S., Gerber, H., Glatt, P.A. and Straub, R. (1985) Equine leucocyte antigens in sarcoid-affected horses. *Equine Veterinarian Journal* 17, 283–286.
Lazary, S., Marti, E., Szalai, G., Gaillard, C. and Gerber, H. (1994) Studies on the frequency and associations of equine leucocyte antigens in sarcoid and summer dermatitis. *Animal Genetics* 25 (Suppl. 1), 75.
Lew, A.M., Bailey, E., Valas, R.B. and Coligan, J. (1986a) The gene encoding the equine soluble class I molecule is linked to the horse MHC. *Immunogenetics* 24, 128–130.
Lew, A.M., Valas, R.B., Maloy, W.L. and Coligan, J.E. (1986b) A soluble class I molecule analogous to mouse Q10 in the horse and related species. *Immunogenetics* 23, 277–283.
Lew, A.M., Maloy, W.L. and Coligan, J.E. (1986c) Characteristics of the expression of the murine soluble class I molecule (Q10). *Journal of Immunology* 163 (1), 254–258.
Lewin, H.A. (1989) Disease resistance and immune response genes in cattle: strategies for their detection and evidence of their existence. *Journal of Dairy Science* 72, 1334–1348.
Lewin, H.A. (1994) Host genetic mechanism of resistance and susceptibility to a bovine retroviral infection. *Animal Biotechnology* 5, 183–191.
Lewin, H.A. (1996) Genetic organization, polymorphism and function of the bovine major histocompatibility complex. In: Shook, L.B. and Lamont, S.J. (eds) *The Major Histocompatibility Complex Region of Domestic Animal Species.* CRC Press, Boca Raton, pp. 65–98.
Lewin, H.A., Clamp, P.A., Beever, J.E. and Schook, L.B. (1991) Mapping genes for resistance to infectious diseases. In: Schook L.B., Lewin H.A. and McLaren, D.G. (eds) *Gene Mapping: Techniques and Applications.* Marcel Dekker, New York.
Lie, O., Solbu, H., Larsen, H.J. and Spooner, R.L. (1986) Possible association of antibody responses to human serum albumin and (T,G)-A--L with the bovine major histocompatibility complex (BoLA). *Veterinary Immunology and Immunopathology* 11, 333–350.
Lie, W.R., Rothschild, M.F. and Warner, C. (1987) Mapping of C2, BF and C4 genes to the swine major histocompatibility complex (swine leucocyte antigen). *Journal of*

Immunology 135, 3388–3394.

Lillehoj, H.S., Ruff, M.D., Bacon, L.D., Lamont, S.J. and Jeffers, T.K. (1989) Genetic control of immunity to *Eimeria tenella*: interaction of MHC genes and genes within the genetic background influence levels of disease susceptibility. *Veterinary Immunology and Immunopathology* 20, 135–148.

Lindberg, P.-G. and Andersson, L. (1988) Close association between DNA polymorphism of bovine major histocompatibility complex class I genes and serological BoLA-A specificities. *Animal Genetics* 19, 245–255.

Longenecker, B.M., Mosmann, T.R. and Shiozawa, C. (1979) A strong, preferential response of mice to polymorphic antigenic determinants of the chicken MHC, analyzed with mouse hybridoma (monoclonal) antibodies. *Immunogenetics* 9, 137–147.

Loudovaris, T., Brandon, M.R. and Fahey, K.J. (1990) The major histocompatibility complex and genetic control of antibody response to sheep red blood cells in chickens. *Avian Pathology* 19, 89–99.

Lumsden, J.S., Kennedy, B.W., Mallard, B.A. and Wilkie, B.N. (1993) The influence of the swine major histocompatibility genes on antibody and cell-mediated immune response to immunization with an aromatic-dependent mutant of *Salmonella typhimurium*. *Canadian Journal of Veterinary Research* 57, 14–18.

Lundén, A., Sigurdardottir, S., Edfors-Lilja, I., Danell, B., Rendel, J. and Andersson, L. (1990) The relationship between bovine major histocompatibility complex class II polymorphism and disease studied by use of bull breeding values. *Animal Genetics* 21, 221–232.

Lunney, J.K. (1994) Current status of the swine leukocyte antigen complex. *Veterinary Immunology and Immunopathology* 43, 19–28.

Lunney, J.K. and Butler, J.E. (1998) Immunogenetics. In: Rothschild, M.F. and Ruvinsky, A. (eds) *The Genetics of the Pig.* CAB International, Wallingford, UK, pp. 163–197.

Lunney, J.K. and Murrell, K.D. (1988) Immunogenetic analysis of *Trichinella spiralis* infections in swine. *Veterinary Parasitology* 29, 179–199.

Lunney, J.K., VanderPutten, D. and Pescovitz, M.D. (1984) Mhc linked immune response gene control of humoral and cellular responses to (T,G)-A--L and lysozyme in miniature swine. *Federation Proceedings* 43, 1821.

MacCubbin, D.L. and Schierman, L.W. (1986) MHC-restricted cytotoxic response of chicken T cells: expression, augmentation, and clonal characterization. *Journal of Immunology* 136, 12–16.

Madden, K.B., Murrell, K.D. and Lunney, J.K. (1990) *Trichinella spiralis*: Major Histocompatibility Complex-associated elimination of encysted muscle larvae in swine. *Experimental Parasitology* 70, 443–451.

Madden, K.B., Moeller, R.F., Douglass, L.W., Goldman, T. and Lunney, J.K. (1993) *Trichinella spiralis*: genetic basis and kinetics of the anti-encysted muscle larval response in miniature swine. *Experimental Parasitology* 77, 23–35.

Maillard, J.C., Chantal, I., DeReynal, M. and Elsen, J.M. (1998) An amino acid motif present in five BoLA-DRB3 alleles is a highly significant marker of susceptibility to bovine dermatophilosis. *Animal Genetics* 29 (*Suppl.* 1) p. 26.

Mäkinen, A., Chowdhary, B., Mahdy, E., Andersson, L. and Gustafsson, L. (1989) Localization of the equine major histocompatibility complex (ELA) to chromosome 20 by *in situ* hybridization. *Hereditas* 110, 93–96.

Mallard, B.A., Leslie, K.E., Dekkers, J.C.M., Hedge, R., Bauman, M. and Stear, M.J. (1995) Differences in bovine lymphocyte antigen associations between immune responsiveness and risk of disease following intramammary infection with *Staphyococcus aureus*. *Journal of Dairy Science* 78, 1937–1944.

Mallard, B.A., Wilkie, B.N. and Kennedy, B.W. (1989a) Genetic and other effects on antibody and cell mediated immune response in swine leucocyte antigen (SLA)-defined miniature pigs. *Animal Genetics* 20, 167–178.

Mallard, B.A., Wilkie, B.N. and Kennedy, B.W. (1989b) Influence of the major histocompatibility genes on serum complement activity in miniature swine. *American Journal of Veterinary Research* 50, 359–363.

Mallard, B.A., Wilkie, B.N. and Kennedy, B.W. (1989c) The influence of the Swine Major Histocompatibility genes (SLA) on variation in serum immunoglobulin (Ig) concentration. *Veterinary Immunology and Immunopathology* 21, 139–151.

Marti, E., Gerber, H. and Lazary, S. (1992) On the genetic basis of equine allergic diseases. II. Insect bite dermal hypersensitivity. *Equine Veterinarian Journal* 24, 113–117.

Marti, E., Szalai, G., Antczak, D.F., Bailey, E., Gerber, H. and Lazary, S. (1996) The equine major histocompatibility complex. In: *The MHC Region of Domestic Animal Species.* CRC Press, Boca Raton, Florida, pp. 245–267.

Martins, C., Mebus, C., Scholl, T., Lawman, M. and Lunney, J. (1988) Virus-specific CTL in SLA-inbred swine recovered from experimental African swine fever virus (ASFV) infection. *Annals of the New York Academy of Science* 532, 462–464.

Meeker, D.L., Rothschild, M.F., Christian, L.L., Warner, C.M. and Hill, H.T. (1987) Genetic control of immune response to pseudorabies and atrophic rhinitis vaccines. I. Heterosis, general combining ability and relationship to growth and backfat. *Journal of Animal Science* 64, 407–413.

Mejdell, C.M., Lie, O., Solbu, H., Arnet, E.F. and Spooner, R.L. (1994) Association of major histocompatibility complex antigens (BoLA-A) with AI bull progeny test results for mastitis, ketosis and fertility in Norwegian cattle. *Animal Genetics* 25, 99–104.

Meredith, D., Else, A.H., Wolf, B., Soma, L.R., Donawick, W.J. and Lazary, S. (1986) Equine leukocyte antigens: relationships with sarcoid tumors and laminitis in two pure breeds. *Immunogenetics* 23, 221–225.

Miller, M.M. (1991) The major histocompatibility complex of the chicken. In: Warr, G. and Cohen, N. (eds) *Phylogenesis of Immune Functions.* CRC Press, Boca Raton, Florida.

Miller, M.M., Goto, R., Young, S., Liu, J. and Hardy, J. (1990b) Antigens similar to major histocompatibility complex B-G are expressed in the intestinal epithelium in the chicken. *Immunogenetics* 32, 45–50.

Milner, C.M. and Campbell, R.D. (1993) The G9a gene in the human major histocompatibility complex encodes a novel protein containing ankyrin-like repeats. *Biochemistry Journal* 290, 811–818.

Müller, V.S., Wanke, R. and Distl, O. (1995) Segregation von bigmentzellanomalien beim münchener miniaturschwein (MMS) troll in Kreuzungen mit der Deutschen Landrasse. *Deutcher Tierärztlich Wokenschrift* 102, 391–394.

Newman, M.J., Truax, R.E., French, D.D., Dietrich, M.A., Franke, D. and Stear, M.J. (1996) Evidence for genetic control of vaccine-induced antibody responses in cattle. *Veterinary Immunology and Immunopathology* 50, 43–54.

Østergard, H., Kristensen, B. and Andersen, S. (1989) Investigations in farm animals of associations between the MHC system and disease resistance. *Livestock Production Science* 22, 49–67.

Park, C.A., Hines, H.C., Mouke, D.R. and Threlfall, W.T. (1993) Association between the bovine major histocompatibility complex and chronic posterior spinal paresis – a form of ankylosing spodylitis – in Holstein bulls. *Animal Genetics* 24, 53–58.

Peelman, L.J., Chardon, P., Vaiman, M., Mattheeuws, M., Van Zeveren, A., Van de Weghe, A., Bouquet, Y. and Campbell, R.D. (1996) A detailed physical map of the porcine major histocompatibility complex (MHC) class III region: comparison with human and mouse MHC class III regions. *Mammalian Genome* 7, 363–367.

Pevzner, I.Y., Nordskog, A.W. and Kaeberle, M.L. (1975) Immune response and the B blood group locus in chickens. *Genetics* 80, 753–759.

Pevzner, I.Y., Trowbridge, C.L. and Nordskog, A.W. (1979) B-complex genetic control of immune response to HSA, (T,G)-A--L, GT and other substances in chickens. *Journal of Immunogenetics* 6, 453–460.

Pharr, G.T., Hunt, H.D., Bacon, L.D. and Dodgson, J.B. (1993) Identification of class II major histocompatibility complex polymorphisms predicted to be important in peptide antigen presentation. *Poultry Science* 72, 1312–1317.

Pickle, J., Chen, C.-L. and Cooper, M. (1990) An avian B lymphocyte protein associated with β_2-microglobulin. *Immunogenetics* 32, 1–7.

Pinard, M.H. and van der Zijpp, A.J. (1993) Effects of major histocompatibility complex on antibody response in F1 and F2 crosses of chicken lines. *Genetics of Selection Evolution* 25, 283–296.

Pinard, M.H., Van Arendonk, J.A.M., Nieuwland, M.G.B. and van der Zijpp, A.J. (1993) Divergent selection for humoral immune responsiveness in chickens: distribution and effects of major histocompatibility complex types. *Genetics of Selective Evolution* 25, 1991–203.

Pink, J.R.L., Droege, W., Hala, K., Miggiano, V.C. and Ziegler, A. (1977) A three-locus model for the chicken major histocompatibility complex. *Immunogenetics* 5, 203–216.

Protais, J., Colin, P., Beaumont, C., Guillot, J.F., Lantier, F., Pardon, P. and Bennejean, G. (1996) Line differences in resistance to *Salmonella enteritidis* PT4 infection. *British Poultry Science* 37, 329–339.

Qureshi, M.A., Dietert, R.R. and Bacon, L.D. (1986) Genetic variation in the recruitment and activation of chicken peritoneal macrophages. *Proceedings of the Society of Experimental Biological Medicine* 181, 560–568.

Qureshi, M.A., Dietert, R.R. and Bacon, L.D. (1988) Chemotactic activity of chicken blood mononuclear leukocytes from $15I_5$-B-congenic lines to bacterially-derived chemoattractants. *Veterinary Immunology and Immunopathology* 19, 351–360.

Rabin, M., Fries, R., Singer, D.S. and Ruddle, F.H. (1985) Assignment of the porcine major histocompatibility complex to chromosome 7 by *in situ* hybridization. *Cytogenetics and Cell Genetics* 39, 206–209.

Rees, M.J. and Nordskog, A.W. (1981) Genetic control of serum immunoglobulin G levels in the chicken. *Journal of Immunogenetics* 8, 425–431.

Renard, Ch., Kristensen, B., Gautschi, C., Hruban, V., Fredholm, M. and Vaiman, M. (1988) Joint report of the first international comparison test on swine lymphocyte alloantigens (SLA). *Animal Genetics* 19, 63–72.

Riegert, P., Andersen, R., Bumstead, N., Dohring, C., Dominguez-Steglich, M., Engberg, J., Salomonsen, J., Schmidt, M., Skjødt, K., Schwager, J. and Kaufman, J. (1996) The chicken β_2-microglobulin gene is located on a non-MHC microchromosome, a small, G + C rich gene with X and Y boxes in the promoter. *Proceedings of the National Academy of Sciences USA* 93, 1243–1248.

Rogel-Gaillard, C., Vaiman, M., Renard, C., Chardon, P. and Yerle, M. (1998) Localization of the β_2-microglobulin gene to pig chromosome 1q17. *Mammalian Genome* 8, 948.

Rose, N.R. (1994) Avian models of autoimmune disease: lessons from the birds. *Poultry Science* 73, 984–990.

Rothschild, M.F., Chen, H.L., Christian, L.L., Lie, W.R., Venier, L., Cooper, M., Briggs, C. and Warner, C.M. (1984a) Breed and swine lymphocyte antigen haplotype differences in agglutination titers following vaccination with *B. bronchiseptica*. *Journal of Animal Science* 59, 643–649.

Rothschild, M.F., Hill, H.T., Christian, L.L. and Warner, C.M. (1984b) Genetic differences in serum neutralization titers of pigs after vaccination with pseudorabies modified live virus-vaccine. *American Journal of Veterinary Research* 45, 1216–1218.

Rubin, L.A., Amos, C.I., Wade, J.A., Martin, J.R., Bale, S.J., Little, A.H., Gladman, D.D., Bonney, G.E., Rabenstein, I.D. and Siminoritch, K.A. (1994) Investigating the genetic basis for ankylosing spondylitis. Linkage studies with the major histocompatibility region. *Arthritis Rheumatism* 37, 1212–1220.

Rubinstein, P., De Haas, L., Pevzner, I.Y. and Nordskog, A.W. (1981) Glyoxalase I (GLO) in the chicken: genetic variation and lack of linkage to the MHC. *Immunogenetics* 13, 493–497.

Sachs, D.H., Leight, G., Cone, J., Schwartz, S., Stuart, L. and Rosenberg, S. (1976) Transplantation in miniature swine I. Fixation of the major histocompatibility complex. *Transpolation* 22, 559–567.

Salomonsen, J., Skjødt, K., Crone, M. and Simonsen, M. (1987) The chicken erythrocyte-specific antigen. Characterization and purification of the B-G antigen by monoclonal antibodies. *Immunogenetics* 25, 373–382.

Salomonsen, J., Dunon, D., Skjødt, K., Thorpe, D., Vainio, O. and Kaufman, J. (1991a) Chicken major histocompatibility complex-encoded B-G antigens are found on many cells that are important for the immune system. *Proceedings of the National Academy of Sciences USA* 88, 1359–1363.

Salomonsen, J., Eriksson, H., Skjødt, K., Lundgren, L., Simonsen, M. and Kaufman, J. (1991b) The 'adjuvant effect' of the polymorphic B-G antigens of the chicken MHC analyzed using purified molecules incorporated in liposomes. *European Journal of Immunology* 21, 649–658.

Schat, K.A. (1994) Cell-mediated immune effector functions in chickens. *Poultry Science* 73, 1077–1081.

Schat, K.A., Taylor, R.L. Jr and Briles, W.E. (1994) Resistance to Marek's disease in chickens with recombinant haplotypes of the major histocompatibility (B) complex. *Poultry Science* 73, 502–508.

Schierman, L.W. and Collins, W.M. (1987) Influence of the major histocompatibility complex on tumor regression and immunity in chickens. *Poultry Science* 66, 812–818.

Schierman, L.W. and McBride, R.A. (1967) Adjuvant activity of erythrocyte isoantigens. *Science* 156, 658–650.

Schierman, L.W. and Nordskog, A.W. (1961) Relationship of blood type to histocompatibility in chickens. *Science* 134, 1008–1009.

Schook, L.B. and Lamont, S.J. (1996) *The Major Histocompatibility Complex in Domestic Animal Species*. CRC Press, Boca Raton, Florida.

Schook, L.B., Rutherford, M.S., Lee, J.-K., Shia, Y.-C., Bradshaw, M. and Lunney, J.K. (1996) The swine major histocompatibility complex. In: Schook, L.B. and Lamont, S.J. (eds) *The Major Histocompatibility Complex of Domestic Animal Species*. CRC Press, Boca Raton, Florida, pp. 212–244.

Shia, Y.-C., Gautschi, C., Ling, M.-S., Beever, J.E., Mclaren, D.G., Lewin, H.A. and Schook, L.B. (1991) RFLP analysis of SLA haplotypes in Swiss Large White and American Hampshire pigs using SLA class I and class II probes. *Animal Biotechnology* 2, 75–91.

Shia, Y.-C., Bradshaw, M., Rutherford, M.S., Lewin, H.A. and Schook, L.B. (1995) Polymerase chain reaction based genotyping for characterization of SLA-DQB and SLA-DRB alleles in domestic pigs. *Animal Genetics* 26, 91–100.

Simonsen, M., Crone, M., Koch, C. and Hala, K. (1982) The MHC haplotypes of the chicken. *Immunogenetics* 16, 513–532.

Singer, D.S., Camerini-Otero, R.D., Satz, M.L., Osborne, B., Sachs, D. and Rudikoff, S. (1982) Characterization of a porcine genomic clone encoding a major histocompatibility antigen: expression in mouse L cells. *Proceedings of the National Academy of Sciences USA* 79, 1403–1407.

Singer, D.S., Ehrlich, R., Golding, H., Satz, L., Parents, L. and Radikoff, S. (1988) Structure and expression of class I MHC genes in the miniature swine. In: Warner, C., Rothschild, M. and Lamont, S. (eds) *The Molecular Biology of the MHC of Domestic Animal Species*. ISU Press, Ames, Iowa, pp. 53–62.

Skow, L.C., Snaples, S.N., Davis, S.K., Tylor, J.F., Huang, B. and Gallagher, D.H. (1996) Localization of bovine lymphocyte antigen (BoLA) DYA and Class I loci to different regions of chromosome 23. *Mammalian Genome* 7, 388–389.

Smith, T.P.L., Rohrer, G.A., Alexander, L.J., Troyer, D.L., Kirby-Dobbels, K.R., Janzen, M.A., Cornwell, D.L., Louis, C.F., Schook, L.B. and Beattie, C.W. (1995) Directed integration of the physical and genetic linkage maps of swine chromosome 7 reveals that the SLA spans the centromere. *Genome Research* 5, 259–271.

Spike, C.A. and Lamont, S.J. (1995) Genetic analysis of 3 loci homologous to human G9a: evidence for linkage of a class III gene with the chicken MHC. *Animal Genetics* 26, 185–187.

Spooner, R.L., Leveziel, H., Grosclaude, F., Oliver, R.A. and Vaiman, M. (1978) Evidence for a possible major histocompatibility complex (BoLA) in cattle. *Journal of Immunogenetics* 5, 335–346.

Spooner, R.L., Oliver, R.A., Sales, D.L., McCoubrey, C.M., Millar, P., Morgan, A.G., Amorena, B., Bailey, E., Bernoco, D., Brandon, M., Bull, R.W., Caldwell, J., Ceik, S., van Dam, R.H., Dodd, J., Gahne, B., Grosclaude, F., Hall, J.G., Hines, H., Leveziel, H., Newman, M.J., Stear, M.J., Stone, W.H. and Vaiman, M. (1979) Analysis of alloantisera against bovine lymphocytes. Joint report of the First International Bovine Lymphocyte Antigen (BoLA) Workshop. *Animal Blood Groups Biochemistry Genetics* 15, 63–68.

Srivastara, R., Ran, B. and Tyle, P. (1991) *Immunogenetics of the Major Histocompatibility Complex*. VCH Publishing, New York.

Steadham, E.M., Lamont, S.J., Kujdych, I. and Nordskog, A.W. (1987) Association of Marek's disease with Ea-B and immune response genes in subline and F2 populations of the Iowa State S1 Leghorn line. *Poultry Science* 66, 571–575.

Stevens, L. (1991) *Genetics and Evolution of the Domestic Fowl*. Cambridge University Press, Cambridge, p. 306.

Sulimova, G.E., Udina, I.G., Shaikhaev, G.O. and Zakharov, I.A. (1995) DNA polymorphism at the BoLA-DRB3 gene of cattle in relation to resistance and susceptibility to leukemia. *Genetika* 31, 1294–1299.

Szalai, G., Bailey, E., Gerber, H. and Lazary, S. (1993) DNA sequence analysis of serologically detected ELA class II haplotypes in the equine DQB locus. *Animal Genetics* 24, 187–190.

Szalai, G., Antczak, D.F., Gerber, H. and Lazary, S. (1994a) Molecular cloning and characterization of equine DQA cDNA. *Immunogenetics* 40, 457.

Szalai, G., Antczak, D.F., Gerber, H. and Lazary, S. (1994b) Molecular cloning and characterization of equine DQB cDNA. *Immunogenetics* 40, 458.

Taracha, E.L.N., Goddeeris, B.M., Teale, A.J., Kemp, S.J. and Morrison, W.I. (1995) Parasite strain specificity of bovine cytotoxic T cell responses to *Theilerin parva* is determined primarily by immunodominance. *Journal of Immunology* 155, 4854–4860.

Taylor, R.L. Jr, Cotter, P.F., Wing, T.L. and Briles, W.E. (1987) Major histocompatibility (b) complex and sex effects on the phytohemagglutinin wattle response. *Animal Genetics* 18, 343–350.

Taylor, R.L. Jr, Ewert, D.L., England, J.M. and Halpern, M.S. (1992) Major histocompatibility (B) complex control of the growth pattern of v-src DNA-induced primary tumors. *Virology* 191, 477.

Tissot, R.G., Beattie, C.W. and Amoss, M.S. Jr (1987) Inheritance of Sinclair swine cutaneous malignant melanoma. *Cancer Research* 47, 5542–5545.

Toivanen, A. and Toivanen, P. (1977) Histocompatibility requirements for cellular cooperation in the chicken: generation of germinal centers. *Journal of Immunology* 118, 431–436.

Uni, Z., Gutman, M., Leitner, G., Landesman, E., Heller, D. and Cahaner, A. (1993) Major histocompatibility complex class IV restriction fragment length polymorphism markers in replicated meat-type chicken lines divergently selected for high or low early immune response. *Poultry Science* 72, 1823–1831.

Väge, D.I., Lingaas, F., Spooner, R.L., Arnet, E.F. and Lie, O. (1992) A study on association between mastitis and serologically defined class I bovine lymphocyte antigens (BoLA-A) in Norwegian cows. *Animal Genetics* 23, 533–536.

Vaiman, M., Renard, Ch., Lafage, P., Ameteau, J. and Nizza, P. (1970) Evidence for a histocompatibility system in swine (SL-A). *Transplantation* 10, 155–164.

Vaiman, M., Hauptmann, G. and Meyer, S. (1978a) Influence of the major histocompatibility complex in the pig (SLA) on serum haemolytic complement levels. *Journal of Immunogenetics* 5, 59–65.

Vaiman, M., Metzger, J.J., Renard, Ch. and Vila, J.P. (1978b) Immune response gene(s) controlling the humoral anti-lysozyme response (Ir-Lys) linked to the major histocompatibility complex SL-A in the pig. *Immunogenetics* 7, 231–238.

Vaiman, M., Chardon, P. and Cohen, D. (1986) DNA polymorphism in the major histocompatibility complex of man and various farm animals. *Animal Genetics* 17, 113–133.

Vaiman, M., Renard, Ch. and Bourgeaux, N. (1988) SLA, the major histocompatibility complex in swine: its influence on physiological and pathological traits. In: Warner, C.M., Rothschild, M.F. and Lamont, S.J. (eds) *The Molecular Biology of Major Histocompatibility Complex of Domestic Animal Species.* Iowa State University Press, Ames, Iowa, pp. 23–37.

Vaiman, M., Chardon, P. and Rothschild, M. (1998) Porcine major histocompatibility complex. *O.I.E. Revue* 17, 95–107.

Vainio, O. and Toivanen, A. (1987) Cellular cooperation in immunity. In: Toivanen, A. and Toivanen, P. (eds) *Avian Immunology: Basis and Practice*, Vol. II. CRC Press, Boca Raton, Florida, pp. 1–12.

Vainio, O., Peck, R., Koch, C. and Toivanen, A. (1983) Origin of peripheral blood macrophages in bursa cell-reconstituted chickens: further evidence for MHC-restricted interactions between T and B lymphocytes. *Scandinavian Journal of Immunology* 17, 193–199.

Vainio, O., Koch, C. and Toivanen, A. (1984) B-L antigens (class II) of the chicken major histocompatibility complex control T-B cell interaction. *Immunogenetics* 19, 131–140.

Vainio, O., Veromaa, T., Eerola, E., Toivanen, P. and Ratcliffe, M. (1988) Antigen presenting cell–T cell interaction in the chicken is MHC class II antigen restricted. *Journal of Immunology* 140, 2864–2868.

Vallejo, R.L., Pharr, G.T., Liu, H.C., Cheng, H.H., Witter, R.L. and Bacon, L.D. (1997) Non-association between Rfp-Y major histocompatibility complex-like genes and susceptibility to Marek's disease virus-induced tumours in $6_3 \times 7_2$ F_2 intercross chickens. *Animal Genetics* 28, 331–337.

van Eijk, M.J.T., Russ, I. and Lewin, H.A. (1993) Order of bovine *DRB3, DYA*, and *PRL* determined by sperm typing. *Mammalian Genome* 4, 113–118.

Velten, F., Renard, C., Rogel-Gaillard, J.C., Vaiman, M. and Chardon, P. (1998) Assignment of BAC and YAC clones spanning the porcine class I region. *Animal Genetics* 29 (*Suppl.* 1) p. 28.

Viza, D., Sugar, J.R. and Binns, R.M. (1970) Lymphocyte stimulation in pigs: evidence for the existence of a single major histocompatibility locus PL-A. *Nature* 227, 949–950.

Wakenell, P.S., Miller, M.M., Goto, R.M., Gauderman, W.J. and Briles, W.E. (1996) Association between the Rfp-Y haplotype and the incidence of Marek's disease in chickens. *Immunology* 44, 242–245.

Warner, C.M. and Rothschild, M.F. (1991) The Swine Major Histocompatibility Complex (SLA). In: Srivastara, R., Ram, B. and Tyle, P. (eds) *Immunogenetics of the Major Histocompatiblity Complex*. VCH Publishers, New York, pp. 368–397.

Weigel, L.A., Freeman, A.E., Kehrli, M.E. Jr, Stear, M.J. and Kelley, D.H. (1990) Association of class I bovine lymphocyte antigen complex alleles with health and production traits in cattle. *Journal Dairy Science* 73, 2538–2546.

Weigend, S. and Lamont, S.J. (1999) Analysis of MHC class II and class IV restriction fragment length polymorphism in chicken lines divergently selected for multitrait immune response. *Poultry Science* 78, 973–982.

Weinstock, D. and Schat, K.A. (1987) Virus specific syngeneic killing of reticuloendotheliosis virus transformed cell line target cells by spleen cells. In: Weber, W.T. and Ewert, D.L. (eds) *Progress in Clinical and Biological Research, Avian Immunology*, Vol. 238. Alan R. Liss, New York, pp. 253–263.

White, E.C., Briles, W.E., Briles, R.W. and Taylor, R.L. Jr (1994) Response of six major histocompatibility (B) complex recombinant haplotypes to rous sarcomas. *Poultry Science* 73, 836–842.

Witter, R.L. and Hunt, H.D. (1994) Poultry vaccines of the future. *Poultry Science* 73, 1087–1093.

Xu, A., McKenna, K. and Lewin, H.A. (1993) Sequencing and genetic analysis of a bovine DQA cDNA clone. *Immunogenetics* 37, 231–234.

Xu, Y., Rothschild, M.F. and Warner, C.M. (1992) Mapping of the SLA complex of miniature swine: mapping of the SLA gene complex by pulsed field gel electrophoresis. *Mammalian Genome* 2, 2–10.

Yonash, N., Kaiser, M.G., Heller, E.D., Cahaner, A. and Lamont, S.J. (1999) Major histocompatibility complex (MHC) related cDNA probes associated with antibody response in meat-type chickens. *Animal Genetics* 30, 92–101.

Yoo, B.H. and Sheldon, B.L. (1992) Association of the major histocompatibility complex with avian leukosis virus infection in chickens. *British Poultry Science* 33, 613–620.

5

Rodent Models of Genetic Resistance to Parasitic Infections

D. Wakelin
School of Biological Sciences, University of Nottingham, Nottingham, UK

Summary

Resistance to disease in general, and to parasitic disease in particular, is a genetically variable characteristic of all animals, including domestic species. Recognition of this fact makes it possible to think in terms of breeding selectively for enhanced resistance or of introducing resistance genes into existing stock, either by breeding programmes or through transgenic technology. Knowledge of the mechanisms through which genetic variation is expressed makes it possible to manipulate immune responsiveness selectively to improve overall breed resistance. Progress in all of these fields depends upon a detailed understanding of the genetic and immunological influences on disease resistance, which can come only from detailed studies in laboratory model systems, particularly those involving mice. Only in these relatively inexpensive, well-defined and manipulable hosts can parameters of infection, response and genotype be rigorously controlled and exploited.

This chapter will describe a range of host–parasite systems selected to cover three of the major groups of parasites responsible for disease in farm animals: Protozoa, Nematoda and Arthropoda. The diseases for which these systems are models include coccidiosis, trypanosomiasis, gastrointestinal helminthiasis and tick infestation. For each system the genetic basis of resistance and susceptibility is defined and the roles of MHC-linked and background genes analysed. The immune and inflammatory responses controlled by these genes, and their functional activities in mediating resistance, are discussed. The data arising from these laboratory studies are considered in relation to selective breeding programmes, to vaccination strategies and to manipulation of host responsiveness designed to reduce immunopathology and prevent parasite-induced immunosuppression.

Introduction

Resistance to disease, and specifically infectious disease, operates at several different levels. Innate resistance reflects those fixed properties of an animal

© CAB *International* 2000. *Breeding for Disease Resistance in Farm Animals*
(eds R.F.E. Axford, S.C. Bishop, F.W. Nicholas and J.B. Owen)

(structural, behavioural, biochemical, physiological) which protect it against the development of a particular disease organism. Acquired resistance comprises those immune and immunologically mediated responses which adaptively protect an initially susceptible animal once it has experienced an infection. Both innate and acquired resistance reflect the phenotypic expression of genetically determined characteristics, and as such can be expected to vary among members of a genetically heterogeneous population.

The empirical observation that individuals within flocks and herds of domestic animals do differ markedly in their resistance to infectious diseases, has a long history. Descriptive studies of this phenomenon extend back for at least 60–70 years, and a number of important papers demonstrating the genetic basis of such differences were published in the 1950s (e.g. Champion, 1954; Whitlock, 1955). Despite the importance of genetic variation in disease resistance, detailed analyses of the ways in which this variation is controlled and expressed in domestic stock have been difficult to undertake, and progress (with some important exceptions) has been relatively slow. There are many reasons for this situation, including the logistic complexity and expense of large-animal research and the long generation times and seasonal reproduction of species such as cattle and sheep. Equally important, however, has been the lack of genetically and immunologically well-defined lines of domestic species, and the paucity of specific immunological reagents for investigating the components of their immune systems. In the years since the publication of the first edition of this book there has been substantial progress in these fields, e.g. inbred and congenic chickens are now available; T-cell, cytokine and immunoglobulin reagents are available for domestic fowl, cattle, pigs and sheep; the structure and function of their immune systems is increasingly well understood, and programmes to map the genomes of domestic animals are under way. All of these facilitate genetic studies of resistance in target species, but considerations of cost and logistics still favour basic research in rodents, and many laboratory models of infectious diseases have now been explored in great detail. In most cases the models have to make use of species related to the target pathogen, rather than the pathogen itself, but this is not a major limitation and such work has led to a detailed understanding of the general principles underlying the expression of genetic variation in resistance (Wakelin and Blackwell, 1988, 1993). This understanding began with the classical studies by Gowen and Webster on bacterial and viral infections in mice (Gowen, 1951; Webster, 1937) and has grown exponentially with the availability of genetically defined strains of mice and with developments in immunology and immunogenetics. It continues with the application of the concepts and approaches of molecular genetics, which have made available sophisticated techniques for the detection and mapping of genes, as well as for the production of genetically modified (transgenic and 'knockout') animals in which the genetic control of specific components of immune and inflammatory responses can be manipulated precisely.

This short review is not intended to be comprehensive. Instead it will focus on a few model systems involving parasitic infections where progress has been substantial and testable hypotheses developed. For a broader and more detailed coverage of the topic the reader is referred to reviews by Abel and Dessein (1997), Hill (1996), Stear and Wakelin (1998), Wakelin and Blackwell (1988, 1993) and Wakelin and Walliker (1996).

Rodent Model Systems

Many of the parasitic infections affecting farm animals can be modelled in the laboratory using rodent hosts. Of these, the most useful from the point of view of the immunogeneticist are those involving mice, rats and guinea-pigs, where well-defined inbred strains are available (Table 5.1). Mice are the best-defined species, both genetically and immunologically, and are the most manipulable experimentally. The range of strains available, and the ease of breeding, makes it possible to locate genes responsible for observed differences in resistance (Tables 5.2 and 5.3) and to identify the ways in which these differences are inherited. The murine genome is increasingly well mapped, with a large number of markers. The use of microsatellite markers for mapping quantitative trait loci makes it possible to identify genomic regions influencing particular phenotypic characteristics, from which individual gene loci can then be pinpointed (see also Chapter 1). Advances in the application of molecular genetic techniques have made it possible to produce mice with specific genes deleted ('knock-out' mice) or with specific genes inserted (transgenic mice). In addition to these genetic advantages, there is a very wide range of immunological reagents available for mice, which makes it possible to identify, delete or augment particular components of immune and inflammatory responses in order to determine their role in resistance.

The murine systems discussed here have been selected to cover three of the major groups of parasites of farm animals – Protozoa, Nematoda and

Table 5.1. Useful rodent models in which genetic influences on resistance to parasites can be studied.

	Parasite	Host	Target disease
Protozoa	*Babesia* spp.	Mouse	Piroplasmosis
	Eimeria spp.	Mouse, rat	Coccidiosis
	Trypanosoma	Mouse	Trypanosomiasis
Platyhelminthes	*Schistosoma* spp.	Mouse, rat	Schistosomiasis
	Taenia taeniaeformis	Mouse	Cysticercosis
	Echinococcus	Mouse	Hydatid
Nematoda	*Heligmosomoides polygyrus*	Mouse	GI nematodiasis
	Nippostrongylus brasiliensis	Mouse, rat	GI nematodiasis
	Trichinella spiralis	Mouse, rat	Trichinellosis
	Trichostrongylus colubriformis	Guinea-pig	GI nematodiasis
	Trichuris muris	Mouse	Trichuriasis
Arthropoda	*Dermacentor variabilis*	Mouse	Tick infestation
	Haemaphysalis longicornis	Mouse	Tick infestation

Table 5.2. Analysis of genetic differences in disease resistance: range of mouse strains available for immunogenetic studies.

Strains	Characteristics
Random-bred	Individuals differ genetically – modelling variation within populations of domestic stock
Inbred	Individuals genetically identical – different strains model variation seen between different individuals within populations
Congenic	Different strains differ only at a specified locus or region (e.g. the MHC loci) – allow influence of specific loci to be determined
Recombinant inbred	Strains produced by crossing two defined inbred strains Distribution pattern of a given trait allows identification of nature of genetic control involved and may indicate chromosomal location of gene/s
Mutant	Strains carrying defined mutation affecting immunological or inflammatory processes
Genetically modified	Strains in which specific genes have been introduced (transgenic) or effectively deleted ('knock-out')

Table 5.3. Analysis of genetic differences in disease resistance: expression of genetic differences.

Gene/s involved in control	
MHC-linked genes (HLA, H-2, BLA, OVA, etc.)	Non-MHC-linked genes (background genes)
Effects expressed in infection	
Qualitative differences between individuals (all-or-nothing responsiveness)	Quantitative differences between individuals (level of response)
seen in:	*seen in:*
Resistance or susceptibility	Degree of resistance of susceptibility
Antigen recognition/non-recognition	Level of antibody response
Lymphocyte responsiveness	Level of inflammatory response

Arthropoda. They illustrate the usefulness of experimental approaches in defining the genetic basis of resistance, identifying the resistance mechanisms concerned and providing predictive models relevant to practical problems.

Coccidia

Coccidiosis has long been a major problem in the poultry industry and is increasingly important in the intensive rearing of calves, pigs and sheep. In chickens, breed variation in resistance to disease has been known for many years, and detailed studies in inbred lines have uncovered the roles of MHC-linked and non-MHC genes (see Chapter 17). A number of species of *Eimeria* infect rodents, providing models that parallel many of the characteristics of coccidial infections in the target host (Wakelin and Rose, 1990), and experimental studies have shown similar degrees of genetically determined variation in resistance and susceptibility. A particularly well-studied model involves infection with *E. vermiformis* in mice. This species is unusual in that there is clear immune control of the duration of primary infections. These are prolonged in susceptible strains of mice, the level and duration of oocyst output providing an easily measured parameter of altered responsiveness. MHC-linked genes play only a small part in determining response phenotype (Fig. 5.1), background (non-MHC) genes exerting the dominant effect.

Resistance is mediated primarily through the activity of CD4+ T cells belonging to the T-helper-1 subset, i.e. cells characterized by the release of the cytokines γ-interferon (γ-IFN) and interleukin-2 (IL-2) (Rose *et al.*, 1988). The first of these cytokines plays a pivotal role (Rose *et al.*, 1989). Comparative

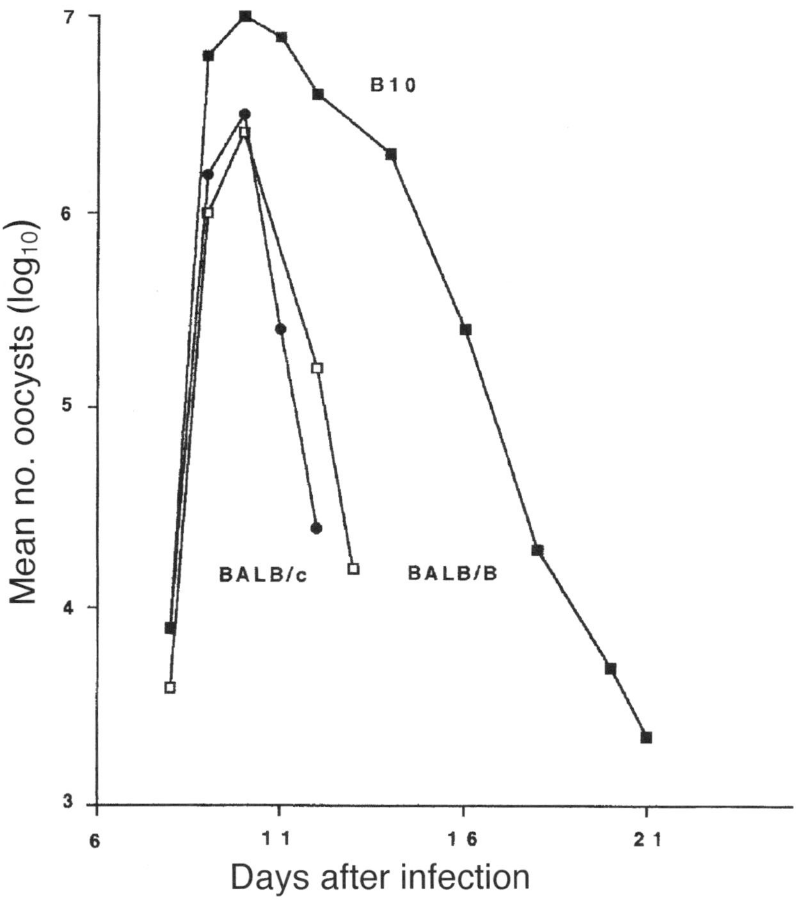

Fig. 5.1. Time course of a primary infection with the coccidian *Eimeria vermiformis* in inbred mice. B10 mice (C57BL/10 – H-2^b) are susceptible and have a high, prolonged oocyst output. Both BALB/c (H-2^d) and BALB/B mice (H-2^b) are resistant, showing that MHC (H-2)-linked genes have little influence on responsiveness. (Redrawn from Joysey *et al.*, 1988.)

studies using resistant and susceptible strains of mice showed that not only do the former show parasite antigen-specific T-cell activation earlier, their T cells also release γ-IFN earlier in infection than cells from susceptible mice (Wakelin *et al.*, 1993a). The role played by γ-IFN in resistance is not yet understood. It clearly influences the capacity of host cells to destroy invading parasites, but this seems not to involve the activity of reactive oxygen or nitrogen intermediates (Ovington *et al.*, 1995). T cells, other than α/β T-cell-receptor-bearing T cells, are involved in resistance to *E. vermiformis*; depletion of γ/δ T cells resulting in greater pathology (Roberts *et al.*, 1996), and thus genetic variation in γ/δ cell function may underlie some of the differences seen between strains in degree of intestinal damage suffered during infection.

One of the long-term aims of research into immunological aspects of coccidiosis has been the development of effective vaccines. Studies in mice have shown that vaccination with oocyst-derived antigens has contrasting effects in mice that are genetically resistant or susceptible to infection (Rose *et al.*, 1996, 1997). In resistant BALB/c mice, infection is enhanced by vaccination, whereas resistance is increased in the susceptible C57BL/6 strain. The mechanisms underlying this paradox are not known, but the phenomenon suggests that vaccination of birds and other domestic species with non-living antigenic preparations may run similar risks, if not in terms of enhancing levels of infections in hosts of certain genotypes, then in terms of exacerbating pathology. It is interesting that the only commercially available anti-*Eimeria* vaccines for use in chickens are based on the use of live attenuated parasites, which elicit the same degree of immunity as normal infections but, because of the truncated life cycle, cause less damage to the intestine (Shirley and Bedrnik, 1997).

Trypanosomes

Trypanosomiasis continues to be a severe problem for cattle production in Africa, despite programmes aimed at controlling the tsetse-fly vector. As discussed in Chapter 9, there is good evidence of marked breed variation in ability to resist trypanosome infections in cattle, some (those that are trypanotolerant) resisting the pathological effects of infection more successfully than others, thus raising the possibility of selective breeding to enhance resistance in herds. Murine models, using several species of trypanosome, have contributed greatly both to our present understanding of trypanotolerance and of genetic variation in responsiveness to this group of parasites.

Much of the basic understanding of genetic variation in resistance to trypanosome infection in mice was laid down 10–20 years ago. Genetic analysis of strain variation in resistance to *T. congolense* (affecting both parasitaemia and survival) led initially to the suggestion that two genes controlled these traits (Morrison and Murray, 1979) whereas later work suggested single-gene control (Blackwell, 1988; Pinder, 1984). Although there is variation in survival times between H-2- (MHC-) congenic strains, it does not appear that MHC genes exert a significant influence on resistance, either to *T. congolense*

(Morrison and Murray, 1979; Pinder *et al.*, 1985) or to *T. rhodesiense* (Levine and Mansfield, 1981). Strains of mice resistant to *T. congolense* (e.g. C57BL/6) produce variant-antigen-specific IgM antibodies earlier in infection than susceptible BALB/c, and this is true also of C57BL/6 nude mice, showing the T-independence of the response (Pinder *et al.*, 1985). In a similar study, Mitchell and Pearson (1985) showed that resistant C57BL/6 mice not only made high IgM responses but produced little specific IgG, whereas susceptible A/J mice showed a preferential switch to IgG isotypes.

Susceptibility to *T. rhodesiense* is a polygenic characteristic showing directional dominance. Susceptible mice, such as C3H, make little or no antibody to the infection, fail to control the initial parasitaemia arising from the infecting variant antigenic type (VAT) and have a short survival time; resistant mice (e.g. B10.BR) control both the infecting VAT and the relapse population, and survive longer; F_1 hybrids between the two strains show short survival times but do clear the initial parasitaemia and make anti-VAT antibody (Seed and Sechelski, 1989). Other strains, and certain of the F_1, F_2 and backcross mice from B10.BR × C3H crosses, have an intermediate status, showing delayed control of the initial parasitaemia but succumbing to the subsequent relapse (Mansfield, 1990). These data strongly suggest that increased survival time may be due to the ability of mice to resist the pathological consequences of infection. The causes of pathology and death in trypanosome infections are complex, but it is clear that immunosuppression induced by the parasites themselves (Sternberg and McGuigan, 1992) and responses to endotoxin-like molecules (Pentreath, 1994) are major components. The mechanisms that are involved in suppression and toxin responses are all likely to be mouse-strain variable and this will influence the overall response phenotype.

A major advance in the immunogenetics of murine trypanosomiasis has been the identification of chromosomal regions containing genes controlling resistance to *T. congolense* (Kemp *et al.*, 1997). Resistant C57BL/6 mice were crossed with susceptible A/J and BALB/c strains to produce F_2 populations. Mice were challenged to establish phenotypes, and those showing the shortest or longest survival times were genotyped for mapped microsatellite markers. The outcome showed that regions on chromosomes 1, 5 and 17 were important in resistance, accounting for most of the genetic variation seen in the F_2 mice, and three regions have been designated on these chromosomes – *Tir3*, *Tir2* and *Tir1*, respectively; *Tir1* was located only in the C57BL/6 × BALB/c cross. At present, no functions have been definitely associated with these regions.

Nematodes

The most important nematode parasites of farm animals are the gastrointestinal (GI) worms. These form a complex of species whose composition and relative importance vary with the host and country concerned. Few of these species can be used directly in laboratory rodents, although guinea-pigs are satisfactory hosts for *Trichostrongylus*. There are, however, several model systems that use related, or other intestinal, species that provide useful data on immunogenetic aspects of these infections. Studies on mouse models have been particularly useful and are facilitated by the detailed information now available on immune responses to GI infections in this host (Finkelman *et al.*, 1997). Three mouse models are discussed here – *Trichinella spiralis*, *Trichuris muris* and *Heligmosomoides polygyrus*. All share a number of features in common. Host-protective immunity results in the physical removal (expulsion) of worms from the intestine. Mouse strains show varying degrees of resistance or susceptibility to infection. This variation is determined genetically and resistance is inherited as a characteristic showing directional dominance. The host-protective immune responses that operate to eliminate worms from the intestine are dependent upon the activity of T-helper (Th) cells belonging to the Th2 subset. Th2 cells regulate immunity through the release of specific cytokines, of which IL-4, IL-9 and IL-13 are of particular importance. The mechanisms of immunity appear to show distinct differences in each system.

Trichinella spiralis

The veterinary importance of *T. spiralis*, although locally significant, is relatively small. However, the low host specificity and ease of maintenance of this species make it an excellent model for studying the responses involved in controlling infections with intestinal worms. Infections generate strong protective responses that have been analysed in some detail (Wakelin *et al.*, 1993b). The time course of infection and the degree of immunity elicited are variable between different strains of mice (Wassom *et al.*, 1983), as are almost all parameters of responsiveness (Table 5.4), and inheritance of resistance shows directional dominance. Resistant strains expel adult worms rapidly from the intestine and in consequence acquire low muscle-larval burdens (Fig. 5.2A). Susceptible strains maintain reproducing worm populations in the intestine for significantly longer periods and develop heavy muscle burdens. Genetic variation is also seen in the timing and expression of immunity to reinfection (Bell *et al.*, 1982).

The genes exerting the strongest influences on immunity lie outwith the MHC but, when H-2 congenic strains are studied, MHC-linked genes are seen

Table 5.4. Models involving mice and GI nematodes: genetic variation in parameters of response to infection.

Parasitological	Immunological
Site of infection in gut	Antibody response (level, isotype)
Size of adult worms	Lymphocyte response (proliferation, cytokine production)
Survival of adult worms	Inflammatory responses
Fecundity of female worms	Degree of immune depression
Response to challenge infection	Response to vaccination

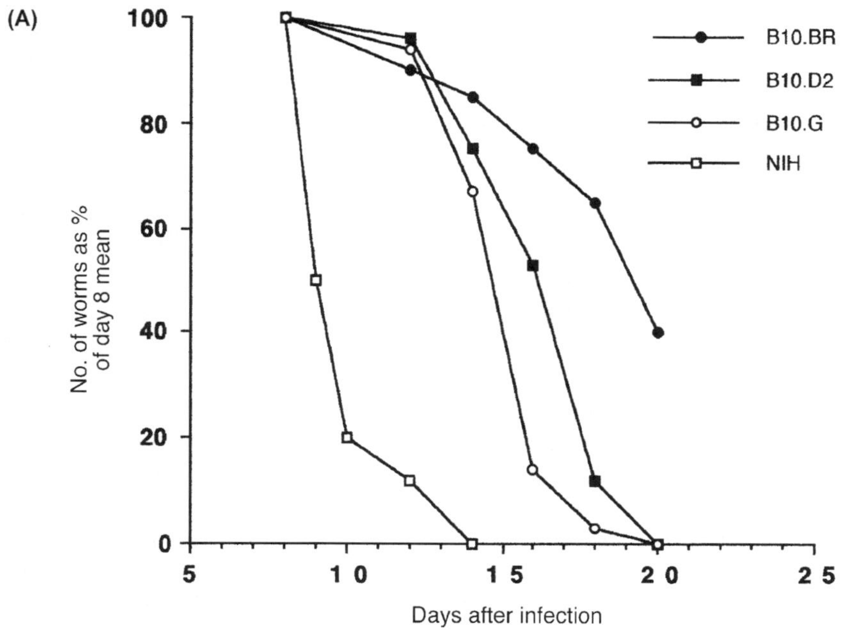

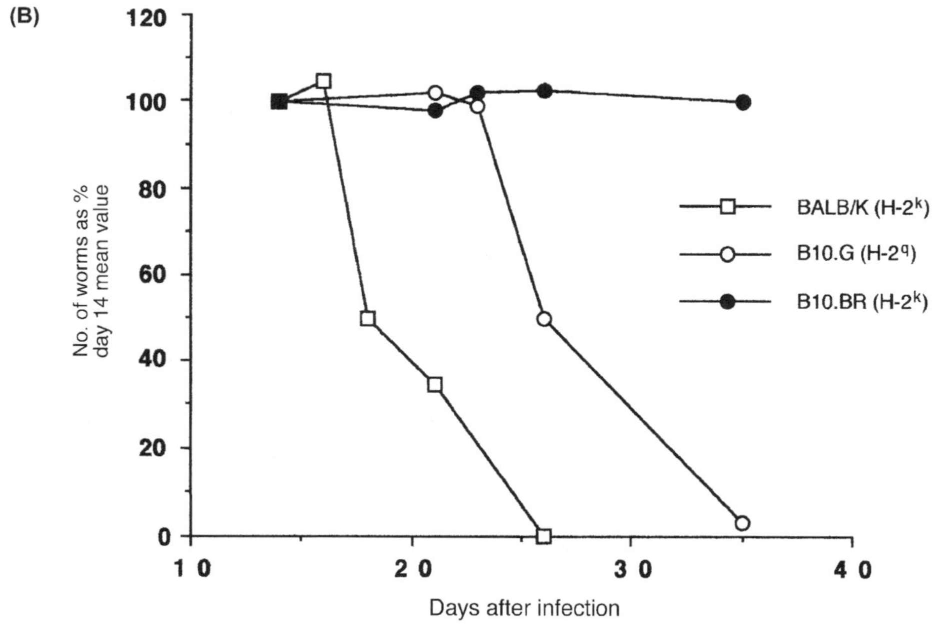

Fig. 5.2. (A) Time course of a primary infection with *Trichinella spiralis* in inbred mice. NIH mice (H-2^q) are very resistant and expel adult worms from the intestine quickly. The B10 congenic strains are more susceptible, but MHC (H-2)-linked genes influence the degree of susceptibility. B10G (H-2^q) are less susceptible than either B10.D2 (H-2^d) or B10.BR (H-2^k). (B) Time course of a primary infection with *Trichuris muris* in inbred mice. NIH (H-2^q) are very resistant and expel larval worms from the large intestine quickly. BALB/K (H-2^k) are more susceptible but still expel worms before they are mature, whereas the MHC compatible B10.BR (H-2^k) are completely susceptible and allow the worms to reach sexual maturity. (Data from Else and Wakelin, 1988.)

to exert a clear effect (Wakelin, 1980; Wassom *et al*., 1983). As with a majority of parasites, therefore, overall resistance appears to reflect the directional dominance associated with background genes, modulated within limits by MHC-linked genes. No genes involved in this control have been identified precisely, but three putative loci in the MHC have been associated with control of immunity to primary infections (Wassom and Kelly, 1990) and a non-MHC gene has been associated with the ability to express the rapid expulsion response to reinfection (Bell *et al*., 1984).

Infection with *T. spiralis* is characterized by marked inflammatory responses which include mucosal mastocytosis, peripheral eosinophilia and elevated IgE, as well as altered mucosal structure and function. There is a strong correlation between response phenotype and these inflammatory changes, which reflects both the involvement of Th2 cells in providing the necessary cytokines, as well as inherent differences in bone-marrow capacity to generate inflammatory cell populations (Wakelin and Grencis, 1992). Although there is controversy in the literature (see, for example, Grencis *et al*., 1991 and Pond *et al*., 1989), there is certainly no clear-cut correlation between response phenotype and Th subset involvement as there is in *Trichuris* infections (see below). However, resistant and susceptible strains do differ in the speed of the T-cell response to infection (Crook and Wakelin, 1994), so that animals with the capacity to express inflammatory changes do so at an earlier stage in infection. Recent work has confirmed this general view by showing that resistant strains also develop more powerful, immune-mediated expulsive peristaltic forces in the infected intestine, and that these forces contribute to worm removal (Vallance and Collins, 1998).

The demonstration that non-immune components can be rate-limiting in the expulsion of *T. spiralis* implies that vaccination is likely to be relatively ineffective in boosting resistance to this infection in animals that have a genetically determined low responsiveness and that this might also apply in

Fig. 5.3. (*opposite*) (A) Resistance to *Trichinella spiralis* in resistant (NIH) and susceptible (C57BL/10–B10) mice vaccinated subcutaneously with muscle larval homogenate antigen presented in Freund's complete adjuvant (FCA) or in an immunostimulatory complex (ISCOMS). The numbers of worms established from an infection of 300 larvae were counted in treated and control groups on day 6 or 7, and the numbers remaining on day 8 (NIH) or 12 (B10) are shown as percentages of the establishment value. NIH mice responded well to either immunization, whereas there was no enhanced resistance in vaccinated B10. ■, Infection-only control; □, antigen + FCA; ▨, antigen + ISCOMS; *, mean significantly lower than control. (Data from Robinson *et al*., 1994.) (B) Resistance to *Trichuris muris* in resistant (BALB/c), susceptible (C57BL/10–B10) and non-responder (B10.BR) mice vaccinated with worm homogenate antigen given subcutaneously in Freund's complete adjuvant (FCA) or orally with cholera toxin (CT). BALB/c and B10 mice were killed 15 days after a challenge infection with 400 eggs and B10.BR mice were killed 21 days after infection. The numbers of worms recovered at those times are expressed as a percentage of the number present in infected-only controls. BALB/c and B10 mice responded well to antigen in FCA, B10.BR responded less well, although the worm burden was reduced significantly. Oral vaccination protected only BALB/c and, to a lesser degree, B10 mice, no protection was seen in B10.BR mice. ■, Antigen + FCA; □, antigen + CT; *, mean significantly lower than control. (Data from Robinson *et al*., 1995c.)

the cases of other worms that are similarly controlled by inflammation-dependent processes. This has been confirmed in a series of comparative studies using resistant and susceptible strains of mice (Wakelin *et al.*, 1986; Robinson *et al.*, 1995a, b). Levels of immunity are increased in vaccinated resistant strains but, although there is increased antibody responsiveness in susceptible mice, there is no enhancement of worm expulsion (Fig. 5.3A). A similar failure of vaccination has been described in guinea-pigs infected with the ovine parasite *Trichostrongylus colubriformis* (Rothwell *et al.*, 1994). Extrapolation of these and similar data to ovine hosts may provide an explanation of the relative unresponsiveness of individual sheep to

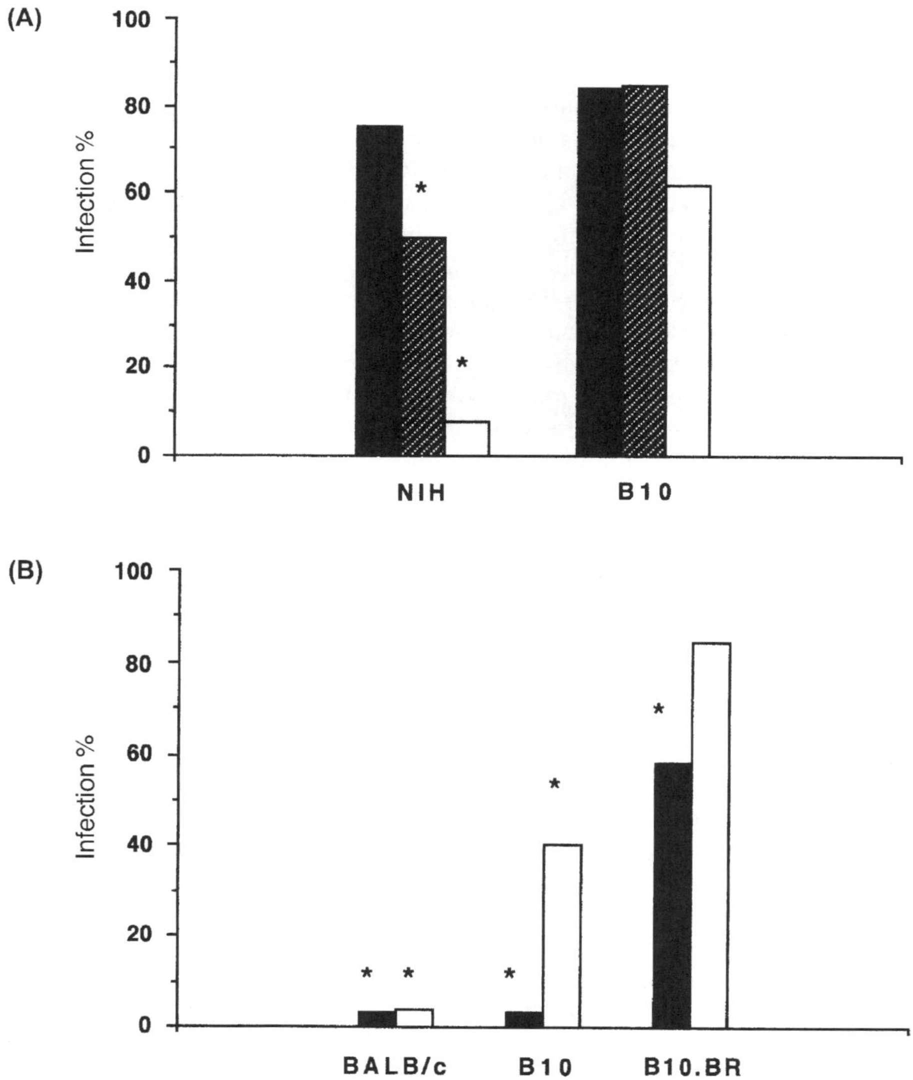

vaccination against species such as *T. colubriformis*, first reported by Dineen and Windon (1980). Although the mechanisms of immunity against these and other trichostrongyle species are not fully understood, there is evidence that implicates the operation of inflammatory mechanisms (Miller, 1996) and, in the guinea-pig model of *T. colubriformis* infection, genetically determined resistance is likewise correlated with greater inflammatory responsiveness (Bendixsen *et al.*, 1991). One conclusion that may be drawn from these studies is that the parameter of inflammatory responsiveness may be a better predictor of antiworm response potential than other genetic and immunological markers (Beh and Maddox, 1996; Douch *et al.*, 1996).

Trichuris muris

Species of *Trichuris* occur in many domestic animals, but are of little clinical importance other than, on occasion, in pigs. Like *T. spiralis*, however, the murine species of this genus – *T. muris* – has been a very valuable model in uncovering aspects of host immunity and in identifying genetic influences. The parasite occupies an epithelial location in the large intestine from which, in resistant mice, it is removed by an immunologically mediated response. Mouse strains differ markedly in the capacity to express this immunity (Else and Wakelin, 1988): some eliminate parasites very quickly, others fail to do so, and sustain long-term sexually mature infections (Fig. 5.2B). This variation is determined by both MHC-linked and non-MHC genes, the latter again playing the most important role (Else *et al.*, 1990). As in *T. spiralis*, inheritance of resistance shows directional dominance, and this has allowed selective breeding from an initially genetically heterogeneous random-bred population to produce resistant and susceptible lines (Wakelin, 1975).

A major advance in understanding this model came from the discovery that the response phenotype was linked to the Th subset response induced by infection and the associated capacity to express type 1 (Th1) or type 2 (Th2) cytokines (Else *et al.*, 1992). A series of elegant experiments (reviewed by Grencis, 1996) has shown clearly that resistance requires expression of Th2 responses, but that, in susceptible strains, the parasite, perhaps through the release of immunomodulatory factors, induces a switch to a non-protective Th1 response. Response phenotype can be manipulated by injection or deletion of the appropriate cytokines (Else *et al.*, 1994).

The precise manner in which host genotype and infection interact to produce one or other Th cell response is still unclear, although there is evidence that parasite load can be a factor (Bancroft *et al.*, 1994). The contribution that the parasite itself makes to determining the host response has been underlined by a comparative study of infections with three isolates of *T. muris* in various strains of mice (Bellaby *et al.*, 1996). Certain strains expressed the resistant phenotype and made Th2-type responses to all three isolates, although the infection patterns differed in terms of the length of survival of the worms. Other strains expressed a resistant phenotype and a Th2 response with one isolate, but a fully susceptible phenotype and a Th1 response with another. Such differences must reflect molecular differences within each isolate. Similar phenomena are likely to occur in any parasite species that is distributed over a wide geographical range, where isolating mechanisms may allow the accumulation of genetic differences to occur. Indeed similar, although less extreme, findings have been described with experimental infections with isolates of *T. spiralis* (Wakelin and Goyal, 1997).

As with *T. spiralis*, vaccination against *T. muris* can induce high levels of protection in mice, but the efficacy of vaccination is determined by the response phenotype of the mouse strain concerned (Robinson *et al.*, 1995c). Resistant BALB/c mice develop almost total (99.2%) resistance to challenge after both parenteral (subcutaneous + Freund's complete adjuvant) and oral (+ cholera toxin) vaccination with a crude whole-worm antigen preparation. In contrast, parenteral vaccination gives only marginal protection (41.4%) in susceptible B10.BR mice and oral vaccination is without effect (Fig. 5.3B).

Heligmosomoides polygyrus

In contrast to some of the model infections studied in rodent hosts, where powerful immune and inflammatory responses eliminate the worms within a short time of infection, most GI nematodes in domestic hosts give rise to long-lasting, persistent infections. Of the reasons put forward to explain this, one of the most convincing is that, as with *T. muris* in susceptible mice, parasites associated with chronic infections have the ability to interfere with the development of protective immunity in hosts of particular genotypes, and that the intensity and frequency of the infection influence the degree to which parasite-induced immune suppression is effective. The trichostrongyle *H. polygyrus*, a natural parasite of wild mice which readily infects laboratory mice, provides a good model for studying this aspect of infection. It produces chronic infections in most mouse strains, and exerts a powerfully suppressive effect on both homologous and heterologous immune responses in the host (Behnke, 1987). Immunity is a Th2-dependent process, and the cytokine IL-4 plays a key role (Finkelman *et al.*, 1997). Immunity is induced by the initial larval stages of infection, but is suppressed by the adult stages, as shown elegantly by Behnke *et al.* (1983). Suppression is associated with the release of low molecular weight immunomodulatory molecules, which have been partially characterized (Monroy *et al.*, 1989; Pritchard *et al.*, 1994). Strains of mice clearly differ in their abilities to express immunity to *H. polygyrus* (or resist immunomodulation) as there is a wide spectrum of responsiveness, some eliminating primary infections after a matter of a few weeks, others retaining worms for many months (Wahid *et al.*, 1989). Both MHC-linked and background genes influence the level of resistance expressed (Behnke and Wahid, 1991; Su and Dobson, 1997).

It has been established that Th2 cell activity and production of the cytokine IL-4 are required for expulsion of adult worms from the intestine

(Finkelman *et al.*, 1997), and it has been proposed that the adult worms bring about a selective modulation of Th2 cytokine production, thereby prolonging their own survival (Behnke *et al.*, 1993). Strains of mice that express the susceptible phenotype to this worm presumably, therefore, have little or no capacity to prevent this immunomodulatory activity.

Heligmosomoides polygyrus in the mouse has proved another valuable model for following the effects of strong selection for resistance to infection. Dobson and colleagues have carried out several series of selective breeding experiments designed to analyse the inheritance of both innate and acquired resistance (see, for example, Brindley and Dobson, 1981; Sitepu and Dobson, 1982; Su and Dobson, 1996, 1997). They have not only established parameters associated with enhanced resistance to the parasite itself, they have also examined concurrent selection of resistance to other organisms, for changes in host characteristics (relevant to the question of 'productivity' versus 'resistance') and changes in parasite biology. All of these aspects are important considerations to be borne in mind when selective breeding for parasite resistance is undertaken in domestic animals.

Arthropoda

Breed and individual variation in resistance of bovines to tick infestation is a well-established fact (see Chapter 7). European breeds (*Bos taurus*) are, in general, more susceptible than *Bos indicus* breeds, and this influences the impact of both tick disease *per se* and the diseases transmitted by these vectors. It is important to understand the genetic and immunological variation that underlies tick resistance, as this information can be used to improve stock performance in areas where ticks are a serious problem. A number of tick species will feed successfully on laboratory rodents, including mice, providing useful models. Considerable mouse strain variation in resistance has been reported by a number of workers (Wakelin, 1988).

One important observation has been that resistance, which affects the ability of the ticks to feed, is associated with IgE antibodies and the development of inflammatory responses at the feeding site (Matsuda *et al.*, 1990). Mast cells play a key role in these responses. Mast-cell-deficient mice (W/W^v strains) develop no resistance to infestation with *Haemaphysalis longicornus*, yet do so if reconstituted with normal bone marrow (Matsuda *et al.*, 1985). The local relevance of functional mast cells in resistance has been demonstrated using elegant grafting experiments with W/W^v mice given skin from compatible mast-cell-sufficient donors (Matsuda *et al.*, 1987). Mast cells also appear to be necessary for the early eosinophil response to infestation in this system (Ushio *et al.*, 1995). Given the evidence from many mouse studies that the ability to raise mast cell and eosinophil responses is genetically determined, differences in activity of these cells between breeds of cattle may be a significant factor in the ability to resist infestation. By the same analogy, the evidence that immune responses to ticks in mice, like those to nematode infections, are Th2 biased (Ganapamo *et al.*, 1995, 1996) also points to the possibility of breed-dependent differences in ability to mount these protective responses effectively.

Conclusions

The value of model systems can be judged in several ways, by their closeness to the situation modelled, by their predictive value and by their capacity to provide fundamental insights. Models for studying the genetic basis of disease resistance fulfil some or all of these criteria to differing degrees. In parasitological terms, few are really close to the host–parasite relationships that exist between farm animals and their parasites. Nevertheless, many model these relationships in their essential details and thus provide data of direct relevance. The predictive value of models has been borne out in several ways. For example, the models show that clear-cut differences in resistance and susceptibility exist almost universally and that these differences are heritable. Most importantly, particularly when comparing inbred lines, they show that inheritance of resistance almost always shows directional dominance. These observations are critical for strategies designed to improve the disease resistance of stock by selective breeding. Identification of genetically controlled differences in inflammatory responses to nematodes in strains of mice, and their correlation with response phenotype, have facilitated the possible use of similar characteristics in predicting resistance to GI nematodes in cattle and sheep. Despite assumptions to the contrary, model systems have shown that relatively few of the complex antigens presented by parasites are significant in terms of functional resistance to infection, and that immunological recognition of such antigens can be tightly regulated. Work during the past decade has also shown quite clearly how important particular Th-cell responses are in determining resistance or susceptibility, and that such responses can be determined genetically. These observations have been critical for the rational design and use of defined vaccines. An excellent example of extrapolation from a mouse model system to a practical vaccine has been the development of a recombinant vaccine against larval tapeworm infections in sheep (Rickard *et al*., 1995), some of the parameters of which were set several years ago by experimental studies on *Taenia taeniaeformis* infections in mice (Mitchell *et al*., 1980). The recent interest in the interrelationship between host and parasite genotype in determining response phenotype will also have significant implications both for production of resistant stock and for the practical use of vaccines as they become available.

The goal of work in the field of host–parasite immunogenetics must be to identify in molecular terms both the genes regulating resistance and the mechanisms through which resistance is expressed. This has yet to be achieved in the majority of experimental systems involving mice; the identification of the *Nramp* gene which confers resistance to *Leishmania donovani* in mice, and understanding of its functions, being one of the most notable successes (Blackwell, 1996). The search for murine genes not only helps to explain host–parasite relationships in that species, but through comparative gene mapping (see Chapter 1), will allow similar genes to be identified more readily in target species.

References

Abel, L. and Dessein, A.J. (1997) The impact of host genetics on susceptibility to human infectious diseases. *Current Opinion in Immunology* 9, 509–516.

Bancroft, A.J., Else, K.J. and Grencis, R.K. (1994) Low-level infection with *Trichuris muris* significantly affects the polarization of the CD4 response. *European Journal of Immunology* 24, 3113–3118.

Beh, K.J. and Maddox, J.F. (1996) Prospects for development of genetic markers for resistance to gastrointestinal parasite infection in sheep. *International Journal for Parasitology* 26, 879–897.

Behnke, J.M. (1987) Evasion of immunity by nematode parasites. *Advances in Parasitology* 26, 1–71.

Behnke, J.M. and Wahid, F.N. (1991) Immunological relationships during primary infection with *Heligmosomoides polygyrus* (*Nematospiroides dubius*): H-2 linked genes determine worm survival. *Parasitology* 103, 157–164.

Behnke, J.M., Hannah, J. and Pritchard, D.I. (1983) *Nematospiroides dubius* in the mouse: evidence that adult worms depress the expression of homologous immunity. *Parasite Immunology* 5, 397–408.

Behnke, J.M., Wahid, F.N., Grencis, R.K., Else, K.J., Ben-Smith, A.E. and Goyal, P.K. (1993) Immunological relationships during primary infection with *Heligmosomoides polygyrus* (*Nematospiroides dubius*): downregulation of specific cytokine secretion (IL-9 and IL-10) correlates with poor mastocytosis and chronic survival of adult worms. *Parasite Immunology* 15, 415–421.

Bell, R.G., McGregor, D.D. and Adams, L.S. (1982) *Trichinella spiralis*: characterization and strain distribution of rapid expulsion in inbred mice. *Experimental Parasitology* 53, 301–314.

Bell, R.G., Adams, L.S. and Ogden, R.W. (1984) A single gene determines rapid expulsion of *Trichinella spiralis* in mice. *Infection and Immunity* 45, 273–275.

Bellaby, T., Robinson, K. and Wakelin, D. (1996) Induction of differential T helper subset responses in mice infected with variants of the parasitic nematode *Trichuris muris*. *Infection and Immunity* 66, 791–795.

Bendixsen, T., Emery, D.L. and Rothwell, T.L.W. (1991) The effect of specific immunization with *Trichostrongylus colubriformis* on production of eosinophil differentiating factor in Guinea-pigs. *International Journal for Parasitology* 21, 883–889.

Blackwell, J.M. (1988) Protozoan infections. In: Wakelin, D. and Blackwell, J.M. (eds) *Genetics of Resistance to Bacterial and Parasitic Infections*. Taylor and Francis, London, pp. 103–151.

Blackwell, J.M. (1996) Genetic susceptibility to leishmanial infections: studies in mice and man. *Parasitology* 112, S67–S74.

Brindley, P.J. and Dobson, C. (1981) Genetic control of liability to infection with *Nematospiroides dubius* in mice: direct and correlated responses to selection of mice for faecal parasite egg counts. *Parasitology* 87, 1113–127.

Champion, L.R. (1954) The inheritance of resistance to cecal coccidiosis in the domestic fowl. *Poultry Science* 33, 670–681.

Crook, K. and Wakelin, D. (1994) Induction of T lymphocyte subsets and levels of interleukin-2 and interleukin-3 after infection with *Trichinella spiralis* are similar in mice of high- and low-responder phenotypes. *International Journal for Parasitology*

24, 119–126.

Dineen, J.K. and Windon, R.G. (1980) The effect of sire selection on the response of lambs to vaccination with irradiated *Trichostrongylus colubriformis* larvae. *International Journal for Parasitology* 10, 189–196.

Douch, P.P.G.C., Green, R.S., Morris C.A., McEwan, J.C. and Windon, R.G. (1996) Phenotypic markers for selection of nematode-resistant sheep. *International Journal for Parasitology* 26, 899–911.

Else, K.J. and Wakelin, D. (1988) The effect of H-2 and non-H-2 genes on the expulsion of the nematode *Trichuris muris* from inbred and congenic mice. *Parasitology* 96, 543–550.

Else, K.J., Wakelin, D., Wassom, D.L. and Hauda, K.M. (1990) The influence of genes mapping within the major histocompatibility complex on resistance to *Trichuris muris* infections in mice. *Parasitology* 100, 479–489.

Else, K.J., Hultner, L. and Grencis, R.K. (1992) Cellular immune responses to the murine nematode parasite *Trichuris muris*. II. Differential induction of Th cell subsets in resistant versus susceptible mice. *Immunology* 75, 232–237.

Else, K.J., Finkelman, F.D., Maliszewski, C.R. and Grencis, R.K. (1994) Cytokine mediated regulation of chronic intestinal helminth infection. *Journal of Experimental Medicine* 179, 347–351.

Finkelman, F.D., Shea-Donohue, T., Goldhill, J., Sullivan, C.A., Morris, S.C., Madden, K.B., Gause, W.C. and Urban J.F. (1997) Cytokine regulation of host defense against parasitic gastrointestinal nematodes: lesson from rodent models. *Annual Reviews in Immunology* 15, 505–533.

Ganapamo, F., Rutti, B. and Brossard, M. (1995) *In vitro* production of IL-4 and IFN-γ by lymph node cells from BALB/c mice infected with nymphs of *Ixodes ricinus* ticks. *Immunology* 85, 120–124.

Ganapamo, F., Rutti, B. and Brossard, M. (1996) Immunosuppression and cytokine production in mice infected with *Ixodes ricinus* ticks – a possible role of laminin and interleukin-10 on the *in vitro* responsiveness of lymphocytes to mitogens. *Immunology* 87, 259–263.

Gowen, J.W. (1951) Genetics and disease resistance. In: Dunn, L.C. (ed.) *Genetics in the 20th Century*. Macmillan, New York, pp. 401–429.

Grencis, R.K. (1996) T cell and cytokine basis of host variability in response to intestinal nematode infections. *Parasitology* 112, S31–S37.

Grencis, R.K., Hultner, L. and Else, K.J. (1991) Host-protective immunity to *Trichinella spiralis* in mice: activation of Th subsets and lymphokine secretion in mice expressing different response phenotypes. *Immunology* 74, 329–332.

Hill, A.V.S. (1996) Genetics of infectious disease resistance. *Current Opinion in Genetics and Development* 6, 348–353.

Joysey, H.S., Wakelin, D. and Rose, M.E. (1988) Coccidiosis: resistance to infection with *Eimeria vermiformis* in mouse radiation chimaeras is determined by donor bone-marrow cells. *Infection and Immunity* 56, 1399–1401.

Kemp, S.J., Iraqi, F., Darvasi, A., Soller, M. and Teale, A.J. (1997) Localization of genes controlling resistance to trypanosomiasis in mice. *Nature Genetics* 16, 194–196.

Levine, R.F. and Mansfield, J.M. (1981) Genetics of resistance to the African trypanosomes. I. Role of the H-2 locus in determining resistance to infection with *Trypanosoma rhodesiense*. *Infection and Immunity* 34, 513–518.

Mansfield, J.M. (1990) Immunology of African trypanosomiasis. In: Wyler, D.J. (ed.) *Modern Parasite Biology. Cellular, Immunological and Molecular Aspects.* W.H. Freeman, New York, pp. 222–246.

Matsuda, H., Fukui, K., Kiso, Y. and Kitamura, Y. (1985) Inability of genetically mast cell deficient W/W^v mice to acquire resistance against larval *Haemaphysalis longicornis* ticks. *Journal of Parasitology* 71, 443–448.

Matsuda, H., Nakano, T., Kiso, Y. and Kitamura, Y. (1987) Normalization of anti-tick response to mast cell deficient W/W^v mice by intracutaneous injection of cultured mast cells. *Journal of Parasitology* 73, 155–158.

Matsuda, H., Watanabe, N., Kiso, Y., Hirota, S., Kanna, Y., Azuma, M., Koyama, H. and Kitamura, Y. (1990) Necessity of IgE antibodies and mast cells for manifestation of resistance against larval *Haemaphysalis longicornis* ticks in mice. *Journal of Immunology* 144, 259–262.

Miller, H.R.P. (1996) Prospects for the immunological control of ruminant gastrointestinal nematodes: natural immunity, can it be harnessed? *International Journal for Parasitology* 26, 801–811.

Mitchell, G.F., Rajasekariah, G.R. and Rickard, M.D. (1980) A mechanism to account for mouse strain variation in resistance to the larval cestode *Taenia taeniaeformis*. *Immunology* 39, 481–489.

Mitchell, L.A. and Pearson, T.W. (1985) Antibody responses during *Trypanosoma congolense* infection in resistance and susceptible mice. *Progress in Leukocyte Biology* 3, 501–515.

Monroy, F.G., Dobson, C. and Adams, J.H. (1989) Low molecular weight immunosuppressors secreted by adult *Nematospiroides dubius*. *International Journal for Parasitology* 19, 125–127.

Morrison, W.I. and Murray, M. (1979) *Trypanosoma congolense*: inheritance of susceptibility to infection in inbred strains of mice. *Experimental Parasitology* 48, 364–374.

Ovington, K.S., Alleva, L.M. and Kerr, E.A. (1995) Cytokines and immunological control of *Eimeria* spp. *International Journal for Parasitology* 25, 1331–1351.

Pentreath, V.W. (1994) Endotoxins and their significance in murine trypanosomiasis. *Parasitology Today* 10, 226–228.

Pinder, M. (1984) *Trypanosoma congolense*: genetic control of resistance to infection in mice. *Infection and Immunity* 578, 185–194.

Pinder, M., Fumous, F. and Roelants, G.E. (1985) Immune mechanisms and genetic control of natural resistance *Trypanosoma congolense*. *Progress in Leukocyte Biology* 3, 495–500.

Pond, L., Wassom, D.L. and Hayes, C.E. (1989) Evidence for differential induction of helper T cell subsets during *Trichinella spiralis* infection. *Journal of Immunology* 143, 4232–4237.

Pritchard, D.I., Lawrence, C.E., Appleby, P., Gibb, L.A. and Glover, K. (1994) Immunosuppressive proteins secreted by the gastrointestinal nematode parasite *Heligmosomoides polygyrus*. *International Journal for Parasitology* 24, 495–500.

Rickard, M.D., Harrison, G.B.L., Heath, D.D. and Lightowlers M.W. (1995) *Taenia ovis* recombinant vaccine – 'quo vadit'. *Parasitology* 110, S5–S9.

Roberts, S.J., Smith, A.L., West, A.B., Wen, L., Findly, R.C., Owen, M.J.and Hayday, A.C. (1996) T-cell $\alpha\beta^+$ and $\gamma\delta^+$ deficient mice display abnormal but distinct phenotypes toward a natural, widespread infection of the intestinal epithelium. *Proceedings of the National Academy of Sciences USA* 93, 11774–11779.

Robinson, K., Bellaby, T. and Wakelin, D. (1994) Vaccination against the nematode *Trichinella spiralis* in high- and low-responder mice. Effects of different adjuvants upon protective immunity and immune responsiveness. *Immunology* 82, 2261–2267.

Robinson, K., Bellaby, T. and Wakelin, D. (1995a) Immune response profiles in vaccinated and non-vaccinated high- and low-responder mice during infection with the intestinal nematode *Trichinella spiralis*. *Parasitology* 110, 71–78.

Robinson, K., Bellaby, T. and Wakelin, D. (1995b) Oral and parenteral immunization against *Trichinella spiralis* infections in high- and low-responder mice. *International Journal for Parasitology* 25, 989–992.

Robinson, K., Bellaby, T. and Wakelin, D. (1995c) Efficacy of oral vaccination against the murine intestinal parasite *Trichuris muris* is dependent upon host genetics. *Infection and Immunity* 63, 1762–1766.

Rose, M.E., Joysey, H.S., Hesketh, P., Grencis, R.K. and Wakelin, D. (1988) Mediation of immunity to *Eimeria vermiformis* in mice by L3T4+ve T cells. *Infection and Immunity* 56, 1760–1765.

Rose, M.E., Wakelin, D. and Hesketh, P. (1989) Gamma interferon controls *Eimeria vermiformis* primary infection in BALB/c mice. *Infection and Immunity* 57, 1599–1603.

Rose, M.E., Wakelin, D. and Hesketh, P. (1996) Immunity to coccidiosis: 1. Responses to vaccination in strains of mice differing in susceptibility to infection with *Eimeria vermiformis*. *Infection and Immunity* 64, 246–252.

Rose, M.E., Wakelin, D. and Hesketh, P. (1997) Oral vaccination against coccidiosis: responses in strains of mice that differ in susceptibility to infection with *Eimeria vermiformis*. *Infection and Immunity* 65, 1808–1813.

Rothwell, T.L.W., Wagland, B.M. and Sangster, N.C. (1994) Expulsion of *Trichostrongylus colubriformis* in high and low responder Guinea-pigs. *International Journal for Parasitology* 24, 527–531.

Seed, J.R. and Sechelski, J.B. (1989) African trypanosomes: inheritance of factors involved in resistance. *Experimental Parasitology* 69, 1–8.

Shirley, M.W. and Bedrnik, P. (1997) Live attenuated vaccines against avian coccidiosis. Success with precocious and egg-adapted lines of *Eimeria*. *Parasitology Today* 13, 481–484.

Sitepu, P. and Dobson, C. (1982) Genetic control of resistance to infection with *Nematospiroides dubius* in mice: selection of high and low immune responder populations of mice. *Parasitology* 85, 73–84.

Stear, M.J. and Wakelin, D. (1998) Parasites. In: *Genetic Resistance to Animal Diseases*. Office International des Epizooties. Revue Scientifique et Technique. 17, 143–153.

Sternberg, J. and McGuigan, K. (1992) Nitric oxide mediates suppression of T cell responses in murine *Trypanosoma brucei* infection. *European Journal of Immunology* 22, 2741–2744.

Su, Z. and Dobson, C. (1996) *Heligmosomoides polygyrus*: resistance in inbred, outbred and selected mice. *Experimental Parasitology* 82, 122–131.

Su, Z. and Dobson, C. (1997) H-2 genes and resistance to infections with *Heligmosomoides polygyrus* in selectively bred mice. *International Journal for Parasitology* 27, 595–600.

Ushio, H., Hirota, S., Jippo, T., Higuchi, S., Kawamoto, K., Kitamura, Y. and Matsuda, H. (1995) Mechanisms of eosinophilia in mice infested with larval *Haemaphysalis longicornis* ticks. *Immunology* 84, 469–475.

Vallance, B.A. and Collins, S.M. (1998) The effect of nematode infection upon intestinal smooth muscle function. *Parasite Immunology* 20, 249–253.

Wahid, F.N., Robinson, M. and Behnke, J.M. (1989) Immunological relationships during primary infection with *Heligmosomoides polygyrus* (*Nematospiroides dubius*): expulsion of adult worms from fast-responder syngeneic and hybrid strains of mice. *Parasitology* 98, 459–469.

Wakelin, D. (1975) Genetic control of immune responses to parasites: selection for responsiveness and non-responsiveness to *Trichuris muris* in random-bred mice. *Parasitology* 71, 377–384.

Wakelin, D. (1980) Genetic control of immunity to parasites. Infection with *Trichinella spiralis* in inbred and congenic mice showing rapid and slow responses to infection. *Parasite Immunology* 2, 85–98.

Wakelin, D. (1988) Arthropod parasites. In: Wakelin, D. and Blackwell, J.M. (eds) *Genetics of Resistance to Bacterial and Parasitic Infections.* Taylor and Francis, London, pp. 225–232.

Wakelin, D. and Blackwell, J.M. (eds) (1988) *Genetics of Resistance to Bacterial and Parasitic Infections.* Taylor and Francis, London.

Wakelin, D. and Blackwell, J.M. (1993) Genetic variation in immunity to parasite infection. In: Warren, K.S. (ed.) *Immunology and Molecular Biology of Parasitic Infections.* Blackwell Scientific Publications, Oxford, pp. 3–22.

Wakelin, D. and Goyal, P.K. (1997) *Trichinella* isolates: parasite variability and host responses. *International Journal for Parasitology* 26, 471–481.

Wakelin, D. and Grencis, R.K. (1992) T cell and genetic control of inflammatory cells. In: Moqbel, R. (ed.) *Allergy and Immunity to Helminths.* Taylor and Francis, London, pp. 108–136.

Wakelin, D. and Rose, M.E. (1990) Immunity to coccidiosis. In: Long, P.L. (ed.) *Coccidiosis of Man and Domestic Animals.* CRC Press, Boca Raton, pp. 281–306.

Wakelin, D. and Walliker, D. (eds) (1996) Genetics of host and parasite: implications for immunity, epidemiology and evolution. *Parasitology*, 112 (supplement).

Wakelin, D., Mitchell, L.A., Donachie, A.M. and Grencis, R.K. (1986) Genetic control of immunity to *Trichinella spiralis* in mice. Response of rapid- and slow-responder strains to immunization with parasite antigens. *Parasite Immunology* 8, 159–170.

Wakelin, D., Rose, M.E., Hesketh, P., Else, K.J. and Grencis, R.K. (1993a) Immunity to coccidosis: genetic influences on lymphocyte and cytokine responses to infection with *Eimeria vermiformis* in inbred mice. *Parasite Immunology* 15, 11–19.

Wakelin, D., Harnett, W. and Parkhouse, R.M.E. (1993b) Nematodes. In: Warren, K.S. (ed.) *Immunology and Molecular Biology of Parasitic Infections.* Blackwell Scientific Publications, Oxford, pp. 496–526.

Wassom, D.L. and Kelly, E.A.B. (1990) The role of the major histocompatibility complex in resistance to parasitic infections. *Critical Reviews in Immunology* 10, 31–52.

Wassom, D.L., Brooks, B.O., Cypess, R.H. and David, C.S. (1983) A survey of suceptibility to infection with *Trichinella spiralis* of inbred mouse strains sharing common H-2 alleles but different genetic backgrounds. *Journal of Parasitology* 69, 1033–1037.

Webster, L.T. (1937) Inheritance of resistance of mice to enteric bacterial and neurotropic virus infections. *Journal of Experimental Medicine* 65, 261–280.

Whitlock, J.H. (1955) A study of inheritance of resistance to trichostrongyliidosis in sheep. *Cornell Veterinarian* 48, 127–133.

Parasites and Vectors

II

6

Genetics of Helminth Resistance

L.C. Gasbarre[1] and J.E. Miller[2]
[1]USDA-ARS, LPSI, Immunology and Disease Resistance Laboratory, Beltsville Agricultural Research Center, Beltsville, USA; [2]Department of Epidemiology and Community Health, School of Veterinary Medicine and Department of Animal Science, Louisiana State University, Baton Rouge, USA

Summary

Gastrointestinal (GI) nematodes disrupt nutrient utilization of grazing ruminants, resulting in reduced growth and productivity by the infected animals. Anthelminthic drug treatment presents a number of problems, including that of drug-resistant parasites. An alternative to complete reliance on drugs is the use of naturally occurring resistance to the parasites. Studies of both small ruminants and cattle indicate that a substantial component of the variance in the level of infection to these parasites is due to host genetic components. The effect of host genetics on resistance to GI nematodes appears to be of the same magnitude across a number of mammalian species, with an estimated heritability of approximately 0.3. However, much variation exists. For example, determinations of the number of parasite eggs passed in the faeces is a good indicator of parasite burdens in sheep in areas where *Haemonchus contortus* is the major problem, but a very poor indicator in cattle in areas where *Ostertagia ostertagi* is the dominant pathogen. Similarly, immune effector mechanisms that affect intestinal-dwelling worms may be less effective on abomasal-dwelling parasites. Economically feasible tools are needed to identify accurately animals with enhanced or diminished resistance to the parasites. Fortunately, the distribution of the parasites is such that control could be accomplished by targeting a small percentage of susceptible animals for manipulation, treatment or removal. Marker-assisted selection could be used to develop genetic markers for susceptibility or resistance and to use these to make breeding and treatment decisions, including introgression of desirable alleles into susceptible populations. A more long-term approach is to identify and characterize the loci associated with immunity through the use of candidate genes and linkage analyses to produce animals of the desired phenotype.

© CAB *International* 2000. *Breeding for Disease Resistance in Farm Animals* (eds R.F.E. Axford, S.C. Bishop, F.W. Nicholas and J.B. Owen)

Introduction

By far the most important helminth infections of livestock are infections of grazing ruminants by nematodes residing within the gastrointestinal (GI) tract of the vertebrate host. Because these are the most economically important helminth parasitoses, they have received the bulk of attention both in terms of research efforts, and in terms of commercial interest in products to control infection. This chapter will focus on the genetics of resistance to these important pathogens, and the potential means by which these infections can be controlled through manipulation of the ruminant genome. Because the biology of the parasites differs substantially between GI nematodes of cattle and those of small ruminants (i.e. sheep and goats), there are significant differences in the way that host genetics could be used to reduce the economic impact of the parasites. For this reason, we will address the genetics of resistance to GI nematodes of small ruminants and of cattle as separate topics.

Biology of GI Nematode Infections

To use host genetics to control infections by GI nematodes it is important to understand the host–parasite interaction in order to identify points of contact between the parasite and its host that can be successfully manipulated to the detriment of the parasite. GI nematode infections of ruminants remain a major constraint to the efficient raising of livestock throughout the world. In the USA alone it is estimated that these parasites cost the American livestock industry in excess of US$2 billion per year in lost productivity, and in increased operating expenses.

Historically, control of these parasites was accomplished by complicated management programmes that kept stocking rates low or attempted to minimize the exposure of susceptible animals to heavily infected pastures. Under favourable conditions these programmes were capable of reducing the occurrence of large outbreaks of clinical illness in livestock herds. But these systems were inefficient, they did not reduce production losses due to subclinical levels of infection and they often failed when conditions for parasite transmission were optimal. This is especially true in small ruminants, where GI nematodes are often one of the most important impediments to the successful raising of the animals (American Association of Veterinary Parasitologists, 1983). Approximately 20–30 years ago anthelmintic agents began to appear that had both low toxicity for the ruminant host and high efficacy against the parasites. As these drugs became available, producers became less and less concerned with traditional parasite control programmes, and more and more dependent on the anthelmintics for control. Although modern anthelmintics have been an important tool for cattle raisers, a number of factors have arisen recently that illustrate that to depend on the drugs as a sole means of parasite control is a serious mistake. The most important of these factors has been the appearance throughout the world of parasites resistant to the anthelmintics. In small ruminants, parasites have developed resistance against all classes of anthelmintics currently employed, and resistance to multiple classes of drugs by the same parasite is commonplace (Waller and Prichard, 1986; Prichard, 1990; Windon, 1991; Waller, 1994). While such resistance is less common in parasites of cattle, there are reports of resistance in New Zealand (Vermunt *et al.*, 1995; Hosking *et al.*, 1996) and in South America (Pinheiro and Echevarria, 1990). A second concern about this dependence on anthelmintics is the growing insistence of consumers that their food and environment be free of chemical residues (Herd *et al.*, 1993). The growing number of 'organic' producers find repeated anthelmintic use to be an unacceptable means of parasite control. The third factor affecting current control programmes is the movement towards more intensive grazing programmes for livestock. Increased grazing intensity is the result of several factors. One impetus is the reduction of the amount of land available for grazing due to forces such as urban development and the setting aside of lands for recreational or conservational use. A second factor is the realization that intensive grazing programmes can result in higher net profits for small farmers, coupled with the perception that grazing systems are more 'sustainable' systems. This move towards more intense use of pastures will necessitate more frequent use of anthelmintics, further increasing the chances for selection of anthelmintic resistance in the parasite populations.

The parasites have a simple life cycle, beginning with adult male and female worms in the alimentary tract of the host. Fertilized eggs are passed in the faeces. The eggs hatch and larval development begins within the protective environment of the faecal pat or droppings. Depending upon ambient temperature, the larvae reach the infective third larval stage approximately 10–14 days after being passed in the faeces. The infective larvae are carried away from the faeces by forces such as rain, and are then found on the grass to await ingestion by the foraging host. Upon ingestion the larvae undergo an additional two moults to the adult stage. In general, the developing larvae are found in more intimate contact with host mucosal surfaces than are the more lumenal-dwelling adult worms.

While on pasture the larvae are very susceptible to desiccation, and transmission is very low to non-existent during very dry periods. In addition, transmission is reduced during periods of persistent cold. For some species this is due to killing of the larvae, but in other cases it is more a reflection of reduced grazing due to dormancy of the grasses. While in the host, most parasites are susceptible to host immune responses. These responses range from those that alter parasite morphology or physiology to those that protect the host from reinfection, presumably by killing of the developing worms or exclusion of the invasive third-stage larvae. Most of the genera of parasites found in ruminants stimulate a relatively effective level of immunity in most animals after several months on pasture. This immunity significantly reduces the number of worms that become established in the grazing animals. The exceptions to this are *Ostertagia ostertagi* in cattle and *Haemonchus contortus* in sheep. Cattle remain susceptible to infection by *Ostertagia* for many months, and immunity that actually reduces the development of newly acquired larvae is usually not evident until the animals are more than 2 years old. In sheep, lambs under the age of about 8 months do not develop significant protective immune responses to infection or immunization (Urquhart *et al.*, 1966; Knight and Rodgers, 1974). This prolonged susceptibility to reinfection is a major reason why these parasites remain the most economically important GI nematodes in their respective species. Although animals may remain susceptible to reinfection for a prolonged period of time, in cattle there is a stunting of newly acquired worms and, in addition, the female worms produce significantly fewer eggs than do female worms in a primary infection. The overall result of either type of immunity is a reduction in parasite transmission.

Given the ubiquitous nature of GI nematodes, the fact that wild ungulates may serve as a source of infection, and that optimal conditions for parasite transmission are the same as optimal conditions for grass growth, it is virtually impossible to eradicate GI nematodes in grazing ruminants. Rather than eradication, the goal of effective nematode control programmes is to protect animals from production losses by reducing parasite transmission and establishment in the host. One such approach is to use genetically superior stock as an adjunct to current methods of control.

Enhancing Helminth Resistance in Small Ruminants

Sources of genetic variation

It may not be possible, or even desirable, to identify animals which have complete genetic resistance to nematode infection, but there is considerable evidence that part of the natural variation in resistance is under genetic control (reviewed by Wakelin, 1978; Barger, 1989; Stear *et al.*, 1990a; Gray and Gill, 1993; Raadsma *et al.*, 1997a; Stear and Wakelin, 1998). Selection of resistant animals within breeds or incorporating resistant breeds into breeding programmes are the available options. Traditional breeding programmes and selection for nematode resistance have been used successfully to establish flocks of sheep with high levels of resistance (Windon and Dineen, 1984; Albers *et al.*, 1987; Windon, 1990, 1991; Woolaston *et al.*, 1990, 1991). A thorough summary of responses achieved in these programmes is given by Morris (1998). Identification of some of the genes involved in regulating resistance will allow earlier selection of genetically superior animals and may increase the rate of selection for resistance.

Within breed

There is variation in nematode burdens between individual sheep in a flock grazing the same pastures. The distribution of burdens (and faecal egg count, FEC) is overdispersed; that is, a minority of the individuals harbour the majority of the nematodes. Evidence for genetic variation in such distributions comes from studies of pedigree populations in which offspring have been assessed for resistance under uniform conditions of exposure and at the same age. Examples of the heritability of resistance to infection in sheep, as measured by FEC, varied from 0.22 to 0.43. Albers *et al.* (1987) challenged 882 lambs with 11,000 *H. contortus* larvae and estimated the heritability of FEC 4 weeks after challenge as 0.34 ± 0.10 (mean ± standard error). Woolaston *et al.* (1991) reported estimated heritabilities, after artificial challenge with *H. contortus*, of FEC in three flocks of Merino sheep as 0.27 ± 0.13, 0.22 ± 0.04 and 0.31 ± 0.03. Cummings *et al.* (1991) reported estimated heritability, after naturally acquired *Ostertagia* infection, of FEC in Merino sheep as 0.42 ± 0.14. Baker *et al.* (1991) reported estimated heritability, after naturally acquired mixed species infection, of FEC in Romney sheep as 0.34 ± 0.19. Other reports of heritable variation in FEC include Bishop *et al.* (1996), Bisset *et al.* (1992), Douch *et al.* (1995), McEwan *et al.* (1992, 1995), Morris *et al.* (1997a, b), Stear *et al.* (1997), Woolaston and Eady (1995) and Woolaston and Piper (1996). Stear *et al.* (1997) showed that for 6-month-old lambs facing *Ostertagia* infection, it was worm size and fecundity rather than worm burden that was the heritable trait.

At these established levels of heritability, there are prospects for selective breeding for improved resistance at least in the sheep industry (Gray, 1991). As a result of these encouraging research findings, there are now structured breeding programmes for resistance to parasites within commercial sheep flocks in both New Zealand and Australia.

Studies conducted in Fiji (humid Pacific island conditions) indicated that the heritability of FEC in Fijian goats from two locations was not significantly different from 0 (Woolaston *et al.*, 1995). On the other hand, heritability of FEC in the crossbred sheep population was estimated at 0.23 ± 0.07. Therefore it was concluded that within-breed genetic improvement for resistance to nematode infection was not feasible for goats, but had potential for sheep.

Between breeds

Some breeds of sheep are more resistant than others to nematode infection. For example, Scottish Blackface sheep are more resistant than Finn-Dorset sheep to *H. contortus* infection (Altaif and Dargie, 1978) and the Red Maasai sheep of Kenya are more resistant than breeds imported into Kenya (Preston and Allonby, 1978, 1979a; Bain *et al.*, 1993; Baker *et al.*, 1993; Miller *et al.*, 1995). In France, Gruner *et al.* (1986) showed that Romanov sheep were more susceptible than Lacaune sheep to infections with *Nematodirus spathiger* and *Ostertagia circumcincta*. In the USA, Radhakrishnan *et al.* (1972) showed that Florida Native lambs were more resistant to *H. contortus* infection than Rambouillet lambs. Courtney *et al.* (1985) subsequently showed that St Croix lambs were more resistant than Florida Native and Barbados Blackbelly lambs, which in turn were more resistant than domestic crossbred (Suffolk, Finn-Dorset, and Rambouillet) lambs. Gamble and Zajac (1992) showed that St Croix sheep were more resistant than Dorset sheep. Bahirathan *et al.* (1996) and Miller *et al.* (1998) have shown that Gulf Coast Native sheep are more resistant than Suffolk sheep. In Indonesia, Gatenby *et al.* (1995) and Romjali *et al.* (1997) found that introduced St Croix ewes were more resistant than the local Sumatra ewes. Moreover, crosses of St Croix and Barbados Blackbelly (also introduced) with local Sumatra sheep were at least as resistant or more resistant than pure Sumatra sheep and could be used to improve production (body size). Further evidence for other breed differences are summarized in Gruner and Cabaret (1988), Gray (1991) and Zajac (1995).

Selection for resistance

For any selection programme to be successful it is essential that the superior individuals can be identified accurately and economically from among the candidate breeding stock. There are direct and indirect means to help make those selections.

Direct selection

There is only one direct way to measure nematode burdens and that is by recovery at necropsy, but it is obvious that this method cannot be employed as a selection criterion. Faecal egg counts have been shown to have moderate (0.61) to high (0.91) correlation with nematode burdens (Baker *et al.*, 1991; Stear *et al.*, 1995a; Bisset *et al.*, 1996), and in essence might be considered a direct measure of infection. The most effective and only practical way to estimate potential pasture contamination is from FEC and, in one sense, this is also direct selection.

Experimental selection programmes for establishing resistant and susceptible lines of sheep have been in place in Australia and New Zealand for several years, where selection has been based on FEC after natural infection and artificial challenge. The rate of selection progress is determined by the flock generation interval, the heritability of resistance and the intensity of selection. It is more efficient if selection of breeding animals can be made at an early age, but it is possible that lambs may need to be 'primed' during their first natural exposure or with initial artificial infections (3–6 months of age) to initiate an immune response (Woolaston *et al.*, 1991). This concept was supported by work demonstrating that no genetic differences were apparent at the sire level or the bloodline level in weaners raised parasite-free and experimentally infected with *H. contortus*. This phenomenon was verified using trickle infections in resistant and susceptible lines (Windon, 1991). Similar observations were noted under natural exposure conditions from birth with Gulf Coast Native and Suffolk sheep (Bahirathan *et al.*, 1996). In this experiment, both Suffolk and Gulf Coast Native lambs had similar increasing *H. contortus* levels until 8–10 weeks of age. Subsequently, the infection level in Gulf Coast Native lambs decreased and in Suffolk lambs it continued to increase. It appeared that resistance was acquired in Gulf Coast Native lambs, but not Suffolk lambs, after being 'primed', but at an earlier age than would be considered normal for an immune response to infection in lambs (Barger, 1988).

Indirect selection

In the context of breeding for resistance, all criteria except those that measure burdens directly must be considered indirect. These indirect selection criteria may include continuously variable physiological or immunological traits, which are referred to as 'indicator traits', or polymorphisms of single genes or small groups of linked genes, which are referred as 'genetic markers'.

INDICATOR TRAITS. Indicator traits usually reflect host response to infection. In living animals, such traits might include blood levels of eosinophils and antinematode antibodies, and blood packed-cell volume (PCV). Dawkins *et al.* (1989) and Buddle *et al.* (1992) reported blood eosinophilia associated with *T. colubriformis* infection in selected lines of resistant compared to susceptible sheep. In contrast, Douch *et al.* (1996) and Woolaston *et al.* (1996) reported several studies indicating that blood eosinophil level was variable and inconsistent for *T. colubriformis* and *H. contortus* resistant and susceptible lines of sheep, which reduced its potential use as an indicator.

Bissett *et al.* (1996) reported that Romney lambs selectively bred for resistance had significantly higher serum *T. colubriformis*-specific antibodies, IgG1 and IgM, than in susceptible lambs. Douch *et al.* (1996) also suggested that antibody levels (particularly IgG1) to excretory or secretory antigens of *T. colubriformis* L3 may be useful in selecting resistant animals. In two trials to test this selection criterion, lambs from rams with high and low antibody levels showed that lines bred true for antibody level but that there was no difference in FEC, which diminished any genetic relationship. Blood PCV may be useful for selecting *H. contortus* resistant animals, as anaemia is a classic sign of haemonchosis, but no studies have been conducted to evaluate this criterion. For *Ostertagia* infections, there is evidence that IgA may be a major regulator of worm length, hence FEC, and therefore a potential indicator trait (Stear *et al.*, 1995a).

Other indicator traits that cannot be evaluated in living animals are accumulation of immune-mediated cell types (specifically mucosal mast cells (MMC), globule leucocytes and eosinophils) in the GI mucosa. At necropsy, increased levels of these cell types have been observed in resistant lines and breeds (Presson *et al.*, 1988; Gamble and Zajac, 1992; Gill *et al.*, 1993).

If immune-mediated indicator traits are to be useful for resistance selection purposes, resistance has to be an acquired phenomenon. Expression of resistance in sheep selected for resistance to *H. contortus* as an acquired immune response has been investigated in Merino sheep. Gill (1991) found that primary infections in parasite-free lambs from resistant and random-bred lines resulted in no significant difference between nematode burdens, serum antibody levels, MMC or circulating and tissue eosinophils between the two groups. However, there was a significant difference in FEC, with lambs of resistant lines having higher levels. Following removal of nematode burdens by anthelmintic treatment, challenge infections resulted in the lambs of resistant lines having significantly lower FEC and nematode burdens and significantly higher serum antibody levels, MMC hyperplasia and mucosal eosinophilia. Additional support for the immunological basis of helminth resistance comes from work done on immunosuppression and ablation of $CD4^+$ T cells. Presson *et al.* (1988) demonstrated that immunosuppression of resistant wethers with dexamethasone resulted in abolishing the differences between them and random-bred susceptible wethers. Gill *et al.* (1993) selectively depleted genetically resistant Merino lambs of their $CD4^+$ T cells, by treatment with mouse monoclonal antibody specific for the determinant, before and during challenge infection with *H. contortus*. Treated lambs lost their expression of genetic resistance as they had higher FEC and nematode burdens compared to control random-bred susceptible lambs. The MMC hyperplasia and tissue

U.W.B. LIBRARY YFRGELL

eosinophilia associated with resistance were also suppressed in treated lambs. These results suggest that CD4 T cells play a role in mediating genetic resistance. It was hypothesized that these CD4 T cells respond to presentation of antigen in the context of class II MHC molecules and produce various cytokines that amplify and regulate the recruitment, differentiation and proliferation of effector cells, such as mast cells, globule leucocytes, eosinophils and antibody-secreting cells. Gill (1994), using lymphocyte proliferation and skin hypersensitivity responses to *H. contortus* antigens, demonstrated a cell-mediated immune component in resistant lambs. Bissett *et al.* (1996) also reported further evidence to support an acquired immune-mediated response in Romney lambs selectively bred for resistance, where numbers of globule leucocytes/MMC were significantly higher in resistant lambs than in susceptible lambs. In contrast, Miller *et al.* (1996) found that depletion of $CD4^{+}$ T cells did not abrogate resistance in Gulf Coast Native sheep, suggesting that there may not be a major acquired immune mechanism involved in resistant breeds of sheep.

GENETIC MARKERS. Recognition of resistance has relied on observed phenotypic variation. It is not known whether this resistance is conferred by a small number of genes or a larger number of genes, each having somewhat smaller effects. In either case, genetic maps are necessary to identify genes directly mediating resistance or markers linked to those genes. Two strategies can be used to identify genes influencing resistance. The first does not require an existing genetic map, and uses the analysis of candidate genes, which are expressed genes that may be expected to play a role in regulating resistance (i.e. genes encoding immunoglobulins, MHC antigens, T-cell receptor molecules, etc.). The second relies upon linkage maps and genome-wide analyses for quantitative trait loci (QTL) detection. This method is based upon the use of polymorphic DNA markers to tag specific genes or regions of the genome carrying resistant genes.

The first genetic marker suggested for use in selection for resistance was haemoglobin type. Sheep have two alleles (*A* and *B*) for haemoglobin and several studies indicated that animals with haemoglobin type AA (HbAA) were more resistant than HbAB, which were more resistant than HbBB to infection with *H. contortus* and its effects (Allonby and Urquhart, 1976; Altaif and Dargie, 1978; Preston and Allonby, 1979b). However, other workers were unable to confirm that association (Radhakrishnan *et al.*, 1972; Riffkin and Yong, 1984; Albers and Gray, 1986; Kassai *et al.*, 1990). Therefore, no general conclusion can be drawn concerning the usefulness of haemoglobin type as a predictive marker for resistance, at least to *H. contortus*.

A second attempt at using the candidate gene approach was the investigation of the major histocompatibility complex (MHC). This is logical because MHC class I and class II genes are involved in immune responses. The MHC regulates resistance to nematode infections in experimental mice (Wassom *et al.*, 1979; Wassom and Kelly, 1990), guinea-pigs (Geczy and Rothwell, 1981) and swine (Bell *et al.*, 1982; Wakelin and Donachie, 1983; Lunney and Murrell, 1988; Madden *et al.*, 1990). Outteridge *et al.* (1985, 1986, 1988) reported an association between class I antigens of the ovine MHC and response to challenge with *T. colubriformis*. Cooper *et al.* (1989), however, found no evidence for an association between class I antigens of the ovine MHC and susceptibility to *H. contortus* infection in the progeny of six rams. Stear *et al.* (1988, 1990b) showed that the bovine MHC is one of the genetic systems that may be involved in regulating resistance to nematode infection. Evidence for class II gene association with parasite resistance in sheep is scarce, but increasing. Hulme *et al.* (1991) reported that restriction fragment length polymorphism (RFLP) analyses using cross-hybridizing human MHC DQA, DQB and DRB cDNA probes showed that genes in or closely linked to the MHC have a significant effect on resistance to *T. colubriformis* infection. Blattman *et al.* (1993) showed no association between MHC class II polymorphism, detected by similar human probes, and resistance to *H. contortus* in a single family of parasite-resistant Merino sheep. Grain *et al.* (1993) demonstrated RFLP of *DQB* and *DRB* class II genes of the ovine MHC using ovine probes specific for the second exons of *Ovar-DQB* and *Ovar-DRB* genes. Miller *et al.* (1995) demonstrated similar polymorphisms in *DQB* and *DRB* genes of the ovine MHC using bovine MHC probes (Muggli-Cockett and Stone, 1991), where allele frequency was significantly different between Red Maasai compared to Dorper breeds of sheep. However, no genetic association with parasite resistance was investigated. Finally, Schwaiger *et al.* (1995) demonstrated a significant association between a *DRB1* allele and low faecal egg counts following natural *Ostertagia* infection. The definition and role of the MHC in livestock species has been reviewed (Stear *et al.*, 1989; Wetherall and Groth, 1992; and, most recently, in this volume).

With the advent of the polymerase chain reaction (PCR) for amplification of genomic DNA, more efficient genetic characterization is possible. This has led to the development of microsatellite markers and, ultimately, genetic linkage maps. Efforts to establish a sheep gene map (by AgResearch, New Zealand; CSIRO and the University of Melbourne, Australia; USDA, USA) has resulted in the derivation of polymorphic markers covering the sheep genome. The first linkage map published (Crawford *et al.*, 1995) included 246 (predominantly microsatellite) markers. Currently there are over 550 such markers (DeGotari *et al.*, 1998). Because of the high degree of homology at the DNA level between cattle and sheep, many cattle microsatellite markers have been positioned in the sheep genome. It is generally accepted that 150–300 markers evenly distributed across the genome are reasonable for a primary linkage map (Teale, 1991), which can be used to detect quantitative trait loci, in this case, for nematode resistance. This led to the reporting of several QTL for nematode resistance at the ISAG meeting in August, 1998.

Ultimately, transgenic technology may be a possibility. If single or multiple genes responsible for expression of nematode resistance can be identified, then transfer of such genes from resistant to susceptible individuals may provide a more long-lasting resistance, since they provide much more of an evolutionary challenge to the parasite.

Objectives of selection for resistance

It is unlikely that a breeding programme would have resistance to parasites as its sole objective. More likely the objective would be expressed in units of production or in the value of production. Therefore, genetic relationships with other traits in the objective, for example liveweight gain, fleece weight or resistance to other diseases, may be as important, or more important, than resistance itself.

Similarly, the usefulness of resistant breeds depends on the commercial requirements of the producer. If resistant breeds are better producers of meat or wool than breeds in current use, then a change to the resistant breed may be advantageous. Comparison of the resistant Red Maasai breed with the susceptible Dorper breed in Kenya has shown that the Red Maasai are just as productive as the Dorpers (Baker *et al.*, 1993, 1998). Not all such comparisons favour the resistant breeds but detailed studies of the breed productivity, including reproductive, survival and production characteristics, all measured in a realistic commercial environment, are required. It should be noted that all reports of resistant breeds do not take into account between-sire differences, and the perceived breed difference may really be due to the effect of a few resistant sires. In addition to their potential as productive replacements for existing breeds, parasite-resistant breeds are of extreme interest for investigating the genetic basis of resistance and how resistance genes may be expressed.

Selection within breed has not resulted in any association with loss of productivity (Windon and Dineen, 1984; Albers *et al.*, 1987; Woolaston *et al.*, 1990). It has been suggested that selection for resistant animals actually results in increased or no loss of production (Baker *et al.*, 1991; Eady *et al.*, 1994; Morris *et al.*, 1995). However, in general there is no consistent pattern in the genetic correlations between FEC and productivity, with published correlations ranging from strongly favourable (i.e. a large negative correlation) (Bishop *et al.*, 1996), through neutral or moderately favourable (Albers *et al.*, 1987; Bisset *et al.*, 1992; Douch *et al.*, 1995; Eady, 1998), to moderately unfavourable (a positive correlation) (McEwan *et al.*, 1992, 1995). The results of Albers *et al.* (1987) imply that the genetic correlation strengthens as the parasite challenge increases. Piper and Barger (1988), using the best-available genetic correlations between parasite resistance and production traits, estimated overall genetic improvement of about 10% per year for a typical Merino flock. However, it should be emphasized that marginal economic gains are not nearly as important as those from preventing devastating losses due to anthelmintic resistance in the parasite population.

There is encouraging evidence that genetic correlations between resistance and more than one nematode parasite is positive (Windon, 1990). Lambs resistant to *Trichostrongylus* also demonstrate significantly reduced FEC after infection with other related species, including *T. rugatus*, *T. axei* and *O. circumcincta* (Windon and Dineen, 1984; Windon *et al.*, 1987). Similarly, resistant *Haemonchus* selection-line lambs had reduced FEC after infection with *Trichostrongylus* spp. and *Ostertagia* spp. (Gray *et al.*, 1992). Heterologous challenge infections of the *Trichostrongylus* and *Haemonchus* selection-line lambs showed that differences between susceptible and resistant lines were not as great after challenge with the heterologous species as that seen after challenge with the homologous species (Windon, 1990; Woolaston, 1990). McEwan *et al.* (1992) also demonstrated a significant positive genetic correlation between *Strongyle* and *Nematodirus* FEC. Perhaps immunological responses differ during infection with each species due to site location, feeding habits, evasion strategies, etc. However, with the exception of the study of Raadsma *et al.* (1997b), genetic correlations between resistance to parasitic disease and other important sheep diseases are lacking. Raadsma *et al.* (1997b) found that resistance to parasites is generally uncorrelated with resistance to other diseases, except for a small positive correlation with resistance to dermatophilosis.

Enhancing Helminth Resistance in Cattle

Effect of genetics on responses to cattle helminths

Any attempt to discern a genetic influence on resistance to GI nematodes requires an accurate means by which to assess parasite burdens in the animals tested. The most commonly used method to assess parasite burdens in ruminants has been the enumeration of parasite eggs in the faeces, and to date most of the work aimed at determining the role of genetics in resistance to this group of parasites has been based on faecal egg counts (FEC). Analyses of FEC values in cattle herds reveals several important facts. First, if calves are sampled on consecutive days, the repeatability of FEC values is approximately 0.6 (Gasbarre *et al.*, 1996); this repeatability decreases dramatically if the time between samples is extended (Stear *et al.*, 1984). Secondly, FEC values decrease with the age of the animal, and change throughout the year based on grazing behaviour of the animals. Thirdly, the sex of the animal influences FEC values, with bulls having higher FEC values than cows (Gasbarre *et al.*, 1990; Stear *et al.*, 1990b). And fourthly, FEC values are not normally distributed within a cattle herd, instead they follow an 'overdispersed' distribution (Crofton, 1971a, b; Genchi *et al.*, 1989; Gasbarre *et al.*, 1993). This 'overdispersed' distribution is one where the value of the standard error of the mean exceeds the value of the mean. In such a distribution most individuals have relatively low FEC values, and a small percentage of animals are responsible for the majority of the eggs released on the pastures. This results in an apparent group of susceptible animals, or 'non-responders', whose number is usually estimated to be between 15 and 25% of the total population (Anderson and May, 1985; Genchi *et al.*, 1989).

Using FEC, the role of host genetics in resistance to GI nematodes has been investigated, both across cattle breeds and within the same breed. Suarez *et al.* (1990) found that Zebu cattle had higher FEC values than did either Hereford or Hereford × Brahman crosses grazed in Argentina. It should be noted that in this study the Zebu steers were maintained on different pastures than the Hereford and Hereford × Brahman cross steers, so that pasture effects would be confounded in this model. Similarly, Almeria *et al.* (1996) reported that FEC values were significantly lower in Pyrenean cattle when compared to Brown Swiss cattle when the animals were co-grazed under a traditional mountain meadow grazing system in the Pyrenees. In contrast, studies in Queensland, Australia, did not indicate significant breed differences in grazing cattle (Barger *et al.*, 1983). Looking at Angus, Hereford, Brahman, Senepol, and Hereford × Senepol and Senepol × Hereford crossed yearling heifers, which were grazed as a single group in the south-eastern USA, we found that environmental factors greatly influenced FEC values. Because all breeds did not exhibit the same grazing behaviour in the winter and summer, group FEC values were strongly affected at these time points. When animals were sampled at times of the year when grazing behaviour among the groups was similar, we found that apparent breed differences in FEC value could be explained by sire effects within the breeds (Gasbarre, unpublished).

There is a greater body of work examining within-breed variation in FEC values. Estimates of the heritability of FEC values have ranged from approximately 0.1 to nearly 0.8 (Stear *et al.*, 1984, 1990b; Leighton *et al.*, 1989; Gasbarre *et al.*, 1990, 1993; Mackinnon *et al.*, 1991; Kloosterman *et al.*, 1992; Gray and Gill, 1993; Suarez *et al.*, 1997). The most commonly accepted estimate of heritability for this trait is in the range 0.3–0.4. In one study covering 4 years, all calves born at the Wye Angus herd in Maryland, USA, were intensively sampled at the time of weaning. The data were subjected to analyses where the model accounted for variation based on age and sex of the calf, year of sampling, potential pasture effects and sire of the calf. Because FEC values do not follow a normal distribution, with a small number of calves accounting for a disproportionate number of eggs released to the pasture, the odds of a given bull producing a calf whose FEC value was in the upper 25th percentile were calculated. Results indicated that the odds of certain bulls producing these calves was approximately 20 times that of other bulls (Gasbarre *et al.*, 1995). The role of BoLA alleles in affecting FEC values is uncertain. Stear *et al.* (1990b) found that two BoLA class I alleles, *W7* and *CA36*, were related to lower FEC values. In other studies, BoLA has been found to exhibit a very minor effect on FEC values (Stear *et al.*, 1984; Gasbarre, unpublished).

There is a serious problem with using FEC values to assess, under field conditions, worm burdens in the host. Rarely, if ever, is a single parasite species found on pastures. Most of the common parasite genera infecting cattle produce eggs that are virtually indistinguishable from each other. Among these different genera, the fecundity of the individual species varies greatly and, even more importantly, host factors may reduce the fecundity in some parasite species but not in others. This is especially true for the important pathogen *Ostertagia ostertagi*. Infections involving this species are characterized by the fact that the worms are less fecund than other commonly encountered species, and this is further compounded by the fact that host immune responses can reduce the fecundity of the worms, without appreciable change in parasite numbers (Smith *et al.*, 1987; Gasbarre, 1994, 1997a). In contrast, in more tropical regions where *Ostertagia* is not an important component of the parasite fauna, this may not be a serious problem (Bryan and Kerr, 1989). In temperate regions of the world, where *Ostertagia* is the dominant pathogen, FEC are a very poor indicator of numbers of *Ostertagia* in the host, because of the effect on fecundity of host immunity. In temperate regions, FEC reflects the numbers of more fecund species, such as *Cooperia* (Gasbarre, 1997b). This has caused investigators to employ additional methods to attempt to estimate more accurately the individual parasite species in the host. The most commonly employed technique is to determine FEC values, and then culture the egg-containing faeces. After an appropriate time interval, larval parasites that have hatched from the eggs are recovered and identified to the genus level. Using these procedures Suarez *et al.* (1997) estimated heritabilities of approximately 0.2 for *Haemonchus*, 0.3 for *Ostertagia* and *Trichostrongylus* and 0 for *Cooperia*. In addition, Schmidt *et al.* (1998) did not find differences in the diversity of parasite genera in calves from different Aberdeen Angus sires, and concluded that the host has limited effect on diversity of parasite genera. This implies that resistance to the different genera of parasites is similar and that resistant individuals will be resistant across the spectra of parasite species. Care should be exercised in interpreting the results of larval cultures, as it has been demonstrated that culture conditions can favour the development of some genera, potentially leading to over- or underestimates of the proportion of the sample made up by each genus (Gasbarre, 1997c; Schmidt *et al.*, 1998).

A second approach to the problem of different fecundities of the parasites is to use artificial infections consisting of a single parasite. Kloosterman *et al.* (1978) used this approach with artificial infections of *Cooperia*. In their experiments they found significant differences between half-sib groups of Dutch Friesian bull calves in FEC values, worm numbers, worm length and antibody responses.

Another approach to addressing the problem of potential erroneous measures of parasite burdens by FEC is to find another measure of parasite burden, such as weight gain of the animals, serum antibody levels, serum pepsinogen levels, blood eosinophil numbers, etc., or to actually slaughter animals and recover, speciate and enumerate the worms themselves. Slaughter, while it is the most accurate, obviously greatly reduces the number of animals that can be studied. Over the past several years we have selectively bred Angus cattle for susceptibility or resistance to infection by gastrointestinal nematodes. These cattle were initially derived from the Wye Angus herd, and as such, all animals can be traced back to 1 of 18 dams or 1 of 19 bulls, with records covering at least seven generations. All calves produced are placed at weaning on pastures containing *Ostertagia ostertagi* and *Cooperia oncophora*. The calves are kept on these pastures for a minimum of 120 days to allow them to express immunity to the parasite exposure (Gronvold *et al.*, 1992). During the test period all animals are monitored weekly, and at the end of the test period some animals are slaughtered for parasitological and immunological assessment. Based upon the slaughter of 49 individuals the following observations have been made. Faecal egg counts in this situation are poorly correlated with *Ostertagia* numbers ($R = 0.1$), and are more an indication of the number of *Cooperia* in the host ($R = 0.6$, $P \geq 0.05$). The numbers of *Ostertagia* and

Cooperia are significantly correlated after 1 month on pasture, but are not significantly correlated after the 120-day test, indicating that immunity to the two parasite genera functions differently over this time frame. Serum antibody levels specific for the parasites (IgG1, IgG2, IgM and IgA) were not correlated with the numbers of either parasite at slaughter. Weight gain over the entire test was significantly correlated with the number of *Ostertagia* at slaughter but not with *Cooperia* numbers. The number of circulating eosinophils did show a relationship to *Ostertagia* numbers, but the only relationship was with peak eosinophil numbers, meaning that animals needed to be sampled repeatedly over time. The best indicator of *Ostertagia* numbers was the serum pepsinogen value of the animal just before slaughter ($R = 0.63$). *Ostertagia* numbers were significantly correlated with measures of anaemia (RBC (red blood cell) counts, haematocrit, haemoglobin levels) but this relationship was less significant than the relationship to serum pepsinogen levels.

Because serum pepsinogen levels appeared to be the best indicator of *Ostertagia* burdens, the analyses done over the 4-year period on the Wye Angus herd were repeated, only this time assessment of heritability was estimated on serum pepsinogen levels of each calf at weaning. The analyses indicated that: (i) serum pepsinogen levels were not normally distributed; (ii) serum pepsinogen levels were significantly affected by sire of the calf; and (iii) the estimated heritability was approximately 0.3. Interestingly, the serum pepsinogen levels were also significantly influenced by sex of the calf, but in this case heifer calves had higher values than bull calves. One note of caution in using serum pepsinogen values is the fact that this relationship has only been demonstrated in young calves. Serum pepsinogen levels are assumed to be a measure of abomasal damage, and the subsequent release of the enzyme precursor into the blood. It is possible that older animals could show hypersensitivity to the parasites, resulting in high levels of abomasal damage with few established worms.

The question becomes: 'How do these observations in an experimental environment compare to those in a more normal field study?' Stear *et al.* (1990b) examined cattle from a more tropical climate, which lacked *Ostertagia* but instead were infected by five species including *Oesophagostomum radiatum*, which is severely pathogenic in young animals. They found that weight gain was negatively correlated with worm numbers, but that only *Oesophagostomum* numbers significantly affected weight gain. No significant relationship was found between parasite burdens and serum antiparasite IgE levels. In tropical cattle, Mackinnon *et al.* (1991) showed that whereas the relationship between growth rate and FEC tended to be negative (favourable) in the wet season when the worm challenge was strong, it was positive (unfavourable) during the dry season. Almeria *et al.* (1996) found that the more resistant Pyrenean cattle had higher RBC, haemoglobin and haematocrit, but lower blood eosinophil values than the more susceptible Brown Swiss cattle. Kloosterman *et al.* (1978) found significant differences in antiparasite antibody levels among half-sib groups experimentally challenged with *Cooperia*, but the relationship between FEC and antibody titre was not clear. Gasbarre *et al.* (1993) found that serum antibody against parasite antigens was strongly influenced by the sire of the calf (heritability 0.7–0.8), but that there was no relationship between serum antibody levels and FEC values.

Methods by which host genetics could be used to control helminth infections in cattle

There is sufficient genetic variation within cattle breeds in terms of their resistance to nematode infection, and the heritability of this trait is high enough that significant genetic progress can be made by classic selection programmes. The question that arises is: 'Do parasites exert sufficient economic pressure to warrant such a selection programme?' The answer to this enquiry is that in some areas there is sufficient pressure, while in others there is not. For example, in the USA, producers in the south-east, north-east and north-west see enough parasitism, and stocking rates are sufficiently high, that genetic improvement in this area would increase their efficiency. In contrast, stockmen in areas of low grazing density may not gain an advantage, although one might still see an advantage in reducing the parasite load of animals in areas of marginal nutrition. In addition, as questions on land use and pesticide application become more important to consumers, programmes that reduce dependence on anthelmintics will likely become more readily adopted by producers.

A potentially more important question is whether it is wise to implement a breeding programme based on resistance to a single disease entity. Resistance to GI nematodes is a complex mixture of many types of immune responses and recent work indicates that protection against the different parasites may involve different types of immune responses (Canals *et al*., 1997; Almeria *et al*., 1998). It is also becoming evident that different types of immune responses, i.e. the so-called Th1 and Th2 response, are counter-regulatory. Thus it is possible that animals that are superior in some types of responses will be inferior in others. This, coupled with the fact that many individuals already question the loss of diversity in agricultural species, makes it unlikely there will be major programmes to breed for helminth resistance in cattle, except in areas of extremely high parasite pressure.

Even though it seems unlikely that there will be major breeding programmes for helminth resistance in cattle, there are several ways host genetics could be used to greatly increase the efficiency of cattle and dairy production. The first is based on the fact that nematodiasis is a 'quantitative' disease. As parasite burdens increase, so does loss, and it is probable that very low parasitaemias induce little or no economic effect in otherwise well-managed animals. With this in mind, the goal of modern parasite control programmes is to keep transmission rates down, and not to eradicate the parasites, which is probably impossible. The second important factor is the 'overdispersed' nature of nematode distributions. This pattern implies that a few members of the herd are responsible for most of the parasite transmission. In an elegant analysis, Anderson and May (1985) calculated that the treatment, removal or vaccination of the susceptibles would be almost as effective in controlling disease as would be treatment of the entire population. From an economic standpoint, rapid and inexpensive identification of these 'susceptible/non-responders' could reduce treatment costs substantially. Such an approach will also lessen the selective pressures placed on parasite populations, and should slow the development of drug resistance in the parasites. The key is the identification of reliable markers for susceptibility and resistance.

In a similar manner, identification of the genes involved in protection should provide producers with the information required to make informed breeding decisions. The past few years have seen cattle producers become more aware of selection for a variety of production traits, and have seen a movement away from selection for a single trait. Cattle producers today judge the genetic potential of bulls for many traits. The use of breeding values is a common subject of interest to cattle producers. Providing producers with reliable information on the performance of their cattle under parasite pressure will allow them to make informed breeding decisions. Producers in areas of high parasite pressure could choose not to use 'susceptible' bulls, while those in areas of less parasite transmission could use those bulls if they provided other traits important in the overall breeding programme.

Finally, as our ability to manipulate the genome becomes more refined it may be possible to genetically engineer cattle to meet the specific needs of a given producer. The ability to do this with parasite resistance will depend upon the exact mechanisms involved, the number of important genes and the precise regulatory processes entailed. In any of these future schemes, it will be important to first identify the gene loci involved in resistance, and to find means to identify the resistance phenotype of individual animals rapidly and reliably.

References

Albers, G.A.A. and Gray, G.D. (1986) Breeding for worm resistance: a perspective. *International Journal of Parasitology* 17, 559–566.

Albers, G.A.A., Gray, G.D., Piper, L.R., Barker, J.S.F., LeJambre, L.F. and Barger, I.A. (1987) The genetics of resistance and resilience to *Haemonchus contortus* in young Merino sheep. *International Journal of Parasitology* 17, 1355–1363.

Allonby, E.W. and Urquhart, G.M. (1976) A possible relationship between haemonchosis and haemoglobin polymorphism in Merino sheep in Kenya. *Research in Veterinary Science* 20, 212–214.

Almeria, S., Gracia, M.J., Llorente, M. and Uriarte, J. (1996) Comparative susceptibility of Pyrenean and Brown Swiss calves to gastrointestinal nematodes in subclinical naturally acquired infections. *Veterinary Parasitology* 63, 345–353.

Almeria, S., Canals, A., Gomez, M.T., Zarlenga, D.S. and Gasbarre, L.C. (1998) Characterization of protective immune responses in local lymphoid tissues after drug attenuated infections with *Ostertagia ostertagi. Veterinary Parasitology* 8, 53–64.

Altaif, K.I. and Dargie, J.D. (1978) Genetic resistance to helminths: the influence of breed and haemoglobin type on the response of sheep to primary infections with *Haemonchus contortus. Parasitology* 77, 161–175.

American Association of Veterinary Parasitologists (1983) Research needs and priorities for ruminant internal parasites in the United States. *American Journal of Veterinary Research* 44, 1836–1847.

Anderson, R.M. and May, R.M. (1985) Herd immunity to helminth infection and implications for parasite control. *Nature* 315, 493–496.

Bahirathan, M., Miller, J.E., Barras, S.R. and Kearnet, M.T. (1996) Susceptibility of Suffolk and Gulf Coast Native suckling lambs to naturally acquired strongylate nematode infection. *Veterinary Parasitology* 65, 259–268.

Bain, R.K., Wanyangu, S.W., Mugambi, J.M., Ihiga, M.A., Duncan, J.L. and Stear, M.J. (1993) Genetic resistance of Red Maasai sheep to *Haemonchus contortus*. *Proceedings of the 11th Scientific Workshop of the Small Ruminant Collaborative Research Support Program, Nairobi, Kenya*, pp. 120–126.

Baker, R.L., Watson, T.G., Bisset, S.A., Vlassoff, A. and Douch, P.G.C. (1991) Breeding sheep in New Zealand for resistance to internal parasites: research results and commercial application. In: Gray, G.D. and Woolaston, R.R. (eds) *Breeding for Disease Resistance in Sheep*. Australian Wool Corporation, Melbourne, pp. 19–32.

Baker, R.L., Reynolds, L.E., Mwanachi, D.M., Audho, J.O., Magadi, M. and Miller, J.E. (1993) Genetic resistance to gastrointestinal parasitism in Dorper and Red Maasai × Dorper lambs in coastal Kenya. *Proceedings of the 11th Scientific Workshop of the Small Ruminant Collaborative Research Support Program, Nairobi, Kenya*, pp. 228–241.

Baker, R.L., Rege, J.E.O., Tembely, S., Mukasa-Mugerwa, E., Anindo, D., Mwamachi, D.M., Thorpe, W. and Lahlou-Kassi, A. (1998) Genetic resistance to gastrointestinal nematode parasites in some indigenous breeds of sheep and goats in East Africa. *Proceedings of the 6th World Congress on Genetics Applied Livestock Production* 25, 269–272.

Barger, I.A. (1988) Resistance of young lambs to *Haemonchus contortus* infections and its loss following anthelmintic treatment. *International Journal of Parasitology* 18, 1107–1109.

Barger, I.A. (1989) Genetic resistance of hosts and its influence on epidemiology. *Veterinary Parasitology* 32, 21–35.

Barger, I.A., Bremner, K.C. and Waller, P.J. (1983) Factors influencing worm populations in cattle. In: Anderson, N. and Waller, P.J. (eds) *The Epidemiology and Control of Gastrointestinal Parasites of Cattle in Australia*. CSIRO, Australia, pp. 44–45.

Bell, R.G., MacGregor, D.D. and Adams, L.S. (1982) *Trichinella spiralis*: genetic basis for differential expression of phase-specific intestinal immunity in inbred mice. *Experimental Parasitology* 53, 315–325.

Bishop, S.C., Bairden, K., McKellar, Q.A., Park, M. and Stear, M.J. (1996) Genetic parameters for faecal egg count following mixed, natural, predominantly *Ostertagia circumcincta* infection and relationships with live weight in young lambs. *Animal Science* 63, 423–428.

Bisset, S.A., Vlassof, A., Morris, C.A., Southey, B.R., Baker, R.L. and Parker, A.G.H. (1992) Heritability of and genetic correlations among faecal egg counts and productivity traits in Romney sheep. *New Zealand Journal of Agricultural Research* 35, 51–58.

Bisset, S.A., Vlassoff, A., Douch, P.G., Jonas, W.E., West, C.J. and Green, R.S. (1996) Nematode burdens and immunological responses following natural challenge in Romney lambs selectively bred for low or high faecal worm egg count. *Veterinary Parasitology* 61, 249–263.

Blattman, A.N., Hulme, D.J., Kinghorn, B.P., Woolaston, R.R., Gray, G.D. and Beh, K.J. (1993) A search for associations between major histocompatibility complex restriction fragment polymorphism bands and resistance to *Haemonchus contortus* infection in sheep. *Animal Genetics* 24, 277–282.

Bryan, R.P. and Kerr, J.D. (1989) The relation between the natural worm burden of steers and the faecal egg count differentiation to species. *Veterinary Parasitology* 30, 327–334.

Buddle, B.M., Jowett, G., Green, R.S., Douch, P.G.C. and Risdon, P.L. (1992) Association of blood eosinophilia with the expression of resistance in Romney lambs to nematodes. *International Journal of Parasitology* 22, 955–960.

Canals, A., Zarlenga, D.S., Almeria, S. and Gasbarre, L.C. (1997) Cytokine profile induced by a primary infection with *Ostertagia ostertagi* in cattle. *Veterinary Immunology and Immunopathology* 58, 63–75.

Cooper, D.W., Van Oorschot, R.A.H., Piper, L.R. and LeJambre, L.F. (1989) No association between the ovine leucocyte antigen (OLA) system in the Australian Merino and susceptibility to *Haemonchus contortus* infection. *International Journal of Parasitology* 19, 695–697.

Courtney, C.H., Parker, C.F., McClure, K.E. and Herd, R.P. (1985) Resistance of exotic and domestic lambs to experimental infection with *Haemonchus contortus*. *International Journal of Parasitology* 15, 101–109.

Crawford, A.M., Dodds, K.G., Ede, A.J., Pierson, C.A., Montgomery, G.W., Garmonsway, H.G., Beattie, A.E., Davies, K., Maddox, J.F., Kappes, S.W., Stone, R.T., Nguyen, T.C., Penty, J.M., Lord, E.A., Broom, J.E., Buitkamp, J., Schwaiger, W., Epplen, J.T., Matthew, P., Matthews, M.E., Hulme, D.J., Beh, K.J., McGraw, R.A. and Beattie, C.W. (1995) An autosomal genetic linkage map of the sheep genome. *Genetics* 140, 703–724.

Crofton, H.D. (1971a) A quantitative approach to parasitism. *Parasitology* 62, 179–193.

Crofton, H.D. (1971b) A model of host–parasite relationships. *Parasitology* 63, 343–364.

Cummings, L.J., Thompson, R.L., Yong, W.K., Riffkin, G.G., Goddard, M.E., Callinan, A.P.L. and Saunders, M.J. (1991) Genetics of *Ostertagia* selection lines. In: Gray, G.D. and Woolaston, R.R. (eds) *Breeding for Disease Resistance in Sheep*. Australian Wool Corporation, Melbourne, pp. 11–18.

Dawkins, H.J.S., Windon, R.G. and Eagleson, G.K. (1989) Eosinophil responses in sheep selected for high and low responsiveness to *Trichostrongylus colubriformis*. *International Journal of Parasitology* 19, 199–205.

DeGotari, M.J., Freking, B.A., Cuthbertson, R.P., Kappes, S.M., Keele, J.W., Stone, R.T., Leymaster, K.A., Dodds, K.G., Crawford, A.M. and Beattie, C.W. (1998) A second-generation linkage map of the sheep genome. *Mammalian Genome* 9, 204–209.

Douch, P.G.C., Green, R.S., Morris, C.A., Bisset, S.A., Vlassof, A., Baker, R.L., Watson, T.G., Hurford, A.P. and Wheeler, M. (1995) Genetic and phenotypic relationships among anti-*Trichostrongylous colubriformis* antibody level, faecal egg count and body weight traits in grazing Romney sheep. *Livestock Production Science* 41, 121–132.

Douch, P.G., Green, R.S., Morris, C.A., McEwan, J.C. and Windon, R.G. (1996) Phenotypic markers for selection of nematode-resistant sheep. *International Journal of Parasitology* 26, 899–911.

Eady, S.J. (1998) Worm resistance in a merino breeding objective – influence of genetic correlations between resistance and production traits. *Proceedings of the 6th World Congress on Genetics Applied to Livestock Production* 27, 141–144.

Eady, S.J., Woolaston, R.R. and Mortimer, S.I. (1994) Internal parasite resistance of Merino flocks selected for production. *Proceedings of the 5th World Congress on Genetics Applied to Livestock Production, Guelph, Ontario*, pp. 289–292.

Gamble, H.R. and Zajac, A.M. (1992) Resistance of St. Croix lambs to *Haemonchus contortus* in experimentally and naturally acquired infections. *Veterinary Parasitology* 41, 211–225.

Gasbarre, L.C. (1994) *Ostertagia ostertagi*: changes in lymphoid populations in the local lymphoid tissues after primary or secondary infection. *Veterinary Parasitology* 55, 105–114.

Gasbarre, L.C. (1997a) Effects of gastrointestinal nematode infection on the ruminant immune system. *Veterinary Parasitology* 72, 327–343.

Gasbarre, L.C. (1997b) Do fecal egg counts accurately reflect worm numbers in cattle? *Veterinary Proceedings* 11, 900–903.

Gasbarre, L.C. (1997c) How do you determine which nematode parasites are infecting cattle? *Veterinary Proceedings* 11, 904–906.

Gasbarre, L.C., Leighton, E.A. and Davies, C. J. (1990) Genetic control of immunity to gastrointestinal nematodes of cattle. *Veterinary Parasitology* 37, 257–272.

Gasbarre, L.C., Leighton, E.A. and Davies, C.J. (1993) Influence of host genetics upon antibody responses against gastrointestinal nematode infections in cattle. *Veterinary Parasitology* 46, 81–91.

Gasbarre, L.C., Leighton, E.A. and Bryant, D. (1995) Genetically controlled resistance to gastrointestinal nematode infections of cattle. *Proceedings of the Beef Improvement Federation* 212–217.

Gasbarre, L.C., Leighton, E.A. and Bryant, D. (1996) Reliability of a single fecal egg per gram determination as a measure of individual and herd values for trichostrongyle nematodes of cattle. *American Journal of Veterinary Research* 57, 168–171.

Gatenby, R.M., Wilson, A.J., Romjali, E., Pandey, V.S., Batubara, L.P. and Bradford, G.E. (1995) Helminth infections of sheep in rubber plantations in Sumatra. In: Gray, G.D., Woolaston, R.R. and Eaton, B.T. (eds) *Breeding for Resistance to Infectious Diseases in Small Ruminants.* Australian Centre for International Agricultural Research, Canberra, pp. 165–171.

Geczy, A.F. and Rothwell, T.L.W. (1981) Genes within the MHC of the guinea pig influence susceptibility to *Trichostrongylus colubriformis* infection. *Parasitology* 82, 281–286.

Genchi, C., Madonna, M. and Traldi, G. (1989) Epidemiology of *Ostertagia ostertagi* in dairy cow from different breeding systems. *Parasitologia* 31, 123–132.

Gill, H.S. (1991) Genetic control of acquired resistance to haemonchosis in Merino lambs. *Parasite Immunology* 13, 617–628.

Gill, H.S. (1994) Cell-mediated immunity in Merino lambs with genetic resistance to *Haemonchus contortus. International Journal of Parasitology* 24, 749–756.

Gill, H.S., Watson, D.L. and Brandon, M.R. (1993) Monoclonal antibody to $CD4^+$ T cells abrogates resistance to *Haemonchus contortus* in sheep. *Immunology* 78, 43–49.

Grain, F., Nain, M.-C., Labonne, M.-P., Lantier, F., Lachopier, P., Gebuhrer, L., Asso, J., Maddox, J. and Betuel, H. (1993) Restriction fragment length polymorphism of DQB and DRB class II genes of the ovine major histocompatibility complex. *Animal Genetics* 24, 377–384.

Gray, G.D. (1991) Breeding for resistance to trichostrongyle nematodes in sheep. In: Axford, R.F.E. and Owen, J.B. (eds) *Breeding for Disease Resistance in Farm Animals.* CAB International, Wallingford, pp. 100–125.

Gray, G.D. and Gill, H.S. (1993) Host genes, parasites and parasitic infections. *International Journal of Parasitology* 23, 485–494.

Gray, G.D., Barger, I.A., Le Jambre, L.F. and Douch, P.G.C. (1992) Parasitological and immunological responses of genetically resistant Merino sheep on pastures contaminated with parasitic nematodes. *International Journal of Parasitology* 22, 417–425.

Gronvold, J., Nansen, P., Gasbarre, L.C., Christensen, C.M., Larsen, M., Monrad, J. and Midtgaard, N. (1992) Development of immunity to *Ostertagia ostertagi* (Trichostrongylidae: Nematoda) in pastured young cattle. *Acta Veterinaria Scandinavica* 33, 305–316.

Gruner, L. and Cabaret, J. (1988) Resistance of sheep and goats to helminth infections: a genetic basis. In: Thomson, E.F. and Thomson, F.S. (eds) *Increasing Small Ruminant Productivity in Semi-arid Areas.* ICARDA, pp. 257–265.

Gruner, L., Cabaret, J., Sauve, C. and Pailhories, R. (1986) Comparative susceptibility of Romanov and Lacaune sheep to GI nematodes and small lungworms. *Veterinary Parasitology* 19, 85–93.

Herd, R.P., Strong, L. and Wardhaugh, K. (eds) (1993) Environmental impact of Avermectin usage in livestock. *Veterinary Parasitology* 48, p. 343.

Hosking, B.C., Watson, T.G. and Leathwick, D.M. (1996) Multigeneric resistance to oxfendazole by nematodes in cattle. *Veterinary Record* 138, 67–68.

Hulme, D.J., Windon, R.G., Nicholas, F.W. and Beh, K.J. (1991) Association between MHC Class II RFLP and *Trichostrongylus* resistance. In: Gray, G.D. and Woolaston, R.R. (eds) *Breeding for Disease Resistance in Sheep.* Australian Wool Corporation, Melbourne, pp. 115–120.

Kassai, T., Fesus, L., Hendrikx, W.M.L., Takats, C., Fok, E., Redl, P., Takacs, E., Nilsson, R., Van Leeuwen, M.A.W., Jansen, J., Bernadina, W.E. and Frankena, K. (1990) Is there a relationship between haemoglobin genotype and the innate resistance to experimental *Haemonchus contortus* infection in Merino lambs? *Veterinary Parasitology* 37, 61–77.

Kloosterman, A., Albers, G.A.A. and Van Den Brink, R. (1978) Genetic variation among calves in resistance to nematode parasites. *Veterinary Parasitology* 4, 353–368.

Kloosterman, A., Parmentier, H.K. and Ploeger, H.W. (1992) Breeding cattle and sheep for resistance to gastrointestinal nematodes. *Parasitology Today* 8, 330–335.

Knight, R.A. and Rodgers, D. (1974) Age resistance of lambs to single inoculation with *Haemonchus contortus. Proceedings of the Helminthological Society of Washington* 41, 116.

Leighton, E.A., Murrell, K.D. and Gasbarre, L.C. (1989) Evidence for genetic control of nematode egg-shedding rates in calves. *Journal of Parasitology* 75, 498–504.

Lunney, J.K. and Murrell, K.D. (1988) Immunogenetic analysis of *Trichinella spiralis* infections in swine. *Veterinary Parasitology* 29, 179–193.

McEwan, J.C., Mason, P., Baker, R.L., Clarke, J.N., Hickey, S.M. and Turner, K. (1992) Effect of selection for productive traits on internal parasite resistance in sheep. *Proceedings of the New Zealand Society of Animal Production* 52, 53–56.

McEwan, J.C., Dodds, K.G., Greer, G.J., Bain, W.E., Duncan, S.J., Wheeler, R., Knowler, K.J., Reid, P.J., Green, R.S. and Douch, P.G.C. (1995) Genetic estimates for parasite resistance traits in sheep and their correlations with production traits. *New Zealand Journal of Zoology* 22, 177.

Mackinnon, M.J., Meyer, K. and Hetzel, D.J.S. (1991) Genetic variation and covariation for growth, parasite resistance and heat tolerance in tropical cattle. *Livestock Production Science* 27, 105–122.

Madden, K.B., Murrell, K.D. and Lunney, J.K. (1990) *Trichinella spiralis*: major histocompatibility complex-associated elimination of encysted muscle larvae in swine. *Experimental Parasitology* 70, 443–451.

Miller, J.E., Cockett, N.E., Baker, R.L. and Stear, M.J. (1995) Susceptibility to nematode infection and genetic variation in the MHC class II region between Dorper and Red Maasai sheep from Kenya. *Journal of Animal Science* 73 (Suppl. 1), 3.

Miller, J.E., Gill, H.S. and Barras, S.R. (1996) Effect of monoclonal antibdy to $CD4^+$ T cells on resistance to *Haemonchus contortus* in Gulf Coast Native lambs. *Animal Genetics* 27 (Suppl. 2), 54.

Miller, J.E., Bahirathan, M., Lemarie, S.L., Hembry, F.G., Kearney, M.T. and Barras, S.R. (1998) Epidemiology of gastrointestinal nematode parasitism in Suffolk and Gulf

Coats Native sheep with special emphasis on relative susceptibility to *Haemonchus contortus* infection. *Veterinary Parasitology* 74, 55–74.

Morris, C.A. (1998) Responses to selection for disease resistance in sheep and cattle in New Zealand and Australia. *Proceedings of the 6th World Congress on Genetics Applied to Livestock Production* 27, 295–302.

Morris, C.A., Watson, T.G., Bisset, S.A., Vlassoff, A. and Douch, P.G.C. (1995) Breeding sheep in New Zealand for resistance or resilience to nematode parasites. In: Gray, G.D., Woolaston, R.R. and Eaton, B.T. (eds) *Breeding for Resistance to Infectious Diseases in Small Ruminants.* Australian Centre for International Agricultural Research, Canberra, pp. 77–98.

Morris, C.A., Vlassof, A., Bisset, S.A., Baker, R.L. and Watson, T.G. (1997a) Direct responses to selection for divergence in faecal nematode egg count in young Romney and Perendale sheep. *Proceedings of the Association for the Advancement of Animal Breeding and Genetics* 12, 413–416.

Morris, C.A., Vlassof, A., Bisset, S.A., Baker, R.L., West, C.J. and Hurford, A.P. (1997b) Responses of Romney sheep to selection for resistance or susceptibility to nematode infection. *Animal Science* 64, 319–329.

Muggli-Cockett, N.E. and Stone, R.T. (1991) Restriction fragment length polymorphisms in bovine major histocompatibility complex class II beta-chain genes using bovine exon-containing hybridization probes. *Animal Genetics* 22, 123–136.

Outteridge, P.M., Windon, R.G. and Dineen, J.K. (1985) An association between a lymphocyte antigen in sheep and the response to vaccination against the parasite *Trichostrongylus colubriformis. International Journal of Parasitology* 15, 121–127.

Outteridge, P.M., Windon, R.G., Dineen, J.K. and Smith, E.F. (1986) The relationship between ovine lymphocyte antigens and faecal egg count of sheep selected for responsiveness to vaccination against *Trichostrongylus colubriformis. International Journal of Parasitology* 16, 369–374.

Outteridge, P.M., Windon, R.G. and Dineen, J.K. (1988) An ovine lymphocyte antigen marker for acquired resistance to *Trichostrongylus colubriformis. International Journal of Parasitology* 18, 853–858.

Pinheiro, A.C. and Echevarria, F.A.M. (1990) Sensitivity of *Haemonchus* spp. in cattle to anthelmintic treatment with albendazole and oxfendazole. *Pesquisa Veterinaria Brasileira* 10, 19–21.

Piper, L.R. and Barger, I.A. (1988) Resistance to gastrointestinal strongyles: feasibility of a breeding programme. *Proceedings of the 3rd World Congress of Sheep and Cattle Breeding*, Vol. 1, pp. 593–611.

Presson, B.L., Gray, G.D. and Burgess, S.K. (1988) The effect of immunosuppression with dexamethasone on *Haemonchus contortus* infections in genetically resistant Merino sheep. *Parasite Immunology* 10, 675–680.

Preston, J.M. and Allonby, E.W. (1978) The influence of breed in the susceptibility of sheep and goats to a single infection with *Haemonchus contortus. Veterinary Record* 103, 509–512.

Preston, J.M. and Allonby, E.W. (1979a) The influence of breed on the susceptibility of sheep to *Haemonchus contortus* infection in Kenya. *Research in Veterinary Science* 26, 134–139.

Preston, J.M. and Allonby, E.W. (1979b) The influence of haemoglobin phenotype on the susceptibility of sheep to *Haemonchus contortus* infection in Kenya. *Research in Veterinary Science* 26, 140–144.

Prichard, R.K. (1990) Anthelmintic resistance in nematodes: extent, recent understanding and future directions for control and research. *International Journal of Parasitology* 20, 515–523.

Raadsma, H.W., Gray, G.D. and Woolaston, R.R. (1997a) Genetics of disease resistance and vaccine response. In: Piper, L. and Ruvinsky, A. (eds) *The Genetics of Sheep.* CAB International, Wallingford, UK, pp. 199–224.

Raadsma, H.W., Nicholas, F.W. and Egerton, J.R. (1997b) Ultimate disease resistance in sheep: what are the major relationships between all major diseases? *Proceedings of the Association for the Advancement of Animal Breeding and Genetics* 12, 63–67.

Radhakrishnan, C.V., Bradley, R.E. and Loggins, P.E. (1972) Host responses of worm-free Florida Native and Rambouillet lambs experimentally infected with *Haemonchus contortus. American Journal of Veterinary Research* 33, 817–823.

Riffkin, G.G. and Yong, W.K. (1984) Recognition of sheep which have innate resistance to trichostrongylid nematode parasites. In: Dineen, J.K. and Otteridge, P.M. (eds) *Immunologic Approaches to the Control of Endoparasites.* CSIRO, Melbourne, pp. 30–40.

Romjali, E., Pandey, V.S., Gatenby, R.M., Doloksaribu, M., Sakul, H., Wilson, A. and Verhulst, A. (1997) Genetic resistance of different genotypes of sheep to natural infections with gastrointestinal nematodes. *Animal Science* 64, 97–104.

Schmidt, E.E., Suarez, V.H. and Cabaret, J. (1998) Nematode genera diversity in cattle: similarity of between-sire progenies. *Veterinary Research* 29, 139–148.

Schwaiger, F.W., Gostomski, D., Stear, M.J., Duncan, J.L., McKellar, Q.A., Epplen, J.T. and Buitkamp, J. (1995) An ovine major histocompatibility complex *DRB1* allele is associated with low faecal egg counts following natural, predominantly *Ostertagia circumcincta* infection. *International Journal of Parasitology* 25, 815–822.

Smith, G., Grenfell, B.T. and Anderson, R.M. (1987) The regulation of *Ostertagia ostertagi* populations in calves: density-dependent control of fecundity. *Parasitology* 95, 373–388.

Stear, M.J. and Wakelin, D. (1998) Genetic resistance to parasitic infection. *Revues Scientifiques et Technicales de l'Office Internationale de Epizooties* 17, 143–153.

Stear, M.J., Nicholas, F.W., Brown, S.C., Tierney, T. and Rudder, T. (1984) The relationship between the bovine major histocompatibility system and faecal worm egg counts. In: Dineen, J.K. and Outteridge, P.M. (eds) *Immunogenetic Approaches to the Control of Endoparasites with Particular Reference to Parasites of Sheep.* CSIRO, Australia, pp. 126–133.

Stear, M.J., Tierney, T.J., Baldock, F.C., Brown, S.C., Nicholas, F.W. and Rudder, T.H. (1988) Class I antigens of the bovine major histocompatibility system are weakly associated with variation in faecal worm egg counts in naturally infected cattle. *Animal Genetics* 19, 115–122.

Stear, M.J., Mallard, B.A., Newman, M.J. and Wilkie, B.N. (1989) The current status of major histocompatibility system definition in cattle, goats, horses, pigs, and sheep. *International Journal of Animal Science* 4, 32–44.

Stear, M.J., Baldock, F.C., Brown, S.C., Gershwin, L.J., Hetzel, D.J.S., Miller, J.E., Nicholas, R.W., Rudder, T.H. and Tierney, T.J. (1990a) The genetic control of nematode infections in ruminants. *Proceedings of the 4th World Congress on Genetics Applied to Livestock Production, Edinburgh, Scotland,* pp. 449–452.

Stear, M.J., Hetzel, D.J.S., Brown, S.C., Gershwin, L.J., Mackinnon, M.J. and Nicholas, F.W. (1990b) The relationships among ecto- and endoparasite levels, class I antigens of the bovine major histocompatibility system, immunoglobulin E levels and weight gain. *Veterinary Parasitology* 34, 303–321.

Stear, M.J., Bishop, S.C., Doligalska, M., Duncan, J.L., Holmes, P.H., Irvine, J., McCririe, L., McKellor, Q.A., Sinski, E. and Murray, M. (1995a) Regulation of egg production, worm burden, worm length and worm fecundity by host responses in sheep infected with *Ostertagia circumcincta. Parasite Immunology* 17, 643–652.

Stear, M.J., Bishop, S.C., Duncan, J.L., McKellar, Q.A. and Murray, M. (1995b) The repeatability of faecal egg counts, peripheral eosinophil counts, and plasma pepsinogen concentrations during deliberate infections with *Ostertagia circumcincta*. *International Journal of Parasitology* 25, 375–380.

Stear, M.J., Bishop, S.C., Bairden, K., Duncan, J.L., Gettinby, G., Holmes, P.H., McKellar, Q.A., Park, M., Strain, S. and Murray, M. (1997) The heritability of worm burden and worm fecundity in lambs following natural nematode infection. *Nature* 389, 27.

Suarez, V.H., Ciminari, O.E., Bedotti, D.O., Busetti, M.R. and Bello, E.M. (1990) Epidemiology, effects and control of nematode infections on Zebu crossbred, Hereford and Hereford × Brahman calves of Argentina's Western Pampas. *Veterinary Parasitology* 35, 79–91.

Suarez, V.H., Lorenzo, R.M., Babinec, F.J. and Schmidt, E.E. (1997) Variabilidad genética en el conteo de huevos de nematodes gastrointestinales en terneros Aberdeen Angus. *Revista de Medicina Veterinaria* 76, 142–146.

Teale, A.J. (ed.) (1991) *Bovine Genome Mapping and Trypanotolerance: Proceedings of a Workshop held at ILRAD, Nairobi, Kenya*, pp. 5–7.

Urquhart, G.M., Jarrett, W.F.H., Jennings, F.W., McIntyre, W.I.M. and Mulligan, W. (1966) Immunity to *Haemonchus contortus* infection: relationship between age and successful vaccination in irradiated larvae. *American Journal of Veterinary Research* 27, 1645–1648.

Vermunt, J.J., West, D.M. and Pomeroy, W.E. (1995) Multiple resistance to ivermectin and oxfendazole in *Cooperia* species in cattle in New Zealand. *Veterinary Record* 137, 43–45.

Wakelin, D. (1978) Genetic control of susceptibility and resistance to parasite infections. *Advances in Parasitology* 16, 219–308.

Wakelin, D. and Donachie, A.M. (1983) Genetic control of immunity to *Trichinella spiralis*: influence of H-2 linked genes on immunity to the intestinal phase of infection. *Immunology* 48, 343–350.

Waller, P.J. (1994) The development of anthelmintic resistance in ruminant livestock. *Acta Tropica* 56, 233–243.

Waller, P.J. and Prichard, R.K. (1986) Drug resistance in nematodes. In: Campbell, W.C. and Rew, R.S. (eds) *Chemotherapy of Parasitic Diseases*. Plenum Press, New York, pp. 339–362.

Wassom, D.L. and Kelly, E.A.B. (1990) The role of the major histocompatibility complex in resistance to parasite infections. *Critical Reviews in Immunology* 10, 31–52.

Wassom, D.L., David, C.S. and Gleich, G.J. (1979) Genes within the MHC influence susceptibility to *Trichinella spiralis* in the mouse. *Immunogenetics* 9, 491–496.

Wetherall, J.D. and Groth, D.M. (1992) The major histocompatibility complex and parasite immunity. In: Yong, W.K. (ed.) *Animal Parasite Control Utilizing Biotechnology*. CRC Press, Boca Raton, Florida, pp. 353–386.

Windon, R.G. (1990) Selective breeding for the control of nematodiasis in sheep. *Revues Scientifiques et Technicales de l'Office Internationale de Epizooties* 2, 555–576.

Windon, R.G. (1991) Resistance mechanisms in the *Trichostrongylus* selection flocks. In: Gray, G.D. and Woolaston, R.R. (eds) *Breeding for Disease Resistance in Sheep*. Australian Wool Corporation, Melbourne, pp. 77–86.

Windon, R.G. and Dineen, J.K. (1984) Parasitological and immunological competence of lambs selected for high and low responsiveness to vaccination with irradiated *Trichostrongylus colubriformis* larvae. In: Dineen, J.K. and Outteridge, P.M. (eds) *Immunogenetic Approaches to the Control of Endoparasites*. CSIRO, Melbourne, pp. 13–28.

Windon, R.G., Dineen, J.K. and Wagland, B.M. (1987) Genetic control of immunological responsiveness against the intestinal nematode *Trichostrongylus colubriformis* in lambs. In: McGuirk, B.J. (ed.) *Merino Improvement Programs in Australia.* Australian Wool Corporation, Melbourne, pp. 371–375.

Woolaston, R.R. (1990) Genetic improvement of resistance to internal parasites in sheep. *Proceedings of the Australian Association of Animal Breeding and Genetics* 8, 163–171.

Woolaston, R.R. and Eady, S.J. (1995) Australian research on genetic resistance to nematode parasites. In: Gray, G.D., Woolaston, R.R. and Eaton, B.T. (eds) *Breeding for Resistance to Infectious Diseases in Small Ruminants.* Australian Centre for International Agricultural Research, Canberra, pp. 53–57.

Woolaston, R.R. and Piper, L.R. (1996) Selection of merino sheep for resistance to *Haemonchus contortus*: genetic variation. *Animal Science* 62, 451–460.

Woolaston, R.R., Barger, I.A. and Piper, L.R. (1990) Response to helminth infection of sheep selected for resistance to *Haemonchus contortus. International Journal of Parasitology* 20, 1015–1018.

Woolaston, R.R., Windon, R.G. and Gray, G.D. (1991) Genetic variation in resistance to internal parasites in Armidale experimental flocks. In: Gray, G.D. and Woolaston, R.R. (eds) *Breeding for Disease Resistance in Sheep.* Australian Wool Corporation, Melbourne, pp. 1–9.

Woolaston, R.R., Manueli, P., Singh, R., Tabunakawai, N. and Le Jambre, L.F. (1995) Breeding to assist control of gastrointestinal parasites of small ruminants in the Pacific Islands. In: Gray, G.D., Woolaston, R.R. and Eaton, B.T. (eds) *Breeding for Resistance to Infectious Diseases in Small Ruminants.* Australian Centre for International Agricultural Research, Canberra, pp. 179–186.

Woolaston, R.R., Manueli P., Eady, S.J., Barger, I.A., Le Jambre, L.F., Banks, D.J.D. and Windon, R.G. (1996) The value of circulating eosinophil count as a selection criteria for resistance of sheep to trichostrongyle parasites. *International Journal of Parasitology* 26, 123–126.

Zajac, A.M. (1995) Genetic resistance to infectious disease in small ruminants: North America and the Carribbean. In: Gray, G.D., Woolaston, R.R. and Eaton, B.T. (eds) *Breeding for Resistance to Infectious Diseases in Small Ruminants.* Australian Centre for International Agricultural Research, Canberra, pp. 153–166.

7

Host Resistance to Ticks and Tick-borne Diseases: Its Role in Integrated Control

B. Minjauw[1] and J.J. de Castro[2]
[1]*International Livestock Research Institute (ILRI), Nairobi, Kenya;*
[2]*FAO Sub Regional Office, Harare, Zimbabwe*

Summary

Host resistance to ticks and tick-borne diseases has been studied increasingly during the past decade and practical applications of the use of host resistance in integrated control strategies are starting to emerge. Different defence mechanisms, including tick avoidance, grooming, skin characteristics and more specific immunological responses, are involved in reducing the numbers of ticks parasitizing cattle. Host resistance inherent in the *Bos indicus* cattle breeds has been used successfully in Australia for the control of *Boophilus microplus*. Culling of heavily infested animals is practised and cattle have been selected and crossed to create new tick-resistant breeds.

In Africa, where several important tick species occur, trials to examine tick resistance in different breeds of cattle and to assess cross-resistance between different tick species are taking place. The results of these trials are important to identify the better-adapted indigenous breeds. At the same time, work is in progress to simplify the assessment of tick burdens on cattle. Preliminary results from Australia indicate the feasibility of selecting cattle for the presence of specific genes responsible for tick resistance.

There is a lack of strong evidence of within breed genetic resistance against tick-borne infections themselves, although there is an increasing number of reports indicating variation of susceptibility among breeds. This requires further investigation if breeding programmes are to be started. The availability of field immunization methods and the prospect of novel subunit vaccines against the major tick-borne diseases, together with a better understanding of the effects of ticks on cattle, have opened the way to improved and more integrated tick control management in Africa. Cattle breeds with resistance qualities exist, the majority of which are zebu genotype. However, the need to increase production efficiency of livestock has resulted in many African countries introducing exotic breeds (of low tick resistance), while at the same time spending large sums of money on tick control. There is a need to plan cattle selection to avoid losing host resistance and to evaluate the productivity and tick and tick-borne disease resistance of the indigenous breeds. The use of better-adapted breeds should reduce the frequency of acaricide application and postpone the development of resistance to acaricides.

© CAB *International* 2000. *Breeding for Disease Resistance in Farm Animals*
(eds R.F.E. Axford, S.C. Bishop, F.W. Nicholas and J.B. Owen)

Introduction

Ticks and tick-borne diseases (TBDs) affect 80% of the world's cattle population and are widely distributed throughout the world, particularly in tropical and subtropical countries (de Castro, 1997). Ticks are responsible for various losses, caused either by the direct effect of attachment (tick-worry), by the injection of toxins, or through morbidity and mortality caused by the diseases transmitted. In a recent review, de Castro (1997) estimated the global costs of ticks and TBD in cattle to be between US$13.9 and US$18.7 billion annually.

Although alternative non-chemical tick-control methods, such as pasture spelling, predators and parasites, sterile male release, tick-resistant cattle and vaccination with tick antigens, exist (Cunningham, 1981), many of these are either at an experimental stage or have shown to be of inadequate efficacy, and tick-control measures are still based mainly on the use of acaricides.

Some breeds of cattle are known to have enhanced tick resistance. Tick resistance in cattle is chiefly associated with zebu (*Bos indicus*) animals, although it is also present in some taurine (*B. taurus*) breeds. It is a widespread tick-control method, present throughout the world and particularly effective in areas where zebu animals are present, such as in pastoralist and smallholder cattle systems in Africa and Australia.

Host resistance to ticks is present in other domestic, wild and laboratory animals, and the reviews by Willadsen (1980), Brown (1985) and, more recently, by Rechav (1992), de Castro and Newson (1993) and Wikel (1996) cover important aspects of this topic.

Host resistance to TBD is less well documented and is not adequately determined. Because TBDs are tick-transmitted, the differences in resistance to the diseases may result in part from differences in tick resistance.

This chapter will review the current knowledge of host resistance to ticks and to TBD and its potential use in integrated control strategies.

Expression of Host Resistance to Ticks

Tick avoidance, grooming behaviour and nature of the skin

The first means by which hosts can express resistance to ticks is by the host being able to avoid contact with the ticks. This was demonstrated by Sutherst *et al.* (1986) for *Boophilus microplus* in Australia. Norval *et al.* (1988b) in Zimbabwe also observed avoidance behaviour with adult *Rhipicephalus appendiculatus*. Although this is commonly attributed to the sighting of the ticks by the animals, it can be also explained by a better knowledge of the indigenous cattle of the infected zone in the grazing area or by morphological differences in the host altering the chance of attachment to the ticks (Meltzer, 1996).

Grooming is another means by which the host can express resistance. Immunological mediators induced by tick antigens introduced into host skin contribute to the itch sensation, which stimulates grooming (Alexander, 1996).

Grooming has been reported, for the case of *B. microplus* infestation, as a response to the release of histamine and other mediators involved in the hypersensitivity response. However, the dislodging of ticks, particularly in the case of two- and three-host ticks by grooming, may simply be a response to the irritation of ticks walking on the skin while seeking their predilection sites before attachment.

Thus, host grooming is an important factor in reducing tick burden, and several workers have studied it. Snowball (1956) and Bennett (1969) carried out experiments to evaluate the importance of grooming. They compared groups of cattle that could groom freely with others fitted with antigrooming devices. Significantly fewer ticks were found on those animals that were able to groom. However, tick numbers decreased over time on animals unable to groom (Bennett, 1969) and the proportion of larvae lost (18–39%) could not be accounted for by grooming alone (Koudstaal *et al.*, 1978). Tatchell and Bennett (1969) demonstrated the role of histamine in host grooming, and further studies (Schleger *et al.*, 1981b) have shown that histamine, liberated by mast-cell degranulation, caused by inflicting cutaneous pain, is important in the initiation of the grooming process.

Under natural tick challenge in the field, tick-infested cattle have been reported to spend more time grooming than tick-free animals, at the expense of other normal behavioural events, suggesting that grooming also plays an important role in reducing the numbers of multihost ticks (de Castro *et al.*, 1985b). Grooming was also important in reducing artificial infestations with *R. appendiculatus* (de Castro *et al.*, 1985a) and Norval *et al.* (1988a) regarded it as the main factor in the reduction in the number of *Amblyomma hebraeum* in cattle.

Skin thickness appears to play an important role on host resistance to ticks. However, there are still contradictory opinions as to whether thick (Bonsma, 1944; Ali and de Castro, 1993) or thin skins (Brown, 1985) are associated with resistance. However, these different observations have been made on different cattle breeds in different locations and other skin characteristics such as coat type, hair density and skin secretions may have influenced the results.

Only a small proportion of ticks are not avoided or removed by grooming by the animals, and go on successfully to attach, becoming the target of other more specific mechanisms of the host's immune system.

Immunological responses to tick feeding

Tick feeding induces a range of immunological mechanisms involving antibodies, complement, cytokines, antigen-presenting cells and T cells (Wikel, 1982). The primary response to the introduction of tick saliva into the skin of the host consists of degranulation of mast cells, which induces the accumulation of basophils and eosinophils (de Castro and Newson, 1993). The onset of tick feeding introduces salivary immunogens to cells in the epidermis and dermis at the attachment site. Proteins and other immunogenic molecules in tick saliva are presented to antigen-specific T cells in the epidermis (Langerhans cells), dermis (macrophages), and draining lymph nodes (macrophages, dendritic cells). Immunogens, antigen-presenting cells, T cells and cytokines contribute to the activation and differentiation of B cells, which produce tick-reactive circulating and homocytotrophic antibodies (Wikel, 1996). Antibody responses to tick infestation in cattle have been described by Brossard (1976), who showed a significant increase in serum gamma-globulin concentration following infestation with *B. microplus*.

Repeated tick infestation exposes the ticks to a more vigorous immune response due to primary introduction of saliva that stimulates generation of T and B cells. This acquired resistance is also expressed by the presence of mast cells in the dermis and basophils presenting surface homocytotrophic antibodies capable of reacting with introduced immunogens from the saliva. Schleger *et al.* (1976) studied the cellular responses of cattle with different levels of resistance to *B. microplus*. Highly resistant animals showed mast-cell degranulation and eosinophil accumulation in the lesion. The level of protective resistance to ticks was related to the degree of hyperaemia which followed larval infestation (Hales *et al.*, 1981), to the number of arteriovenous anastomoses in the hair-follicular layer above the sebaceous-gland level of cattle skin (Schleger *et al.*, 1981a), and to the concentration and degranulation of eosinophils at the attachment site of the larvae (Schleger *et al.*, 1981b).

The level of expression of immunity depends on the host and tick species concerned. The effects vary and range from simple rejection of the tick, reduced engorged weight of all instars, reduced number or viability of eggs, to death of the tick on the host (Willadsen, 1980).

Breeding for Host Resistance to Ticks

Individual resistance assessment and resistance markers

Cattle resistance to the one-host tick *B. microplus* has been measured by either counting and evaluating the level of engorgement of the ticks after a known larval challenge or by ranking the animals while undergoing presumed equal natural tick-challenge (Sutherst *et al.*, 1979).

Scoring of cattle resistance to the three-host tick *R. appendiculatus* was described by de Castro *et al.* (1985a) and this was the technique used by Latif (1984), Norval *et al.* (1988b) and de Castro *et al.* (1989). Ranking of cattle for two- and three-host tick infestations is also possible by assessing the number of adult ticks attached to the cattle (Kaiser *et al.*, 1982).

Since these techniques are often time consuming, the use of simpler tests has been proposed (Hewetson, 1978). Visual assessments of tick numbers have been used in Africa (de Castro, 1991). In this method, three independent observers assessed tick loads on cattle into five classes, namely very high, high, medium, low and very low tick burdens. When the results were compared with the measured tick counts on the animals, they were found to be highly correlated (0.80–0.93). Also, Scholtz *et al.* (1989) defined a simplified method for the assessment of tick resistance in cattle by counting all ticks under the tail and on the scrotum or udder only. This method gave a very good correlation (0.85) with the entire body count and seems to be adequate for practical application and for screening large number of animals.

No easily recognizable host characteristics have yet been found associated with host resistance to ticks but cattle with more arteriovenous anastomoses in the skin showed higher resistance to *B. microplus* (Schleger *et al.*, 1981a). This observation may help to explain the positive relationship between hyperaemia and host resistance to *B. microplus* reported by Hales *et al.* (1981). Thick hide and high vascularity was one of the criteria adopted in the selection of cattle for tick resistance (Bonsma, 1944). This author stressed that ticks avoided animals with sleek coats and high sebum secretion as well as areas of the body where subcutaneous muscles were present. Sleek coats were later associated with animals in good condition and hence cattle of unimpaired tick resistance due to stress (Turner and Schegler, 1960). Not only was the appearance of the coat related to resistance but rectal and skin temperatures were also shown to be lower in animals that showed the highest tick resistance (O'Kelly and Spiers, 1983).

A relationship between phenotypes of the bovine major histocompatibility system and resistance to *B. microplus* has been described (Stear *et al.*, 1984). However, although significant, its role in the resistance to *B. microplus* is small. Nokoe *et al.* (1993) developed a mathematical index to discriminate between susceptible and resistant cattle on the basis of the weights of engorged female and nymphal instars.

More recently, Frish (1994) identified the presence of a major gene for resistance to cattle ticks in the Belmont Adaptaur cattle (Hereford × Shorthorn). Segregation analyses of data, such as those performed by the FINDGENE program (Kerr *et al.*, 1994), could facilitate the screening of populations for suitable candidates to be used in test matings for quantitative trait loci (QTL) detection.

Breed characteristics

Several workers have compared *B. taurus* with *B. indicus* cattle with regard to their relative capacity to develop resistance to *B. microplus* (Table 7.1). Seifert (1971a) analysed 3000 tick counts made on over 1000 animals of several breeds and found no differences between Afrikander (*B. taurus* × *B. indicus*) and Brahman crosses, though these carried significantly fewer ticks than British taurine breeds studied (Hereford and Shorthorn). These results were supported by the findings of Utech *et al.* (1978) and Sutherst *et al.* (1988).

The phenomenon of varying resistance to tick infestation in different cattle breeds has been much less studied in parts of the world other than Australia (Sutherst *et al.*, 1988). Australian scientists have successfully exploited the differences in susceptibility to parasites by developing the Belmont Adaptaur breed and the Australian Friesian Sahiwal (Alexander *et al.*, 1984; Angus, 1998). Furthermore, the isolation of extremely resistant individual

Table 7.1. Cattle breed comparisons for resistance to tick infestations.

Resistant breed	Comparison breed	Trait	Type of infection	Tick species	Reference
AMZ[a]	Friesian	TC[b]	A[e]	B.micr[g]	Hewetson (1968)
Zebu crosses	Hereford ×	TC	NI[f]	B.micr	Wharton *et al.* (1970)
Zebu	Friesian	TC	N	R.app[h]	Latif *et al.* (1991)
Zebu	Friesian	TC	A	R.app	Nokoe *et al.* (1993)
Horro	Horro × Friesian	TC	N	A.coh[i]	Ali and de Castro (1993)
Gaudali	Wakwa	TC	N	Mixed	Tawah (1992)
N'Dama	Zebu N'dama × Zebu	TC, PCV[c]	N	A.var.[j]	Mattioli *et al.* (1993)
Sanga Nkoni	Friesian × Hereford	TC, LWG[d]	A	R.app.	Norval *et al.* (1988b)

[a]Australian Milking Zebu; [b]tick count; [c]packed cell volume; [d]liveweight gain; [e]artificial; [f]natural; [g]*Boophilus microplus*; [h]*Rhipicephalus appendiculatus*; [i]*Amblyomma cohaerens*; [j]*Amblyomma variegatum.*

animals in the Belmont Adaptaur (half Hereford, half Shorthorn) led to the demonstration of a putative major gene responsible for host resistance (Kerr *et al.*, 1994). This is an important development that undoubtedly will enhance the use of host resistance in the control of ticks.

In Africa also, many cases of breed resistance to ticks have been reported. Bonsma (1944) in South Africa demonstrated that Afrikander cattle were more resistant to *Amblyomma hebraeum* and heartwater than imported exotic breeds. Of the total number of ticks naturally infesting the cattle, the former carried a mean of 9.6% of the ticks compared with 90.4% for the latter. The tick resistance of Criollo cattle (*B. taurus*) has also been reported (Ulloa and de Alba, 1957) as well as differences observed between *B. indicus* and *B. taurus* and Kenana and Butana (*B. indicus*) breeds in the Sudan (Latif, 1984). In western Ethiopia, Ali and de Castro (1993) found Horro and Boran (*B. indicus*) to be more resistant than their crosses with Jersey, Simmental and Friesian (*B. taurus*). Horro cattle invariably carried fewer ticks of all species than the other breeds, including Boran. Jersey crosses showed relatively high levels of resistance and Friesian crosses very low resistance levels. Also in Ethiopia, Solomon and Kaaya (1996) found Arssi (*B. indicus*) and Boran to be more resistant than crossbreds and showed that cattle resistant to one species of tick were also resistant to other tick species. Other studies have shown that *B. indicus*, including Sanga cattle (*B. indicus* × *B. taurus*), carry significantly fewer ticks than exotic *B. taurus* cattle (Spickett *et al.*, 1989; Rechav *et al.*, 1990). Recently, Mattioli *et al.* (1993, 1995) and Mattioli and Cassama (1995) identified N'dama breeds (*B. taurus*) as more resistant breeds to *Amblyomma variegatum* and *Hyalomma* sp. than Gobra zebu (*B. indicus*).

In conclusion, even if there are good indications to believe that resistance seems to be directly related to the presence of zebu genes (Utech *et al.*, 1978), more studies involving a larger number of cattle with indication of previous history and early parasitic state of the parents, are needed to establish the cause of the differences and to show conclusively that some breeds have genetically better innate and/or acquired resistance against ticks (Willadsen, 1980).

Selecting for host resistance

Resistance to ticks is a heritable character and several workers have estimated the heritability for resistance to *B. microplus* (Table 7.2). Cattle have been selected for tick resistance and significant progress has been made in Australia with the development of breeds of cattle that are resistant to ticks, and at the same time, show good productivity (Frisch and Vercoe, 1978; Burrow *et al.*, 1991; Vercoe and Frisch, 1992).

Hewetson (1972) showed that the heritability of tick resistance ranged between 40 and 50%, and Utech *et al.* (1978) concluded that resistance in Brahman cattle appeared to be a dominant trait when interbreeds are observed. Over 17 years of selection, resistance increased from 89.2% to 99% in an Australian Illawarra Shorthorn herd. Concurrently, in this herd the resistance of the progeny increased from 93.7% to 97.7%, demonstrating that the selection and breeding of cows and bulls resulted in genetic improvement in the resistance of the progeny (Utech and Wharton, 1982). Tick-resistant breeds of cattle have now been developed in an effort to find animals that are productive (particularly for milk) under tick challenge and in a tropical environment (Reason, 1983; Alexander *et al.*, 1984). The development of these breeds started with the parallel development of cows with acceptable levels of milk production and bulls with high tick resistance and from there excellent

Table 7.2. Heritability ($h^2 \times 100$) estimates for resistance to *Boophilus microplus* in cattle.

References	Challenge	Breed of cattle	$h^2 \times 100$
Hewetson (1968)	Artificial	Australian Milking Zebu	28–42
Seifert (1971a) and	Artificial	AX	59
Davis (1993)		BX	10
Davis (1993), Seifert (1971b)		HS	10
Seifert (1971b)	Natural	F_2–F_3 European × Zebu	82
		HS	48
Madalena *et al.* (1985)	Natural	Holstein/Friesian	20
Wharton *et al.* (1970)	Natural	Australian Illawarra Shorthorn	48
Mackinnon *et al.* (1991)	Natural	AX	37
		AX BX	35
		AX, AXBX	34
Burns (1991)	Natural	AX	45

AX, Belmont Red: 50% Afrikander, 50% HS.
BX: 50% Brahman, 50% HS.
HS, Belmont Adaptaur: 50% Hereford, 50% Shorthorn.

dairy breeds, such as the Australian Friesian Sahiwal, have been created (Reason, 1983). Mackinnon (1990) also showed that tick resistance can be readily changed by selection in tropical beef breeds.

Despite the progress made in Australia in selecting for tick resistance, where they are concerned with only one tick species of economic importance in cattle, namely *B. microplus*, there is less enthusiasm from other parts of the world, especially Africa, where mixed species infestations are prevalent. The pioneering work of Bonsma (1944) set the criteria for breeding of 'tick-repellent' cattle, which also had good heat tolerance, thick hides, high vascularity and specific coat colours. He also recommended the culling of those cattle that are heavily infested, in agreement with Wilkinson (1955) and Kaiser *et al.* (1982), but this has not been widely applied in the field. Despite the importance of ticks to the African cattle industry, very little work followed on the selection of cattle and/or identification of adapted breeds for host resistance in Africa. Further investigations are needed to define whether cattle can develop resistance to several tick species and what level of tick burdens are necessary for the development and maintenance of the resistance (Kaiser *et al.*, 1982; Tatchell, 1986; de Castro *et al.*, 1991; Solomon and Kaaya, 1996).

Host Resistance to Tick-borne Disease Infections

Research in the past decade shows some evidence for host genetic resistance to TBD in cattle but it is often anecdotal and sometimes conflicting (Spooner and Brown, 1991). The differences in resistance to TBD are difficult to distinguish from resistance to ticks and therefore may just be a consequence of the number of ticks that the individual or group of animals is carrying, and the infection prevalence and intensity within ticks. The difficulties of showing evidence of resistance against TBD in controlled studies lie in the very complex relation between genetic susceptibility, immune responses and parasite diversity.

Genetic resistance can be innate and defined in terms of non-specific immunity to infection (Rumyantsev, 1998) or acquired and dependent on an adaptive immune response involving immunological mechanisms such as antibodies, T cells and immune-mediated inflammatory responses (Stear and Wakelin, 1998). The ability of an animal to respond to a given immunogen depends upon the genetic capacity to process immunogens and present them to immunocompetent T cells in the context of major histocompatibility complex (MHC) antigens (Wakelin and Blackwell, 1993). Genetic variation is expected in the aptitudes of different animals to develop and express resistance to tick feeding or to any infectious agent (Wikel *et al.*, 1994). A genetic component might be also responsible for the ability to control the severity of pathology. Some animals show the ability to control leucopenia and anaemia, and thus their white cell concentration and packed cell volume (PCV) are not affected by the infection (Spooner and Brown, 1991).

The magnitude and type of the immune responses define the balance between pathological change and protection. This fragile equilibrium is influenced directly by the immunological history of the animal, including previous exposure to other infections and also the overall condition, the age and nutritional status of the animal. The immune response is also influenced by interactions with the kinetics and the duration of the infection.

The host immunological reaction will depend on the antigenic characteristics of the parasite through its different stages (i.e. membrane antigens and antigen combinations). The parasite diversity (i.e. allelic polymorphisms, antigenic variation, sexual reproduction) and phenotypic characteristics (i.e. multiplication rate, specificity of cytoadherence, serotype) can further complicate the issue and may complicate any experiment trying to demonstrate genetic resistance against disease. Also, the severity of the infection depends on the quantity of infectious material injected and therefore the intensity and the duration of the challenge can be decisive factors influencing pathological change and protection.

Nevertheless, natural immunity or disease resistance have been described in different instances, as follows.

Age-related immunity

In the light of the limited amount of work that has been carried out on resistance to TBD and the fact that we are often dealing with extremely severe diseases, it is not often possible to clearly separate the effects of maternal immunity from those due to genetic resistance.

Available evidence indicates that young animals show reduced susceptibility to TBD when compared with adult animals undergoing a similar challenge (Koch *et al.*, 1990). Young ruminants are relatively resistant to babesiosis and anaplasmosis, and passive protection from the infected cow to the calf reinforces this natural immunity, which is most effective between 4 and 7 months of age. In endemic areas for babesiosis, young stock are protected for about 2 months by colostral antibodies and by limited innate resistance, and there is reverse age immunity in susceptible stock (Losos, 1986), whereby calves, for instance, can undergo mild infections with little or no mortality. However, in the case of East Coast fever, Cunningham *et al.* (1989) concluded that 1–2-month-old taurine calves from artificially immunized dams were not protected from experimental *Theileria parva* sporozoite challenge and that there was no evidence of inherent calfhood resistance to East Coast fever. In the case of heartwater the reverse age immunity is much shorter than with babesiosis and anaplasmosis, and young calves under 6 weeks of age, and lambs of less than 1 week, are fairly resistant and may recover spontaneously from natural and experimental infection (Losos, 1986).

A feature allowing some protection is the lower tick burden reported in calves, mainly due to behavioural and management practices, such as maternal grooming and tethering. With lower tick burdens resulting in low numbers of parasites injected, calves are more likely to acquire immunity without suffering acute clinical disease (Norval *et al.*, 1992).

The need for further study on age-related and maternal immunity has been recognized (Spooner and Brown, 1991; Norval *et al.*, 1992) but unfortunately not much work has taken place over the past 10 years.

Species resistance

Species resistance is best illustrated by the difference in susceptibility of wildlife and cattle to the common TBDs. The African buffalo, *Syncerus caffer*, is resistant to *Theileria lawrencei* infection while the water buffalo, *Bubalis bubalis*, is highly susceptible to *Theileria lawrencei* but resistant to *Theileria annulata* infection (Spooner and Brown, 1991). The major issue of resistance to infection by wildlife species resides in their ability to become carriers that tolerate the presence of the pathogen and thus become a source of infection to the very susceptible cattle population. The importance of wildlife as a carrier in the epidemiology of *Cowdria ruminantium* infection, responsible for heartwater, has been shown by Peter *et al.* (1998).

Breed resistance

There is less evidence of breed differences in resistance to TBD than there is for tick-resistant breeds. Nevertheless, there is evidence, often anecdotal, that there are breed differences in resistance, with local breeds offering more resistance than exotic breeds imported to the tropics (Spooner and Brown, 1991). Some evidence of varying resistance to tropical theileriosis and to East Coast fever in different cattle breeds has been reported (Paling *et al.*, 1991). Bonsma (1944) reported resistance to *A. hebraeum* and heartwater in South African cattle. Genetic resistance to heartwater in indigenous goats was reported in South Africa and in Guadeloupe (Matheron *et al.*, 1987). Less unambiguous evidence exists for breed differences in resistance to TBD as it is often indistinguishable from tick resistance or acquired resistance due to previous infection. Further investigation is required, as there would be real advantages if one could breed for resistance to TBD (Spooner and Brown, 1991). Different native breeds should be tested for their parasite resistance but also simultaneously for their productivity aptitude in the tropics (Frisch and Vercoe, 1978; Mackinnon, 1990; Burrow *et al.*, 1991). The N'Dama breed, known for its trypanotolerant ability, has been reported by Mattioli and Dempfle (1995) as having high levels of tick and tick-borne diseases resistance. This could be very promising and requires further investigation. Recently, Bock *et al.* (1997) found some indication that *B. indicus* is relatively more resistant against *Babesia bovis*, *B. bigemina* and *Anaplasma marginale* infections than their crosses with *B. taurus* or pure *B. taurus* breeds in Australia.

In conclusion, there is a lack of strong evidence for genetic variation in susceptibility, and further investigations are needed to be able to breed for TBD resistance. Thus, the future in controlling TBD may remain more in controlling the vector and the use of TBD vaccines. While a range of live vaccines exists for most of the TBDs, research is focusing on the production of new vaccines, which are safer and cheaper. The prospects for these subunit vaccines against TBD have been reviewed recently by Musoke *et al.* (1996).

Conclusions

As a result of many years of research, Australia has tackled the problem of acaricide resistance in *B. microplus* successfully by increasing the proportion of zebu blood in the cattle of northern Australia, and by regularly culling those animals with heavy tick burdens. Additionally, new breeds of highly productive, tick-resistant cattle, better adapted to the environment, have been developed, strengthening the role of host resistance in tick control even further and reducing the need for chemical methods (Alexander *et al.*, 1984; Mackinnon *et al.*, 1991). Further, the development of an anti-tick vaccine against *B. microplus*, based on the tick antigen Bm86 (Willadsen *et al.*, 1996), might also play an important role in integrated control strategies.

The situation in Africa with regard to ticks and cattle is very different. Although a number of crossbred animals exist, the great majority of cattle in Africa today are of *B. indicus* stock, kept under traditional management with low productivity (in some circumstances because of these management systems) but with high levels of host resistance to ticks. However, regular and intensive acaricide tick control is still practised, due to the implementation of schemes based on policies dictated many years ago and directed to the eradication of TBD.

The ever-increasing need for human food, particularly milk, due to population growth exerts great pressure on African governments to improve their livestock. This is achieved most commonly through breeding schemes, which use artificial insemination, crossing indigenous animals with European breeds. As an example, the Friesian is often selected, which rapidly improves milk production but with the unwanted consequence of lowering resistance to ticks and tick-borne diseases, and making acaricide use still more frequent and continuous. Most governments, have to spend increasing amounts of scarce foreign exchange on measures to protect these ‘improved’ cattle. This situation has created a high dependency on acaricides. The grave consequences of a break in this practice have been clearly described (Norval, 1978).

In areas where no tick control is carried out, endemic stability between ticks, tick-borne diseases and cattle exists (Young *et al.*, 1981; Norval *et al.*, 1992). With immunization against *T. parva* available (Young, 1985) and with cattle that can tolerate relatively high tick burdens (de Castro *et al.*, 1985a, b; Norval *et al.*, 1988b, Pegram *et al.*, 1989; Minjauw *et al.*, 1997), the reduction of acaricide application and the establishment of endemic stability in many areas of Africa becomes possible. In a situation of endemic stability tick control can be greatly influenced by the level of host resistance of the cattle. Therefore, the reduction of tick resistance by the steady introduction of crossbreeds could be detrimental in the long run, if not carefully performed.

The recent findings in the field of host resistance to ticks, and to some extent to tick-borne diseases, should progressively convert the prevailing control methods characterized by the intensive use of acaricides, to a more integrated tick-control strategy where genetic resistance and immunology will play an important role (Brossard, 1998). The African livestock industry needs to evaluate the different African cattle breeds through genetic research, for both productivity and parasite resistance. As in Australia, the African livestock industry might benefit from the genetic resistance of indigenous breeds or develop new breeds better adapted to the African conditions (Angus, 1996).

While there is evidence of genetic variation for tick resistance, there is a need for further investigation into genetic variation in susceptibility to tick-borne diseases. In the meantime, vaccination against TBD remains an important component of integrated tick and TBD control. Since the current live vaccines against TBD are the only ones available, scientists should continue to improve them while novel vaccines are developed. The innate and acquired host resistance to parasites remains an underexploited control strategy which can also be significantly exploited. The latter approach, integrated with seasonal application of acaricide to control the tick burdens and the use of specific vaccines against ticks and TBD, could revolutionize the economics of tick and TBD control strategies.

Unfortunately, resources for research in this area are decreasing rapidly and studies on ticks and TBD in Africa are becoming scarce. Although at the present time research on ticks and TBD control is not looked on as a priority, progress in the understanding of the genetics of host resistance and its potential use in ticks and TBD control should emphasize the importance of ticks and TBD research as a component of animal health progress in Africa. Since ticks and TBDs are major factors reducing animal production, particularly in the small-scale farming sector, the authors believe that research on ticks and TBD will have a significant positive long-term impact on food security and alleviating poverty.

Acknowledgements

The authors are grateful to R. Leyden Baker and Brian Perry from the International Livestock Institute for their helpful criticism of the manuscript. This is ILRI publication number 98055.

References

Alexander, J.O.D. (1996) The physiology of itch. *Parasitology Today* 2 (12), 345–350.

Alexander, J.O.D., Reason, G.K. and Clark, C.H. (1984) The development of the Australian Friesian Sahiwal. *World Animal Review* 51, 27–34.

Ali, M. and De Castro, J.J. (1993) Host resistance to tick in different breeds of cattle at Bako, Ethiopia. *Tropical Animal Health Production* 25, 215–222.

Angus, B.M. (1996) The history of the cattle tick *Boophilus microplus* in Australia and achievement in its control. *International Journal for Parasitology* 26 (12), 1341–1355.

Angus, B.M. (1998) *Tick Fever and the Cattle Tick in Australia.* Department of Primary Industries, Brisbane, Queensland, Australia.

Bennett, G.F. (1969) *Boophilus microplus* (Acarina:Ixodidae): experimental infestations on cattle restrained from grooming. *Experimental Parasitology* 26, 323–328.

Bock, R.E., De Vos, A.J., Kingston, T.G. and McLellan, D.J. (1997) Effect of breed of cattle on innate resistance to infection with *Babesia bovis, B. bigemina* and *Anaplasma marginale. Australian Veterinary Journal* 75 (5), 337–340.

Bonsma, J.C. (1944) Hereditary heartwater-resistant characters in cattle. *Farming in South Africa* 19, 71–96.

Brossard, M. (1976) Relations immunologiques entre bovins et tiques, plus particulierement entre bovins et *Boophilus microplus. Acta Tropica* 33, 15–36.

Brossard, M. (1998) The use of vaccines and genetically resistant animals in tick control. In: Muller, M. and Brem, G. (eds) Genetic Resistance to Animal Diseases. *Revues Scientifiques et Technicales de l'Office Internationale des Epizooties,* 17 (1), 188–199.

Brown, S.J. (1985) Immunology of acquired resistance to ticks. *Parasitology Today* 1, 166–171.

Burns, B.M. (1991) Relationships between adaptive traits and growth in different cattle breeds in a subtropical environment. M.Sc. thesis, Graduate School of Tropical Veterinary Sciences, James Cook University of North Queensland.

Burrow, H.M., Seifert, G.W. and Hetzel, J.S. (1991) Consequences of selection for weaning weight in Zebu, *B. taurus* and Zebu × *B. taurus* cattle in the tropics. *Australian Journal of Agricultural Research* 42, 295–307.

Cunningham, M.P. (1981) Biological control of ticks with particular reference to *Rhipicephalus appendiculatus.* In: Irvin, A.D., Cunningham, M.P. and Young, A.S (eds) *Advances in the Control of Theileriosis.* Nijhoff, The Hague.

Cunningham, M.P., Brown, C.G.D., Burridge, M.J. and Morzaria, S.P. (1989) *Theileria parva*: the immune status of calves born of dams immunised against ECF. *Research in Veterinary Science* 46, 90–94.

Davis, G.P. (1993) Genetic parameters for tropical beef cattle in northern Australia: a review. *Australian Journal of Agricultural Research* 44, 179–198.

De Castro, J.J. (1991) Resistance to ixodid ticks in cattle with an assessment of its role in tick control in Africa. In: Owen, J.B. and Axford, R.F.E. (eds) *Breeding for Disease Resistance in Farm Animals.* 1st edn. CAB International, Wallingford.

De Castro, J.J. (1997) Sustainable tick and tickborne disease control in livestock improvement in developing countries. *Veterinary Parasitology* 71, 77–97.

De Castro, J.J., Cunningham, M.P., Dolan, T.T., Dransfield, R.D. and Young, A.S. (1985a) Effects on cattle of artificial infestations with the tick *Rhipicephalus appendiculatus. Parasitology* 90, 21–33.

De Castro, J.J., Young, A.S., Dransfield, R.D., Cunningham, M.P. and Dolan, T.T. (1985b) Effects of tick infestation on Boran (*Bos indicus*) cattle immunised against theileriosis in an endemic area of Kenya. *Research in Veterinary Science* 39, 279–288.

De Castro, J.J., Newson, R.M. and Herbert, I.H. (1989) Resistance in cattle against *Rhipicephalus appendiculatus* Neumann, 1901 with an assessment of cross-resistance to Rhipicephalus pulchellus Gerstacker, 1873 (Acari:Ixodidae). *Experimental and Applied Acarology* 6, 237–244.

De Castro, J.J., Capstick, P.B., Nokoe, S., Kiara, H., Rinkanya, F., Slade, R., Okello, O. and Bennun, L. (1991) Towards the selection of cattle for tick resistance in Africa. *Experimental Applied Acaralogy* 12, 219–227.

De Castro, J.J. and Newson, R.M. (1993) Host resistance in cattle tick control. *Parasitology Today* 9 (1), 13–17.

Frisch, J.E. (1994) Identification of a major gene for resistance to cattle ticks. *Proceedings of the 5th World Congress on Genetic Applied Livestock Production* 20, 293–295.

Frisch, J.E. and Vercoe, J.E. (1978) Utilizing breed differences in growth of cattle in the tropics. *World Animal Review* 25, 8–12.

Hales, J.R.S., Schleger, A.V., Kemp, D.H. and Fawcett, A.A. (1981) Cutaneous hyperaemia elicited by larvae of the Cattle Tick, *Boophilus microplus*. *Australian Journal of Biological Sciences* 34, 37–46.

Hewetson, R.W. (1968) Resistance of cattle to cattle tick *Boophilus microplus*. II. The inheritance of resistance to experimental infestations. *Australian Journal of Agricultural Research* 19, 497–505.

Hewetson, R.W. (1972) The inheritance of resistance by cattle to cattle tick. *Australian Veterinary Journal* 48, 299–303.

Hewetson, R.W. (1978) Selection of cattle for resistance against *Boophilus microplus*. In: Wilde, J.K.H. (ed.)*Tickborne Diseases and their Vectors*. University of Edinburgh, Centre for Tropical Veterinary Medicine, Edinburgh, pp. 258–261.

Kaiser, M.N., Sutherst, R.W. and Bourne, A.S. (1982) Relationship between ticks and zebu cattle in southern Uganda. *Tropical Animal Health and Production* 14, 63–74.

Kerr, R.J., Frisch, J.E. and Kinghorn, B.P. (1994) Evidence for a major gene for tick resistance in cattle. *Proceedings of the 5th World Congress on Genetic Applied Livestock Production* 20, 265–268.

Koch, H.T., Kambeva, L., Norval, R.A.I., Ocama, J.G.R., Masaka, S., Munatswa, F.C., Hornhold, N. and Irvin, A.D. (1990) Age resistance to *Theileria parva* bovis infection in calves. *Veterinary Parasitology* 37, 197–206.

Koudstaal, D., Kemp, D.H. and Kerr, J.D. (1978) *Boophilus microplus*: rejection of larvae from British breed cattle. *Parasitology* 76, 379–386.

Latif, A.A. (1984) Resistance to natural tick infestations in different breeds of cattle in the Sudan. *Insect Science and its Application* 5, 95–97.

Latif, A.A., Nokoe, S., Punyua, D.K. and Capstick, P.B. (1991) Tick infestation on Zebu cattle in Western Kenya: quantitative assessment of host resistance. *Journal of Medical Entomology* 28, 122–126.

Losos, G.J. (1986) *Infectious Tropical Diseases of Domestic Animals*. Longman Scientific and Technical, UK.

Mackinnon, M.J. (1990) Genetic relationships between parasite resistance, growth and fertility in tropical beef cattle. In: *Proceedings of the 8th Australian Association of Animal Breeding and Genetics,* pp. 155–161.

Mackinnon, M.J., Meyer, K. and Hetzel, D.J.S. (1991) Genetic variation and covariation for growth, parasite resistance and heat tolerance in tropical cattle. *Livestock Production Science* 27, 105–122.

Madalena, F.E., Teodero, R.L., Lemos, A.M. and Oliveira, G.P. (1985) Causes of variation of field burdens of cattle tick (*Boophilus microplus*). *Revue Brasilia Genetics* 8, 361–375.

Matheron, G., Barre, N., Camus, E. and Gogue, J. (1987) Genetic resistance of Guadeloupe native goats to heartwater. *Onderstepoort Journal of Veterinary Research* 54, 337–340.

Mattioli, R.C. and Cassama, M. (1995) Comparison of characteristics of life cycle in female ticks collected on N'Dama and Zebu cattle. *Tropical Animal Health Production* 27, 150–154.

Mattioli, R.C. and Dempfle, L. (1995) Recent acquisitions on tick and tick-borne disease resistance in N'Dama (*B. taurus*) and Gobra zebu (*B. indicus*) cattle. *Parasitologia* 37, 63–67.

Mattioli, R.C., Bah, M., Faye, J., Kora, S. and Cassama, M. (1993) A comparison of field tick infestation on N'Dama, Zebu and N'Dama × Zebu crossbred cattle. *Veterinary Parasitology* 47, 139–148.

Mattioli, R.C., Bah, M., Kora, S., Cassama, M. and Clifford, D.J. (1995) Susceptibility to different tick genera in Gambian N'Dama and Gobra zebu cattle exposed to naturally occurring tick infestations. *Tropical Animal Health Production* 27, 95–105.

Meltzer, M.I. (1996) A possible explanation of the apparent breed-resistance in cattle to bont tick (*A. hebraeum*) infestations. *Veterinary Parasitology* 67, 275–279.

Minjauw, B., Otte, J., James, A.D., de Castro, J.J. and Sinyangwe, P. (1997) Effect of different ECF control strategies on fertility, milk production and weight gain of Sanga cattle in Central Province of Zambia. *Experimental and Applied Acarology*, 21, 715–730.

Musoke, A.J., Palmer, G.H., McElwain, T.F., Nene, V. and McKeever, D. (1996) Prospects for subunit vaccines against tick-borne diseases. *British Veterinary Journal* 152, 321–639.

Nokoe, S., Capstick, P.B., Latif, A.A. and Punyua, D.K. (1993) Derivation of an index for assessing the resistance of zebu cattle to *Rhipicephalus appendiculatus* Neumann (Acarina: Ixodidae). *Insect Sciences Application*, 14 (1), 15–19.

Norval, R.A.I. (1978) The effects of partial breakdown of dipping in African areas in Rhodesia. *Rhodesian Veterinary Journal* 9, 9–16.

Norval, R.A.I., Floyd, R.B. and Kerr, J.D. (1988a) Ability of adults of *Amblyomma hebraeum* (Acarina:Ixodidae) to feed repeatedly on sheep and cattle. *Veterinary Parasitology* 29, 351–355.

Norval, R.A.I., Sutherst, R.W., Kurki, J., Gibson, J.D. and Kerr, J.D. (1988b) The effect of the Brown Ear tick (*Rhipicephalus appendiculatus*) on the growth of sanga and European breed cattle. *Veterinary Parasitology* 30, 149–164.

Norval, R.A.I., Perry, B.D. and Young, A.S. (1992). *The Epidemiology of Theileriosis in Africa.* Academic Press, London.

O'Kelly, J.C. and Spiers, W.G. (1983) Observations on body temperature of the host and resistance to the tick *Boophilus microplus* (Acari: Ixodidae). *Journal of Medical Entomology* 20, 498–505.

Paling, R.W., Mpangala, C., Luttikhuizen, B. and Sibomana, G. (1991) Exposure of Ankole and crossbred cattle to theileriosis in Rwanda. *Tropical Animal Health Production* 23, 203–214.

Pegram, R.G., Lemche, J., Chizyuka, H.G.B., Sutherst, R.W., Kerr, J.D. and McCosker, P.J. (1989) Effect of tick control on liveweight gain of cattle in Central Zambia. *Medical and Veterinary Entomology* 3, 313–320.

Peter, T.F., Anderson, E.C., Burridge, M.J. and Mahan, S.M. (1998) Demonstration of a carrier state for *Cowdria ruminantium* in wild ruminants from Africa. *Journal Wildlife Disease* 34, 567–575.

Reason, G.K. (1983) Dairy cows with tick resistance: twenty years of the Australian Friesian Sahiwal. *Queensland Agricultural Journal* 109, 135–138.

Rechav, Y. (1992) Naturally acquired resistance to ticks – a global view. *Insect Science and its Application* 13 (4), 495–504.

Rechav, Y., Dauth, J. and Els, D.A. (1990) Resistance of Brahman and Simmentaler cattle to Southern African ticks. *Onderstepoort Journal of Veterinary Research* 57, 7–12.

Rumyantsev, S.N. (1998) Constitutional and non-specific immunity to infection. In: Muller, M. and Brem, G. (eds) Genetic Resistance to Animal Diseases. *Revues Scientifiques et Technicales de l'Office Internationale des Epizooties* 17 (1), 71–83.

Schleger, A.V., Lincoln, D.T., McKenna, R.V., Kemp, D.H. and Roberts, J.A. (1976) *Boophilus microplus*: cellular responses to larval attachment and their relationship to host resistance. *Australian Journal of Biological Sciences* 29, 499–512.

Schleger, A.V., Lincoln, D.T. and Bourne, A.S. (1981a) Arteriovenous anastomoses in the dermal vasculature of the skin of *Bos taurus* cattle, and their relationship with resistance to the tick, *Boophilus microplus*. *Australian Journal of Biological Sciences* 34, 27–35.

Schleger, A.V., Lincoln, D.T. and Kemp, D.H. (1981b) A putative role for eosinophils in tick rejection. *Experientia* 37, 49–50.

Scholtz, M.M., Lombard, P.E., de Bruin, D.S. and Enslin, C.B. (1989) Eenvoudige metode vir die beraming van bosluisbestandheid by beeste. *South African Journal Animal Science* 19 (3), 121–124.

Seifert, G.W. (1971a) Variations between and within breeds of cattle in resistance to field infestations of the Cattle Tick (*Boophilus microplus*). *Australian Journal of Agricultural Research* 22, 159–168.

Seifert, G.W. (1971b) Ecto- and endoparasitic effects on the growth rates of zebu crossbred and British cattle in the field. *Australian Journal of Agricultural Research* 22, 839–850.

Snowball, G.J. (1956) The effect of self-licking by cattle on infestations of Cattle Tick *Boophilus microplus* Canestrini. *Australian Journal of Agricultural Research* 7, 227–232.

Solomon, G. and Kaaya, G.P. (1996) Comparison of resistance in three breeds of cattle against African Ixodid ticks. *Experimental Applied Acaralogy* 20, 223–230.

Spickett, A.M., De Klerk, D., Enslin, C.B. and Scholtz, M.M. (1989) Resistance of Nguni, Bonsmara and Hereford cattle to ticks in a bushveld region of South Africa. *Onderstepoort Journal of Veterinary Research* 56, 245–250.

Spooner, R.L. and Brown, D. (1991) Theileriosis and evidence for genetic resistance. In: Owen, J.B. and Axford, R.F.E. (eds) *Breeding for Disease Resistance in Farm Animals*, 1st edn. CAB International, Wallingford, pp. 235–243.

Stear, M.J. and Wakelin, D. (1998) Genetic resistance to parasitic infection. In: Muller, M. and Brem, G. (eds) Genetic Resistance to Animal Diseases. *Revues Scientifiques et Technicales de l'Office Internationale des Epizooties* 17 (1), 143–153.

Stear, M.J., Newman, M.J., Nicholas, F.W., Brown, S.C. and Holroyd, R.G. (1984) Tick resistance and the major histocompatibility system. *Australian Journal of Experimental Biology and Medical Science* 62, 47–52.

Sutherst, R.W., Wharton, R.H., Cook, I.M., Sutherland, I.D. and Bourne, A.S. (1979) Long-term population studies on the Cattle Tick (*Boophilus microplus*) on untreated cattle selected for different levels of tick resistance. *Australian Journal of Agricultural Research* 30, 353–368.

Sutherst, R.W., Floyd, R.B., Bourne, A.S. and Dallwitz, M.J. (1986) Cattle grazing behaviour regulates tick populations. *Experientia* 42, 194–196.

Sutherst, R.W., Maywald, G.F., Bourne, A.S., Sutherland, I.D. and Stegeman, D.A. (1988) Ecology of the cattle tick (*Boophilus microplus*) in Subtropical Australia. II Resistance of different breeds of cattle. *Australian Journal of Agricultural Research* 39, 299–308.

Tatchell, R.J. (1986) Interactions between ticks and their hosts. In: Howell, M.J. (ed.) *Parasitology – Quo Vadit?* Australian Academy of Science, Canberra, pp. 597–606.

Tatchell, R.J. and Bennett, G.F. (1969) *Boophilus microplus*: antihistaminic and tranquilising drugs and cattle resistance. *Experimental Parasitology* 26, 369–377.

Tawah, C.L. (1992) Comparative study of tick burdens in Guadali and Wakwa cattle under natural infestation in the sub-humid highlands of Wakwa, Cameroun. *Revue*

d'elevage et de Medecine Veterinaire des Pays Tropicaux 45, 310–313.
Turner, H.G. and Schleger, A.V. (1960) The significance of coat type in cattle. *Australian Journal of Agricultural Research* 11, 645–663.
Ulloa, G. and de Alba, J. (1957) Resistance to ectoparasites in some races of cattle. *Turrialba* 7, 8–12.
Utech, K.B.W. and Wharton, R.H. and Kerr, J.D. (1978) Resistance to *Boophilus microplus* (Canestrini) in different breeds of cattle. *Australian Journal of Agricultural Research* 29, 885–895.
Utech, K.B.W. and Wharton, R.H. (1982) Breeding for resistance to *Boophilus microplus* in Australian Illawarra Shorthorn and Brahman × Australian Illawarra Shorthorn cattle. *Australian Veterinary Journal* 58, 41–46.
Vercoe, J.E. and Frisch, J.E. (1992) Genotype (breed) and environment interaction with particular reference to cattle in the tropics, a review. *Asian–Australasian Journal of Animal Science* 5 (3), 401–409.
Wakelin, D. and Blackwell, J.M. (1993) Genetic variation in immunity to parasite infection. In: Warren, K.S. (ed.) *Immunology and Molecular Biology of Parasitic Infections.* Blackwell Scientific Publications, Oxford, pp. 3–22.
Wharton, R.H., Utech, K.W.B. and Turner, H.G. (1970) Resistance to the tick *Boophilus microplus* in a herd of Australian Illawara Shorthorn cattle. Its assessment and heritability. *Australian Journal of Agricultural Research* 21, 163–181.
Wikel, S.K. (1982) Immune responses to arthropods and their products. *Annal Revue Entomology*, 27, 21–48.
Wikel, S.K. (1996) Host immunity to ticks. *Annal Revue Entomology* 41, 1–22.
Wikel, S.K., Ramachandra, R.N. and Bergman, D.K. (1994) Tick-induced modulation of the host immune response. *International Journal for Parasitology* 24 (1), 59–66.
Wilkinson, P.R. (1955) Observations on infestations of undipped cattle of British breeds with the Cattle Tick, *Boophilus microplus* (Canestrini). *Australian Journal of Agricultural Research* 6, 655–665.
Willadsen, P. (1980) Immunity to ticks. *Advances in Parasitology* 18, 293–313.
Willadsen, P., Cobon, G.S. and McKenna, R.V. (1996) Comparative vaccination of cattle against *Boophilus microplus* with recombinant antigen Bm86 alone or in combination with recombinant Bm91. *Parasite Immunology* 18 (5), 241–246.
Young, A.S. (1985) Immunization of cattle against theileriosis in the TransMara Division of Kenya. A comparison of trials under traditional Maasai management with trials on a ranch development. In: Irvin, A.D. (ed.) *Immunization Against Theileriosis in Africa.* ILRAD, Nairobi, pp. 64–68.
Young, A.S., Leitch, B.L. and Newson, R.M. (1981) The occurrence of a *Theileria parva* carrier state in cattle from an East Coast fever endemic area of Kenya. In: Irvin, A.D., Cunningham, M.P. and Young, A.S. (eds) *Advances in the Control of Theileriosis.* Nijhoff, The Hague, pp. 60–62.

Genetic Aspects of Resistance to Ovine Cutaneous Myiasis

8

H.W. Raadsma
Centre for Sheep Research and Extension, The University of Sydney, Camden, Australia

Summary

Myiasis of sheep remains a serious health problem posing significant constraints on production efficiency. Myiasis flies of economic significance in sheep production belong to the superfamily Oestroidae, characterized by three major families of flies: (i) Oestridae (bots and warbles); (ii) Sacrophagidae (flesh flies); and (iii) Calliphoridae (blowflies). In sheep the nasal bot fly is possibly the most prevalent form of myiasis, but can be easily controlled and poses limited impact on production. Infestation with screw-worm flies on the other hand pose enormous constraints on sheep production. Fortunately, most of the important sheep production countries in the world are free from this parasite. In combination, the greatest impact of myiasis in sheep is caused by *Lucilia* spp. (*L. cuprina* and to a lesser extent *L. sericata*) worldwide.

An extensive range of control methods are being used to either reduce blowfly populations or increase resistance of the host. Within the current range, use of management practices such as tail docking, mulesing, shearing, and internal parasite control are highly effective. In addition, the strategic use of insecticides still forms the backbone of fly strike control programmes despite the widespread development of resistance by blowflies to a wide range of chemicals, and reduced consumer acceptance of chemical residues in wool. In the case of relatively susceptible Merino sheep, selection of resistant genotypes has proven to be effective to reduce susceptibility to body strike. Genetic improvement of host resistance to breech strike in non-Merinos has not yet been demonstrated, but is thought to be feasible. Additional strategies to reduce the population of blowflies through trapping, biological or genetic control have limited scope on a large scale, or have been shown not to be cost effective. Vaccination against blowfly larvae or the main predisposing agents of blowfly strike is not sufficiently reliable to offer widespread protection in the field.

In isolation each of the current control strategies has limitations for high-level sustainable control of blowflies. The integration of our current

© CAB *International* 2000. *Breeding for Disease Resistance in Farm Animals* (eds R.F.E. Axford, S.C. Bishop, F.W. Nicholas and J.B. Owen)

knowledge of blowfly control-strategies to deliver sustainable integrated pest management (IPM) programmes can now be achieved. Although interactive control is currently in place, it occurs on an *ad hoc* basis rather than part of a designed IPM programme. The management of insecticide resistance, strategic use of chemicals and manipulation of host resistance, in combination with suppressing blowfly populations should all be important components as well as breeding for increased resistance.

Introduction – Myiasis of Sheep

Myiasis is the infestation of organs or tissues of host animals by the larval stages (maggots or grubs) of dipterous flies. The fly larvae feed directly on the host's necrotic or living tissue. Toure (1994), Hall and Wall (1995), and Wall and Shearer (1997) describe myiasis of veterinary and economic significance in considerable detail.

Classification of myiasis

Myiasis is often classified according to the anatomical location of infestation on the host, and is broadly speaking described as dermal, subdermal or cutaneous, nasopharyngeal, ocular, intestinal/enteric or urogenital. As outlined by Wall and Shearer (1997), it is often more appropriate to classify myiasis in terms of relationships between host and parasite, since this provides insight into the biology of the fly species, its likely pathological effect, host responses and methods of control. The terms 'obligatory', 'facultative' and 'accidental' cover possible relationships between host and parasite. Obligatory ectoparasites must have a living host (often quite specific) to complete their development and are unable to survive in the absence of the host. Facultative species, on the other hand, can develop both in living and dead organic matter, and can be further subdivided into primary and secondary facultative ectoparasites. The major distinction within this classification is that primary species are capable of initiating myiasis, but will also live as saprophages in carrion. Secondary species normally live as saprophages, and are normally unable to initiate myiasis but invade pre-existing infestations. The detail is of significance, since the major species of blowflies in sheep are primary, facultative ectoparasites, as discussed below. Accidental blowfly species, do not normally live on a specific host, but will do so once any host is suitably predisposed, through other myiases or wounds.

Myiasis flies of importance in sheep production

Myiasis flies of economic significance in sheep production belong to the superfamily Oestroidae, which is characterized by three major families of flies: (i) Oestridae (bots and warbles); (ii) Sacrophagidae (flesh flies); and (iii) Calliphoridae (blowflies). Table 8.1 summarizes myiasis flies of potential significance in sheep.

Table 8.1. Summary of myiasis flies of potential significance in sheep production systems.

Species	Classification	Host specificity	Site of infection	Occurrence	Impact on sheep production
Oestridae					
Oestrus ovis	Obligate	High	Nasopharyngeal	All sheep-producing areas	Mild/medium
Calliphoridae					
Cochliomyia hominivorax	Obligate screw worm	Low	Wounds, body orifices	Neartic and neotropical	Potentially very high
Chrysomya bezziana	Obligate screw worm	Low	Wounds, body orifices	Afrotropical/oriental	Potentially very high
Lesser *Chrysomya* spp. *C. megacephala* *C. rufifaces* *C. albiceps*	Facultative saprophagous	Low	Secondary myiasis	Australasian/oriental	Low
Lucilia spp. *Lucilia cuprina* *Lucilia sericata*	Facultative	High	Primary cutaneous myiasis agent All predisposed sites	Worldwide	High–very high
Lesser *Calliphora* spp. *C. augur* *C. vicina* *C. vomitoria* *C. albifrontalis* *C. nociva* *C. stygia*	Saprophytic	Low	Primary/secondary myiasis	Australasian	Low–moderate
Sarcophagidae					
Wolfarthia magnifica	Obligate screw worm	Low	Wounds, body orifices	Eastern palaeartic	Potentially high

Bot flies

The oestrid species of greatest economic significance in sheep production is the sheep nasal bot fly, *Oestrus ovis*, an obligate parasite to sheep and goats. Larvae of *O. ovis* develop in the head sinuses and nasal passages during all three major larval stages. At larval maturation, larvae are sneezed out. In general, infestations are relatively light (2–20 larvae) in the frontal sinus of infested animals. Mucoid nasal discharge, sneezing, nose rubbing, or head shaking are clinical symptoms of infestation. Dead larvae in the sinuses can cause allergic reaction and inflammatory responses, followed by secondary bacterial infections. Death following infestation is rare. Production losses in the form of reduced weight gain, loss of body condition and loss of wool and milk production have been attributed to infestation with *O. ovis*. On balance, such losses can be considered mild. Prevalence in individual flocks varies greatly, ranging from 0 to 100%, and infestations have been reported from all sheep production countries worldwide. Treatment with a single oral dose of ivermectin or rafoximide is systemically active and highly effective. No reports describing genetic variation in resistance between sheep have been found. In the light of highly effective chemotherapy and relatively mild production losses due to infestation, it is unlikely that such investigations are warranted.

Screw-worm flies

The three species of obligate screw-worm flies that can cause potentially severe myiasis in sheep are *Cochliomyia hominivorax*, *Chrysomya bezziana* and *Wolfarthia magnifica*. All three species will infect almost all warm-blooded animals and therefore do not have a host-specific relationship with sheep. Screw-worm myiasis occurs largely as a consequence of oviposition at sites of skin damage due to trauma, shearing, tail docking or castration wounds. Oviposition also occurs near body orifices such as nostrils, eyes, mouth, ears, anus and vagina. Larvae from these three species hatch within 24 hours, and are characterized by aggressive gregarious feeding behaviour, causing deep, cavernous tissue wounds. Wounds rapidly become very extensive, attracting other female screw-worm flies and secondary myiasis agents. If left unattended, repeated infestation within the wounds rapidly leads to death of the host. Distribution of the New World screw-worm fly, *C. hominivorax*, is largely restricted to the southern areas of USA, Central America and northern South America. The Old World screw-worm fly, *C. bezziana*, is largely restricted to much of Africa, India, the Arabian peninsula, South-East Asia and New Guinea. *Wolfarthia magnifica* is located primarily around the Mediterranean Basin, in eastern and central Europe, and in Asia Minor. Australia, as one of the major sheep-producing countries in the world, is free from all screw-worm flies. However, introduction from neighbouring countries could pose a A$500 million per annum threat to the pastoral industries of Australia. The Merino sheep industry would be highly exposed as a consequence of routine management practices that leave widespread surgical and shearing wounds in almost all sheep. There have been no documented reports on genetic variation between sheep in their predisposition to these flesh-eating flies.

Lesser *Chrysomya* spp. (*C. megaphala*, *C. rufifaces* and *C. allbiceps*) occur throughout Australasia and oriental regions. Although these species are mainly saprophagous, and carrion breeders, they occasionally act as secondary agents in myiasis, infecting pre-existing myiasis wounds. On balance, the economic impact of myiasis attributed to these species is considered less important than that of screw-worms or *Lucilia* spp. (described below).

Blowflies

Sheep myiases attributed to *Lucilia sericata* and *Lucilia cuprina* are considered of greatest economic significance worldwide: these species are the primary agents of cutaneous myiasis. Although both species can cause myiasis in a range of domestic animals, they are of particular importance in sheep, and are known as 'sheep blowflies'. *L. cuprina* and *L. sericata* are facultative ectoparasites. Larvae develop following hatching of the egg clusters deposited on sites that are suitably predisposed. Their larvae infest and feed superficially on the living tissues (epidermis), lymphatic exudate or necrotic tissue of the host. The mouth hooks are used to macerate tissues, and digestion occurs extraorally by means of amylase in the saliva and proteolytic enzymes in larval excreta.

Blowfly strike by both *L. cuprina* and *L. sericata* occurs most commonly in the perianal area of sheep, and is strongly associated with faecal and urine soiling of the breech/tail region. In addition, bacterial dermatoses as consequence of prolonged wetting of the skin, with or without additional involvement of *Dermatophilosis congolensis* or infections attributable to footrot, are major predisposing factors in blowfly strike involving *L. cuprina*. There is little recorded involvement of dermatitis in the predisposition of sheep to body strike by *L. sericata*. Microbial agents producing attractants for gravid females encourage oviposition, with females laying eggs in batches of 200–250. Under suitable conditions, eggs hatch within 24 hours, allowing first-instar larvae to feed on soluble extraneous protein in the immediate environment. On development of mouth hooks in later stages, second- and third-instar larvae will feed more aggressively. Oviposition sites, and pre-existing strikes, are attractive to gravid females, and it is common for extensive development of strikes over the body if left untreated (Fig. 8.1B). Sheep that are not suitably predisposed will not allow attraction, initiation and the development of blowfly strike.

DISTRIBUTION OF *L. CUPRINA* AND *L. SERICATA*. In combination, the effects of *L. cuprina* and *L. sericata* have been reported from all major sheep-producing countries worldwide. In general it is believed that *L. sericata* is the most important agent of sheep myiasis throughout the cool–temperate habitats of northern Europe and New Zealand. *L. cuprina*, on the other hand, is widespread in warm—temperate and subtropical regions, including Australia, southern Spain, and northern and southern Africa. In Australia *L. cuprina* is present in over 90% of fly-strike cases. *L. sericata* is believed to be a primary agent in 75% of all cases of fly-strike in New Zealand. The recent importation and rapid spread of *L. cuprina* (Fig. 8.1A) in that country, shows that this species is almost synonymous with blowfly strike throughout the most important sheep-production countries in the world. A notable exception of the import-

(A)

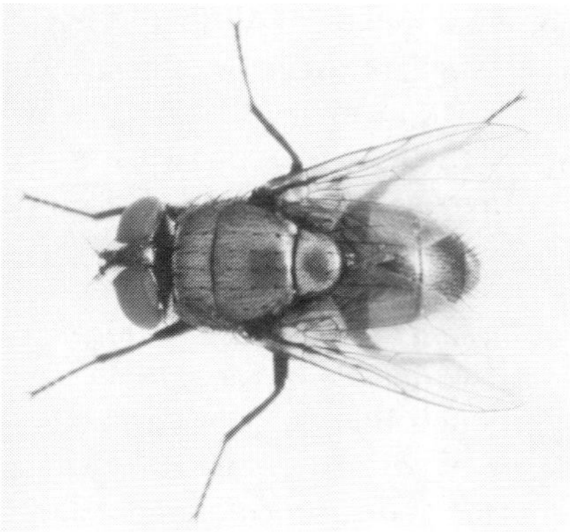

(B)

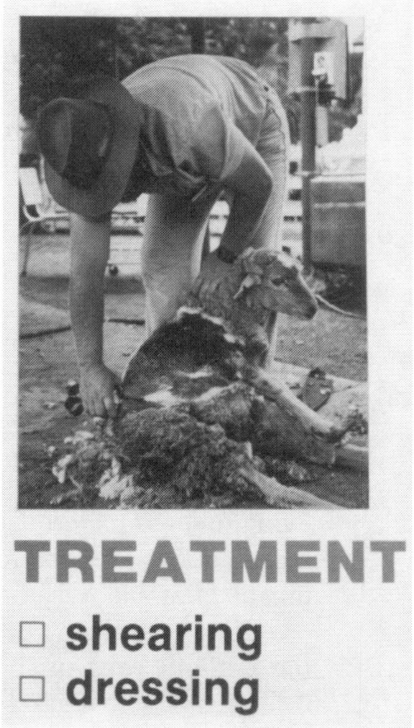

Fig. 8.1. (A) *Lucilia cuprina*. The Australian green blowfly is responsible for 90% of all primary fly strikes in Australian sheep production. (B) Body strike in sheep. Extensive damage and animal suffering can occur as a consequence of *Lucilia cuprina* myiasis. Affected animals often perish if left untreated. Inspection and treatment of affected sheep ('flying sheep') is labour intensive and is part of routine husbandry on many Australian sheep farms.

ance of *L. cuprina* in sheep myiasis appears in the USA, where it is known to be present but is considered unimportant.

Given the widespread importance of *L. cuprina*, and to a lesser extent *L. sericata*, in sheep myiasis, this chapter will restrict itself to these species. Other *Calliphora* spp. of interest in sheep myiasis include *C. augur*, *C. vicina*, *C. vomitoria*, *C. albifrontalis*, *C. nociva* and *C. stygia*. Although these are primarily saprophytic and carrion breeders, their role as agents of occasional primary and mostly secondary myiasis warrant their mention in this chapter. These species are widely distributed throughout the Australasian region, but on balance their economic significance is low when accounting for the economic impact of the essential primary myiasis caused by other species, in particular *L. cuprina*.

Summary of the aetiology and pathogenesis of myiasis

The pathological effects seen in sheep affected by myiasis vary considerably, depending on species of ectoparasites, number of larvae and site of infestation. Small infestations restricted to confined sites may have little or no discernible effects on the host. Large infestations result in rapid increase of body temperature and respiratory rate, accompanied by anorexia and weight loss. In heavy and extensive infections animals are likely to suffer significant tissue damage, and bacterial infections, resulting in septicaemia, toxaemia, dehydration and anaphalaxis. Severe infestations, if left untreated, are likely to result in the death of the animal.

Economic impact of sheep myiasis

The economic impact of sheep myiasis is difficult to ascertain. Although the proportion of farms likely to be affected by fly strike is high (>80%), the prevalence in a national flock may appear relatively low at 1.5–2.0% of all sheep farmed (UK, Australia, New Zealand). However, this does not reflect the substantial cost associated with widespread adoption of husbandry practices such as crutching, shearing, tail docking and mulesing to minimize the impact of fly strike. Furthermore, the sporadic occurrence of severe outbreaks with high prevalence (>10%) often attracts high mortality and high labour costs, with continued surveillance of flocks at risk. The covert costs associated with decreased reproductive efficiency, and lifetime reduction in meat, wool and milk production as a consequence of severe fly strike, and decreased genetic gain in diseased flocks, are seldom accounted for.

McCleod (1995) shows the relative cost associated with blowfly strike control in Australia, which is predominantly a wool-producing country. Methods of control represented 80% of the annual cost attributable to fly strike, and labour represented the greatest single cost in this (88% of cost of control). Production losses contributed the remaining 20% to the annual cost of A$161 million to fly strike, with mortality the single greatest cost component at 38%.

The predicted impact of myiasis on sheep production is even more difficult to estimate if relative risk scenarios need to be incorporated that allow for decreased reliance on current and highly effective control methods. Such scenarios are realistic in the light of increased resistance to current chemicals and insecticides, decreased reliance on use of chemicals as dictated by consumer preferences, and decreased reliance on proven husbandry techniques such as mulesing and tail docking, on the grounds of animal welfare.

Genetic Improvement in Resistance to Blowfly Strike in Sheep

Incentive for genetic improvement in resistance to fly strike

High labour costs and reduced reliance on insecticides for control often form the main incentive for breeding for resistance to fly strike as a long-term and permanent control option. The widespread development of resistance by blowflies to a broad range of insecticides (Hughes and McKenzie, 1987; Levot 1993, 1995), and the limited development of new chemical compounds, provided the stimulus for much of the earlier research in developing alternative methods of control, including breeding for increased resistance. In addition, chemical residues in wool, meat and environmental by-products (wool scouring) restricts broad-based use of chemicals as a main control agent. Despite these restrictions, the use of cyromazine (Hart *et al.*, 1982; Levot 1995) has proven to be highly effective in controlling fly strike, but should be used strategically in order to prolong its effective life span. Development of new chemical compounds to control fly strike is discussed below.

It is projected that in combination with other forms of blowfly control, resistant sheep will require less frequent preventive and therapeutic treatments, reduced labour input in flock management, and will incur lower production penalties associated with fly strike in affected flocks. However, genetic change in resistance is not without cost, in the sense that it will need to be combined with other breeding objectives, and is likely to reduce the rate of genetic gain in these traits. It will also make breeding programmes more expensive and complicated. In well-designed breeding programmes, all traits of economic importance are considered, and programmes are designed to optimize profitability through selective breeding. In this context, it is important to have a clear and well-defined perspective of how resistant genotypes will be managed in a production system, in order to calculate financial benefits of inclusion of resistance in a breeding programme.

Most of the research evaluating breeding for resistance has assumed that proven and highly effective management practices, such as mulesing and tail docking, will continue to be available to industry. However, concerns based on grounds of animal welfare may also limit the use of these tools in the control of fly strike. Removal of these highly effective tools will radically change current thinking on the control of blowfly strike, as detailed below.

Feasibility of breeding for resistance to fly strike

In the previous edition of this text, Raadsma (1991a) detailed the feasibility of breeding for resistance to blowfly strike. Under conditions of continued mulesing and tail docking, body strike was the main form of fly strike which warranted consideration in a genetic improvement programme. The state of knowledge as it was available then showed that:

1. Considerable genetic variation existed within and between flocks in resistance to body strike, and in its main predisposing factor, the bacterial dermatitis – fleece rot.
2. Major genetic differences between strains and bloodlines (stud-lines) in resistance to fleece rot and body strike were a major avenue for rapid genetic change in resistance.
3. Heritability of resistance to body strike was sufficiently high (>0.25) that direct selection through culling of affected sheep under high disease expression should be effective (direct selection).
4. Indirect selection for reduced susceptibility to body strike would be more effective through reducing susceptibility to fleece rot, based on the heritability (0.3–0.4), its high genetic correlation with body strike resistance (>0.9) and higher level of expression of this trait (Raadsma, 1991c).
5. Indirect selection for resistance to fleece rot and body strike on the basis of fleece and skin traits was calculated to have high potential. However, feasibility based on a limited range of known fleece traits showed indirect selection to be less effective than direct selection using fleece rot. Skin traits still required evaluation.
6. The genetic basis of resistance to both fleece rot and body strike was not understood.

7. The genetic relationship between resistance to fleece rot and body strike and other diseases of economic importance such as internal parasites and foot rot, was unknown.
8. The genetic relationships between resistance to fleece rot and body strike and other economically important production traits such as fleece weight, fibre diameter, body weight and reproductive performance were poorly understood.
9. Projected advantages of genetically resistant sheep over susceptible sheep could be demonstrated in single-trait selection flocks selected for and against resistance to fleece rot.

Recent developments in our genetic understanding of resistance

Mechanisms of resistance

Recently investigations into the mechanisms of resistance to fleece rot and blowfly strike have shifted from the factors associated with fleece characteristics, as reviewed by Raadsma (1991a, 1993), to skin-based and immunological factors. Despite the strong range of immunological and inflammatory responses to larval attack of the host, repeated infections generate no substantial acquired resistance. Multiple repeated infections by larvae showed a weak and poorly sustained protective immune response (Sandeman *et al.*, 1986; Eisseman *et al.*, 1990). The weak acquired resistance responses result in slightly fewer larvae surviving and reductions in larval growth rate. During larval challenge, sheep generate strong antibody responses (humoral, primarily IgG1) to a wide range of larval products, in particular excretory enzymes, gut and salivary gland proteins (Bowles *et al.*, 1987; Seaton *et al.*, 1992; Tellam *et al.*, 1994). In addition, there are major cellular responses evoked during fly strike, particularly in the skin. These are characterized by the rapid influx of neutrophils, eosinophils, macrophages and lymphocytes (CD4-helper cells and $\gamma\delta$ cells) (Bowles *et al.*, 1992). In conjunction with the cellular responses, a wide repertoire of T-cell cytokine and inflammatory agents has been demonstrated, including IL-1a, IL-1b, IL-6 and IL-8. The inflammatory cytokine TNF-α remained unchanged during infection. The T-cell-dependent cytokines IL-2 and γ-IFN also increase, but not IL-4. The acute inflammatory response is characteristic of strikes, and does not differ markedly between repeated infestations. The capacity of *L. cuprina* larvae to degrade and digest antibody (Sandeman *et al.*, 1995) may actually be enhanced by strong inflammatory and antibody responses. It is unlikely that these immune responses are natural mechanisms of resistance.

Sheep selected for increased and decreased resistance to fleece rot and body strike, as described by Raadsma (1991a, 1992), provide powerful experimental resources with which to understand the genetic basis of resistance. Investigations by O'Meara *et al.* (1992, 1995) showed that sheep from the resistant line showed substantially greater inflammatory skin responses after intradermal injection with excretory–secretory larval products compared with sheep from the susceptible line. Investigations by Colditz *et al.* (1992, 1994), on the other hand, showed that resistant sheep showed lower plasma leakage and skin inflammatory responses when challenged with specific and characterized endogenous inflammatory agents. Despite these differences between the selection lines in inflammatory responses, no difference was observed in antibody titres and specificity directed against *L. cuprina* larval antigens (O'Meara and Raadsma, 1995; O'Meara *et al.*, 1997), or larval survival and growth rate between sheep from the resistant and susceptible lines following experimental implantation (Colditz *et al.*, 1996). Although these results appear, at face value, to be in conflict with the large and consistent difference in prevalence of body strike under natural and induced conditions between these lines (Raadsma, 1992), they can be ratified when considering that the lines were selected primarily for and against fleece rot, and the latter condition provides the major opportunity for fly strike to develop. The differences in resistance are thought to be partly due to wool and skin differences in response to moisture and bacterial agents, as detailed by Raadsma (1991a, b), and not specific mechanisms directly targeted against larvae operate in these flocks. This hypothesis is further supported by the higher antibody responses to bacterial antigens and complement activation in resistant sheep, compared with responses from susceptible sheep (Chin and Watts, 1991).

Use of indicator traits

For genetic improvement of resistance to body strike, there is little doubt that direct selection against the main predisposing dermatitis – fleece rot – is highly effective. From known genetic parameters, we can calculate the co-heritability as described by Raadsma (1991a) from $r_g \times h_1 \times h_2$, where r_g is the genetic correlation between resistance to body strike and the indicator trait, and h_1 and h_2 the square root of the heritability estimates of the two traits. The use of fleece rot as an indirect selection trait has further advantages in that the prevalence is usually higher than that of body strike, hence more effective selection differentials can be achieved; it is easy and cheap to score; and it does not incur the costly production penalties associated with direct selection against body strike. A detailed and effective scoring system to assess susceptibility to fleece rot and hence body strike as used in commercial practice is presented in Fig. 8.2.

Under some circumstances, direct selection against fleece rot suffers from the same problems as direct selection against body strike, in that the prevalence is too low to allow effective selection. In this case considerable effort has focused on the identification of skin and fleece traits that could be used as indirect selection traits. Raadsma (1991a, b, 1993) and O'Meara and Raadsma (1995) reviewed the range of indirect selection traits that may be appropriate for selection to increase resistance to fleece rot. Recent developments have shown an on-farm skin test, based on weal development following intradermal injection with blowfly larval excretory enzymes, to have high genetic and practical potential for use as an indirect selection trait. The co-heritability with resistance to fleece rot was estimated at 0.25, and the test would allow many animals to be screened at relatively low cost (Raadsma *et al.*, 1992). The test requires further development and evaluation under field conditions, and appropriate genetic parameters with production traits need to be obtained.

It should be noted that indirect selection as discussed here is unlikely to be effective against breech strike. Provided mulesing and tail docking remain acceptable, this should not pose a problem. Should genetic resistance to breech strike be required, indirect selection on the basis of breech-fold development may be one avenue for evaluation.

Use of gene markers

The search for genetic markers, expressed either as gene products (cell surface or immunoregulatory proteins) or polymorphisms at the DNA level, for resistance to fly strike or its predisposing dermatoses has been limited. Raadsma *et al.* (1992) and Engwerda *et al.* (1996) examined differences in frequency of polymorphic variants of IgE, TNF-α, IL-1β, IL-4 and γ-IFN gene polymorphisms between flocks selected for resistance and susceptibility to fleece rot and fly strike. No obvious difference consistent with resistance or susceptibility was observed. It should be noted that such association studies are at best indicative and may not have sufficient power to detect linkage between candidate genes and resistance. Primarily as a function of insufficient research activity, there is no conclusive evidence either way that gene markers can have a role in selective breeding for resistance to blowfly strike. Possible research to evaluate genetic markers for resistance is discussed below.

Resistance and other breeding objectives

As discussed in the previous text (Raadsma, 1991a), resistance to blowfly strike is not likely to be the sole breeding objective, and breeders are required to combine resistance with other economically important breeding objectives. From preliminary data we have some understanding of the genetic and phenotypic relationships between important production traits and resistance (Raadsma, 1991a). For wool sheep it is likely that selection for increased resistance will lead to a slight reduction in fleece weight (unfavourable), decrease in fibre diameter (favourable) and no effect on body weight. Preliminary correlated responses in selection flocks for resistance and susceptibility to fleece rot and blowfly strike, as presented by Mortimer *et al.* (1998), confirm these initial observations. However, more robust and reliable information on appropriate genetic parameters and relative economic weightings are warranted to develop comprehensive breeding programmes that incorporate resistance to blowfly strike.

Resistance to other diseases

There is now a considerable body of evidence on genetic aspects of disease resistance in Merino sheep, as reviewed by Gray *et al.* (1995) and Raadsma *et al.* (1997a, b, 1998). The primary focus of such studies has been on resistance to internal parasites, footrot, blowfly strike (and its major predisposing dermatoses – fleece rot and dermatophilosis), showing that it is feasible to change resistance to each disease through selective breeding. However, there is a paucity of information for prediction of the likely consequences of selecting for resistance to one disease on correlated responses in resistance to other diseases. This is of importance, given that the important diseases identified

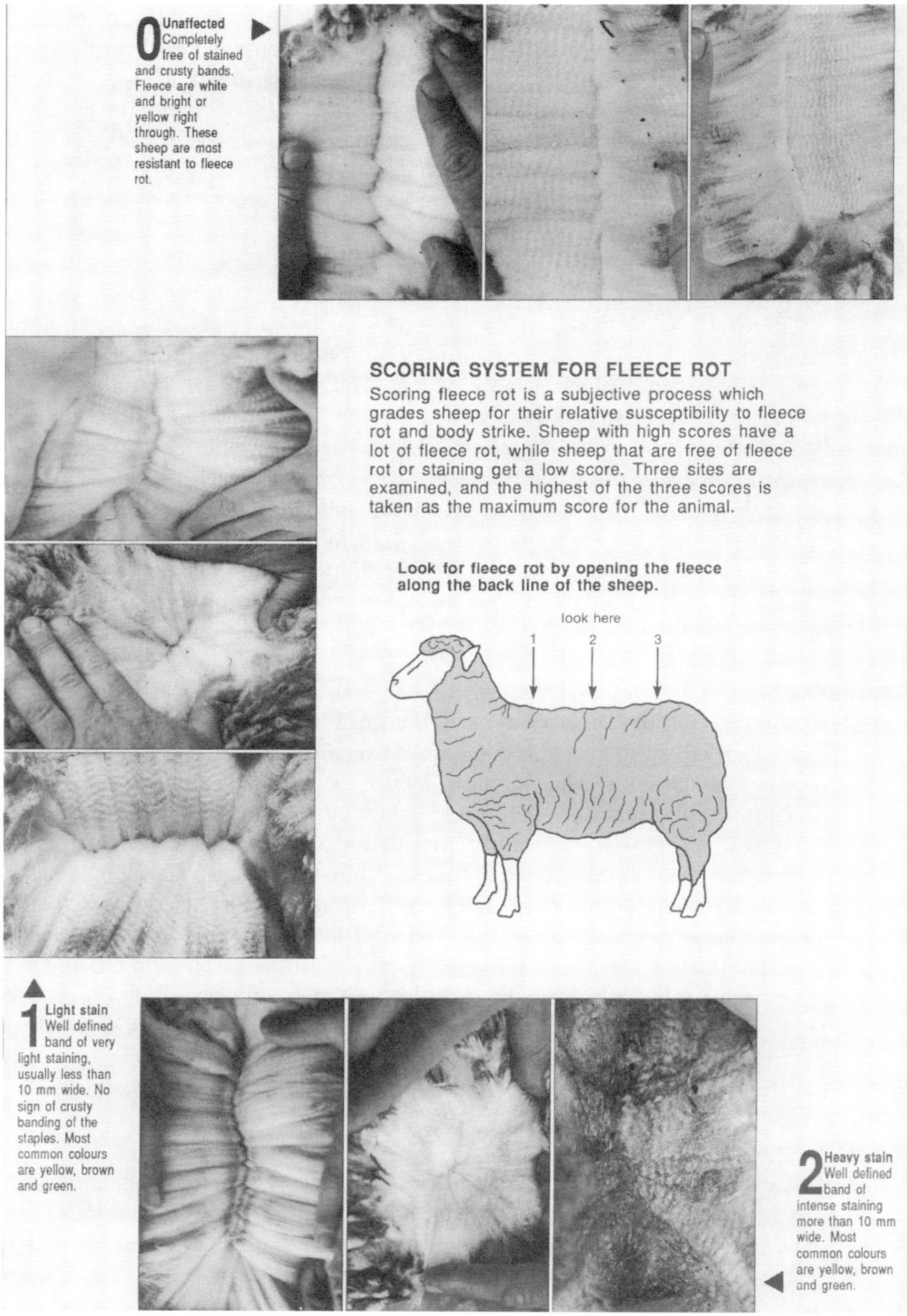

Fig. 8.2. Fleece rot dermatitis is the major predisposing factor to body strike. Resistance can be scored quickly and cheaply using a graded scoring system to assess young sheep after exposure to periods of wet weather. (Source: NSW Agriculture.)

5 Heavy crusty fleece rot
Heavy bands of crusty fleece rot more than 10 mm thick, or many bands with a total thickness of more than 10 mm. These sheep are the most susceptible to body strike. Animals which have previously been struck on the body are also classed in this grade.

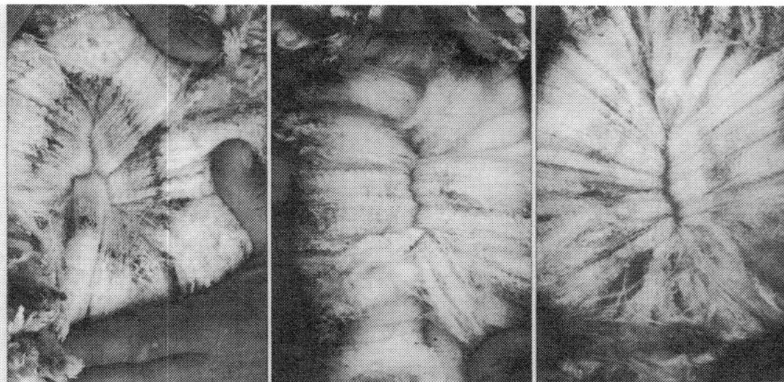

4 Moderate crusty fleece rot
A well defined band of crusty fleece rot which is easily felt in the fingers when the wool is opened. Bands are 5–10 mm wide. Associated staining can be common, but is less important than the degree of crustiness.

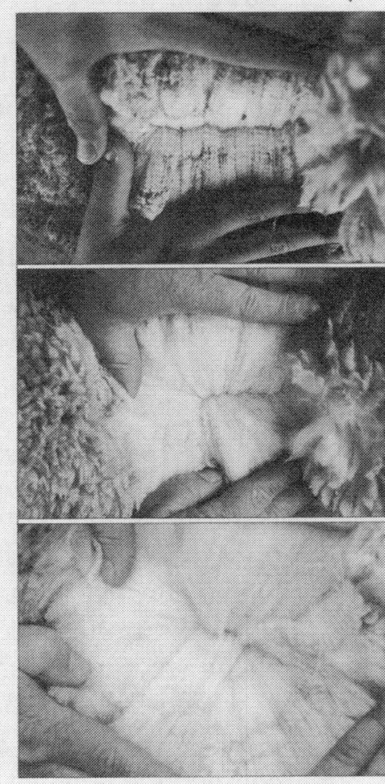

What to do with the results

Take scores for individual animals and record them for a later date. Calculate the average by summing all scores and dividing by the number of sheep scored. All sheep can also be ranked from 0 to 5 and selected or culled depending on what is required; for instance, all ewes with a score of 5 can be culled and only rams with scores of 0 or 1 selected.

- For a simple light–moderate–heavy classification:
  grades 0 and 1 are grouped as light or 'resistant';
  grades 2 and 3 are grouped as moderate or 'intermediate';
  grades 4 and 5 are grouped as heavy or 'susceptible'.
- For the 'incidence' of crusty fleece rot:
  grades 3, 4 and 5 are classed as 'affected';
  grades 0, 1 and 2 are classed as 'unaffected'.

  Divide the number of animals affected by the number of animals scored and multiply by 100 to find the percentage incidence of crusty fleece rot in the flock. For example, if 36 sheep in a flock of 420 are affected with fleece rot, the incidence of fleece rot is $\frac{36}{420} \times 100 = 8.5$ per cent.
- For a simple selection in the yard, inspect all sheep and raddle and cull only those which are unsuitable (say score 5). If not very much fleece rot is present and there are no sheep with a score of 5, all sheep with scores of, say, 3 and 4 can be raddled and culled.
- For a detailed description of all fleece rot along the back line, register the score for each site inspected. Normally three sites are examined: the shoulder/wither, the back, and the loin/rump.

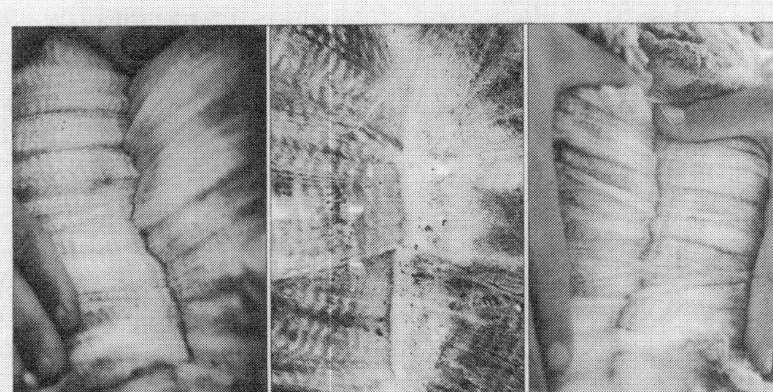

3 Slight crusty fleece rot
Few narrow bands of crusty, matted material in the fleece, less than 5 mm thick. Staining is often present but is not a necessary feature.

Fig. 8.2. *Continued*

above have common environmental (climatic) risk factors and are likely to occur in the same flock. Furthermore, strong favourable or unfavourable genetic correlations between resistance to multiple diseases may provide insight into the nature of disease-resistance genes. Raadsma *et al.* (1997b) report on an investigation in which resistance to all major diseases was investigated in a single experiment, and predictions could be made on likely changes in disease resistance following selection for any of the major diseases. In the case of predisposition to fly strike, resistance to fleece rot had no genetic relationship with resistance to dermatophilosis, showing that genes controlling these two important risk factors act independently. Predisposition to blowfly strike was also unlikely to lead to favourable or unfavourable changes in resistance to internal parasites or footrot. Selection for resistance to multiple diseases simultaneously will thus be considerably slower than selection for resistance to any one disease alone, since there are no genetic synergies between resistance to the most important diseases in sheep (Raadsma *et al.*, 1997a, b).

Breeding for Resistance in the Face of Other Control Options

Current control strategies

Use of insecticides

The use of insecticides has been a core component in the control of blowflies since the early 1900s. Insecticides are generally applied to key areas on the body in the prevention of strike, or used in dressings of existing strikes to prevent restrikes. The development of new chemical compounds has been highly effective in fly-strike prevention, with some formulations offering up to 16 weeks' protection, sufficient to take sheep through a high-risk period. One of the striking features of the history of insecticide use for fly-strike control has been the enormous capacity of the primary blowfly to develop genetic resistance to almost all chemical compounds. Insecticide resistance to multiple compounds is generally widespread and greatly reduces the efficacy of this control option. Resistance to available insecticides has also forced the chemical industries to develop new compounds, often at great expense, with potentially little return on investment if the compound shows efficacy against blowflies only. For a comprehensive review of the use of insecticides for fly-strike control, the reader is referred to Levot (1993, 1995), who provides both an historical account and outline of management strategies to deal with insecticide resistance. Although the repertoire of chemicals has changed considerably over the 100-year history, the application technology has changed relatively little. The relatively crude methods of insecticide application in wool often lead to unpredictable amounts of chemical being deposited on the animal. Significant residual levels of chemical at the time of wool harvest can lead to consumer resistance for wool, and can limit the time periods during which chemicals may be used. Despite the potentially negative aspects of chemicals in blowfly-strike control, where effective compounds are available, the cost of controlling blowfly strike by chemical means may well be lower than adopting long-term control strategies through selective breeding. However, this scenario assumes that chemical compounds will remain unaltered in their efficacy over time, hardly a realistic expectation.

Management strategies

There is no doubt that corrective surgical procedures such as tail docking and removal of skin folds in the breech area (mulesing) are amongst the most effective methods to reduce the predisposition of sheep to blowfly strike. Such measures, although questioned on the basis of animal welfare (French *et al.*, 1994), offer life-long protection against fly strike in sheep. For this reason, the use of the surgical control methods is widespread in Australia, where the ratio of sheep (DSE, dry sheep equivalent) per labour unit now exceeds 6000 in economically viable units (Abbot, personal communication). The use of mulesing and tail docking is less widespread in developing countries, or countries where sheep production is less extensive, allowing for a much lower ratio of sheep/labour unit, and hence greater scope for individual animal attention. Strategic removal of wool from the main areas inherently susceptible to fly strike (breech, pizzle and poll in rams) also greatly reduces predisposition to fly strike, but protection is generally not long lasting.

The Australian wool industry has now become reliant on a combination of effective management practices, opportunistic use of insecticides and the use of suitable genotypes for the management of fly strike. The emphasis or need for additional or replacement blowfly control options will certainly increase in the future. Use of insecticides will decrease as consumer demands favour lower chemical residues and contamination in wool, and blowflies are likely to develop genetic resistance to current insecticides. The use of surgical modifications as we know them today, will eventually cease on grounds of animal welfare, driven by consumers.

Future and upcoming control strategies

Significant research effort has been devoted to develop alternative or complementary forms of blowfly-strike control. Those that have received the most attention are described below.

Genetic or biological control of blowfly populations

The manipulation of pest populations through biological or genetic means has worked remarkably well for some arthropod pests (i.e. screw-worm control). In the case of blowfly populations, genetic control through irradiated or genetically modified release stocks has been shown to be technically possible, but economically unattractive. For a full review of this topic, the reader is referred to reviews by Foster (1989) and Foster *et al.* (1992). Biological control of *L. cuprina* populations through predation by the microsporidium *Octosporea muscaedomesticae* (Smallridge *et al.*, 1995), *Hymenoptera* spp. (Bishop *et al.*, 1996) or entomopathic agents delivered through engineered fleece microflora (*Bacillus thuringiensis* toxins (Bt): Pinnock, 1994; Lyness *et al.*, 1994) have been demonstrated to be efficacious on a limited scale. However, these control strategies have to overcome significant practical limitations before being used on a large scale.

Immunological control of blowflies

Vaccination against the larval stages of blowflies, or the main predisposing dermatoses of blowfly strike (namely fleece rot and dermatophilosis infection) are potentially highly attractive control strategies. Where vaccination has been shown to be an effective means of disease control (primarily microbial infectious agents), there is little doubt that such means are cost effective and amenable to widespread and sustained use as a single control option. In the case of parasitic diseases, vaccination is less effective and often requires additional methods of disease control. The development of immunological means of blowfly-strike control has been a long and arduous route (Sandeman, 1990). Initial stimulation for the work was derived from the observation that sheep with repeated infection developed a weak but non-sustained immunological response to blowfly strike. In addition, the control of the cattle tick, *Boophilus microplus*, through vaccination against antigens not normally seen by the host (concealed antigens) has been shown to be effective. Despite 15 years of intensive research, there is currently no vaccine against blowfly products that offers sufficient protection. At best, blowfly vaccines show a marginal reduction in larval growth rates under highly artificial conditions of fly strike. For a comprehensive review of the development and current status of such vaccines, the reader is referred to Tellam and Bowles (1997). In similar vein, vaccines with partial efficacy against the microbial agents that cause fleece rot and dermatophilosis have been developed (Burrell and MacDiarmid, 1983; Burrell, 1985; Sutherland *et al.*, 1991), but provide insufficient protection against blowfly strike. Not until we have a clear and comprehensive understanding of the aetiology of blowfly strike, associated immune responses, and identity of the key immunogens, can we develop effective vaccines.

Other forms of blowfly control

These have focused on physical trapping of blowflies to reduce fly density (Anderson *et al.*, 1990; Wall and Smith, 1997; Morris *et al.*, 1998; Smith and Wall, 1998). Like biological control programmes, this form of control has not been shown to be feasible or effective on a large scale. The development of transgenesis in sheep has presented researchers with the theoretical option to develop hosts with novel parasite control genes such as chitinase or Bt production in sweat or wax glands (Ward *et al.*, 1993). At present such research is largely conceptual, and far removed from practical application.

Breeding for Resistance as Part of Integrated Pest Management (IPM)

It is clear from our understanding of blowfly strike in sheep that many forms of control, when used in isolation, offer at best partial protection. It is also clear that sheep producers are able to integrate, often on an *ad hoc* basis, the various forms of fly-strike control that may give them the best form of protection. Based on our current knowledge of blowfly-strike control, we could enhance the efficacy of individual control strategies by combining them in an integrated form of blowfly pest management control. This section will deal briefly with IPM to describe its advantages and prerequisite characteristics to make it successful. Most of the material comes from Bram (1994), who describes integrated control of ectoparasites.

Definition and general components of IPM

In the livestock context, IPM consists of the systematic application of two or more technologies in an environmentally compatible and cost-effective manner, to control arthropod pest populations that adversely affect livestock production. In addition, it is now recognized that such integrated control programmes need to be sustainable in their efficacy. In designed IPM programmes, technologies are applied deliberately and systematically (rather than at random) to optimize efficacy of the combined control strategy.

Chemo-prophylactic control is likely to be the backbone of any pest management system. Not only should novel compounds be considered but also novel modes of delivery to allow judicious use of insecticides at any stage of production. For some forms of pests, biological and mechanical control can provide strong complementary components of an IPM programme. In addition, genetic manipulation of the pest and the host through conventional selection may provide useful tools. With the addition of immunological control, a combined arsenal of technologies can be applied to pest control. The key features of an IPM programme is the strategic use of such combinations.

Feasibility of control of blowfly strike through IPM

From population studies involving both field data and modelling on both *L. cuprina* and *L. sericata*, it is evident that the two critical determinants affecting prevalence of blowfly strike are population density of blowflies at low densities, and sheep susceptibility once a threshold of blowfly population density has been exceeded (Vogt and Woodburn, 1980; Wardhaugh *et al.*, 1989; Wardhaugh and Morton, 1990; Wall *et al.*, 1993b, 1995; French and Morgan, 1996; Gleeson and Heath, 1997; Fenton *et al.*, 1998, 1999). The climatic variables of temperature and rainfall are of overriding importance in the determination of blowfly population and sheep susceptibility. Integrated control strategies which can manipulate blowfly populations, sheep susceptibility and which can predict optimum interactive use of the various control strategies should lead to effective IPM for blowfly control for a wide range of environments under which sheep are farmed.

From details presented above, it is obvious that many of our current control strategies for blowfly strike can be readily applied to an IPM programme. It is unlikely that the core components in blowfly-strike control will change from those currently in place, rather the manner in which they should be used in combination may change. There is a paucity of information on the critical interaction of the variables that control blowfly populations. IPM requires a clear understanding of the degree of control that is possible through chemo-prophylaxis, the rate of development of resistance, the extent of protection offered by resistant hosts (through management, genetic or immunological means) and environmental factors that influence blowfly populations. We have knowledge of the individual components but no formal evaluation or prediction of an integrated approach. There is now reliable and consistent evidence that selective breeding of sheep for resistance to body strike, poll strike in rams and possibly breech strike can become a major tool in IPM (Raadsma and Rogan 1985, 1987; Raadsma 1987, 1991a, b; Raadsma *et al.*, 1998).

Scope for Further Research in Breeding

There are a few areas that warrant further consideration in breeding for resistance to fly strike. These are outlined briefly below.

Across-flock predictions

Although the Merino is relatively susceptible to fly strike, extensive variation has been observed between stud-lines in their relative susceptibility. Sheep from some stud-lines require minimal intervention for fly-strike control and would provide a rapid means for sheep breeders to access superior genetic seedstock. The difficulty lies in finding a reliable and convenient means to predict the relative susceptibility of the many stud-lines available to breeders. Comparative benchmarking with appropriate reference links across environments is one avenue to overcoming this difficulty. Identification of resistant studs on the basis of indicator traits or genetic markers for resistance genes may provide additional avenues (Raadsma, 1991b).

Molecular characterization of resistance

With the development of a high-resolution genetic linkage map of highly polymorphic microsatellite DNA markers (Montgomery and Crawford, 1997) it is now feasible to identify chromosomal locations for genes contributing to resistance. Should genes with a sufficiently large effect be identified, indirect selection for resistance with the aid of markers (marker-assisted selection, MAS) can provide a means for increasing resistance without the need for direct challenge. Although many experiments are currently in progress to identify such markers for production traits and disease resistance, no such experiments have been designed for resistance to blowfly strike.

Selection as an alternative to mulesing and tail docking

Tail docking and mulesing are the most powerful tools in the control of blowfly strike. Despite the many advantages of these surgical management tools, there is considerable misgiving from public and farmers alike for the procedures. It is highly likely that these procedures will be phased out in the future. Raadsma and Rogan (1985) reviewed potential options as an alternative control strategy. Breeding was considered the most promising means by which to control breech strike in the period preceding the development of the mules operation (Seddon *et al.*, 1931; MacKerras, 1936; Belschner, 1937).

Modelling benefits of inclusion of breeding in IPM programmes

There are now a number of models that can simulate or predict blowfly abundance and activity. Predictions can be made of flystrike prevalence through incorporation of environmental, seasonal and known risk factors leading to fly strike (Wardhaugh *et al.*, 1989; Wall *et al.*, 1993a; Fenton and Wall, 1997; Fenton *et al.*, 1999). Although such models are extremely useful in predicting blowfly activity and potential impact, they do not extend far enough to model optimum use of IPM. Incorporation of pesticide management strategies and inclusion of genetic selection for increased host resistance are two additional components that warrant serious consideration. In addition, selection decisions for increased resistance need to be compatible with the overall breeding objective for sheep production under high risk of blowfly strike.

References

Anderson, J.M.E., McLeod, L.J., Shipp, E., Swan, A. and Kennedy, J.P. (1990) Trapping sheep blowflies using bait-bins. *Australian Veterinary Journal* 67, 93–97.

Belschner, H.G. (1937) Studies on the sheep blowfly problem. *Department of Agriculture, New South Wales, Science Bulletin*, No. 54.

Bishop, D.M., Heath, A.C.G. and Haack, N.A. (1996) Distribution, prevalence and host associations of Hymenoptera parasitic on Calliphoridae occurring in flystrike in New Zealand. *Medical and Veterinary Entomology* 10, 365–370.

Bowles, V.M., Carnegie, P.R. and Sandeman, R.M. (1987) Immunization of sheep against infection with larvae of the blowfly, *Lucilia cuprina*. *International Journal of Parasitology* 17, 759–765.

Bowles, V.M., Grey, S.T. and Brandon, M.R. (1992) Cellular immune responses in the skin of sheep infected with larvae of *Lucilia cuprina*, the sheep blowfly. *Veterinary Parasitology* 44, 151–162.

Bram, R.A. (1994) Integrated control of ecto-parasites. *Revues Scientifiques et Technicales, Office International des Epizooties* 13, 1357–1365.

Burrell, D.H. (1985) Immunisation of sheep against experimental *Pseudomonas aeruginosa* dermatitis and fleece rot associated body strike. *Australian Veterinary Journal* 62, 55–57.

Burrell, D.H. and MacDiarmid, J.A. (1983) The role of *Pseudomonas* in pathogenesis of fleece rot and indirect control of body strike by immunization. In: Raadsma, H.W. (ed.) *Second National Symposium on Sheep Blowfly and Flystrike in Sheep.* NSW Department of Agriculture, Sydney, pp. 283–291.

Chin, J.C. and Watts, J.E. (1991) Dermal and serological response against *Pseudomonas aeruginosa* in sheep bred for resistance and susceptibility to fleece-rot. *Australian Veterinary Journal* 68, 28–31.

Colditz, I.G., Woolaston, R.R., Lax, J. and Mortimer, S.I. (1992) Plasma leakage in skin of sheep selected for resistance or susceptibility to fleece rot and flystrike. *Parasite Immunology* 14, 587–594.

Colditz, I.G., Lax, J., Mortimer, S.I., Clarke, R.A. and Beh, K.J. (1994) Cellular inflammatory responses in skin of sheep selected for resistance or susceptibility to fleece rot and flystrike. *Parasite Immunology* 16, 289–296.

Colditz, I.G., Eisemann, C., Tellam, R.L., McClure, S.J., Mortimer, S.I. and Husband, A.J. (1996) Growth of *Lucilia cuprina* larvae following treatment of sheep divergently selected for fleece rot and flystrike with monoclonal antibodies to T lymphocyte subsets and interferon gamma. *International Journal for Parasitology* 26, 775–782.

Eissemann, C.H., Johnston, L.A.Y., Broadmeadow, M., O'Sullivan, B.M., Donaldson, R.A., Pearson, R.D., Vuocolo, T. and Kerr, J.D. (1990) Acquired resistance of sheep to larvae of *Lucilia cuprina*, assessed *in vivo* and *in vitro*. *International Journal of Parasitology* 20, 299–305.

Engwerda, C.R., Dale, C.J. and Sandeman, R.M. (1996) IgE, TNF alpha, IL1 beta, IL4 and IFN gamma gene polymorphisms in sheep selected for resistance to fleece rot and flystrike. *International Journal for Parasitology* 26, 787–791.

Fenton, A. and Wall, R. (1997) Sensitivity analysis of a stochastic model for the sheep blowfly *Lucilia sericata. Journal of Applied Ecology* 34, 1023–1031.

Fenton, A., Wall, R. and French, N. (1998) The incidence of sheep strike by *Lucilia sericata* on sheep farms in Britain: a simulation model. *Veterinary Parasitology* 76, 211–228.

Fenton, A., Wall, R. and French, N.P. (1999) The effect of farm management strategies on the incidence of sheep strike in Britain: a simulation analysis. *Veterinary Parasitology* 79, 341–357.

Foster, G.G. (1989) The sheep blowfly [*Lucilia cuprina*] genetic control program in Australia. *Insect and Pest Control Newsletter* No. 43, 23–26.

Foster, G.G., Weller, G.L., James, W.J., Paschalidis, K.M. and McKenzie, L.J. (1992) Advances in sheep blowfly genetic control in Australia. In: Management of Insect Pests: Nuclear and Related Molecular and Genetic Techniques. *Proceedings of the International Atomic Energy Agency Symposium, Vienna*, pp. 299–312.

French, N.P. and Morgan, K.L. (1996) A model of ovine cutaneous myosis using the predicted abundance of *Lucilia sericata* and a pattern of sheep susceptibility. *Preventive Veterinary Medicine* 26, 143–155.

French, N.P., Wall, R. and Morgan, K.L. (1994) Lamb tail docking: a controlled field study of the effects of tail amputation on health and productivity. *Veterinary Record* 134, 463–467.

Gleeson, D.M. and Heath, A.C.G. (1997) The population biology of the Australian sheep blowfly, *Lucilia cuprina*, in New Zealand. *New Zealand Journal of Agricultural Research* 40, 529–535.

Gray, G.D., Woolaston, R.R. and Eaton, R.T. (eds) (1995) *Breeding for Resistance to Infectious Diseases of Small Ruminants.* Watson Ferguson Pty, Brisbane, Australia, ACIAR Monograph Series, No. 34.

Hall, I.M. and Wall, R. (1995) Myiasis of human and domestic animals. *Advances in Parasitology* 35, 257–334.

Hart, R.J., Cavey, W.A., Ryan, K.J., Strong, M.B., Moore, B., Thomas, P.L., Boray, J.C. and Orelli, M. von (1982) A new sheep blowfly insecticide. *Australian Veterinary Journal* 59, 104–109.

Hughes, P.B. and Mckenzie, J.A. (1987) Insecticide resistance in the Australian sheep blowfly, *Lucilia cuprina*: speculation, science and strategies. In: Ford, M.G., Holloman, D.W., Khambay, B.P.S. and Sawicki, R.M. (eds) *Combating Resistance to Xenobiotics. Biological and Chemical Approaches.* Ellis Horwood, London, pp. 162–177.

Levot, G.W. (1993) Insecticide resistance: new developments and future options for fly and lice control on sheep. *Wool Technology and Sheep Breeding,* 41 (2), 108–119.

Levot, G.W. (1995) Resistance and the control of sheep ecto parasites. *International Journal of Parasitology,* 25, 1355–1362.

Lyness, E.W., Pinnock, D.E. and Cooper, D. (1994) Microbial ecology of sheep fleece. *Agriculture Ecosystems and Environment* 49, 103–112.

McCleod, R.S. (1995). Costs of major parasites to the Australian livestock industries. *International Journal for Parasitology* 25, 1363–1367.

MacKerras, I.M. (1936) *The Sheep Blowfly Problem in Australia.* Council for Industrial and Scientific Research Pamphlet, No. 66.

Montgomery, G.W. and Crawford, A.M. (1997) The sheep linkage map. In: Piper, L. and Ruvinsky, A. (eds) *The Genetics of Sheep.* CAB International, Wallingford, pp. 297–352.

Morris, M.C., Morrison, L., Joyce, M.A. and Rabel, B. (1998) Trapping sheep blowflies with lures based on bacterial cultures. *Australian Journal of Experimental Agriculture* 38, 125–130.

Mortimer, S.I., Atkins, K.D. and Raadsma, H.W. (1998) Responses to selection for resistance and susceptibility to fleece rot and body strike in Merino sheep. *Proceedings of the 6th World Congress on Genetics Applied to Livestock Production* 27, 283–286.

O'Meara, T.J. and Raadsma, H.W. (1995) Phenotypic and genetic indicators of resistance to ectopathogens. In: Gray, G.D., Woolaston, R.R. and Eaton, B.T. (eds) *Breeding for Resistance to Infectious Diseases of Small Ruminants.* ACIAR Monograph Series, No. 34, pp. 187–218.

O'Meara, T.J., Nesa, M., Raadsma, H.W., Saville, D.G. and Sandeman, R.M. (1992) Variation in skin inflammatory responses between sheep bred for resistance or susceptibility to fleece rot and blowfly strike. *Research in Veterinary Science* 52, 205–210.

O'Meara, T.J., Nesa, M., Seaton, D.S. and Sandeman, R.M. (1995) A comparison of inflammatory exudates released from myiasis wounds on sheep bred for resistance or susceptibility to *Lucilia cuprina. Veterinary Parasitology* 56, 207–223.

O'Meara, T.J., Nesa, M. and Sandeman, R.M. (1997) Antibody responses to *Lucilia cuprina* in sheep selected for resistance or susceptibility to *L. cuprina. Parasite Immunology* 19, 535–543.

Pinnock, D.E. (1994) The use of *Bacillus thuringiensis* for control of pests of livestock. *Agriculture Ecosystems and Environment* 49, 59–63.

Raadsma, H.W. (1987) Flystrike control: an overview of management and breeding options. *Wool Technology and Sheep Breeding* 35 (3), 174–185.

Raadsma, H.W. (1991a) Genetic variation in resistance to fleece rot and flystrike. In: Owen, J.B. and Axford, R.F.E. (eds) *Breeding for Disease Resistance in Farm Animals.* Proceedings of the International Symposium, Bangor, Wales, Sept. 1990. CAB International, Wallingford, pp. 264–290.

Raadsma, H.W. (1991b) Genetics of resistance to fleece rot and footrot. *Proceedings Australian Association of Animal Breeding and Genetics* 9, 67–72.

Raadsma, H.W. (1991c) Fleece rot and body strike in Merino sheep. V. Heritability of liability to body strike in weaner sheep under flywave conditions. *Australian Journal of Agricultural Research* 42, 279–253.

Raadsma, H.W. (1992) The susceptibility to body strike under high rainfall conditions of flocks selected for and against fleece rot. *Australian Journal of Experimental Agriculture* 31, 757–759.

Raadsma, H.W. (1993) Fleece rot and body strike in Merino sheep. VI. Experimental evaluation of some fleece and body traits for indirect selection against fleece rot. *Australian Journal of Agricultural Research* 44, 915–933.

Raadsma, H.W. and Rogan, I.M. (1985) Potential for genetic improvement in resistance to blowfly strike. In: Moore, B.L and Chenoweth, P.J. (eds) *Grazing Animal Welfare Symposium.* Queensland branch of the Australian Veterinary Association, pp. 48–56.

Raadsma, H.W. and Rogan, I.M. (1987) Genetic variation in resistance to blowfly strike. In: McGuirk, B.J. (ed.) *Merino Improvement Programs in Australia.* Proceedings of a National Symposium, Leura, New South Wales. Australian Wool Corporation, Melbourne, pp. 321–340.

Raadsma, H.W., Sandeman, R.M., Sasiak, A.B., Engwerda, C.R. and O'Meara, T.J. (1992) Genetic improvement in resistance to body strike in Merino sheep: where are we at with indirect selection? *Proceedings Australian Association of Animal Breeding and Genetics* 10, 143–146.

Raadsma, H.W., Gray, D.G. and Woolaston, R.R. (1997a) Genetics of disease resistance and vaccine response. In: Piper, L.R. and Ruvinsky, A. (eds) *Genetics of the Sheep.* CAB International, Wallingford, UK, pp. 199–224.

Raadsma, H.W., Nicholas, F.W. and Egerton, J.R. (1997b) Ultimate disease resistance in sheep: what are the relationships between all diseases? *Proceedings Australian Association of Animal Breeding and Genetics* 12, 63–67.

Raadsma, H.W., Gray, D.G. and Woolaston, R.R. (1998) Breeding for disease resistance in Merino sheep in Australia. In: Brem, G. (ed.) Genetic Resistance to Animal Diseases. *Revues Scientifiques et Technicales, Office International des Epizooties* 17, 315–326.

Sandeman, R.M. (1990) Prospects for the control of sheep blowfly strike by vaccination. *International Journal for Parasitology* 20, 537–541.

Sandeman, R.M., Chandler, R.A., Turner, N. and Seaton, D.S. (1995) Antibody degradation in wound exudates from blowfly infections on sheep. *International Journal for Parasitology* 25, 621–628.

Sandeman, R.M., Bowles, V.M., Stacey, I.W. and Carnegie, P. (1986) Acquired resistance in sheep to infection with larvae of the blowfly, *Lucilia cuprina. International Journal for Parasitology* 16, 69–75.

Seaton, D.S., O'Meara, T.J., Chandler, R.A. and Sandeman, R.M. (1992) The sheep antibody response to repeated infection with *Lucilia cuprina. International Journal for Parasitology* 22, 1169–1174.

Seddon, H.R., Belschner, H.G. and Mulhearn, C.R. (1931) Studies on cutaneous myiasis of sheep (sheep blowfly attack). *Department of Agriculture, New South Wales, Science Bulletin*, No. 37.

Smallridge, C.J., Cooper, D.J. and Pinnock, D.E. (1995) The effect of the microsporidium *Octosporea muscaedomesticae* on adult *Lucilia cuprina* (Dipthera: Calliphoridae). *Journal of Invertebrate Pathology* 66, 196–197.

Smith, K.E. and Wall, R. (1998) Suppression of the blowfly *Lucilia sericata* using odour-baited triflumuron-impregnated targets. *Medical and Veterinary Entomology* 12, 430–437.

Sutherland, S.S., Ellis, T.M. and Edwards, J.R. (1991) Evaluation of vaccines against Dermatophilosis. *Veterinary Micobiology* 27, 91–100.

Tellam, R.L. and Bowles, W.M. (1997) Control of blowfly strike in sheep: current strategies and future prospects. *International Journal for Parasitology* 27, 261–273.

Tellam, R.L., Eisemann, C.H. and Pearson, R.D. (1994) Vaccination of sheep with purified serine proteases from the secretory and excretory material of *Lucilia cuprina* larvae. *International Journal for Parasitology* 24, 757–764.

Toure, S.M. (1994) Les myiases d'importance economique. *Revues Scientifiques et Technicales, Office International des Epizooties* 13, 1053–1073.

Vogt, W.G. and Woodburn, T.L. (1980) The influence of temperature and moisture on the survival and duration of the egg stage of the Australian sheep blowfly, *Lucilia cuprina* (Wiedemann) (Diptera: Calliphoridae). *Bulletin of Entomological Research* 70, 665–671.

Wall, R. and Shearer, D. (1997) *Veterinary Entomology: Arthropod Ectoparasites of Veterinary Importance.* Chapman & Hall, London.

Wall, R. and Smith, K.E. (1997) The potential for control of the blowfly *Lucilia sericata* using odour-baited targets. *Medical and Veterinary Entomology* 11, 335–341.

Wall, R., French, N.P. and Morgan, K.L. (1993a) Sheep blowfly population control: development of a simulation model and analysis of management strategies. *Journal of Applied Ecology* 30, 743–51.

Wall, R., French, N.P. and Morgan, K.L. (1993b) Predicting the abundance of the blowfly *Lucilia sericata* (Diptera: Calliphoridae). *Bulletin of Entomological Research* 83, 431–436.

Wall, R., French, N.P. and Morgan, K.L. (1995) Population suppression for control of the blowfly *Lucilia sericata* and sheep blowfly strike. *Ecological Entomology* 20, 91–97.

Ward, K.A., Brownlee, A.G., Leish, Z. and Bonsing, J. (1993) Genetic manipulation for disease control. *Proceedings, VII World Conference on Animal Production, Edmonton, Alberta, Canada, June 28–July 2, 1993*, Vol. 1. University of Alberta, Edmonton, pp. 267–280.

Wardhaugh, K.G. and Morton, R. (1990) The incidence of flystrike in sheep in relation to weather conditions, sheep husbandry, and the abundance of the Australian sheep blowfly, *Lucilia cuprina* (Wiedemann) (Diptera: Calliphoridae). *Australian Journal of Agricultural Research* 41, 1155–1167.

Wardhaugh, K.G., Vogt, W.G., Dallwitz, R. and Woodburn, T.L. (1989) The incidence of flystrike in relation to sheep susceptibility and the abundance of the blowfly *Lucilia cuprina* (Wiedemann) (Diptera: Calliphoridae). *General and Applied Entomology* 21, 11–16.

9

Exploitation of Resistance to Trypanosomes and Trypanosomosis

G.D.M. d'Ieteren[1], E. Authié[2], N. Wissocq[1] and M. Murray[3]

[1]International Livestock Research Institute (ILRI), Nairobi, Kenya;
[2]Programme Santé Animale, CIRAD-EMVT, Montpellier, France;
[3]Department of Veterinary Clinical Studies, University of Glasgow Veterinary School, Glasgow, UK

Summary

While an increasing number of examples of innate resistance to disease are being identified in domestic livestock, resistance to the effect of trypanosomes is one of the best-recognized and most thoroughly investigated. Experimental and field studies reviewed in this chapter are providing the basic tools with which the trypanotolerance trait can be identified and exploited. Comprehensive evaluation of the degree of genetic determination of the different disease resistance traits, their heritability and their genetic correlations with each other and with animal performance traits should allow progress to be made in the development of breeding programmes and policies.

There is increasing recognition that Africa possesses animal genetic resources probably unparalleled in any other continent. Evidence that these resources can provide sustainable and environmentally sound solutions for some of the vast disease problems currently confronting Africa is now being found. Thus, the natural innate resistance possessed by breeds of cattle, such as the N'Dama and the West African Shorthorn, to trypanosomosis and to several other important infectious diseases is now accepted as an important component of national and regional disease-control programmes. The fact that these breeds also possess considerable production potential and that their disease-resistance traits could be exploited in crossbreeding offers an unparalleled opportunity to improve livestock production in the vast areas of Africa dominated by the tsetse fly, ticks and helminths, particularly as more market-orientated production evolves.

Introduction

The African continent is faced with the challenge of satisfying a dramatic increase in demand for livestock products, in particular milk and meat.

© CAB *International* 2000. *Breeding for Disease Resistance in Farm Animals*
(eds R.F.E. Axford, S.C. Bishop, F.W. Nicholas and J.B. Owen)

Domesticated species play an important role in supporting human populations and in generating income and economic activity. The areas with the greatest potential for significant increases in livestock population and livestock productivity are the subhumid and the non-forested parts of the humid zones. Large areas of natural grassland could be exploited better to support the increasing demand for livestock products if constraints to, and opportunities for, their increased contribution to market economies were understood and overcome. Animal diseases, particularly those caused by parasites, are major constraints to animal production in these areas, and trypanosomosis is arguably the most important of these. Jahnke *et al.* (1988) considered that a total increase in cattle of 33 million heads might be possible and would lead to an additional production of 495,000 t of meat per year (assuming productivity of 15 kg $head^{-1}$ $year^{-1}$) and an increase in milk production of 1.26 Mt $year^{-1}$ (using estimates of 38.3 kg milk $head^{-1}$ $year^{-1}$) if eradication or sustainable control of trypanosomosis were achieved over the entire tsetse fly-affected area in the subhumid and humid zones (7 million km^2). The potential benefit of sustainable control of trypanosomosis is considerable for the 40 countries in Africa affected by this disease. Non-tsetse-transmitted trypanosomosis appears also to be a growing problem in Latin America (Davila *et al.*, 1998).

Trypanosomes, Major Hindrance to Sustainability of Livestock Production in the Tropics

This chapter draws its major components from a recent review by d'Ieteren *et al.* (1998). Other relevant recent reviews that should be consulted include Murray *et al.* (1990, 1991). Pathogenic species of salivarian trypanosomes are present throughout vast areas of Africa, Asia, Latin America and the Middle East, and cause disease in cattle, sheep, goats, water buffalo, pigs, equidae, camels, wildlife and humans. In Africa, the major pathogenic trypanosome species for livestock are transmitted by the tsetse fly (genus *Glossina*) and include *Trypanosoma congolense*, *T. vivax*, *T. brucei brucei* and *T. simiae*. The subspecies of *T. brucei*, *T. b. rhodesiense* and *T. b. gambiense* cause sleeping sickness in humans. Forms of trypanosomosis not transmitted by tsetse flies also occur in Africa, as well as in the Middle East, Asia and Latin America. The most important pathogen, under these circumstances, is *T. evansi*. This parasite can cause severe disease in horses and camels and can lead to significant losses in production and performance in cattle and water buffalo (Murray *et al.*, 1990). The fact that domestic animals and wildlife also act as reservoir hosts for the human pathogens *T. b. rhodesiense* and *T. b. gambiense*, and that trypanosomosis in humans is an important health constraint to rural development in large areas of Africa, must also be noted. No other continent appears to be dominated by one disease to the same extent that Africa is dominated by trypanosomosis transmitted by tsetse flies. This disease not only results in severe losses in production in domestic livestock due to poor growth, weight loss, low milk yield, reduced capacity for work, infertility and abortion, but also impairs the development of animal agriculture in zones that constitute 41% of the land but which carry only 26% of the ruminant population. The annual loss in meat production alone was estimated at US$5 billion in 1984 (Murray and Gray, 1984), but this figure excludes milk, hides and mixed agriculture. In Africa, 80% of traction power is non-mechanized. A sixfold increase in agricultural output as a result of the availability of a draught ox to a family unit has been calculated (McDowell, 1977). Furthermore, the manure provided by livestock is essential for the production of food and cash crops and is a potential source of energy in the form of biogas. In view of this, and despite the continental impact of the disease, assessment of the global impact of trypanosomosis is difficult and costly, due to the need to carry out impact studies at the production system level taking into consideration the many complex interactions that affect farm outputs of which trypanosomosis risk is only one component: hence the scarcity of available, reliable, quantitative indicators of the direct and indirect impact of trypanosomosis and of its control.

Indeed, many factors contribute to the magnitude of the problem of African trypanosomosis. One major factor is the complexity of the disease itself: for example, the genetic diversity of pathogenic trypanosomes and the frequent occurrence of multiple infections. These trypanosomes are transmitted cyclically by the tsetse fly, of which there are at least 36 species and subspecies, each adapted to different climatic and ecological conditions (Ford, 1971). While the tsetse fly is not the only vector of African trypanosomes, cyclical transmission of infection represents the most important problem because the tsetse fly, once infected, remains infective for a long period, in contrast to the ephemeral nature of non-cyclical transmission. At the same time, trypanosomes infect a wide range of hosts, including wild and domestic animals. The former do not usually suffer severe clinical disease but become carriers and constitute an important reservoir of infection. The success of the trypanosome as a parasite is due largely to its ability to undergo antigenic variation, i.e. to change a single glycoprotein (called variant surface glycoprotein (VSG); Cross, 1975), thereby enabling evasion of host immune responses and the establishment of persistent infections. In addition to multiplicity of variable antigen types (VATs) expressed by a single parasite, each trypanosome species comprises an unknown number of different strains and clones, all capable of elaborating a different repertoire of VATs (Van Meirvenne *et al.*, 1977).

Trypanosomosis Control

Besides the complexity of the disease and its epidemiology, other factors contributing to the failure to contain and reduce the problem of trypanosomosis include the enormous geographical area affected and the limitations of methods currently available for extensive control. The sustained public efforts and resources required to achieve definitive solutions for large-scale vector eradication are paradoxical in an era when the public sector is less involved than ever before (as a result of the privatization of animal health and veterinary services), and when sustainability is increasingly believed to rely on private initiative and resources. Recent reviews of trypanosomosis control elaborated on the possible causes of the difficulties encountered in the sustainable economic development of trypanosomosis-affected areas (Murray *et al.*, 1990, 1991; Peregrine, 1994; Holmes, 1997; d'Ieteren *et al.*, 1998). While the eradication of trypanosomosis is an unrealistic goal for most of Africa, considerable effort has been invested in control of this disease by the use of trypanocidal drugs, management of the vector and exploitation of the genetic resistance exhibited by indigenous breeds, such as N'Dama cattle and Djallonké sheep.

The use of trypanocidal drugs is well established and represents the most widely adopted approach to control trypanosomosis. However, estimates available in 1984 revealed that only 25–30 million curative or prophylactic treatments are given every year for the 48 million cattle exposed to trypanosomosis; thus there is scope for increased use (Murray and Gray, 1984). Considering the number of recent reports of trypanosome resistance to current drugs, drug resistance can be assumed to be developing faster than generally thought. Not only are individual cases recognized (Peregrine, 1994), but regional distribution is increasingly reported in East and West Africa (Clausen *et al.*, 1992; d'Ieteren *et al.*, 1997; Rowlands *et al.*, 1993). As there appears to be little hope for developing new trypanocidal drugs to benefit smallholder farmers in the short term, major consideration is now being given to better use of the existing drugs, with appropriate guidelines for delaying the development of drug resistance (Geerts and Holmes, 1999). In the past two decades a considerable pool of knowledge and expertise has been built up on trypanocidal drug control. Information technology offers today novel and effective methods of structuring and providing access to such knowledge and expertise. Decision support systems for more efficient management of livestock disease are becoming available, particularly for evaluation of most effective and cost-efficient drug regimes for treatment and prophylaxis of bovine trypanosomosis. They include information on trypanocidal drug resistance (Perry *et al.*, 1998). However, given the actual or potential problem of drug resistance in many areas, drug usage clearly cannot be relied upon continuously as the sole method of trypanosomosis control.

Attempts to control tsetse flies have been made over many decades. Initially, these attempts included eradication of wildlife, clearing of fly barriers to prevent the advance of the vector, widespread bush clearing to destroy breeding habitats, and ground and aerial insecticide spraying. The release of sterile males has been used for the eradication of tsetse flies in isolated areas such as on the island of Zanzibar. The main method currently employed to control tsetse flies is the use of synthetic pyrethroid insecticides to impregnate traps and screens, baited or not with odour attractants (Jordan, 1986). More recently, live animals have been used as targets by spraying, dipping or by pour-on treatments (Bauer *et al.*, 1992). This approach offers a major possibility of reducing maintenance costs and of combining tick control with tsetse control (Barrett, 1997) and may be well accepted by farmers as a mixed public–private good (Swallow *et al.*, 1995). However, the limits of these techniques still need to be assessed before wider adoption is promoted. Holmes (1997) reviewed some of the factors that determine the probability of success:

for example, cattle densities required to ensure effective control, the scope for treating only a proportion of the cattle, the level of fly–cattle interactions as measured by fly blood-meal analyses, etc.

One of the major components of sustainability of the recently developed tsetse fly control methods is the active participation of the majority of communities contributing to a relevant production system in a given environment or region. Major economic incentives are thus required for these techniques to be accepted by farmers for collective action, as compared to methods of a more private nature, such as the use of curative or prophylactic drugs or trypanotolerant livestock. The very private nature of breed choices and breeding strategies for parasite or disease resistance traits can explain the major success of trypanotolerant cattle as a component of, or as a single option for, livestock production under trypanosomosis risk. The exploitation of disease resistance traits to trypanosomosis and dermatophilosis – two very important diseases of humid Central and West Africa – has certainly been seen by farmers as an integral part of their production approaches to the alleviation of disease constraints for many decades (d'Ieteren *et al.*, 1994).

A potential vaccine against trypanosomosis is an attractive option. However, there is little hope that a conventional, anti-infection vaccine will be produced in the near future, due to the antigenic variation of trypanosomes and the complexity of their antigenic repertoire. The concept of antidisease vaccines for parasitic diseases is one that has gained acceptance in recent years and has received attention in the area of trypanosomosis research. Searching for parasite components that mediate pathology and attempting to inhibit these 'toxins' through immunization is a promising avenue for improving trypanosomosis control. In this context, trypanotolerant cattle are an invaluable research resource.

Trypanotolerance: a Major Asset for Sustainable Livestock Production Under Trypanosomosis Risk

Trypanotolerance, the ability of some livestock species and breeds to survive, reproduce and remain productive under trypanosomosis risk without the aid of trypanocidal drugs, was recognized and exploited by farmers long before research on trypanotolerance began. The exploitation of trypanotolerant breeds is practised as a major (if not the only) option for sustainable livestock production in 19 countries in the most humid parts of West and Central Africa. In 11 countries, trypanotolerant cattle (mainly N'Dama) were either moved into the highest risk areas or were imported from other countries. There are now N'Dama herds in nearly all West and Central African countries, and this could be the source of genetic material for further dissemination (d'Ieteren, 1994). Trypanotolerance in cattle is well documented, particularly in N'Dama cattle, the most numerous trypanotolerant breed, and in the West African Shorthorn. While significant differences in resistance to trypanosomosis occur also among various zebu (*Bos indicus*) types (Ismael and Njogu, 1985; Njogu *et al.*, 1985; Dolan *et al.*, 1994; Mwangi *et al.*, 1994), most *B. indicus* cattle in tsetse fly infested areas require regular treatment or are found only on the fringes of fly belts. Exotic breeds cannot be maintained even in areas of low tsetse fly risk without intensive trypanocidal drug therapy and veterinary care.

There is a continued perception that because of their small size, trypanotolerant livestock are less productive than other breeds (Holmes, 1997). The International Livestock Centre for Africa (ILCA) demonstrated, however, that herd productivity of N'Dama cattle raised under very high trypanosomosis risk was equal to the herd productivity of Boran cattle also maintained under high disease risk but with permanent chemo-prophylaxis (Feron *et al.*, 1988; Trail *et al.*, 1985). Trypanotolerant N'Dama cattle thus compare very well with Boran cattle, which are regarded as one of the best beef cattle breeds in Africa, with the added advantage that N'Dama cattle are not dependent on trypanocidal drugs, whereas Boran cattle would not survive without the drugs. Similarly, Agyemang *et al.* (1994) demonstrated that when milk extracted from N'Dama cattle for human consumption was taken into consideration, their overall productivity was superior to that of zebu breeds maintained under similar traditional systems in the absence of tsetse fly challenge.

While it is generally accepted that trypanotolerance as a breed characteristic is under genetic control, there is evidence that the stability of trypanotolerance can be affected by environmental factors, such as overwork, intercurrent disease and repeated bleeding, pregnancy, parturition, suckling and lactation, the single most important factor being nutrition. In the first large-scale attempt to evaluate the effect of trypanosomosis risk (defined rather subjectively, but using relevant information available at that time) on performance of N'Dama and West African Shorthorn cattle at 30 different locations, ILCA, The Food and Agriculture Organization and the United Nations Environment Programme demonstrated that although these breeds remain productive under trypanosomosis risk, their outputs were affected by increasing risk (ILCA, 1979). Later studies, reviewed by d'Ieteren *et al.* (1998), attempted to provide more accurate evaluation of the links between trypanosomosis risk components and animal performance. It was concluded that many interactions between risk, drugs and treatment regimes, livestock breeds and levels of disease resistance can influence farm outputs and the profitability of interventions.

There is no one solution that will be valid for all production systems, ecological zones, or regional or national markets. However, the decreasing efficacy of the trypanocidal drugs available and the difficulties of sustaining tsetse fly control increase the imperative need for enhancing trypanotolerance through selective breeding, either within breed or through cross-breeding. This last option is possibly the most appropriate for dairy-orientated production systems (Cunningham and Syrstad, 1987; Dempfle, 1993). These conclusions support the analysis by Holmes (1997) that more integrated strategies need to be developed. Thus livestock production under trypanosomosis risk will have to focus increasingly on those integrated control strategies that are more reliant on trypanotolerant livestock, on methods for increasing disease resistance and/or on improved vector control techniques, and possibly on less (or more careful) use of trypanocidal drugs.

Biology of Trypanotolerance

Trypanotolerance has been defined as the relative capacity of an animal to control the development of the parasites and to limit their pathological effects, the most prominent of which is anaemia (Murray *et al.*, 1982; Murray and Dexter, 1988). While a general relationship exists between the parasite load and the severity of anaemia, the mechanisms involved in each component of trypanotolerance – control of parasite development and mitigation of anaemia – appear to be distinct and to operate independently (Murray *et al.*, 1990, 1991). Understanding these mechanisms has been a major research goal over the past 20 years, first with the aim of identifying markers that could be used in selection for disease resistance, and secondly, in the hope to improve disease control. Conferring to susceptible cattle a similar capacity to that identified in resistant cattle, and optimizing this capacity in trypanotolerant breeds, are indeed challenging options to complement conventional control methods.

Considerable insight into the immunological events associated with trypanosome infection (reviewed by Mansfield, 1990; de Baetselier, 1996; Sternberg, 1997; Taylor, 1998) and into the genetic basis for trypanoresistance (Kemp *et al.*, 1997; Iraqi *et al.*, 1998; Iraqi and Teale, 1999) has been gained from the study of rodent models. However, trypanosomosis in mice and cattle differ in many aspects and no experimental model has led so far to identification of a mechanism that was also confirmed to contribute to bovine trypanotolerance. Thus, while tumour necrosis factor (TNF-α) plays a key role in trypanoresistance of mice (Iraqi *et al.*, 1998; Iraqi and Teale, 1999), studies in cattle provided conflicting results and failed to indicate the involvement of this cytokine in trypanotolerance (Sileghem *et al.*, 1994; Mertens *et al.*, 1999). A detailed review of the immunological events in trypanosome-infected mice and cattle has been published recently by Taylor (1998); the current section will focus on responses to infection in trypanotolerant and trypanosusceptible cattle.

Early studies on trypanotolerance were mostly carried out in field situations. Experiments in which cattle of unknown history and different physiological status were infected with uncharacterized parasites yielded conflicting results. More recently, comparative studies in African Zebu and Taurine cattle were carried out during controlled experimental infections with cloned, cyclically transmitted *T. congolense*. While parasitaemia and anaemia do not differ greatly between the two types of cattle in the initial stages of a primary infection, the capacity of Taurine cattle to limit parasitaemia and to resist anaemia becomes patent after 30–50 days of infection (Duvallet *et al.*, 1988; Paling *et al.*, 1991a, b), and increases upon rechallenge (Paling *et al.*, 1991a, b; Williams *et al.*, 1991).

Dissecting the cellular component of the bovine immune response to trypanosomes, addressing both T-cell (Lutje *et al.*, 1995a, b, 1996) and monocyte/macrophage functions (Taylor *et al.*, 1996a, 1998), has greatly improved our understanding of trypanosome-associated immunosuppression, but has failed to identify cellular events that differ fundamentally between susceptible and resistant cattle. Both types of cattle undergo T-cell suppression and their cytokine patterns are similar, except for IL-4, whose synthesis appears to be increased in trypanotolerant N'Dama cattle (Mertens *et al.*, 1999).

Since African trypanosomes are mainly intravascular and extracellular parasites, the major immunological differences associated with trypanotolerance are more likely to exist at the level of humoral immunity. Indeed, the most prominent immunological feature that has been identified so far is the capacity of trypanotolerant cattle to generate sustained antibody responses to trypanosome antigens. Following infection, both susceptible and trypanotolerant cattle develop transient trypanosome-specific IgM (Authié *et al.*, 1993a; Williams *et al.*, 1996). A distinct population of IgM, of low specificity, reacts with both trypanosome and non-trypanosome antigens (Williams *et al.*, 1996). Interestingly, these 'polyspecific' antibodies, which may be responsible for pathology rather than protection, have been detected in susceptible Boran but not in trypanotolerant N'Dama cattle (Williams *et al.*, 1996). The trypanosome-specific IgG response is elicited almost coincidentally with the IgM response and is predominantly of the IgG1 isotype in both types of cattle (Authié *et al.*, 1993a). However, trypanotolerant cattle develop higher and more sustained levels of specific IgG1 than susceptible Zebu cattle (Authié *et al.*, 1993a; Williams *et al.*, 1996). The difference is not the mere reflection of differential absorption of antibody by circulating antigen, but lies at the B-cell level (Taylor *et al.*, 1996b). The difference in specific IgG levels increases during rechallenge infections (Authié *et al.*, 1993a), and thus it is likely to be a prominent feature of cattle maintained under field challenge.

The failure of susceptible cattle to develop a sustained IgG response to trypanosome antigens is one manifestation of immunosuppression, the causes and consequences of which remain unknown. Of particular interest is the possibility that some of this IgG may play a role in the control of infection. Early studies indicated that more resistant cattle might develop higher levels of neutralizing antibodies to trypanosomes during the first peak of parasitaemia than fully susceptible cattle (Akol *et al.*, 1986; Pinder *et al.*, 1988). Studies in N'Dama and Boran cattle failed to confirm these findings and inferred that, in primary infections, the superior control of parasitaemia in N'Dama cattle is unlikely to operate through anti-VSG responses (Williams *et al.*, 1996). In contrast to primary infections, the superior control of parasitaemia in trypanotolerant cattle following secondary (homologous) infections is associated with more efficient anamnestic responses to VATs (Paling *et al.*, 1991b; Williams *et al.*, 1991). Thus, in agreement with early field observations (Desowitz, 1959), acquired immunity probably contributes to exacerbating resistance and the capacity to mount highly efficient secondary responses must be considered as an important determinant of the overall level of resistance that trypanotolerant cattle express in the field.

The superior humoral response of N'Dama cattle may also mediate neutralization of parasite products that are responsible for pathology, therefore contributing to increased resistance to the disease.

It is widely speculated that trypanosomes release immunomodulatory molecules that may depress T-cell responses. However, if in the mouse model, soluble extracts of trypanosomes have shown a suppressive effect *in vitro* (Bakhiet *et al.*, 1993; Sternberg *et al.*, 1996), no immunosuppressive factor has been confirmed to play a role in bovine trypanosomosis.

Degradative enzymes, such as proteases (Coombs and Mottram, 1997) and sialidases, have been considered as potentially harmful to the host (Tizard *et al.*, 1978). The erythrocytes of N'Dama cattle were found to contain more sialic acid than those of zebu cattle, and may thus be less sensitive to damage by parasite sialidases (Esievo *et al.*, 1986, 1990). However, no sialidase has been characterized from bloodstream forms of African trypanosomes so far. Congopain, a cysteine protease expressed in *T. congolense* bloodstream and metacyclic forms, has been well characterized (Authié *et al.*, 1992; Mbawa *et al.*, 1992; Chagas *et al.*, 1997). This enzyme is of particular interest due to its hydrolytic activity over a wide range of protein substrates at physiological pH, its presence in the circulation of infected cattle (Authié *et al.*, 1993b; Jaye, 1993) and the fact that it was identified through comparative studies of immune responses in resistant and susceptible cattle. Congopain is highly antigenic during primary infections of N'Dama cattle and in other trypanotolerant breeds, whereas cattle of susceptible breeds develop very low or undetectable levels of specific IgG (Authié *et al.*, 1993b). IgG elicited by trypanotolerant cattle during infection exerts an inhibitory effect on congopain (Authié *et al.*, unpublished). The pathogenicity of this enzyme, the role of antibody-mediated protease inhibition in bovine trypanotolerance and the potential of congopain-specific antibody to modulate the disease are being assessed in immunization trials with recombinant congopain.

Besides having a greater ability to develop specific humoral responses, trypanotolerant cattle have been found to maintain higher complement levels than susceptible zebu cattle, which suffer from severe hypocomplementaemia during trypanosome infection (Authié and Pobel, 1990). Moreover, a correlation was found between an individual's ability to maintain complement levels, its ability to maintain packed-cell volume (PCV) values and survival under natural challenge infection (Authié and Pobel, 1990). The control of complement levels might be of relevance to trypanoresistance, since C3 plays crucial roles in the generation of immune responses, particularly in antigen processing and presentation (Dempsey *et al.*, 1996), in the isotype switch mechanism and in the generation of memory B cells (Böttger and Bitter-Suermann, 1987).

The evidence of a role for immune mechanisms in trypanotolerance does not rule out the contribution of other non-immunological mechanisms (reviewed by Authié, 1993). As suggested by studies on resistance to trypanosomosis in wildlife, serum factors other than antibodies may conceivably modulate the viability, multiplication or differentiation of the parasites (Mulla and Rickman, 1988). However, no such factor has been identified in the serum of either susceptible or tolerant cattle that could explain the difference in parasite loads between the two breeds. The intrinsic capacity of the bone marrow to respond to anaemia by efficient haematopoiesis should also be considered. Recent studies in *T. congolense*-infected cattle have provided evidence of an early bone marrow impairment in susceptible Boran cattle and also indicated that N'Dama cattle may compensate more effectively than

Boran cattle for the degree of anaemia (Andrianarivo *et al.*, 1995, 1996). Finally, the differential attractivity of cattle for tsetse flies remains a pending question (Filledier *et al.*, 1988; Leak *et al.*, 1994).

In conclusion, the mechanisms underlying bovine trypanotolerance remain mostly unravelled; however, humoral immune responses to trypanosomes differ greatly between susceptible and resistant cattle, a feature that might be exploited to improve trypanosomosis control. Enhancing disease resistance through immunization with appropriate antigens, and assessment of specific IgG responses and complement factors as indicators of trypanotolerance, are research areas that are worth exploring.

Finally, the possibility of finding a solution to the enigma of bovine trypanotolerance through the molecular genetic approach should be mentioned (Kemp and Teale, 1998). The search for genetic markers, and subsequently for the genes controlling resistance, is hampered by the difficulty of phenotyping trypanotolerance accurately under experimental conditions. PCV, parasitaemia and performance, the currently available indicators that have proved useful in the field, might not be powerful enough to reveal, during a single experimental primary infection of limited severity, the maximum of the variation observed in the field. However, recent research provides encouraging preliminary results suggesting that at least three chromosomal regions carry quantitative trait loci associated with trypanotolerance (ILRI, 1998).

Continued efforts towards understanding bovine trypanotolerance remain vital, since better exploitation of trypanotolerant cattle, as well as alternative control strategies for trypanosomosis, in the long term, will be likely to arise from both immunological and genetic approaches.

Exploitation of Trypanotolerance Traits

Further to understanding the mechanisms underlying trypanotolerance, the exploitation of resistance traits relies on the characterization of these traits in the field and their practical measurement. The successful use of any criteria for identification of trypanotolerant breeds of cattle or superior animals within these breeds depends on the practicality of their measurement; on the strength of the linkage of the criteria with the economically important production traits such as viability, reproductive performance and growth, and on the associated genetic parameters.

As indicated above, experimental and field studies have clearly demonstrated that trypanotolerance is associated with the capacity to control parasitaemia and the capacity to resist the development of severe anaemia. In field studies, the degree of anaemia can be quantified easily by measuring the haematocrit or packed-cell volume per cent (PCV). Although the measurement of PCV is very accurate, the biological interpretation of PCV variation in the field is meaningful only if other factors affecting PCV are well identified, quantified or controlled. In contrast, the degree of parasitaemia is not so easily quantified, as quantification depends on demonstration of trypanosomes in peripheral blood by parasitological techniques. The most sensitive practical field approach has been to detect the presence of trypanosomes by the dark-ground or phase-contrast buffy coat technique (BCT) (Murray *et al.*, 1977) and to quantify the intensity of the infection as a parasitaemia score (Paris *et al.*, 1982). Although this method has contributed to major progress in characterizing trypanotolerance in the field, all parasite detection techniques have limitations. A high proportion of infections go undetected, as many are chronic. In addition, parasitaemia often fluctuates markedly and may be below the limit of detection, particularly in trypanotolerant livestock.

Relation between measurements of trypanotolerance and livestock performance

Results in recent years have shown that PCV in particular and parasitaemia, the two principal indicators of trypanotolerance, are strongly correlated with animal performance, especially post-weaning growth, reproductive performance and overall cow productivity (d'Ieteren, 1994; Trail *et al.*, 1992b, 1993, 1994b). The inclusion of trypanosome species in the measurement of parasite control capability highlighted the importance of determining trypanosome species for studying relationships between infection, anaemia and performance (Wissocq *et al.*, 1993).

Initially, control of anaemia development (as measured by average PCV value) appeared to be the criterion of trypanotolerance most closely linked to overall productivity, in production systems where other possible causes of anaemia were systematically controlled. However, although the direct effects of trypanosome infections on PCV and growth are obvious, a more sensitive method than BCT for reflecting parasite control is required so that individual animals can be reliably categorized for parasite control capability. The contribution of parasite burden for the assessment of a component of resistance has been well documented in tick (Spickett, 1998) and helminth (Baker, 1997) resistance research, with a high level of genetic control sometimes demonstrated (Stear *et al.*, 1997a).

Based on these relative difficulties, recent work reviewed by d'Ieteren *et al.* (1998) explored other approaches to the characterization of parasite control using newly developed diagnostic techniques. One assumption was that trypanosome antigen detection tests would give a better indication of an ongoing infection than the BCT analysis. The possible contribution of antigen detection to measure the infection control component of trypanotolerance more accurately had been evaluated initially by Trail *et al.* (1992a, b). It was emphasized that animals that gave positive results in the antigen-ELISA, but which were able to control parasites below levels detected by BCT, were assumed to be more tolerant than those that gave positive results for both tests. Another justification of the work came from the demonstration that, while PCV can be measured accurately in the field, the precise use of PCV as a trypanotolerance criterion (after other possible causes of anaemia had been controlled systematically) depended greatly on the accuracy of measurement of trypanosome infection status.

Wissocq *et al.* (1999a) determined the relationships between parasitaemia and antigen tests and demonstrated further that trypanosome antigen tests carried out over a 21-week period proved to be useful in trypanotolerance research. The trypanosome antigen detection patterns of animals were shown to reflect better their infection status and thus provided possible practical measurement allowing a more accurate ranking of animals' capability to control trypanosomes. These patterns were also associated to a significant extent with anaemia, the other trypanotolerance criterion, and with performance as indicated by growth rate. Subsequent work will explore the importance of the determination of trypanosome species antigen detection as a predictor of tolerance and for estimating anaemia control more accurately. In addition, as more data become available the interactions between parasitaemia, antigenaemia, anaemia control and growth can be explored more fully. The results suggest thus that, despite their limitations as diagnostic tools (Masake and Minja, 1998), antigen test information, serially recorded, can contribute to characterizing infection status and animals' trypanosomosis resistance capability more accurately. They support the justification for further development of more reliable diagnosis tools for trypanosomosis.

The evolution of the pattern of trypanosome infections over time was evaluated in calves from birth to maturity and thereafter in the different age groups of their dams (Trail *et al.*, 1994a). There was strong evidence for the ability of N'Dama cattle to acquire significant control of the development of parasitaemia following *T. vivax* infection, but apparently not following *T. congolense* infections. However, this finding needs to be confirmed in more than one area. The data also confirmed that pre-weaner calves grazing with their dams are more protected from, or more resistant to, *T. vivax* and *T. congolense* infections.

The importance of the ability to acquire resistance to trypanosome infection and the possible linkage between this ability, anaemia control and growth of trypanotolerant N'Dama cattle was further assessed by Wissocq *et al.* (1999b). The results indicated that with increasing age, the effect of *T. vivax* parasitaemia on animal growth was reduced significantly in comparison to the effect of *T. congolense* parasitaemia. Thus, not only is the ability to acquire some control of the development of parasitaemia over time following a *T. vivax* infection confirmed, but the deleterious influence of this parasite on animal performance also appears to be reduced. Further research will have to clarify whether, and how, these findings could be put to practical use in selection decisions for trypanotolerance levels.

Genetics of trypanotolerance measurements

The genetic components of these trypanotolerance measurements are being determined progressively using a quantitative approach in longitudinal studies of well-monitored production systems, in which the necessary identification and quantification of environmental and genetic effects are possible. Heritability values for, and genetic and phenotypic correlations between, growth and different PCV estimates have been reported by Trail and co-workers for N'Dama cattle under moderate challenge (Trail *et al.*, 1991b). These values, coupled with the higher heritabilities of the PCV measures, indicate some possibility of selection on PCV for control of anaemia development. The fact that evidence of genetic variation for the same components of trypanotolerance has been provided in susceptible *Bos indicus* cattle in East Africa should also be noted (Dolan, 1993; Rowlands *et al.*, 1995; Mwangi *et al.*, 1998).

In later studies using the same antigen detection technique as reported above, antigen-positive, parasite-negative animals were assumed to be classified as having more ability to control parasite growth than parasitaemic animals. A significant sire effect was found, suggesting that a degree of genetic control was involved. Thus it was concluded that more sensitive trypanosomosis diagnostic techniques, such as those detecting trypanosome antigens, might offer a practical possibility for selection of trypanotolerant animals based on infection criteria (Trail *et al.*, 1992a).

Preliminary genetic parameters provide evidence that trypanotolerance is not only a breed characteristic but is also a heritable trait within the N'Dama population. The genetic variation identified within the N'Dama breed has opened new opportunities for improved productivity through selection for trypanotolerance.

With the selection criteria for trypanotolerance already available, or in the process of becoming so, the design of selection programmes is becoming possible. Having established how such criteria relate to all other economically important traits for given production systems, the next step will be to compute the most appropriate relative weightings between these criteria and all economically important traits, based on economic importance, heritability and phenotypic and genetic correlations, in order to develop appropriate and relevant selection indices.

Quantification of the relative importance of trypanotolerance indicators is being pursued in longitudinal studies in well-characterized production systems, with the simultaneous recording of all possible criteria along with the relevant production parameters (reviewed by d'Ieteren *et al.*, 1998). A key finding of these studies is the major contribution of each of the indicators simultaneously evaluated to the overall trypanotolerance variance. Once the genetic aspects of each indicator have been estimated reliably, the results may well demonstrate that all must be handled simultaneously for optimal progress. This would require use of new diagnostics for assessing parasite control capability and precise identification of trypanosome alongside of anaemia control traits as measured by PCV (d'Ieteren *et al.*, 1998).

Future exploitation

Identification and utilization of superior animals in breeding programmes will depend on accurate measurement of these criteria of trypanotolerance in a 'trypanotolerance test' (Trail *et al.*, 1991a, c). The practical characteristics of such a test still need to be established. The impact will depend on the proportion of the variation associated with the trypanotolerance indicators that can be manipulated. However, genetic correlation between these indicators, the production traits and resistance to other economically important parasites and diseases must still be determined. New practical immunological markers (if and when available) could contribute to more comprehensive characterization of trypanotolerance in the field. Likewise, molecular markers would greatly assist and expedite selection decisions based on the resistance traits (Teale, 1997), provided that these markers are associated with a sufficiently large proportion of the genetic variation observed in the field. Stear *et al.* (1997a, b) identified one such influential gene, associated with resistance of sheep to the nematode *Ostertagia circumcincta*, which is the most significant gene identified for resistance to any parasite species, within or around the MHC.

Multiple Attributes of Trypanotolerant Livestock

In addition to their resistance to trypanosomosis, trypanotolerant cattle, and the N'Dama breed in particular, have other genetic advantages that contribute to their potential for use in livestock development programmes in the tropics (reviewed by d'Ieteren *et al.*, 1998). These cattle are reported to be resistant to several other important infectious diseases (Murray *et al.*, 1991), including a number of tick-borne infections such as dermatophilosis (Stewart, 1937), heartwater, anaplasmosis and babesiosis (Epstein, 1971). Lower tick burdens have been reported recently in N'Dama cattle in comparison with zebu cattle (Claxton and Leperre, 1991; Mattioli *et al.*, 1993, 1995; Mattioli and Casama, 1995) as well as lower prevalence of strongyle infections (Mattioli *et al.*, 1992). The success of the introduction of N'Dama cattle in the most humid parts of West and Central Africa must be recognized as being attributed equally to their resistance to trypanosomosis and to dermatophilosis (d'Ieteren *et al.*, 1998).

A major advantage of trypanotolerant livestock, particularly N'Dama cattle, is thus the resistance or adaptation to many of the important pathogens prevailing in the subhumid and humid tropics. Research on practical indicators of resistance to these conditions will lead to more integrated strategies based on disease-resistant livestock. Selective breeding will have to integrate the traits that farmers see as important for their production systems. More recent studies have indicated how highly farmers, who have knowledge about trypanotolerant cattle, rank disease-resistance traits in choosing a cattle breed (Jabbar *et al.*, 1997; Tano *et al.*, 1998).

Acknowledgements

The work of the authors and the many studies carried out by ILRI and its partners, which have been reported extensively in this chapter, were supported by the governments of Belgium, France, Germany, Switzerland, the UK and the EU through the European Development Fund.

References

Agyemang, K., Little, D.A. and Dwinger, R.H. (1994) Salvaging the image of the N'Dama breed: productivity evidence from village production systems in the Gambia. In: Rowlands, G.J. and Teale, A.J. (eds) *Towards Increased Use of Trypanotolerance: Current Research and Future Directions.* Proceedings of a Workshop Organised by the International Laboratory for Research on Animal Diseases (ILRAD) and the International Livestock Centre for Africa (ILCA), 26–29 April 1993, Nairobi, Kenya. ILRAD/ILCA, Nairobi, pp. 123–133.

Akol, G.W.O., Authié, E., Moloo, S.K., Murray, M., Pinder, M. and Roelants, G.E. (1986) Susceptibility and immune responses of Zebu and Taurine cattle of West Africa to infection with *Trypanosoma congolense* transmitted by *Glossina morsitans centralis. Veterinary Immunology and Immunopathology* 11, 361–373.

Andrianarivo, A.G., Muiya, P., Opollo, M. and Logan-Henfrey, L.L. (1995) *Trypanosoma congolense*: comparative effects of a primary infection on bone-marrow progenitor cells from N'Dama and Boran cattle. *Experimental Parasitology* 80, 407–418.

Andrianarivo, A.G., Muiya, P. and Logan-Henfrey, L.L. (1996) *Trypanosoma congolense*: high erythropoietic potential in infected yearling cattle during the acute phase of anaemia. *Experimental Parasitology* 82, 104–111.

Authié, E. (1993) Contribution à l'étude des mécanismes immunologiques impliqués dans la trypanotolérance des taurins d'Afrique. PhD thesis, Department of Biochemistry and Cellular Biology, University of Bordeaux II, Bordeaux, France.

Authié, E. and Pobel, T. (1990) Serum haemolytic and C3 levels in bovine trypanosomiasis under natural conditions of challenge – early indication of individual susceptibility to disease. *Veterinary Parasitology* 35, 43–59.

Authié, E., Muteti, D.K., Mbawa, Z., Lonsdale-Eccles, J., Webster, P. and Wells, C. (1992) Identification of a major antigen of *Trypanosoma congolense* as a cysteine protease. *Molecular and Biochemical Parasitology* 56, 103–116.

Authié, E., Muteti, D.K. and Williams, D.J.L. (1993a) Antibody responses to non-variant antigens of *Trypanosoma congolense* in cattle of differing susceptibility to trypanosomiasis. *Parasite Immunology* 15, 101–111.

Authié, E., Duvallet, G., Robertson, J. and Williams, D.J.L. (1993b) Antibody responses to a cysteine protease of *Trypanosoma congolense* – relationship to trypanotolerance in cattle. *Parasite Immunology* 15, 465–474.

Baker, R.L. (1997) Résistance génétique des petits ruminants aux helminthes en Afrique. *INRA Production Animale* 10 (1), 99–110.

Bakhiet, M., Olsson, T., Edlund, C., Höjeberg, B., Holmberg, K., Lorentzen, J. and Kristensson, K. (1993) A *Trypanosoma brucei brucei*-derived factor that triggers CD8+ lymphocytes to interferon-gamma secretion: purification, characterization and protective effects *in vivo* by treatment with a monoclonal antibody against the factor. *Scandinavian Journal of Immunology* 37, 165–178.

Barrett, J. (1997) Control strategies for African trypanosomiasis: their sustainability and effectiveness. In: Hide, G., Mottram, J.C., Coombs, G.H. and Holmes, P.H. (eds) *Trypanosomiasis and Leishmaniasis Biology and Control.* CAB International, Farnham Royal, pp. 347–359.

Bauer, B., Kabore, I., Liebisch, A., Meyer, F. and Petrich-Bauer, J. (1992) Simultaneous control of ticks and tsetse flies in Satiri, Burkina Faso, by the use of flumethrin pour-on for cattle. *Tropical Medicine and Parasitology* 43, 41–46.

Böttger, E.C. and Bitter-Suermann, D. (1987) Complement and the regulation of humoral immune responses. *Immunology Today* 8, 261–264.

Chagas, J., Authié, E., Serveau, C., Lalmanach, G., Juliano, L. and Gauthier, F. (1997) A comparison of the enzymatic properties of the major cysteine proteinases from *Trypanosoma congolense* and *Trypanosoma cruzi*. *Molecular and Biochemical Parasitology* 88, 85–94.

Clausen, P.H., Sidibe, I., Kabore, I. and Bauer, B. (1992) Development of multiple drug resistance of *Trypanosoma congolense* in Zebu cattle under high natural tsetse fly challenge in the pastoral zone of Samorogouan, Burkina Faso. *Acta Tropica* 51, 229–236.

Claxton, J. and Leperre, P. (1991) Parasite burdens and host susceptibility of Zebu and N'Dama cattle in village herds in Gambia. *Veterinary Parasitology* 40, 293–304.

Coombs, G.H. and Mottram, J.C. (1997) Proteinases of trypanosomes and *Leishmania*. In: Hide, G., Mottram, J.C., Coombs, G.H. and Holmes, P.H. (eds) *Trypanosomiasis and Leishmaniasis: Biology and Control*. CAB International, Farnham Royal, pp. 177–198.

Cross, G.A.M. (1975) Identification, purification and properties of clone-specific glycoprotein antigens constituting the surface coat of *Trypanosoma brucei*. *Parasitology* 71, 393–417.

Cunningham, E.P. and Syrstad, O. (1987) *Crossbreeding* Bos indicus *and* Bos taurus *for Milk Production in the Tropics*. Animal Production and Health Paper No. 68. Food and Agriculture Organisation, Rome.

Davila, A.M.R., Ramirez, L. and Silva, R.A.M.S. (1998) Trypanosomoma in the America: morphometry and host range. *Revue d'Elevage et de Medecine Veterinaire des Pays Tropicaux* 51, 29–35.

De Baetselier, P. (1996) Mechanisms underlying trypanosome-induced T-cell immunosuppression. In: Mustafa, A.S., Al-Attiyah, R.J., Nath, I. and Chugh, T.D. (eds) *T-Cell Subsets and Cytokines Interplay in Infectious Diseases*. Karger, Basel, pp. 124–139.

Dempfle, L. (1993) *Strategies for Improving the Genetic Performance of Local Breeds*. Food and Agriculture Organization (FAO) Production and Health Technical Paper No. 110. FAO, Rome, pp. 3–10.

Dempsey, P.W., Allison, M.E.D., Akkaraju, S., Goodnow, C.C. and Fearon, D.T. (1996) C3d of complement as a molecular adjuvant: bridging innate and acquired immunity. *Science* 271, 348–350.

Desowitz, R.S. (1959) Studies on immunity and host–parasite relationships. I. The immunological response of resistant and susceptible breeds of cattle to trypanosomal challenge. *Annals of Tropical Medicine and Parasitology* 53, 293–313.

d'Ieteren, G. (1994) Trypanotolerant livestock, a sustainable option for increasing livestock production in tsetse affected areas. In: Rowlands, G.J. and Teale, A.J. (eds) *Towards Increased Use of Trypanotolerance: Current Research and Future Directions*. Proceedings of a Workshop Organised by the International Laboratory for Research on Animal Diseases (ILRAD) and the International Livestock Centre for Africa (ILCA), 26–29 April 1993, Nairobi, Kenya. ILRAD/ILCA, Nairobi, pp. 3–11.

d'Ieteren, G.D.M., Coulibaly, L., Atse, P.A., Hecker, P.A., Krebs, H.A., Rowlands, G.J., Leak, S.G.A. and Nagda, S.M. (1997) Trypanocidal drug resistance in four regions of Côte d'Ivoire. Importance and possible impact on sustainability of integrated strategies for trypanosomiasis control. In: *Proceedings of the 23rd Meeting of the International Scientific Council for Trypanosomiasis Research and Control, 11–15 September 1995, Banjul, the Gambia*. Organisation of African Unity, Scientific, Technical and Research Commission, Nairobi, Kenya, pp. 233–247.

d'Ieteren, G.D.M., Authié, E., Wissocq, N. and Murray, M. (1998) Trypanotolerance, an option for sustainable livestock production in areas at risk from trypanosomosis.

Revues Scientifique et Technicales de l'Office Internationale des Epizooties 17 (1), 154–175.

Dolan, R.B. (1993) *Trypanotolerance in Orma Boran Cattle on Galana Ranch in Kenya.* International Livestock Centre for Africa (ILCA) Consultancy Report. ILCA, Nairobi.

Dolan, R.B., Alushula, H., Munga, L., Mutugi, M., Mwendia, C., Okech, G., Sayers, P.D., Stevenson, P.G.W., Baker, R.L. and Magadi, M. (1994) The Orma Boran – ten years of field observations. In: Rowlands, G.J. and Teale, A.J. (eds) *Towards Increased Use of Trypanotolerance: Current Research and Future Directions.* Proceedings of a Workshop Organised by the International Laboratory for Research on Animal Diseases (ILRAD) and the International Livestock Centre for Africa (ILCA), 26–29 April 1993, Nairobi, Kenya. ILRAD/ILCA, Nairobi, pp. 71–79.

Duvallet, G., Ouedraogo, A., Pinder, M. and Van Melick, A. (1988) Observations following the cyclical infection with *Trypanosoma congolense* of previously uninfected Baoule and Zebu cattle. In: *Livestock Production in Tsetse Affected Areas of Africa.* International Laboratory for Research on Animal Diseases/International Livestock Centre for Africa, Nairobi, pp. 318–325.

Epstein, H. (1971) *The Origin of the Domestic Animals of Africa,* Vols I and II. Africana, New York.

Esievo, K.A.N., Saror, D.I., Kolo, M.N. and Eduvie, L.O. (1986) Erythrocyte surface sialic acid concentrations in N'Dama and Zebu cattle. *Journal of Comparative Pathology* 96, 95–99.

Esievo, K.A.N., Jaye, A., Andrews, J.N., Ukoha, A.I., Alafiatayo, R.A., Eduvie, L.O., Saror, D.I. and Njoku, C.O. (1990) Electrophoresis of bovine erythrocyte sialic acids: existence of additional band in trypanotolerant N'dama cattle. *Journal of Comparative Pathology* 102, 357–361.

Feron, A., Sheria, M., Mulungo, M., Pelo, M., Kakiese, O., d'Ieteren, G.D.M., Durkin, J., Itty, P., Maehl, J.H.H., Nagda, S.M., Rarieya, J.M., Thorpe, W., Trail, J.C.M. and Paling, R.W. (1988) Productivity of ranch N'Dama cattle under trypanosomiasis risk. In: *Livestock Production in Tsetse Affected Areas of Africa.* International Laboratory for Research on Animal Diseases/International Livestock Centre for Africa, Nairobi, pp. 246–250.

Filledier, F., Duvallet, G. and Mérot, P. (1988) Comparaison du pouvoir attractif des bovins Zébu et Baoulé pour *Glossina tachinoides* Westwood, 1850 et *Glossina morsitans submorsitans* Newstead, 1910 en savane soudano-guinéenne, Burkina Faso. *Revue d'Elevage et de Medecine Veterinaire des Pays Tropicaux* 41, 191–196.

Ford, J. (1971) *The Role of the African Trypanosomiases in African Ecology: a Study of the Tsetse Fly Problem.* Clarendon Press, Oxford.

Geerts, S. and Holmes, P.H. (1999) Drug management and parasite resistance in animal trypanosomiasis in Africa. In: *Proceedings of the 24th Meeting of the International Scientific Council for Trypanomosiasis Research and Control (ISCTRC), Maputo, Mozambique.* Organisation of African Unity/ISCTRC, Nairobi, Kenya.

Holmes, P.H. (1997) New approaches to the integrated control of trypanosomosis. *Veterinary Parasitology* 71, 121–135.

International Livestock Centre for Africa (ILCA) (1979) *Trypanotolerant Livestock in West and Central Africa.* Vol. 1. Monograph No. 2. ILCA, Addis Ababa, Ethiopia.

International Livestock Research Institute (ILRI) (1998) *Annual Project Progress Report.* ILRI, Nairobi, Kenya, in press.

Iraqi, F. and Teale, A. J. (1999) Polymorphisms in the TNF-α gene of different inbred mouse strains. *Immunogenetics* 49, 242–245.

Iraqi, F., Sekikawa, K. and Teale, A. (1998) TNF-α-deficient mice are highly susceptible to

Trypanosoma congolense infection. *12th International Mouse Genome Conference, Garmisch, Germany, 30 September–4 October, and 2nd Louis Pasteur Conference on Infectious Diseases, Paris, France, 8–10 October, 1998.*

Ismael, A.A. and Njogu, A.R. (1985) Susceptibility of Orma and Galana cattle to infection with bloodstream forms of *Trypanosoma congolense* and *T. vivax*. In: *Proceedings of the 18th Meeting of the International Scientific Council for Trypanosomiasis Research and Control, 4–9 March, Harare, Zimbabwe.* Organisation of African Unity, Scientific, Technical and Research Commission, Nairobi, Kenya, pp. 176–181.

Jabbar, M.A., Swallow, B.M., d'Ieteren, G.D.M. and Busari Adesina (1997) *Farmer Preferences and Market Values of Cattle Breeds of West Africa and Central Africa.* Social and Policy Research Working Paper No. 21. International Livestock Research Institute, Addis Ababa, Ethiopia.

Jahnke, H.E., Tacher, G., Keil, P. and Rojat, D. (1988) Livestock production in tropical Africa, with special reference to the tsetse-affected zone. In: *Livestock Production in Tsetse Affected Areas of Africa.* International Livestock Centre for Africa/ International Laboratory for Research on Animal Diseases, Nairobi, Kenya, pp. 430–432.

Jaye, A.B. (1993) Characterization of a *Trypanosoma (Nannomonas) congolense*-specific antigen: identification as a thiol-protease precursor. PhD Thesis, Department of Biology and Biochemistry, Brunel University, Uxbridge, UK.

Jordan, A.M. (1986) *Trypanosomiasis Control and African Rural Development.* Longman, London.

Kemp, S.J. and Teale, A.J. (1998) Genetic basis of trypanotolerance in cattle and mice. *Parasitology Today* 14, 450–454.

Kemp, S.J., Iraqi, F., Darvasi, A., Soller, M. and Teale, A.J. (1997) Localisation of the genes controlling resistance to trypanosomiasis in mice. *Nature Genetics* 16, 194–196.

Leak, S.G.A., d'Ieteren, G. and Rowlands, G.J. (1994) Factors affecting estimation of tsetse challenge and the expression of trypanotolerance. In: Rowlands. G.J. and Teale, A.J. (eds) *Towards Increased Use of Trypanotolerance: Current Research and Future Directions.* Proceedings of a Workshop Organised by International Livestock Centre for Africa/International Laboratory for Research on Animal Diseases, 26–29 April 1993, Nairobi, Kenya. ILRAD/ILCA, Nairobi, pp. 15–22.

Lutje, V., Mertens, B., Boulangé, A., Williams, D.J.L. and Authié, E. (1995a) *Trypanosoma congolense*: proliferative responses and interleukin production in lymph node cells of infected cattle. *Experimental Parasitology* 81, 154–164.

Lutje, V., Taylor, K., Boulangé, A. and Authié, E. (1995b) *Trypanosoma congolense*: tissue localisation of long term T-cell and B-cell responses in N'Dama cattle. *Immunology Letters* 48, 29–34.

Lutje, V., Taylor, K.A., Kennedy, D., Authié, E., Boulangé, A., Logan-Henfrey, L.L. and Gettinby, G. (1996) *Trypanosoma congolense*: a comparison of T-cell mediated responses in lymph nodes of trypanotolerant and trypanosusceptible cattle during primary infection. *Experimental Parasitology* 84, 320–329

McDowell, R.E. (1977) *Ruminant Products: More Meat than Milk.* Winrock International Livestock and Training Centre, Marilton, Arkansas.

Mansfield, J.M. (1990) Immunology of African trypanosomiasis. In: Wyler, D.J. (ed.) *Modern Parasite Biology: Cellular, Immunological and Molecular Aspects.* W.H. Freeman and Co., New York, pp. 222–246.

Masake, R.A. and Minja, S.H. (1998) Antigen-ELISA for detection of trypanosome. *Proceedings of a Workshop on Evaluation of Antigen ELISAs for Trypanosomes, 9–11 December, 1996, Nairobi, Kenya,* pp. 21–30.

Mattioli, R.C. and Cassama, M. (1995) Comparison of characteristics of life cycle in female ticks collected on N'Dama and Zebu cattle. *Tropical Animal Health and Production* 27, 150–154.

Mattioli, R.C., Cassama, M. and Kora, S. (1992) A comparative study of gastrointestinal nematode egg output in N'Dama, Zebu and N'Dama × Zebu crossbred cattle. *Parasitología* 34, 109–113.

Mattioli, R.C., Bah, M., Faye, J., Kora, S. and Cassama, M. (1993) A comparison of field tick infestation on N'Dama, Zebu and N'Dama × Zebu crossbred cattle. *Veterinary Parasitology* 47, 139–148.

Mattioli, R.C., Bah, M., Kora, S., Cassama, M. and Clifford, D.G. (1995) Susceptibility to different tick genera in Gambian N'Dama and Bobra zebu cattle exposed to naturally-occurring tick infestations. *Tropical Animal Health and Production* 79, 1–17.

Mbawa, Z.R., Gumm, I.D., Shaw, E. and Lonsdale-Eccles, J.D. (1992) Characterisation of a cysteine proteinase from bloodstream forms of *Trypanosoma congolense*. *European Journal of Biochemistry* 204, 371–379.

Mertens, B., Taylor, K., Muriuki, C. and Rocchi, M. (1999) Cytokine mRNA profiles in trypanotolerant and trypanosusceptible cattle infected with the protozoan parasite *Trypanosoma congolense*: protective role for IL-4? *Interferon Cytokine Reseach* 19, 59–65.

Mulla, A.F. and Rickman, L.R. (1988) How do African game animals control trypanosome infections? *Parasitology Today* 4, 352–354.

Murray, C., Murray, M., Murray, P.K., Morrison, W.I., Pyne, C. and McIntyre, W.I.M. (1977) Diagnosis of African trypanosomiasis in cattle: improved parasitological and serological techniques. In: *Proceedings of the 15th Meeting of the International Scientific Council for Trypanosomiasis Research and Control, 25–30 April, Banjul, the Gambia.* Organisation of African Unity, Scientific, Technical and Research Commission, Nairobi, Kenya, pp. 247–254.

Murray, M. and Gray, A.R. (1984) The current situation on animal trypanosomosis in Africa. *Preventive Veterinary Medicine* 2, 23–30.

Murray, M. and Dexter, T.M. (1988) Anaemia in bovine African trypanosomiasis. *Acta Tropica* 45, 389–432.

Murray, M., Morrison, W.I. and Whitelaw, D.D. (1982) Host susceptibility to African trypanosomiasis: trypanotolerance. *Advances in Parasitology* 21, 1–68.

Murray, M., Trail, J.C.M. and d'Ieteren, G.D.M. (1990) Trypanotolerance in cattle and prospects for the control of trypanosomiasis by selective breeding. *Revue Scientifique et Technique de l'Office Internationale des Epizooties* 9 (2), 369–386.

Murray, M., Stear, M.J., Trail, J.C.M., d'Ieteren, G.D.M., Agyemang, K. and Dwinger, R.H. (1991) Trypanosomiasis in cattle: prospects for control. In: Axford, R.F.E. and Owen, J.B. (eds) *Breeding for Disease Resistance in Farm Animals.* CAB International, Wallingford, pp. 203–223.

Mwangi, E.K., Stevenson, P., Gettinby, G. and Murray, M. (1994) Variation in susceptibility to tsetse-borne trypanosomiasis among *Bos indicus* cattle breeds in East Africa. In: Rowlands, G.J. and Teale, A.J. (eds) *Towards Increased Use of Trypanotolerance: Current Research and Future Directions.* Proceedings of a Workshop Organised by the International Livestock Centre for Africa and the International Laboratory for Research on Animal Diseases, 26–29 April 1993, Nairobi, Kenya. ILRAD/ILCA, Nairobi, pp. 81–86.

Mwangi, E.K., Stevenson, P., Gettinby, G., Reid, S.W.J. and Murray, M. (1998) Susceptibility to trypanosomosis in *Bos indicus* cattle breeds in areas of differing tsetse fly challenge. *Veterinary Parasitology* 79, 1–17.

Njogu, A.R., Dolan, R.B., Sayer, P.D., Wilson, A.J. and Alushula, H. (1985) Trypanotolerance in East African Orma Boran cattle. *Veterinary Record* 117, 632–636.

Paling, R.W., Moloo, S.K., Scott, J.R., Gettinby, G., McOdimba, F.A. and Murray, M. (1991a) Susceptibility of N'Dama and Boran cattle to sequential challenges with tsetse-transmitted clones of *Trypanosoma congolense. Parasite Immunology* 13, 427–445.

Paling, R.W., Moloo, S.K., Scott, J.R., Logan-Henfrey, L.L., Murray, M. and Williams, D.J.L. (1991b) Susceptibility of N'Dama and Boran cattle to tsetse-transmitted primary and rechallenge infections with a homologous serodeme of *Trypanosoma congolense. Parasite Immunology* 13, 413–425.

Paris, J., Murray, M. and McOdimba, F. (1982) An evaluation of the sensitivity of current parasitological techniques for the diagnosis of bovine African trypanosomiasis. *Acta Tropica* 39, 307–316.

Peregrine, A.S. (1994) Chemotherapy and delivery systems: haemoparasites. *Veterinary Parasitology* 54, 223–248.

Perry, B., Dolan, T., Morzaria, S., Peregrine, A., Teale, A., Lessard, P., Thorpe, B., Gettinby, G., Revie, C., McKendrick, I., McDermott, J. and Gu, Y. (1998) Generic decision-support systems to improve the control of trypanosomosis and theileriosis in cattle. In: *DFID and the Consultative Group on International Agriculture Research (CGIAR): The Competitive Research Facility 1990–1997.* Department of International Development (DFID), UK, pp. 83–85.

Pinder, M., Bauer, J., Van Melick, A. and Fumoux, F. (1988) Immune responses of trypanoresistant and trypanosusceptible cattle after cyclic infection with *Trypanosoma congolense. Veterinary Immunology and Immunopathology* 18, 245–257.

Rowlands, G.J., Woudyalew Mulatu, Leak, S.G.A., Authié, E., d'Ieteren, G.D.M., Nagda, S.M. and Peregrine, A.S. (1993) Epidemiology of bovine trypanosomiasis in the Ghibe valley, Southwest Ethiopia. 2. Factors associated with variations in trypanosome prevalence, incidence of new infections and prevalence of recurrent infections. *Acta Tropica* 53, 135–150.

Rowlands, G.J., Woudyalew Mulatu, Nagda, S.M., Dolan, R.B. and d'Ieteren, G.D.M. (1995) Genetic variation in packed red cell volume and frequency of parasitaemia in East African Zebu cattle exposed to drug-resistant trypanosomes. *Livestock Production Sciences* 43, 75–84.

Sileghem, M., Flynn, J.N., Logan-Henfrey, L. and Ellis, J. (1994) Tumor necrosis factor production by monocytes from cattle infected with *Trypanosoma (Duttonella) vivax* and *Trypanosoma (Nannomonas) congolense*: possible association with severity of anaemia associated with the disease. *Parasite Immunology* 16, 51–54.

Spickett, A.M. (1998) Host genetic resistance to ticks and tick-borne diseases. *Proceedings of the 16th International Conference of the World Association for the Advancement of Veterinary Parasitology, 10–15 August 1997, Sun City, South Africa*, in press.

Stear, M.J., Bairden, K., Bishop, S.C., Buitkamp, J.L., Duncan, J.L., Gettinby, G., McKellar, Q.A., Park, M., Parkins, J.J., Reid, S.W.J., Strain, S. and Murray, M. (1997a) The genetic basis of resistance to *Ostertagia circumcinata* in lambs. *Veterinary Journal* 154, 111–119.

Stear, M.J., Bairden, K., Duncan, J.L., Holmes, P.H., McKellar, Q.A., Park, M., Strain, S., Murray, M., Bishop, S.C. and Gettinby, G. (1997b) How hosts control worms. *Nature* 389, 27.

Sternberg, J. (1997) Immunobiology of African trypanosomiasis. *Chemical Immunology* 70, 186–199.

Sternberg, J. and Mabbot, N. (1996) Nitric oxide-mediated suppression of T-cell responses during *Trypanosoma brucei* infection: soluble trypanosome products and interferon-α are synergistic inducers of nitric oxide synthase. *European Journal of Immunology* 26, 539–543.

Stewart, J.L. (1937) The cattle of the Gold Coast. *Veterinary Record* 49, 1289–1297.

Swallow, B.M., Woudyalew Mulatu and Leak, S.G.A. (1995) Potential demand for a mixed public–private animal health input: evaluation of a pour-on insecticide for controlling tsetse-transmitted trypanosomiasis in Ethiopia. *Preventive Veterinary Medicine* 24, 265–275.

Tano, K., Kamuanga, M., Swallow, B., Feminow, M. and d'Ieteren, G. (1999) Demand for trypanotolerant cattle and their traits: evidence from southern Burkina Faso. In: *Proceedings of the 24th Meeting of the International Scientific Council for Trypanosomiasis Research and Control, 29 September–3 October 1997, Maputo, Mozambique.* Organisation of African Unity, Scientific, Technical and Research Commission, Nairobi, Kenya (addendum).

Taylor, K.A. (1998) Immune responses of cattle to African trypanosomes: protective or pathogenic? *International Journal for Parasitology* 28, 219–240.

Taylor, K., Lutje, V. and Mertens, B. (1996a) Nitric oxide synthesis is depressed in *Bos indicus* cattle infected with *Trypanosoma congolense* and *T. vivax* and does not mediate T-cell suppression. *Infection and Immunity* 64, 4115–4122.

Taylor, K., Lutje, V., Kennedy, D., Authié, E., Boulangé, A., Logan-Henfrey, L.L., Gichuki, B. and Gettinby, G. (1996b) B-lymphocyte responses differ between trypanotolerant and trypanosusceptible cattle during primary infection with *Trypanosoma congolense*. *Experimental Parasitology* 83, 106–116.

Taylor, K., Mertens, B., Lutje, V. and Saya, R. (1998) *Trypanosoma congolense* infection of trypanotolerant N'Dama (*Bos taurus*) cattle is associated with decreased secretion of nitric oxide by interferon-gamma-activated monocytes and increased transcription of interleukin 10. *Parasite Immunology* 20, 421–429.

Teale, A. (1997) Biotechnology: a key element in the CGIAR's livestock research programme. *Outlook on Agriculture* 26 (4), 217–225.

Tizard, I., Nielsen, K.H., Seed, J.R. and Hall, J.E. (1978) Biologically active products from African trypanosomes. *Microbiological Review* 42, 661–681.

Trail, J.C.M., Sones, K., Jibbo, J.M.C., Durkin, J., Light, D.E. and Murray, M. (1985) *Productivity of Boran Cattle Maintained by Chemoprophylaxis under Trypanosomiasis Risk.* Research Report No. 9, International Livestock Centre for Africa, Addis Ababa, Ethiopia.

Trail, J.C.M., d'Ieteren, G.D.M., Colardelle, C., Maille, J.C., Ordner, G., Sauveroche, B. and Yangari, G. (1991a) Evaluation of a field test for trypanotolerance in young N'Dama cattle. *Acta Tropica* 48, 47–57.

Trail, J.C.M., d'Ieteren, G.D.M., Maille, J.C. and Yangari, G. (1991b) Genetic aspects of control of anaemia development in trypanotolerant N'Dama cattle. *Acta Tropica* 48, 285–291.

Trail, J.C.M., d'Ieteren, G.D.M. and Murray, M. (1991c) Practical aspects of developing genetic resistance to trypanosomiasis. In: Axford, R.F.E. and Owen, J.B. (eds) *Breeding for Disease Resistance in Farm Animals.* CAB International, Wallingford, pp. 224–234.

Trail, J.C.M., d'Ieteren, G.D.M., Maille, J.C., Yangari, G. and Nantulya, V.M. (1992a) Use of antigen-detection enzyme immunoassays in assessment of trypanotolerance in N'Dama cattle. *Acta Tropica* 50, 11–18.

Trail, J.C.M., d'Ieteren, G.D.M., Viviani, P., Yangari, G. and Nantulya, V.M. (1992b) Relationships between trypanosome infection measured by antigen-detection enzyme immunoassays, anaemia and growth in trypanotolerant N'Dama cattle. *Veterinary Parasitology* 42, 213–223.

Trail, J.C.M., d'Ieteren, G.D.M., Murray, M., Ordner, G., Yangari, G., Maille, J.C., Viviani, P., Colardelle, C. and Sauveroche, B. (1993) Measurement of trypanotolerance criteria and their effect on reproductive performance of N'Dama cattle. *Veterinary Parasitology* 45, 241–255.

Trail, J.C.M., Wissocq, N.M., d'Ieteren, G.D.M., Kakiese, O., Mulungo, M. and Murray, M. (1994a) Patterns of *Trypanosoma vivax* and *Trypanosoma congolense* infection differ in young N'Dama cattle and their dams. *Veterinary Parasitology* 55, 175–183.

Trail, J.C.M., Wissocq, N., d'Ieteren, G.D.M., Kakiese, O. and Murray, M. (1994b) Quantitative phenotyping of N'Dama cattle for aspects of trypanotolerance under field tsetse challenge. *Veterinary Parasitology* 55, 185–195.

Van Meirvenne, N., Magnus, E. and Verwoort, T. (1977) Comparison of variable antigen types produced by trypanosome strains of the sub-genus *Trypanozoon*. *Annale de la Societe Belge de Médecine Tropicale* 57, 409–423.

Williams, D.J.L., Naessens, J., Scott, J.R. and McOdimba, F.A. (1991) Analysis of peripheral leukocyte populations in N'Dama and Boran cattle following a rechallenge infection with *Trypanosoma congolense*. *Parasite Immunology* 13, 171–185.

Williams, D.J.L., Taylor, K.A., Newson, J. and Gichuki, B. (1996) The role of anti-variable surface glycoprotein antibody responses in bovine trypanotolerance. *Parasite Immunology* 18, 209–218.

Wissocq, N., Trail, J.C.M., Kakiese, O., d'Ieteren, G.D.M., Pelo, M. and Mulungo, M. (1993) Importance of trypanosome species in relationship between infection, anaemia and reproductive performance in N'Dama cattle. In: *Proceedings of the 21st Meeting of the International Scientific Council for Trypanosomiasis Research and Control, 21–25 October 1991, Yamoussoukro, Côte d'Ivoire.* Organisation of African Unity, Scientific, Technical and Research Commission, Nairobi, Kenya, p. 278.

Wissocq, N., d'Ieteren, G.D.M., Trail, J.C.M., Masake, R., Nantulya, V. and Monsengo, B. (1999a) Trypanosome antigen test to characterise infection status in N'Dama cattle. In: *Proceedings of the 24th Meeting of the International Scientific Council for Trypanosomiasis Research and Control, 29 September–3 October, Maputo, Mozambique.* Organisation of African Unity, Scientific, Technical and Research Commission, Nairobi, Kenya, pp. 310–314.

Wissocq, N., d'Ieteren, G., Trail, J.C.M. and Monsengo, B. (1999b) Ability to acquire resistance to trypanosome infections and grown of N'Dama cattle. In: *Proceedings of the 24th Meeting of the International Scientific Council for Trypanosomiasis Research and Control, Maputo, Mozambique, 29 September–3 October 1997.* OAU/STRC, Nairobi, Kenya, pp. 306–309.

Bacteria III

10

Genetic Aspects of Resistance to Ovine Footrot

H.W. Raadsma
Centre for Sheep Research and Extension, The University of Sydney, Camden, Australia

Summary

A primary requirement for inclusion of resistance to footrot in breeding programmes is knowledge of the extent of genetic variation in resistance and the genetic correlations between resistance and all other important traits, including production and resistance to other diseases. At the moment sufficient preliminary information is available on all relevant genetic parameters to include resistance to footrot in Merino breeding programmes. Similar information for other breeds is more scant, but there is no reason to expect the parameters to be greatly different in other breeds. The main challenge for animal breeders is to decide on the appropriate weighting for resistance to footrot.

Finally, it is not sensible to consider selective breeding programmes to improve resistance to footrot, or other important diseases for that matter, without also considering non-genetic forms of disease control. An overall animal-health programme will need to consider all avenues that are economically feasible, and may need to improve management as well as incorporating disease resistance in selective breeding programmes.

With the development of high-grade veterinary vaccines, there is a need for animals to respond to a large number of clearly defined immunogens. Genetic restrictions that prevent an adequate response to multivalent vaccines need to be identified and investigated for possible modification through selective breeding. The use of footrot as a model host–vaccine response system may show how this could be done for other species and other veterinary vaccines.

Introduction: Footrot of Sheep

It is not within the scope of this chapter to provide a detailed account of all aspects of footrot. The reader is referred to an excellent overview of footrot in

© CAB *International* 2000. *Breeding for Disease Resistance in Farm Animals* (eds R.F.E. Axford, S.C. Bishop, F.W. Nicholas and J.B. Owen)

ruminants by Egerton *et al.* (1989), and in particular the comprehensive and detailed chapter on footrot of sheep by Stewart (1989). For those with a specific interest in footrot, the benchmark publication by Beveridge (1941) is compulsory reading, and a recent review by Allworth (1995) also provides an excellent overview of the subject.

Definition of footrot

Footrot of sheep is an infectious and, on specific occasions, an exceptionally contagious disease resulting from invasion of epidermal tissue of the hooves by a mixed group of bacteria (Egerton *et al.*, 1969; Roberts and Egerton, 1969). An essential component of this mixture is *Dichelobacter nodosus*, formerly *Fusiformis nodosus* (Beveridge, 1941) or *Bacteroides nodosus* (Dewhirst *et al.*, 1990). *Dichelobacter nodosus* is a Gram-negative anaerobic bacterium which, so far as is known, occurs naturally only in the feet of ruminants affected by footrot. The disease is characterized by infection of the interdigital skin (IDS) and may, under certain conditions, progress to separation of the sole, soft and hard horn from the underlying hoof matrix.

Foot conditions not classified as footrot

Stewart (1989) provided a comprehensive description of differential diagnosis of foot conditions and diseases associated with lameness in sheep. The most common infections and conditions include ovine interdigital dermatitis (OID, a necrotizing infection by *F. necrophorum* in the absence of *D. nodosus* (Parsonson *et al.*, 1967)), foot and toe abscess, and 'shelly toe'. Less common conditions and infections of feet include strawberry footrot, post-dipping lameness, toxic laminitis, scabby mouth, ulcerative dermatosis, bluetongue and foot-and-mouth disease. None of these conditions is associated with *D. nodosus* and, with the exception of OID, experienced clinical examiners can distinguish these conditions readily from ovine footrot.

Summary of the aetiology and pathogenesis of footrot

Infection of normal, dry healthy interdigital skin of sheep with *D. nodosus* alone is insufficient for the development of footrot. Predisposition of feet through water maceration and activity of normal environmental skin microflora, including *Corynebacterium pyogenes* and various surface-located diphtheroid bacteria, is an essential prerequisite for the development of footrot. Footrot is therefore a mixed bacterial infection, and *Fusiformis necrophorum*, a faecal organism, is essential to the disease. *F. necrophorum* is often considered to be the main pathogen in the predisposition and early stages of initiation of footrot. Predisposition causes sufficient inflammation and damage of the stratum corneum for infection to proceed. The synergistic relationship between *F.*

necrophorum and *D. nodosus* leads to progressive and destructive infection of the epidermis of the hoof, causing in severe cases a separation of the horn from the dermis. The underlying dermis may become inflamed but is not invaded, and the stratum germinativum is not destroyed. The extent of tissue damage is dependent on bacterial and host-resistance factors, described later on.

Transmission of the disease is facilitated directly by transfer of infected material containing *D. nodosus* from exposed lesions into the environment, and thereby contaminating the feet of other sheep. Successful transmission is only possible under wet and warm conditions, in sheep that have been sufficiently predisposed. *D. nodosus*, as the obligate parasite, is thus considered to be the essential transmitting agent of footrot, but development of footrot requires the symbiotic relationship with *F. necrophorum*, as a normal environmental inhabitant, and the action of prolonged wetting of feet.

Expression of Footrot

Clinical signs in individual sheep

Following infection with *D. nodosus* and in the successful development of footrot, a range of clinical signs may be evident. Inflammation, characterized by diffuse superficial necrosis and erythema of the interdigital skin, is evident during predisposition and initial onset of footrot. In more severe cases, a break at the skin–horn junction is visible approximately 1 week after infection. Extensive separation, which commences at the heel and the posterior region of the sole, may progress along the sole to the toe. In extremely severe cases, separation will extend to the abaxial wall of the hoof.

Chronic infection will cause the horn to be overgrown and misshapen, with extensive necrotizing damage to underlying soft tissues. Apparent self-cure is possible under dry conditions, particularly in cases where the infection has been confined to the interdigital skin.

Scoring of lesions

In order to standardize the description of severity and progression of lesions, a number of scoring systems have been developed for the subjective assessment of footrot. Egerton and Roberts (1971) were the first to propose a scoring system for footrot lesions, as follows:

- Score 0: normal dry or wet foot
- Score 1: limited interdigital dermatitis
- Score 2: more extensive interdigital dermatitis
- Score 3: severe interdigital dermatitis and underrunning (separation) of the horn of the heel and sole
- Score 4: as for score 3, but with underrunning extended to the walls of the hoof

Other scoring systems have evolved from this system in an attempt to

Normal foot. There is normal skin between the claws, with no reddening or inflammation and no loss of hair. No exudate is present.

Score 1. Slight to moderate inflammation with some erosion between the claws. There is no underrunning or erosion of the skin or horn.

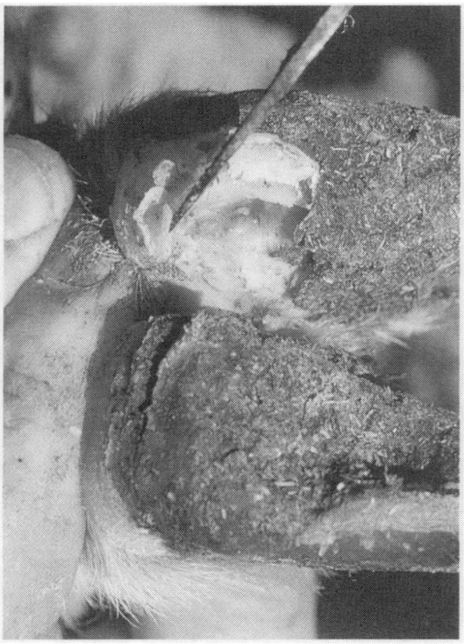

Score 3b. Underrunning no more than halfway across the heel or sole.

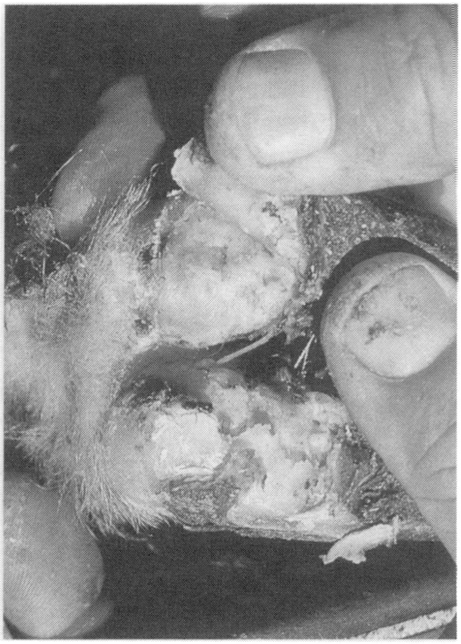

Score 3c. More extensive underrunning of the heel or sole, but not extending to the outside edge of the sole of the claw.

Fig. 10.1. Examples of footrot in the scoring system used in NSW (Source: NSW Agriculture).

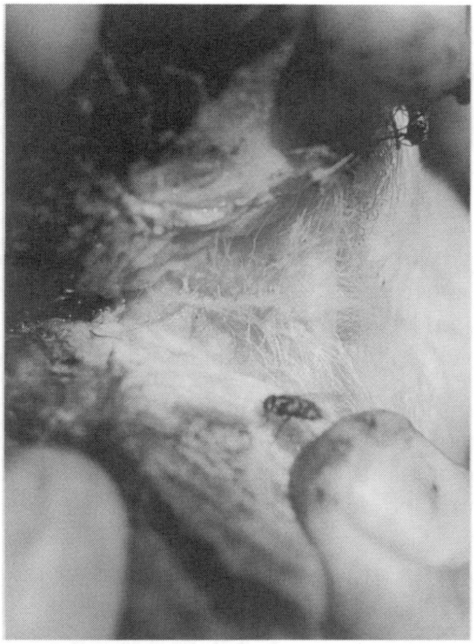

Score 2. The skin between the claws is inflamed and raw. This condition may involve part, or all, of the soft horn of the inside of the claws. There is no underrunning of the horn.

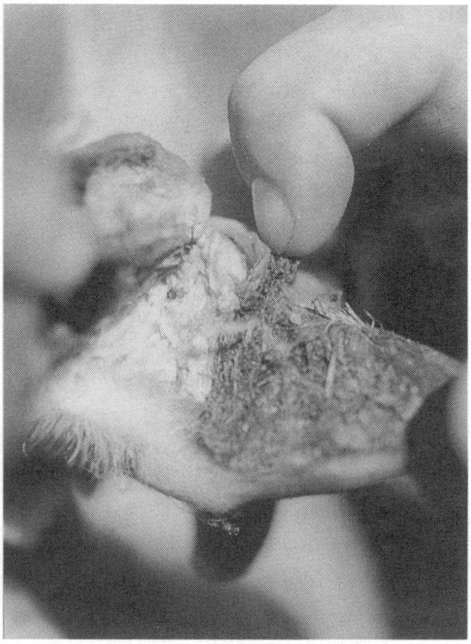

Score 3a. Separation of the skin–horn junction, with underrunning extending no more than 5 mm.

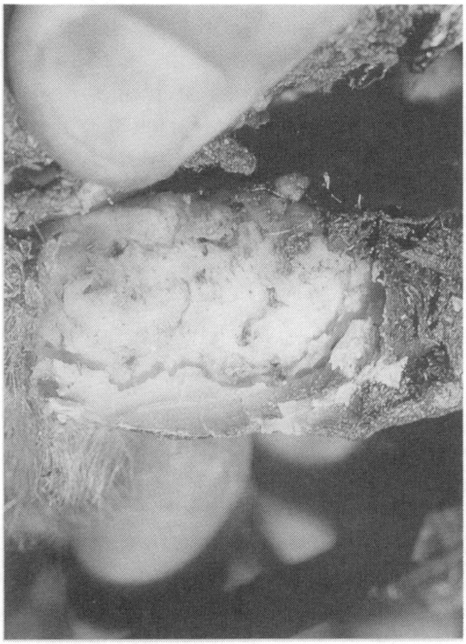

Score 4. The underrunning extends to the outside edge of the sole of the claw and involves hard horn.

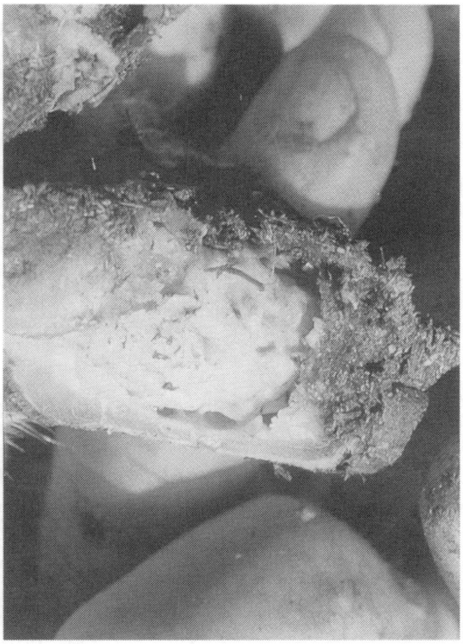

This is a severe form of the disease involving the sole, with extensive inflammation and underrunning of the hard horn of the hoof.

Fig. 10.1. *Continued*

differentiate more clearly between various levels of progression of footrot infection. An example of a clinical scoring system for use in the New South Wales sheep industry is shown in Fig. 10.1.

Expression on a flock basis

The prevalence of sheep affected with footrot in different environments exposed to agents of varying virulence can range from negligible (<5%) to extremely high (>95%), as illustrated by Egerton and Raadsma (1991). For convenience, footrot at the flock level can be classified as described by Egerton and Raadsma (1991):

1. Severe footrot – high proportion (>10%) of animals with severe lesions (score 3 or score 4); rapid development; severe production losses; little evidence of self-cure. This is a severe and debilitating disease with significant lameness and welfare implications.
2. Moderate footrot – disease expressed at a severity in between severe and mild footrot. A low proportion of sheep may express severe lesions which are usually confined to underrunning of the sole of the hoof (score 3); self-cure is evident; few sheep remain chronically infected. Fewer sheep show lameness, although the few sheep with severe lesions may show acute lameness, and may remain chronically infected.
3. Mild footrot – very low proportion (<1%) of animals with severe lesions, with most lesions confined to the interdigital skin (score 1 or score 2). Most lesions resolve spontaneously with onset of dry conditions. Little effect on production. Associated lameness is less than in severe or moderate footrot. Expression of mild footrot is further complicated in that it closely resembles ovine interdigital dermatitis (OID).

It should be noted that these categories are subjective, and a wide range of expression of footrot is usually recognized in the field. The expression of footrot based on the severity and economic impact of the disease in a flock is, in fact, a continuum.

Factors affecting expression of footrot

The expression of footrot in a flock of sheep is governed by three factors: (i) virulence of *D. nodosus*; (ii) the suitability of the environment for predisposition of the host and transmission of the organism; and (iii) inherent susceptibility of the host. These factors have been described in detail by Egerton and Raadsma (1991). Their influence on the expression of the disease is shown in Fig. 10.2. It should be recognized that control of the disease is feasible through direct manipulation of any of the three 'windows' of opportunity. This chapter will focus on those factors affecting host susceptibility only.

Variation in resistance to infectious disease is the consequence of a combination of innate and acquired resistance. In the case of footrot, innate

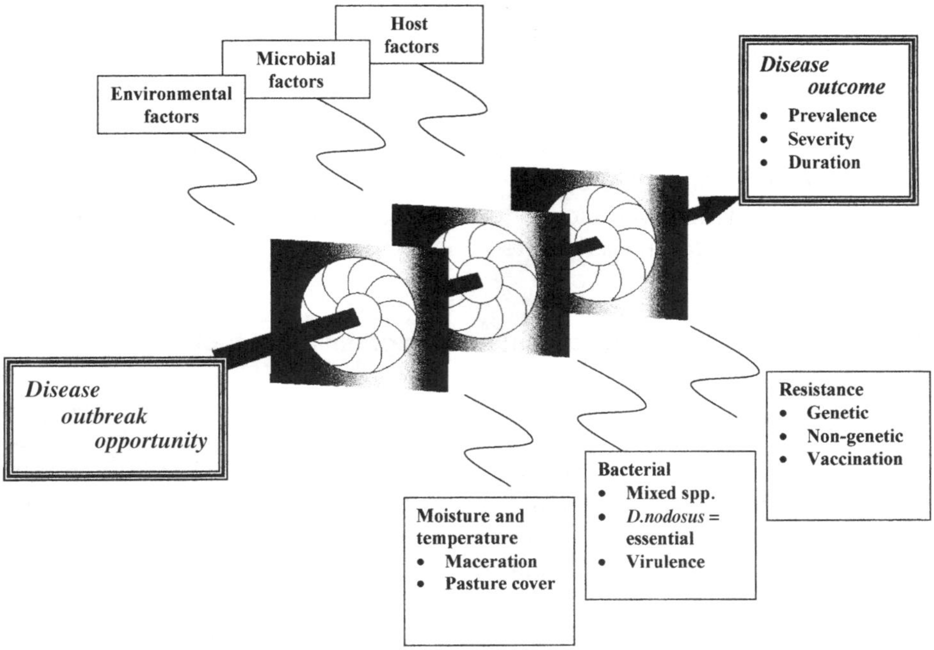

Fig. 10.2. The three major factors that influence the prevalence, severity and duration of footrot shown as windows of opportunity. The window which allows for the lowest opportunity will determine the overall expression of footrot.

resistance may be responsible for preventing invasion of the epidermis by the bacteria responsible for the disease. O'Meara and Raadsma (1995) provide an overview of the physical and immunological factors of the host that aid or arrest the development of footrot.

The influence of non-genetic effects such as age, birth or rearing type, age of dam and sex of sheep on susceptibility to footrot has been reported. Adult sheep are more susceptible than lambs, and rams more susceptible than ewes (Beveridge, 1941; Littlejohn, 1961; Raadsma *et al.*, 1993, 1995; Woolaston, 1993).

In general terms, acquired resistance is a sequel to naturally acquired infection, colostral transfer of immunity or immunization. The identification of the principal causative bacterial agent, *D. nodosus*, and the immunogenic properties of *D. nodosus* fimbriae, have made it possible to use vaccination as a method to control footrot. Following the initial development of whole-cell vaccines (Egerton and Burrell, 1970; Egerton and Roberts, 1971), recombinant DNA techniques have now made it possible to produce effective vaccines comprised largely of *D. nodosus* fimbrial antigens (Egerton *et al.*, 1987; O'Meara *et al.*, 1993). To establish the importance of these antigens in vaccine formulations, the measurement of K-agglutinating antibody titres can be used as an indication of the ability of the vaccine to protect against infection with *D. nodosus* (Raadsma *et al.*, 1994b).

Vaccines need to induce antibody titres against fimbrial antigens from each of the nine major serogroups of *D. nodosus*, since there is little or no cross-protection between serogroups, and field infections often involve infection with more than one serogroup. The problem of reduced titres, possibly due to antigenic competition in the host immune response (Raadsma *et al.*, 1994b, c), limits the efficacy of multicomponent footrot vaccines. The existence of genetic variation in response to active immunization against footrot is reviewed later in this chapter.

Accepted control methods to minimize expression of footrot

Control of footrot has focused largely on elimination (eradication) of non-benign isolates of *D. nodosus* from flocks; prophylactic control through vaccination; or control of affected sheep through topical and systemic therapy, together with vaccination. None of these measures offers a long-term easy-care approach to disease management. The full range of control methods of footrot in sheep, goats and deer has been extensively reviewed by Stewart (1989) and Egerton *et al.* (1989). Only aspects relevant to exploiting genetic variation in innate resistance or the protective response to vaccination will be discussed further in this review.

Scope for Genetic Improvement of Resistance to Footrot

Incentive for genetic improvement in resistance to footrot

In the absence of regulatory control, the occurrence of footrot on a farm does not always justify treatment. Egerton and Raadsma (1991) presented estimates of losses likely to arise when owners take one of three options: (i) take no action; (ii) implement control; or (iii) proceed through control to eradication. A number of important points were made:

1. For mild footrot, the cost of control or eradication would exceed that which could be directly attributed to the disease if left untreated.
2. The losses due to severe and moderate footrot could be halved through conventional control techniques, but the cost of control would be recurrent from season to season.
3. The economic loss from severe or moderate footrot under a control management option is greater than that from uncontrolled mild footrot.

Breeding strategies could aim to reduce the impact of infection with virulent or intermediate isolates of footrot to that experienced with benign isolates of footrot, so that no specialized control strategies are warranted and the disease has minimum impact on production. An alternative strategy could be to improve the responsiveness of sheep to vaccination, so that footrot could be managed similarly to the clostridial diseases, with annual booster vaccinations offering effective protection.

Resistance to footrot defined in general terms

It is appropriate to think of resistance as a response to challenge. Clear clinical signs and diagnostic techniques have been established for bacterial diseases to classify animals according to their level of resistance after a disease outbreak.

Indicators based on clinical scores

Numerous measurement systems have been used to describe differences between sheep in their clinical response to footrot. Most of these systems are based on a simple binary scale indicating whether a foot or a sheep is affected or not. Sometimes the scale is extended beyond the two classes, in order to describe the severity of footrot as indicated by the extent of underrunning of soft and hard horn of the foot. A number of footrot severity indices have been derived from the individual foot scores. Raadsma *et al.* (1993) evaluated 22 primary and derived indicators, including footrot scores incorporating healing, for their utility to describe differences between feet, between sheep and between flocks. For differences between sheep, all indicators were highly correlated (Raadsma *et al.*, 1993). Those traits with an ordered scale, such as the number of feet affected or underrun, are inherently more useful than scores or grades that do not reflect incremental levels of severity, as used, for example, by Skerman *et al.* (1988).

Indicators and liability to footrot

Resistance is often measured as an all-or-none trait. However, in reality, resistance is a multifactorial trait. The polygenic nature of the trait and all the non-genetic factors that influence expression of disease can be readily accommodated by adopting an underlying scale of liability, as proposed by Falconer (1965). This allows disease to be treated as a trait with quantitative characteristics similar to those of other production traits for which genetic parameters can be estimated. The genetic implication that all-or-none traits are due to single-gene effects, where resistance behaves as a trait with Mendelian inheritance, is oversimplistic and incorrect for most bacterial diseases.

Of particular interest is how traits with multiple categories, such as number of feet affected or underrun, fit a threshold model. Similarly, the relationship between the threshold for becoming affected, and the threshold for severe footrot (i.e. underrunning) can also be examined under a multi-threshold model of liability. Raadsma *et al.* (1993) showed that for those indicators with several categories on an ordered scale (number of feet affected, or number of feet underrun), increasing grades of severity reflect a single underlying variable. Figure 10.3 shows how thresholds for the number of feet underrun (0, 1, 2, 3, 4) were 'mapped' on to a single underlying scale of liability for varying levels of expression of resistance to footrot. On the basis of this analysis, it is recommended to use 'number of feet affected with underrun footrot' as a simple indicator to describe differences between sheep in resistance to footrot after challenge.

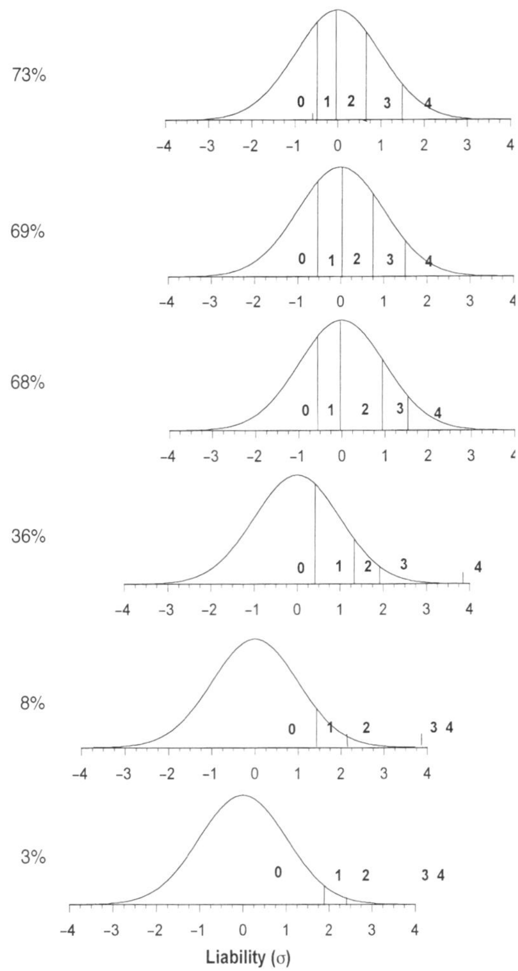

Fig. 10.3. Thresholds for the number of feet with underrun footrot (0–4) mapped on to the underlying scale of liability for a range of expression (prevalence) of severe footrot (%). The area under the curve represents the proportion of animals in each category. (Source: Raadsma *et al.*, 1993.)

Serological indicators of resistance to footrot

It has been shown that antibodies specific to *D. nodosus* are generated during chronic and severe footrot infection (Egerton and Roberts, 1971; Egerton and Merritt, 1973; Fahey *et al.*, 1983; Emery *et al.*, 1984; Ferrier *et al.*, 1986). Such antibodies may be directed to pilus or outer-membrane components. In addition, haemagglutination and protease-inhibiting antibodies have been detected following chronic infection (Egerton and Merritt, 1973). The results of these studies suggest that groups of sheep with more severe levels of infection develop higher antibody titres as a consequence of that infection. Raadsma *et al.* (1993) were the first to report positive correlations between *D. nodosus*-specific antibodies generated during infection, and the susceptibility to footrot, on an individual sheep basis. Although antibody titre was correlated with severity and duration of infection (Fig. 10.4A), this relationship was not sufficiently strong ($r = 0.3–0.6$) to replace clinical scores as a phenotypic indicator to describe resistance in individual sheep.

Acquired immunity following vaccination with native or recombinant pili preparations is an important tool in the control of footrot. Immunity in this case is reflected by the absence of footrot following challenge, and by the level of antibodies directed against the protective pilus antigen, which is traditionally measured by K-agglutination (Egerton and Merritt, 1973). Raadsma *et al.* (1994b) showed a strong relationship between K-agglutinating antibody titre following vaccination, and resistance following challenge (Fig. 10.4B).

The appropriate measures by which resistance in individuals can be measured are clinical scores of footrot lesions and K-agglutinating antibody levels following vaccination. The value of K-agglutinating antibodies in selection for innate resistance or responsiveness to vaccination is discussed in detail below (p. 234).

Repeatability of resistance

For both clinical assessment of footrot and measurement of K-agglutinating antibody titres, repeatability between and within operators is very high ($r > 0.9$; Raadsma, unpublished), demonstrating that experienced operators can make consistent assessment of these two major indicators of resistance to footrot and response to vaccination, respectively, at any point in time.

As a consequence of the changes in footrot status of individuals following challenge and subsequent vaccination, clinical scores are moderately correlated when inspections are made at intervals of 2–3 weeks ($r = 0.31–0.70$ prior to vaccination, and $r = 0.02–0.31$ after vaccination) (Raadsma *et al.*, 1993, 1994a), showing that timing of inspections is of critical importance in the assessment of footrot, and hence resistance. It is recommended to make at least two inspections at least 3 weeks apart during an outbreak of footrot, to get a better assessment of footrot in individual sheep.

Using repeated measures of footrot status during an outbreak in the assessment of resistance has the additional advantage that repeatability models can be used in the estimation of genetic variation and prediction of breeding values. Raadsma *et al.* (1994a) showed that genetic correlations between footrot status at consecutive inspections were almost unity, thus describing the same genetic trait. Genetic correlations between footrot assessed before and after vaccination within the same outbreak were slightly lower (0.8, Raadsma *et al.*, 1994). To quantify resistance over time adequately, repeated assessment of individual sheep may thus be necessary.

Also of relevance is the relationship between the method of footrot challenge, and resistance following challenge with different serogroups. Unfortunately relevant data are scarce. Raadsma *et al.* (1994a) obtained phenotypic, genetic and environmental correlations between resistance following induced and natural challenge, which was (through necessity of the experimental design) confounded with the serogroup of challenge strains. Phenotypic correlations between the responses to the two challenges were low (<0.10). Low negative

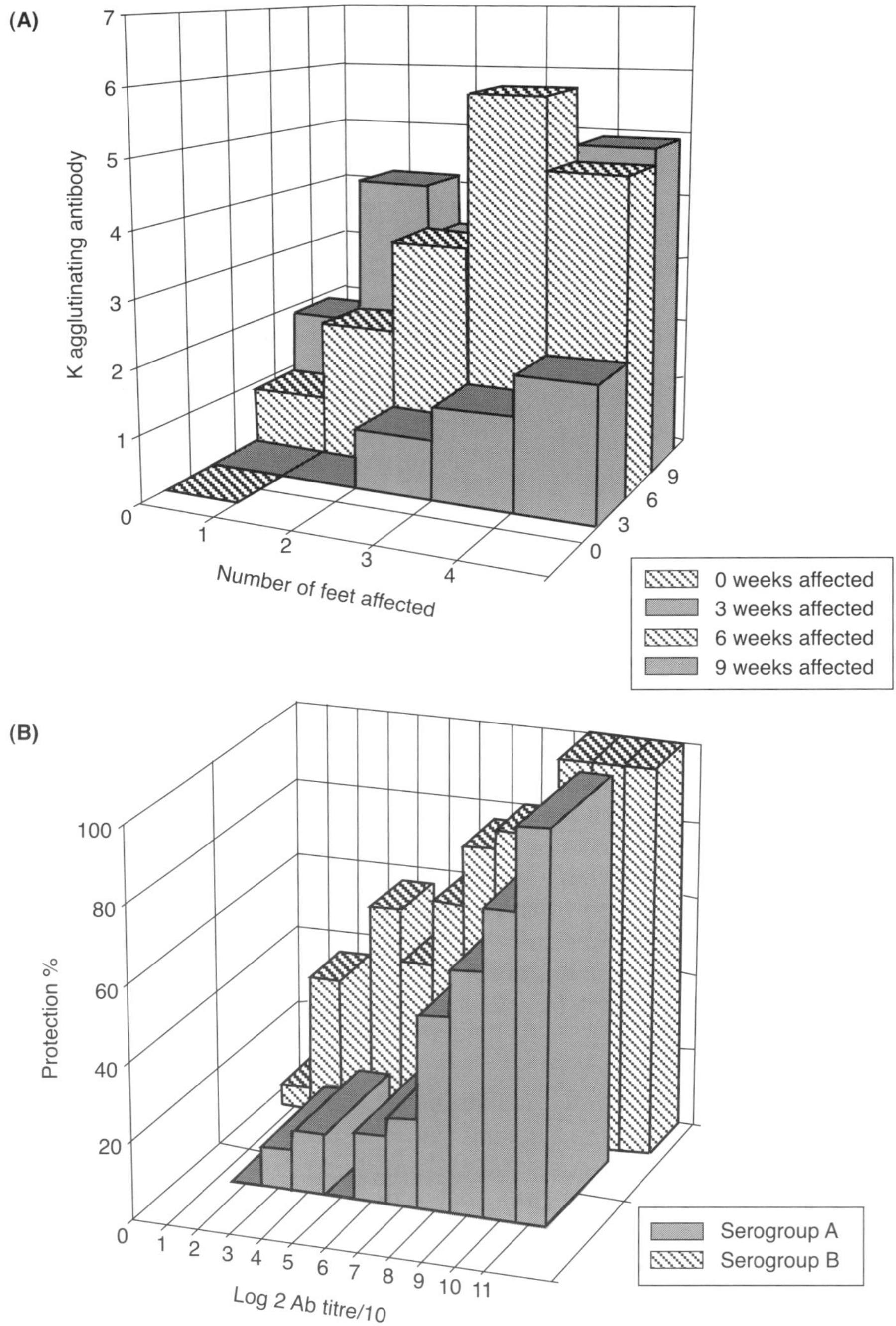

Fig. 10.4. (A) Relationship between footrot severity (number of feet affected) and duration of infection (0, 3, 6, 9 weeks), and serological response measured as serogroup-specific K-agglutinating antibody titres. (B) Relationship between protection following challenge with virulent isolates of *D. nodosus*, and K-agglutinating antibody titre following vaccination with pilus immunogens from two homologous serogroups.

environmental correlations (0.0 to –0.14) suggested that some carry-over effects may exist between sheep exposed to repeated challenge. Corresponding genetic correlations were moderate (0.37–0.67), suggesting that resistance under different challenge conditions may not be completely the same trait. Further data are necessary to determine whether this is due to method of challenge, or if resistance is specific for different serogroups of *D. nodosus*.

Documented genetic variation in resistance to footrot

Although for most major production traits, the extent of genetic variation is reasonably well known, for footrot there is a paucity of information, except for Merinos (Egerton and Raadsma, 1991).

Differences between breeds

Limitations of the available literature describing breed differences in resistance to footrot were identified by Egerton and Raadsma (1991) and will not be covered further. No further additional information of significance has become available since then.

Differences between strains and bloodlines

As detailed by Egerton and Raadsma (1991), within the Australian Merino breed there are special distinct strains and bloodlines within strains. These strains and bloodlines within strains vary considerably in major production characteristics (such as fleece weight, fleece length, body weight) and in resistance to fleece rot, and susceptibility to fly strike (Atkins and McGuirk, 1979; Mortimer and Atkins, 1987; Raadsma and Rogan, 1987; Raadsma, 1991). In contrast, there is relatively little variation between strains or between bloodlines within strains for resistance to internal parasites (Eady *et al.*, 1996)

Estimates of differences between major industry bloodlines of Merinos in their susceptibility to footrot are limited (Raadsma *et al.*, 1994a). One relatively small investigation conducted by Raadsma, Swan and Purvis (unpublished) involved 400 wethers from 11 fine- and medium-wool Merino bloodlines after separate exposure to an intermediate and then a virulent isolate of *D. nodosus*. Although no differences between the 11 flocks in resistance to the intermediate isolate was observed, substantial differences were observed following challenge with the virulent isolate. Repeated inspections over a 27-week period, following a challenge protocol similar to that used by Raadsma *et al.* (1994a), showed that the most resistant flock had 34% of sheep affected with severe footrot (score 3 or score 4) compared with 79% for the most susceptible flock (Fig. 10.5). These results are interesting in that they highlight potential differences between bloodlines in susceptibility to severe footrot, which has not been recorded previously, presumably because in most flocks, managers are actively trying to minimize the expression of footrot.

Woolaston (1993) observed no difference in resistance to footrot in single-trait selection flocks selected for different levels of resistance to *Haemonchus contortus*.

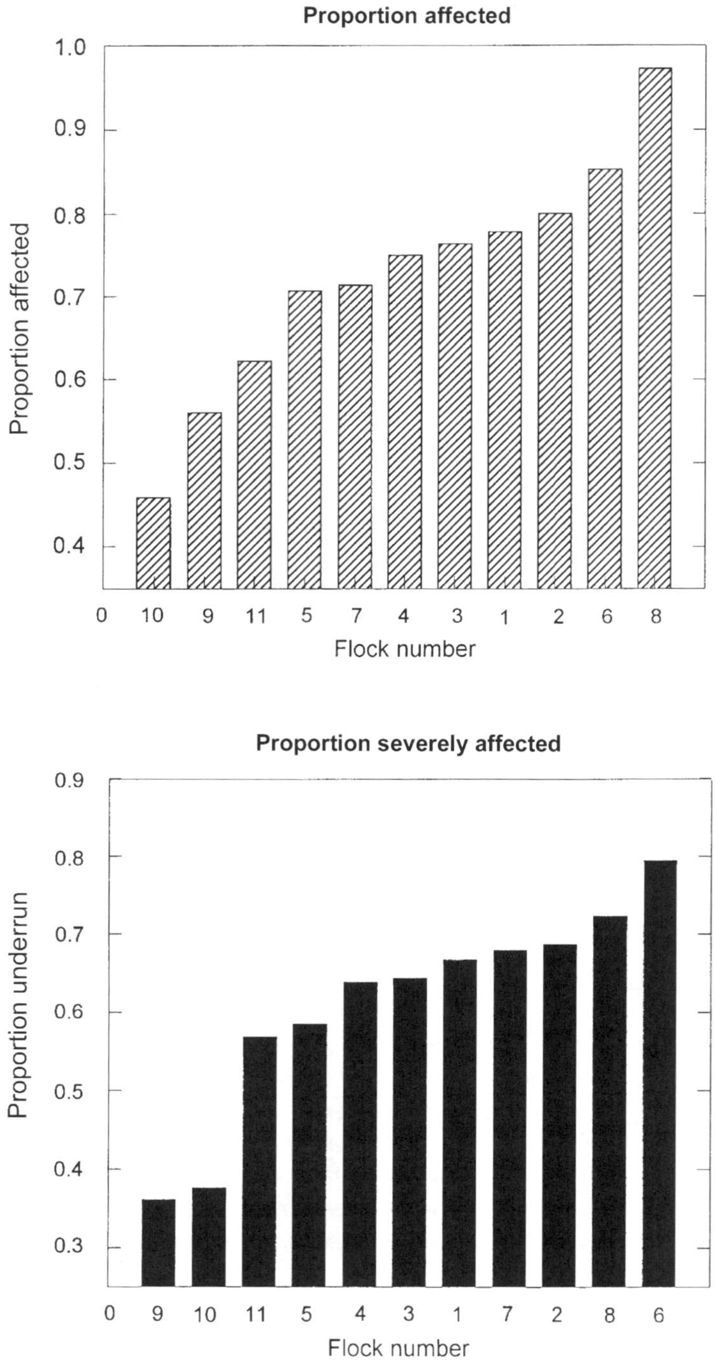

Fig. 10.5. Distribution of susceptibility to footrot expressed as prevalence of affected and severely affected sheep from 11 distinct fine and super-fine bloodlines of Merino sheep. (Source: Raadsma *et al.*, unpublished.)

Differences between sheep within flocks

The estimation of additive genetic variance (heritability) of resistance to footrot has received relatively little scientific attention compared with other important traits in sheep. Early estimates of heritability of resistance to footrot in various breeds have been summarized by Egerton and Raadsma (1991), who also discussed some of the problems associated with traits describing resistance to footrot. First, to diagnose 'footrot' only when underrunning of soft horn occurs, and 'footscald' when only interdigital skin inflammation has been recorded, is erroneous, unless individual intradigital lesions were specifically confirmed to be free from *D. nodosus* infection. Failure to recognize this mars some studies. Secondly, most of the early estimates are from binomial data. In such cases, heritability estimates are dependent on the prevalence of the condition, which means that differences in heritability estimates could reflect differences in prevalence, rather than differences in magnitude of genetic variation. However, it is possible to obtain estimates of the heritability of liability to footrot, independent of prevalence of the condition, through transformation of estimates on the observed scale (Falconer, 1989) or directly on the underlying scale (Gilmour *et al.*, 1985). The difficulty in studying infectious diseases under uncontrolled conditions was highlighted by Woolaston (1993), who reported genetic differences between sire-lines in prevalence and severity of footrot in Merino lambs following natural challenge. However, sire effects were confounded with paddock effects, which were of the same magnitude as the paddock effects in adult ewes allocated at random to sires.

Raadsma *et al.* (1994a) reported heritability estimates for eight clinical indicators of resistance to footrot (five describing the extent of clinical signs, and three describing the extent of subsequent healing). Resistance was assessed in 1562 Merino sheep, representing the progeny from 162 sires in four major bloodlines, following exposure to virulent isolates of *D. nodosus* under both an experimental challenge in which footrot was induced, and a separate natural (field) challenge involving a different isolate of *D. nodosus*. Resistance was assessed on seven occasions following induced challenge, and on five occasions following natural challenge. All sheep were vaccinated with primary and booster injections of an homologous rDNA pilus after initiation of the induced and natural challenge. Half-sib heritability estimates of resistance to footrot were low to moderate for single observations recorded pre-vaccination (0.07–0.22), and slightly lower for inspections made after vaccination (0.07–0.15). Genetic correlations among footrot indicators recorded at repeat inspections were high for observations pre-vaccination (range 0.87–1.00) and slightly lower for observations made after vaccination (0.52–1.00). Heritability estimates derived from repeat measurements approached 0.30 for most indicators, except for indicators describing healing, which had a heritability of almost zero. Heritability estimates of liability to footrot ranged between 0.09 and 0.41, depending on the time after challenge when the inspections were made. The genetic correlation between induced and natural footrot ranged from 0.14 to 0.95, depending on the period over which inspections were made, with an average of 0.67. It was concluded that there is substantial genetic variation within flocks of Merino sheep in resistance to challenge with virulent isolates of *D. nodosus*, especially if resistance is assessed on the basis of preferably three inspections at 3-week intervals.

Skerman (1985) and Skerman and Moorhouse (1987) reported the development of a bloodline with increased resistance to footrot, in each of the Romney Marsh and Corriedale breeds. Both bloodlines evolved through direct selection under natural outbreaks of footrot and extensive use of sires whose progeny showed increased resistance over their contemporaries. Both studs claim now that footrot is an insignificant problem in their flocks (Skerman and Moorhouse, 1987; Warren *et al.*, 1990). Although reports of this nature highlight the potential for genetic control of this disease, formal genetic comparisons such as those described by Skerman and Moorhouse (1987) are needed, but are usually lacking, in on-property experiments. The follow-on benefit of increased usage of breeding stock from more resistant bloodlines in the Merino industry still awaits evaluation, but has already prompted certain stud breeders to place heavy selection pressure on resistance (Patterson and Patterson, 1989, 1991).

Genetic variation in immunological response to *D. nodosus* antigens

Immunological responsiveness to *D. nodosus* antigens is relevant in three distinct scenarios. First, immunological host responses generated during the course of an infection are a valuable objective indicator of resistance during or immediately after infection. Secondly, response following vaccination in non-challenged animals is potentially a valuable means of indirect selection for resistance without the need for direct challenge. Thirdly, vaccination with *D. nodosus* immunogens represents an effective method of footrot control (as described above), and response to vaccination may be amenable to genetic improvement to enhance vaccine efficacy.

At the time of the review presented by Egerton and Raadsma (1991), little evidence was available indicating genetic differences in the ability of the host to respond to *D. nodosus* antigens. Since then, Raadsma *et al.* (1994a, b, 1995, 1996) have addressed each of the three potential applications of immunological responsiveness in breeding for resistance to footrot. A summary of their findings is presented below.

Immune response as an alternative clinical indicator during infection

Despite the consistently positive phenotypic correlations between antibody titre and footrot score, estimates of the genetic correlations and heritability suggested that measures of antibody level during infection were not suitable alternative indicators to footrot scores. Efficiency of selection based on serological indicators during infection would be only 50% of that based on clinical scores.

Vaccine response as an indirect selection criterion

Vaccine responses to *D. nodosus* antigens measured in sheep free from challenge show low genetic correlations with resistance to footrot in non-vaccinated sheep under either homologous (challenge serogroup same as

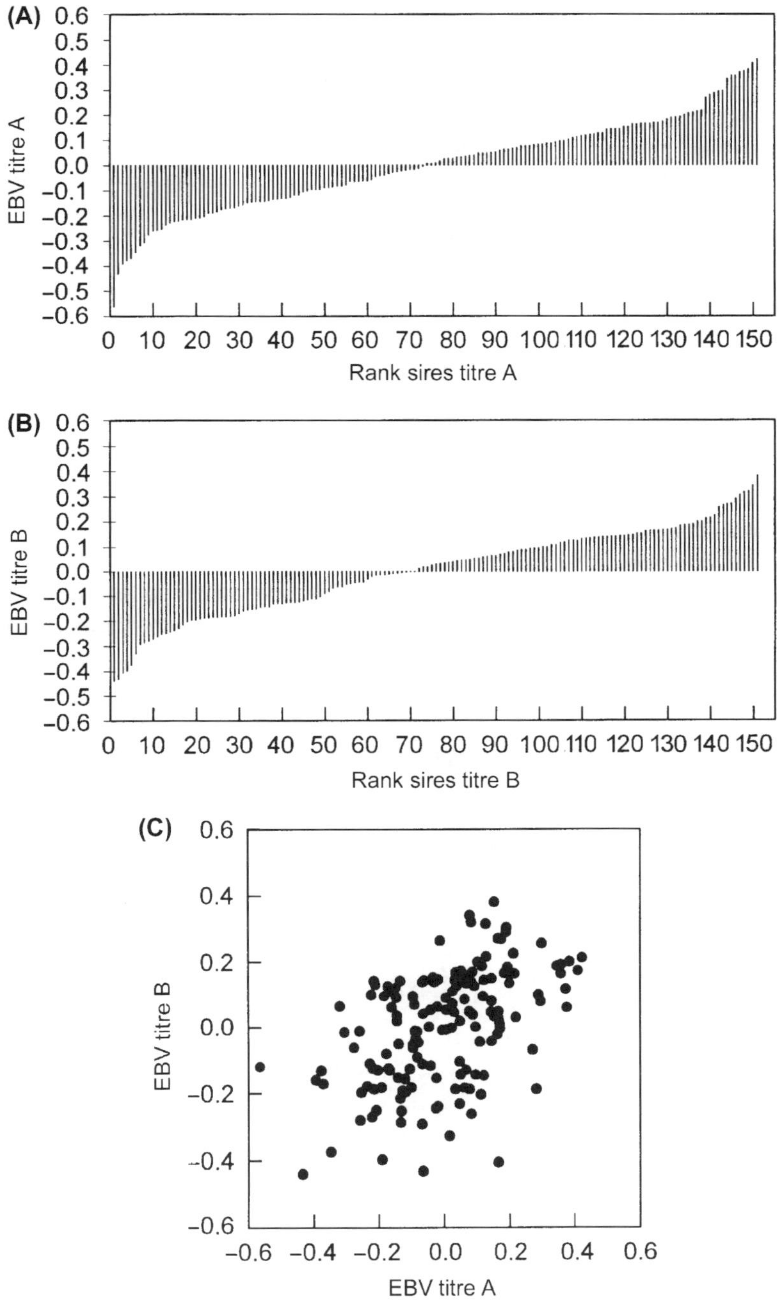

Fig. 10.6. Distribution of estimated breeding values (EBV) for antibody responsiveness measured in 130 sire-progeny groups vaccinated with a multivalent *D. nodosus* vaccine. (A) Response to serogroup A; (B) response to serogroup B; (C) relationship between EBV for response to serogroups A and B. (Source: Raadsma *et al.*, 1995.)

vaccine antigen) or heterologous serogroups. Despite the higher heritability for vaccine response than innate resistance, indirect selection for innate resistance based on vaccine response in non-challenged sheep has little potential because of the low and inconsistent genetic correlations between the two traits. Efficiency of selection based on response to vaccination in the absence of challenge is expected to be less than 20% of that based on direct selection.

O'Meara and Raadsma (1995) reviewed other immunological indicators of resistance to footrot with a view to assessing their potential as alternative indicators, but found none which were more consistent than K-agglutinating antibody levels.

Protective immune response following vaccination

Large phenotypic and genetic differences were recorded for antibody responses to all nine major serogroups of *D. nodosus* commonly used in multivalent footrot vaccines. Based on a sample of 120 half-sib sire families comprising a total of 1200 progeny, considerable variability in heritability estimates was observed for antibody responses to each of the nine serogroups. The estimates ranged from 0.24 to 0.58 (standard error 0.08–0.12) and were independent of mean antibody level. The range in estimated breeding values (EBV) for a sample of sire groups against two of the nine serogroups is shown in Figs 10.6A and 10.6B, where the difference between the lowest- and highest-ranked sire represents a greater than twofold difference in mean antibody titre following vaccination. Results suggest that genotype of the host confers a differential capacity to respond to individual vaccine antigens when multiple antigens are presented simultaneously in the same vaccine.

Genetic heterogeneity in vaccine response

Despite 34–78% homology of the amino acid sequence among the fimbrial subunits of the nine *D. nodosus* fimbrial proteins, each elicits a serogroup-specific antibody response that offers almost no cross-protection between serogroups. Examination of genetic correlations among antibody response to each serogroup revealed considerable heterogeneity, ranging from moderately negative (–0.40) to strong positive (+0.87). An example showing the genetic correlation between vaccine responsiveness for serogroup A and B is shown in Fig. 10.6C. Utilizing procedures described by Raadsma *et al.* (1996), genetic control of antibody response to *D. nodosus* antigens was partitioned into that which was shared (common) and that which was serogroup-specific. On average, 32% of antibody response was under shared genetic control, but the actual genetic correlations ranged from 0 to 87%, depending on the combination of antigens. It was concluded that significant additive genetic variation exists in Merino sheep for response to *D. nodosus* vaccines. However, there was no strong evidence for common genetic control of immune response to different antigens, and each serogroup may require a unique combination of immune-response genes in the host for optimum vaccine performance.

Genetic markers for resistance and vaccine responsiveness

The option of exploiting genetic differences in footrot resistance through direct selection is of limited practical value, as it requires challenging all animals before selection. The possibility of using indirect selection strategies is therefore appealing. The limited utility of serological responses as indirect selection traits was discussed above. No other physiological responses have been identified that show potential as indirect selection criteria (O'Meara and Raadsma, 1995). Other options for indirect selection strategies include the use of genetic markers linked to resistance genes. The role of the major histocompatibility complex (MHC) in modulating immune responses and subsequently disease resistance is well documented for a number of species (see Chapter 4). MHC gene products are glycoproteins, which are present on the surface of some cells and are divisible into two types, class I and class II histoglobulins. Genetic polymorphism within both the class I and class II regions has been investigated in relation to innate resistance to footrot, and response after vaccination with *D. nodosus* antigens (Outteridge *et al.*, 1989; Litchfield *et al.*, 1993; Escayg *et al.*, 1997; Raadsma and Stear, unpublished data). Overall, the conclusion is that associations between resistance or antibody response and MHC polymorphism are not sufficiently strong or consistent to justify their application in the field. Obviously, the search for suitable markers should be extended to other regions within (class III) or outside the MHC. The use of genome-wide screens with high-density coverage of polymorphic microsatellite markers in crosses between resistant and susceptible populations (breeds or selection lines), as conducted by Crawford *et al.* (personal communication), may provide evidence for major genes that affect resistance to footrot. Identification of such major genes may then make it possible to use marker-assisted introgression, or marker-assisted selection within flocks.

Raadsma *et al.* (1998) reported preliminary findings from genome screens in resource flocks held by Crawford, McEwan and colleagues at AgResearch (New Zealand), in relation to the response to vaccination with *D. nodosus* antigens. The results suggested that a region in the MHC plays a role in regulating serogroup-specific responses, and a region within chromosome 1 contributes to variation in generalized vaccine response. Further studies will identify the chromosomal regions, and, where possible, specific genes, that affect vaccine responsiveness.

Resistance and other breeding objectives

Should it prove to be feasible to improve resistance to footrot through selective breeding, it is unlikely to be the sole breeding objective in any well-designed programmes. It is important, therefore, to have accurate information on the correlations between resistance to footrot and all the major production traits that are recommended as breeding objectives or selection criteria for sheep and wool production. In high-rainfall areas, resistance to other important diseases, such as fly strike and internal parasites, may also need to be considered.

No published estimates of phenotypic or genetic correlations between resistance to footrot and production traits are available. Raadsma *et al.* (unpublished) obtained such estimates in the study described above (Raadsma *et al.*, 1993, 1994a). Estimates of genetic correlations between resistance and production were obtained for production traits both under challenged and non-challenged conditions, to account for the environmental influence of disease on production performance. The results indicate no strong undesirable genetic correlations. In fact, all estimates of correlations between the important production traits (clean fleece weight and mean fibre diameter) were neutral (range –0.1 to +0.2). These data suggest that selection for important production traits will have no adverse or desirable effect on resistance to footrot. Similarly, selection for increased resistance to footrot will not adversely affect important production traits, and there should be sufficient scope to improve production and resistance simultaneously, albeit at slower rates than if either objective was taken as a sole breeding objective.

Resistance to other diseases

The question of broad-based resistance (resistance to multiple diseases) is relevant here, since the important diseases in sheep production are often influenced by common environmental factors. Raadsma *et al.* (1997) reported a unique study, where resistance to all important diseases that affect production was examined in the same flock under the same environmental conditions. Genetic variation for resistance to each of the diseases existed within the flock, but genes conferring resistance to one disease did not, in general, have any effect on resistance to another disease – the genetic correlations between resistance to different diseases were low. The only possible exception was a moderate (undesirable) genetic correlation between resistance to fleece rot (a major predisposing factor to blowfly strike) and resistance to footrot.

The observation on a neutral genetic correlation between resistance to footrot and resistance to internal parasites confirms earlier observations by Woolaston (1993) who reported a neutral genetic correlation (0.02 ± 0.20) between resistance to footrot and resistance to a major internal parasite in sheep, namely *Haemonchus contortus*. Similarly, Gray and Wollaston (1991) did not observe significant differences in K-agglutinating antibody levels following vaccination with a whole-cell *D. nodosus* vaccine, in flocks with various levels of resistance to *H. contortus*.

It is likely that breeding programmes aimed at resistance to multiple diseases will need to consider each relevant disease separately. It may be feasible to exploit and combine resistance to multiple diseases, including footrot, from different flocks that have been selected specifically for resistance to just one disease. This type of breeding exercise would be greatly assisted by gene markers.

References

Allworth, M.B. (1995) Investigations of the eradication of footrot. PhD thesis, University of Sydney, Australia.

Atkins, K.D. and McGuirk, B.J. (1979) Selection of Merino sheep for resistance to fleece rot and body strike. *Wool Technology and Sheep Breeding* 27 (3), 15–19.

Beveridge, W.I.B. (1941) *Foot-rot in Sheep: A Transmissible Disease due to Infection with* Fusiformis nodosus (*n. sp*). Bulletin 140, Council for Scientific and Industrial Research, Melbourne.

Dewhirst, F.E., Paster, B.J., La Fontaine, S. and Rood, J.I. (1990) Transfer of *Kingella indologenes* (Snell and Lapage 1976) to the genus *Suttonella* gen. nov. as *Suttonella indologenes* comb. nov.: transfer of *Bacteroides nodosus (Beveridge 1941) to the genus Dichelobacter nodosus* comb. nov.: and assignment of the genera *Cardiobacterium, Dichelobacter*, and *Suttonnella* to *Cardiobacteriaceae* fam. nov. in the gamma division of *Proteobacteria* based on 16S ribosomal ribonucleic acid sequence comparisons. *International Journal of Systematic Bacteriology* 40, 426–433.

Eady, S.J., Woolaston, R.R., Mortimer, S.I., Lewer, R.P., Raadsma, H.W., Swan, A.A. and Ponzoni, R.W. (1996) Resistance to nematode parasites in Merino sheep: sources of genetic variation. *Australian Journal of Agricultural Research* 47, 895–915.

Egerton, J.R. and Burrell, D.H. (1970) Prophylactic and therapeutic vaccination against ovine footrot. *Australian Veterinary Journal* 46, 517–522.

Egerton, J.R. and Merritt, G.C. (1973) Serology of footrot: antibodies against *Fusiformis nodosus* in normal affected vaccinated and passively immunised sheep. *Australian Veterinary Journal* 49, 139–145.

Egerton, J.R. and Raadsma, H.W. (1991) Breeding sheep for resistance to footrot. In: J.B. Owen, J.B. and Axford, R.F.E. (eds) *Breeding for Disease Resistance in Farm Animals.* CAB International, Wallingford, pp. 347–370.

Egerton, J.R. and Roberts, D.S. (1971) Vaccination against ovine footrot. *Journal of Comparative Pathology* 81, 179–185.

Egerton, J.R., Roberts, D.S. and Parsonson, I.M. (1969) The aetiology and pathogenesis of ovine footrot. I. A histological study of the bacterial invasion. *Journal of Comparative Pathology* 79, 207–216.

Egerton, J.R., Cox P.T., Anderson, B.J., Kristo, C., Norman, M. and Mattick, J.S. (1987) Protection of sheep against footrot with a recombinant DNA-based fimbrial vaccine. *Veterinary Microbiology* 14, 393–409.

Egerton, J.R., Yong, W.K. and Riffkin, G.G. (1989) *Footrot and Foot Abscess of Ruminants.* CRC Press, Boca Raton, Florida.

Emery, D.L., Stewart, D.J. and Clark, B.L. (1984) The structural integrity of pili from *Bacteroides nodosus* is required to elicit protective immunity against foot-rot in sheep. *Australian Veterinary Journal* 61, 237–238.

Escayg, A.P., Hickford, J.G.H. and Bullock, D.W. (1997) Association between alleles of the ovine major histocompatibility complex and resistance to footrot. *Research in Veterinary Science* 63, 283–287.

Fahey, K.J., McWaters, P.G., Stewart, D.J., Peterson, J.E. and Clark, B.L. (1983) Quantitation by ELISA of pili and sheep antibodies to the pili of *Bacteroides nodosus. Australian Veterinary Journal* 66, 111–116.

Falconer, D.S. (1965) The inheritance of liability to certain diseases, estimated from the incidence among relatives. *Annals of Human Genetics (London)* 29, 51–76.

Falconer, D.S. (1989) *Introduction to Quantitative Genetics*, 3rd edn. Longman, London.

Ferrier, G.R., Smith, S.C. and Spencer, T.L. (1986) Serological responses to antigens of

Bacteroides nodosus in footrot infected and vaccinated sheep. In: Stewart, D.J., Peterson, J.D., McKern, N.M. and Emery, D.L. (eds) *Footrot in Ruminants.* Proceedings of a Workshop, Melbourne, 1985. CSIRO, A.W.C., pp. 161–163.

Gilmour, A.R., Anderson, R.D. and Rae, A.L. (1985) The analysis of binomial data by a generalised linear mixed model. *Biometrika* 72, 593–599.

Gray, G.D. and Woolaston, R.R. (1991) *Breeding for Disease Resistance.* Australian Wool Corporation, Melbourne.

Litchfield, A.M., Raadsma, H.W., Hulme, D.J., Brown, S.C., Nicholas, F.W. and Egerton, J.R. (1993) Disease resistance in Merino sheep. II. RFLPs in Class 2 MHC and their association with resistance to footrot. *Journal of Animal Breeding and Genetics* 110, 321–334.

Littlejohn, A.I. (1961) Field trials of a method for the eradication of footrot. *Veterinary Record* 32, 773–780.

Mortimer, S.I. (1987) Australian estimates of genetic parameters for wool production and quality traits. In: McGuirk, B.J. (ed.) *Merino Improvement Programs in Australia.* Proceedings of the National Symposium, Leura, NSW, pp. 159–173.

O'Meara, T.J. and Raadsma, H.W. (1995) Phenotypic and genetic indicators of resistance to ectopathogens. In: Gray, G.D., Woolaston, R.R. and Eaton, B.T. (eds) *Breeding for Resistance to Infectious Diseases in Small Ruminants.* ACIAR Monograph Series, No. 34, pp. 187–218.

O'Meara, T.J., Egerton, J.R. and Raadsma, H.W. (1993) Recombinant vaccines against ovine footrot. *Immunology and Cell Biology* 71, 473–488.

Outteridge, P.M., Stewart, D.J., Skerman, T.M., Dufty, J.H., Egerton, J.R., Ferrier, G. and Marshall, D.J. (1989) A positive association between resistance to ovine footrot and particular lymphocyte antigen types. *Australian Veterinary Journal* 66, 175–179.

Parsonson, I.M., Egerton, J.R. and Roberts, D.S. (1967) Ovine interdigital dermatitis. *Journal of Comparative Pathology* 77, 309–313.

Patterson, R.G. and Patterson, H.M. (1989) A practical approach to breeding footrot resistant Merinos. *Journal of New Zealand Mountain Lands Institute* 46, 64–75.

Patterson, R.G. and Patterson, H.M. (1991) The selection and breeding of Merino sheep for foorot resistance. *Proceedings of the New Zealand Society of Animal Production* 51, 283–286.

Raadsma, H.W. and Rogan, I.M. (1987) Genetic variation in resistance to blowfly strike. In: McGuirk, B.J. (ed.) *Merino Improvement Programs in Australia.* Proceedings of the National Symposium, Leura, NSW, pp. 321–340.

Raadsma, H.W., Egerton, J.R. and Nicholas, F.W. (1991) Investigations in genetic variation for resistance to footrot. In: Gray, G.D. and Woolaston, R.R. (eds) *Breeding for Disease Resistance in Sheep.* Proceedings of a Workshop, UNE, Armidale, Nov. 1990, pp. 41–50.

Raadsma, H.W., Egerton, J.R., Nicholas, F.W. and Brown, S.C. (1993) Disease resistance in Merino sheep. I. Traits indicating resistance to footrot following experimental challenge and subsequent vaccination with an homologous rDNA pilus vaccine. *Journal of Animal Breeding and Genetics* 110, 281–300.

Raadsma, H.W., Egerton, J.R., Wood, D., Kristo, C. and Nicholas, F.W. (1994a) Disease resistance in Merino sheep. III. Genetic variation in resistance following challenge and subsequent vaccination with an homologous rDNA pilus vaccine under both induced and natural conditions. *Journal of Animal Breeding and Genetics* 111, 367–390.

Raadsma, H.W., O'Meara, T.J., Egerton, J.R., Lehrbach, P.R. and Schwartzkoff, C.L. (1994b) Protective antibody titres and antigenic competition in multivalent

Dichelobacter nodosus fimbrial vaccines using characterised rDNA antigens. *Veterinary Immunology and Immunopathology* 40, 253–274.

Raadsma, H.W., O'Meara, T.J., Egerton, J.R., Nicholas, F.W. and Attard, G. (1994c) Genetic factors in protective antibody response to ovine footrot vaccines. *Proceedings of the 5th World Congress of Genetics Applied to Livestock Production* 20, 296–299.

Raadsma, H.W., Attard, G.A., Nicholas, F.W. and Egerton, J.R. (1995) Disease resistance in Merino sheep. IV. Genetic variation in immunological responsiveness to fimbrial *Dichelobacter nodosus* antigens and its relationship with resistance to footrot. *Journal of Animal Breeding and Genetics* 112, 349–372.

Raadsma, H.W., Attard, G.A., Nicholas, F.W. and Egerton, J.R. (1996) Disease resistance in Merino sheep. V. Genetic heterogeneity in response to vaccination with *Dichelobacter nodosus* and clostricial antigens. *Journal of Animal Breeding and Genetics* 113, 181–199.

Raadsma, H.W., Nicholas, F.W. and Egerton, J.R. (1997) Ultimate disease resistance in sheep: what are the relationships between all diseases? *Proceedings of the Australian Association of Animal Breeding and Genetics* 12, 63–67.

Raadsma, H.W., MacDonald, P.A., Attard, G., Wright, C.S., Knowler, K.J., Beattie, A.E., Dodds, K.G., McEwan, J.C. and Crawford, A.M. (1998) Genetic markers and vaccine response to immunologically defined foot rot (*Dichelobacter nodosus*) antigens in sheep. *Proceedings of the 6th World Congress of Genetics Applied to Livestock Production,* Armidale, 27, 327–330.

Roberts, D.S. and Egerton, J.R. (1969) The aetiology and pathogenesis of ovine footrot. II. The pathogenic association of *Fusiformis nodosus* and *F. necrophorus. Journal of Comparative Pathology* 79, 217–226.

Skerman, T.M. (1985) Isolation and identification of *Bacteroides nodosus.* In: Stewart, D.J., Petersen J.E., McKern, N.M. and Emery, D.L. (eds) *Footrot in Ruminants.* Proceedings of a Workshop, Melbourne, AWC, pp. 77–78.

Skerman, T.M. and Moorhouse, S.R. (1987) Broomfield Corriedales: a strain of sheep selectively bred for resistance to footrot. *New Zealand Veterinary Journal* 35, 101–106.

Skerman, T.M., Johnson, D.L., Kane, D.W. and Clarke, J.N. (1988) Clinical footscald and footrot in a New Zealand Romney Flock; phenotypic and genetic parameters. *Australian Journal of Agricultural Research* 39, 907–916.

Stewart, D.J. (1989) Vaccination against footrot and foot abscess. In: Egerton, J.R., Yong, W.K. and Riffkin, G.G. (eds) *Footrot and Foot Abscess of Ruminants.* CRC Press, Boca Raton, Florida, pp. 5–45.

Warren, H., Daniel, W. and Parker, T. (1990) Where are we at in sheep breeding? *Wool Technology and Sheep Breeding* 38 (1), 27–28.

Woolaston, R.R. (1993) Factors affecting the prevalence and severity of footrot in a merino flock selected for resistance to *Haemonchus contortus. Australian Veterinary Journal* 70, 365–369.

Mastitis in Dairy Cattle

11

J.B. Owen[1], R.F.E. Axford[1] and S.C. Bishop[2]
[1]School of Agricultural and Forest Sciences, University of Wales, Bangor, UK; [2]Roslin Institute (Edinburgh), Roslin, UK

Summary

During the last decade of the twentieth century much new evidence accumulated on the question of whether and how mastitis resistance should be incorporated into dairy cattle breeding. Although unease about the poor udder health of the modern cow, especially its welfare implications, has increased, there have been several positive developments that enable breeders to reduce the rate of deterioration without adversely affecting overall profitability, maybe even increasing profitability.

The availability of somatic cell counts (SCC) has given breeders an additional convenient, routine, indirect selection criterion, associated with both clinical and subclinical manifestations of mastitis. The relevant estimated genetic parameters for SCC confirm that it is more heritable, particularly after the first lactation, than clinical assessments of mastitis. Additionally, SCC and measures of clinical mastitis show a reasonably high genetic correlation.

A promising development from molecular biology, that of mapping the cattle genome, is proceeding apace and this review indicates that SCC will soon be significantly supported by marker-assisted selection to an extent that we cannot easily assess.

Computer simulation evaluations, based on the most recent estimates of genetic parameters and appropriate economic weightings, confirm the economic justification for including emphasis on mastitis resistance in modern dairy cattle breeding schemes. At present, however, this may only arrest rather than reverse the trend in mastitis occurrence.

Considering that the benefit in future cow welfare cannot be assessed merely in money terms, it must surely be the aim to enhance mastitis resistance so that, with other improvements in milking equipment and hygiene, the major curse of the dairy industry is put in retreat.

© CAB *International* 2000. *Breeding for Disease Resistance in Farm Animals* (eds R.F.E. Axford, S.C. Bishop, F.W. Nicholas and J.B. Owen)

Introduction

Mastitis, inflammation of the udder, results from the invasion of the udder by a wide variety of pathogenic bacteria. Clinical mastitis is widespread throughout dairy herds, with a UK incidence, about 50%, which has not changed greatly over the past 20 years. There are three types of mastitis, defined by the route of infection, each associated with particular pathogens (Blowey, 1988). Contagious mastitis is spread from cow to cow during the milking process; environmental mastitis is acquired from fouled bedding; and the separate condition, summer mastitis, is insect transmitted. Control in each case depends on preventing the deposition of the pathogens on the teats, their transfer through the teat canals and their establishment in the udder tissue.

Well-established specific routines such as teat dipping, antibiotic therapy, dry cow therapy, the avoidance of teat damage, good milking machine maintenance and fly control are combined with the practice of good hygiene to limit infection rates (Blowey and Edmonson, 1995). These have reduced the proportion of new cases of contagious mastitis so that most surveys (e.g. Hillerton *et al.*, 1995) now find environmental mastitis to be the main manifestation of the condition. Prevention of environmental mastitis requires avoidance of contact between the teat of the freshly milked cow and faecal material, which is difficult to maintain reliably in housed cattle. The continuing high incidence of mastitis suggests that practical husbandry methods alone give inadequate control, and that additional benefit could be obtained from increasing the resistance and reducing the susceptibility of cows to udder infection.

Since the publication of the first edition of this book (Eriksson, 1991) the economic importance of mastitis to the world's dairy industry has become even more evident. In the meantime research has tended to dispel the gloom about the economic justification for including mastitis resistance as an objective within dairy cattle breeding schemes. The full cost of clinical and subclinical mastitis, in terms of milk yield and quality as well as its cost in terms of cow welfare, is becoming ever clearer. Wells *et al.* (1998) have analysed the factors contributing to US dairy cow health and have placed mastitis as the single most important problem in this respect. In their analysis the authors considered, in addition to traditional evaluation of the impact of health problems on animal productivity, zoonotic risks, international trade implications and animal welfare. On both methods of evaluation mastitis was the top-ranked disease problem affecting dairy cattle.

In terms of cost to the UK dairy industry (1996 figures), output losses ranged from £57–185 million, treatment cost £45–78 million and prevention cost £4 million per annum. Excluding BSE, which was not evaluated, mastitis is far and away the economically most important animal disease in the UK, in any species (Bennett *et al.*, 1999; further information on the web page http://www.rdg.ac.uk/AcaDepts/ae/AEM/livestockdisea/).

Concurrently, much greater confidence is now attached to estimates of genetic parameters for mastitis resistance and the best predictors of resistance in on-farm recording. The rapid expansion of molecular studies has added to the armoury for explaining the genetic basis of resistance and to provide efficient DNA markers to underpin marker-assisted selection (MAS).

This chapter summarizes some of the pertinent studies that have extended the scope of genetic means to help combat mastitis during the period 1990–1999.

Rationale for Genetic Improvement

Subsequent sections detail some of the evidence for the unfavourable genetic correlation between mastitis resistance and milk yield. This confirms the widely held suspicion that the modern dairy cow, hitherto bred mainly for yield in early lactations, is more susceptible to mastitis than would be expected purely on the basis of her higher yield.

Jones *et al.* (1994) in a long-term selection experiment commenced in 1964, compared 236 cows sired by top AI bulls, selected only on milk yield, with 227 sired by unselected control bulls. Cows sired by the selected bulls had significantly higher health expenses than those sired by the control bulls (approximately US$50 per lactation). This difference was mainly due to expenses for mastitis in the first 20 days of lactation. There is now also a greater appreciation of the widespread nature of the effect of mastitis on herd profitability and on the cow's welfare (Juga and Voutilainen, 1998). Both overt clinical mastitis and its subclinical manifestation contribute to extra costs of treatment and prevention, to direct payment penalties and higher replacement rates, as well as to the inevitable loss of milk production and unfavourable changes in milk composition stemming from the disease.

Genetic Parameters of Mastitis Resistance with Special Reference to Somatic Cell Counts and Type Traits as Selection Criteria

Clinical mastitis events are difficult and costly to record as quantitative traits in cattle breeding schemes, so that SCC, combined with a few type traits, has become a useful additional, or the preferred, selection criterion in most practical breeding schemes. Somatic cell counts enumerate the concentrations of udder epithelial cells and white blood cells, which are shed in milk, as a single total, and are used routinely in milk quality assessment. These cells are normal constituents of milk, but are found in increased concentrations, particularly the neutrophils, in milk from cows with udder damage. The justification for the use of SCC depends not only on the ease and lower cost of measurement but also on the relative heritability of SCC compared with measures of clinical mastitis. Additionally, the justification also depends on the genetic correlation between the objectives, i.e. the number and severity of infections in challenged cows, and the recorded criterion (i.e. SCC).

Heringstad *et al.* (1996) have emphasized the interrelation of the heritability of clinical mastitis, progeny group size and genetic correlation of clinical mastitis with SCC in deciding whether dairy cattle selection criteria should include clinical mastitis, SCC or both. Mrode and Swanson (1996) in a similar analysis have concluded that, in relation to the then current estimates of genetic parameters, SCC was the most suitable indirect trait for reduction of mastitis through selection and that it had a promising role to play under the right circumstances.

Groen *et al.* (1994), in a study of 3617 cows sired by 224 bulls of the Dutch Black and White breed, showed that the genetic correlation of clinical mastitis problems with milk yield was unfavourable (–0.26). Boettcher *et al.* (1992) undertook a genetic evaluation of Holstein bulls for SCC in the milk of their first-lactation daughters from five US data processing centres. The heritability of SCC overall was 0.10 and genetic correlations with important economic yield traits were significantly unfavourable. Miglior *et al.* (1995) gave an estimate of 0.16 for the heritability of SCC, in a study of 65,491 lactation mean SCC values, for first-lactation Canadian Holsteins. In the Finnish study, involving data on 23,196 cows, reported by Poso and Mantysaari (1996), the heritability of SCC was 0.14–0.19, distinctly higher than the 0.02–0.05 estimated for clinical mastitis. Reents *et al.* (1995) in their Canadian study, involving 235,000 test-day records of lactations 1–3 for 15,922 cows in 143 herds, reported heritability of 0.09–0.11 for SCC in Holstein cows. For 31,180 American Jersey cows, sired by 411 bulls, Rogers *et al.* (1995) reported heritability estimates for SCC of 0.05–0.15. Schutz *et al.* (1994) report estimates of heritability for lactation mean SCC varying from 0.07 to 0.11 for six common breeds, including a value of 0.09, based on 1,135,752 Holsteins.

An Israeli study involving 32,448 mean lactation records for 19,764 Israeli Holsteins in 54 herds (Weller *et al.*, 1992) reported genetic parameters for SCC, udder bacterial infection and clinical mastitis. The heritabilities of these three traits were respectively 0.13–0.27, 0.02–0.04 and 0.01 (for field mastitis). The genetic correlation between bacterial infection and SCC was near unity but for clinical mastitis and SCC it was only 0.3. Lund *et al.* (1994), using first-lactation data from the Danish young sire sampling programme, showed that the heritability of mastitis was low (0.025) and for SCC was 0.18. Their estimate of the genetic correlation of SCC with mastitis was high at 0.97, making SCC a feasible indicator for clinical mastitis. Shook and Schutz (1994) have noted the increased genetic susceptibility to mastitis that has accompanied the rapid genetic increase in US dairy cattle. Estimating the heritability of SCC in US dairy cattle to be 0.10 and the genetic correlation between SCC and clinical mastitis around 0.6–0.8, they concluded that using SCC as a selection criterion would marginally decrease genetic gain for milk yield but enhance total economic merit. Optimal selection indices that include SCC would slow down the rate of increase in mastitis although they were not likely to decrease its incidence.

A Swedish study reported by Philipsson *et al.* (1995) involved 750,000 effective daughters of 1462 Red and White and 911 Friesian progeny-tested bulls. Estimates of 0.79 and 0.71 for the two daughter groups, respectively, were obtained for the genetic correlation of SCC and clinical mastitis. Pryce *et al.* (1998) in a study of 10,569 records from 4642 cows of all parities in 33 herds, over a 6-year period, confirmed the general finding that the incidence of mastitis increased with lactation number. The heritability of SCC across lactations was 0.15. All genetic correlations with production traits were antagonistic. However the genetic correlation between SCC and mastitis incidence was 0.65, confirming the value of SCC as a selection criterion for resistance to mastitis.

VanDorp *et al.* (1998) studied 4368 records from 30 herds in Canada. They estimated that the heritability of mastitis in first-lactation Holsteins was less than 0.05 and that the genetic correlation between mastitis and udder conformation traits ranged from 0 to 0.37 (unfavourable).

A British study by Mrode *et al.* (1998) involved the average lactation SCC record from the first three lactations of 63,424 Holstein/Friesian, 7966 Ayrshire and 14,509 Jersey cows. The heritabilities of log SCC for first lactations were 0.11, 0.12 and 0.09 for the Holstein/Friesian, Jersey and Ayrshire breeds, respectively. Genetic correlations with milk fat and protein yield were unfavourable in the Holstein/Friesian breed. Significant negative correlations were observed in Holsteins and Ayrshires between sire progeny transmitting ability (PTA) for log SCC and daughters' foot angle (–0.14), fore-udder attachment (–0.19) and udder depth (–0.19) and a positive correlation for teat length (0.15).

Detilleux *et al.* (1995) reported an average heritability value of 0.10 for mastitis indicators, including SCC and clinical mastitis, in Holsteins and showed an interesting negative genetic correlation between the number of quarters infected with major pathogens and those infected with minor pathogens. Koenen *et al.* (1994) in their study of the Swedish Friesian breed, based on 9516 observations, gave an estimate of 0.018 for the heritability of registered veterinary-treated mastitis. However, the inclusion of culling reasons in the analysis gave a significant increase in the heritability estimate (0.083). Rogers *et al.* (1991) estimated genetic correlations between linear type traits (see below) and SCC in US Holsteins in datasets ranging from 4294 daughters of 216 sires to 58,235 daughters of 301 sires. Their observations included correlations between this category of traits when both were measured in the first lactation and also between linear type traits measured in the first lactation and SCC measured in the second and third lactations. Overall genetic correlations between SCC and each of udder depth, fore-udder attachment and teat placement were negative, that is favourable, whereas the genetic correlation of SCC with teat length tended to be positive. The highest genetic correlation was that with udder depth (–0.35) thus indicating that udder depth, and possibly also closer teat placement, could be useful criteria in selecting for resistance to mastitis in dairy cattle.

The estimates described above confirm that the heritability of SCC is higher than that of more directly available measures of clinical mastitis in cattle. The genetic correlation with milk yield will have been responsible for a significant fall in mastitis resistance at a time when higher yields have placed considerably more strain on the modern dairy cow.

A consequence of the lowish heritabilities for mastitis or SCC is that progeny group sizes in progeny testing schemes may have to be increased in order to estimate accurately breeding values (BVs) for bulls for these traits. It is evident that the Scandinavian countries have done just this, accounting for their success in limiting the rate of increase in mastitis. Heringstad *et al.* (1996) give a more detailed account of these Nordic developments that have pioneered the development of multitrait selection for mastitis resistance in dairy cattle. A next step in terms of SCC data-handling is to find ways of obtaining and utilizing SCC data such that it is most informative about the actual mastitis status and the consequences of mastitis for the cow.

Molecular Basis for Mastitis Resistance

Recent molecular studies have revealed a number of quantitative trait loci (QTLs), as well as some possible candidate genes involved in mastitis resistance. Some of these are promising possibilities for MAS, as well as aiding the better understanding of the underlying basis of the mechanism of resistance.

Sharif *et al.* (1998) showed that there was an association ($P < 0.05$) between the *BoLA-DRB3.2*16* allele (in part of the MHC complex) and lower SCC in Holsteins. There was also an association ($P < 0.05$) between the *BoLA-DRB3.2*23* allele and the occurrence of severe mastitis, mainly involving coliform bacteria, important agents in environmental mastitis. It seems that *BoLA* loci may have a potential as markers in breeding for mastitis resistance. Ashwell *et al.* (1998), working with US Holsteins, reported a significant association between both udder depth and SCC on the one hand and markers 513 and BM1258, thus indicating the presence of relevant loci on chromosome 23. Dietz *et al.* (1997), in a study of US Holsteins, identified the *DRB3.2*16* allele as being associated with acute intramammary infection in cows classified as having acute, as opposed to chronic, elevated levels of SCC. The authors suggest the possibility of a useful marker being based on DNA from milk samples. Kelm *et al.* (1997) studied the association of various measures of mastitis resistance and parameters of innate resistance in periparturient US Holsteins. They found that molecular markers accounted for up to 40% of the variation in estimated breeding values (EBVs) for measures of mastitis. The presence of allele *DRB3..2*16*, in contrast with Sharif *et al.* (1998) (above), was associated with higher values of SCC. Alleles *DRB3..2*8*, *IgG2* (β) and the normal *CD18* were associated with increased EBV for clinical mastitis. Alleles *DRB3..2*11*, *IgG2* (α) and the recessive allele for bovine leucocyte adhesion deficiency were associated with lower levels of clinical mastitis. Apparently, in contrast to the findings of Sharif *et al.* (1998), allele *DRB3..2*23* was also associated with decreased clinical mastitis. Allele *DRB3.2*24* and EBV for intramammary infection (IMI) by major pathogens showed a positive genetic association, as did allele *DRB3.2*3* and IMI by minor pathogens. Several of the EBVs for immunological assays showed significant correlations with some EBVs for mastitis measures.

Some recent reports on QTLs with potential application in marker-assisted selection for mastitis resistance are contained in the *Proceedings of the XXVIth International Conference on Animal Genetics* held in Auckland, New Zealand in 1998. These include QTLs influencing SCC, located on chromosomes 5, 22 and 23 ($P < 0.005$) (Heyen *et al.*, 1998) and on chromosomes 1 and 18 ($P < 0.05$) (Kalm *et al.*, 1998).

Due to the apparent contradictions in *BoLA* loci marker effects, noted above (a not uncommon occurrence in MHC loci in mammals generally) caution has to be exercised when interpreting *BoLA* loci results. In spite of this, immunological parameters, including physiological and molecular markers, may well become useful aids to understanding the genetics of resistance to mastitis and should contribute useful selection criteria in breeding schemes.

Cost–Benefit of Inclusion of Mastitis Resistance in Multitrait Selection

Several economic evaluation studies, many involving computer simulation, have helped in the assessment of the net value of including mastitis resistance in cattle breeding schemes. Christensen (1998) has discussed the 'Nordic profile' concept relating to the inclusion of non-production traits in multitrait selection of young dairy bulls. Deterministic computer simulation of alternative breeding schemes showed that, in contrast to multitrait selection, single-trait selection for milk yield led to substantial deterioration in most functional traits, including resistance to mastitis. Response in money terms from single-trait selection was 26% less than from multitrait selection. In all the multitrait schemes studied, 20–25% of the total genetic gain was due to improvement in the non-production traits. The potential importance of this conclusion for cow welfare is obvious!

Colleau and Lebihanduval (1995) have reported the results from using computer simulation to assess the value of methods to improve mastitis resistance in dairy cows under French conditions. Annual genetic gains for the conventional breeding scheme were estimated using statistical parameters from the literature. Selection on yield and log SCC, with or without mastitis culling rate, increased genetic gains for overall profitability by 0.7 or 0.9% respectively, compared with selection on yield only. Increases of log SCC and mastitis problems were reduced by 40–60% in multitrait as against single-trait selection for milk yield.

Philipsson *et al.* (1994) report the results from a simulation study on the efficiency of index selection for overall merit in Swedish dairy cows. They show that when the real breeding goal includes functional non-production traits, such as mastitis resistance, in addition to production traits, their omission from the selection index decreases efficiency by 15–25%.

These studies and simulations are positive indications for the value of including mastitis resistance (mainly through SCC as the selection criterion) as an objective in dairy cattle breeding schemes. Despite its relatively low heritability, major economic progress can be made by selecting for mastitis resistance, with additional benefits accruing to the welfare of the cow.

Acknowledgement

SCB wishes to thank BBSRC C59.

References

Ashwell, M.S., Da, Y., Vanraden, P.M., Rexroad, C.E. and Miller, R.H. (1998) Detection of putative loci affecting conformational type traits in an elite population of United States Holsteins using microsatellite markers. *Journal of Dairy Science* 81 (4), 1120–1125.

Bennett, R., Christiansen, K. and Clifton-Hadley, R. (1999) Preliminary estimates of the direct costs associated with endemic diseases of livestock in Great Britain. Paper presented at the workshop on 'Economic assessment of livestock disease problems', University of Reading, 2 March, 1999.

Blowey, R.W. (1988) *A Veterinary Book For Dairy Farmers*, 2nd edn. Farming Press, Ipswich, UK.

Blowey, R. and Edmonson, P. (1995) *Mastitis Control in Dairy Herds.* Farming Press, Ipswich, UK.

Boettcher, P.J., Hansen, L.B., Vanraden, P.M. and Ernst, C.A. (1992) Genetic evaluation of Holstein bulls for somatic cells in milk of daughters. *Journal of Dairy Science* 75 (4), 1127–1137.

Christensen, L.G. (1998) Possibilities for genetic improvement of disease resistance, functional traits and animal welfare. *Acta Agriculturae Scandinavica Section A – Animal Science* S29, 77–89.

Colleau, J.J. and Lebihanduval, E. (1995) A simulation study of selection methods to improve mastitis resistance of dairy cows. *Journal of Dairy Science* 78 (3), 659–671.

Detilleux, J.C., Jacquinet, E., Harvengt, A. and Leroy, P.L. (1995) Genetic selection for resistance to mastitis. *Annales de Médecine Vétérinaire* 141 (3), 199–204.

Dietz, A.B., Cohen, N.D., Timms, L. and Kehrli, M.E. (1997) Bovine lymphocyte antigen class II alleles as risk factors for high somatic cell counts in milk of lactating dairy cows. *Journal of Dairy Science* 80 (2), 406–412.

Eriksson, J.A. (1991) Mastitis in cattle. In: Owen, J.B. and Axford, R.F.E. (eds) *Breeding for Disease Resistance in Farm Animals.* CAB International, Wallingford, pp. 394–411.

Groen, A.F., Hellinga, I. and Oldenbroek, J.K. (1994) Genetic correlations of clinical mastitis and feet and legs problems with milk yield and type traits in Dutch Black-and-White Dairy-Cattle. *Netherlands Journal of Agricultural Science* 42 (4), 371–378.

Heringstad, B., Klemetsdal, G. and Ruane, J. (1996) Selection for mastitis resistance in dairy cattle – a review with special consideration of the situation in the Nordic countries. *47th Annual Meeting of the EAAP, Lillehammer, Norway, 25–29 August 1996.*

Heyen, D.W., Weller, J.I., Ron, M., Band, M., Feldmesser, E., Da, Y., Wiggans, G.R., Van Raden, P.M. and Lewin, H.A. (1998) Genome scan for QTL influencing milk production and health traits in dairy cattle. *Proceedings of the XXVIth International Conference on Animal Genetics, E006, 9–14 August, 1998, Auckland, New Zealand.*

Hillerton, J.E., Bramley, A.J., Staker, R.T. and McKinnon, C.H. (1995) Patterns of intramammary infection and clinical mastitis over a 5-year period in a closely monitored herd applying mastitis control measures. *Journal of Dairy Research* 62, 39–50.

Jones, W.P., Hansen, L.B. and Chesterjones, H. (1994) Response of health-care to selection for milk-yield of dairy cattle. *Journal of Dairy Science* 77 (10), 3137–3152.

Juga, J. and Voutilainen, U. (1998) The key elements for genetic response in Finnish dairy cattle breeding. *Agricultural and Food Science in Finland* 7 (2), 207–217.

Kalm, E., Reinsch, N., Xu, H. *et al.* (1998) Mapping QTL on cattle chromosome 2, 5, 10, 16, 18 and 23. *Proceedings of the XXVIth International Conference on Animal Genetics, E009, 9–14 August, 1998, Auckland, New Zealand.*

Kelm, S.C., Detilleux, J.C., Freeman, A.E., Kehrli, M.E., Dietz, A.B., Fox, L.K., Butler, J.E., Kasckovics, I. and Kelley, D.H. (1997) Genetic association between parameters of innate immunity and measures of mastitis in periparturient Holstein cattle. *Journal of Dairy Science* 80 (8), 1767–1775.

Koenen, E., Berglund, B., Philipsson, J. and Groen, A. (1994) Genetic parameters of fertility disorders and mastitis in the Swedish Friesian breed. *Acta Agriculturae Scandinavica Section A – Animal Science* 44 (4), 202–207.

Lund, T., Miglior, F., Dekkers, J.C.M. and Burnside, E.B. (1994) Genetic relationships between clinical mastitis, somatic-cell count, and udder conformation in Danish Holsteins. *Livestock Production Science* 39 (3), 243–251.

Miglior, F., Burnside, E.B. and Dekkers, J.C.M. (1995) Non-additive genetic effects and inbreeding depression for somatic-cell counts of Holstein Cattle. *Journal of Dairy Science* 78 (5), 1168–1173.

Mrode, R.A. and Swanson, G.J.T. (1996) Genetical and statistical properties of somatic cell count and its suitability as an indirect means of reducing the incidence of mastitis in dairy cattle. *Animal Breeding Abstracts* 64 (11), 847–857.

Mrode, R.A., Swanson, G.J.T. and Winters, M.S. (1998) Genetic parameters and evaluations for somatic cell counts and its relationship with production and type traits in some dairy breeds in the United Kingdom. *Animal Science* 66 (3), 569–576.

Philipsson, J., Banos, G. and Arnason, T. (1994) Present and future uses of selection index methodology in dairy cattle. *Journal of Dairy Science* 77 (10), 3252–3261.

Philipsson, J., Ral, G. and Berglund, B. (1995) Somatic-cell count as a selection criterion for mastitis resistance in dairy cattle. *Livestock Production Science* 41 (3), 195–200.

Poso, J. and Mantysaari, E.A. (1996) Relationships between clinical mastitis, somatic cell score, and production for the first three lactations of Finnish Ayrshire. *Journal of Dairy Science* 79 (7), 1284–1291.

Pryce, J.E., Esslemont, R.J., Thompson, R., Veerkamp, R.F., Kossaibati, M.A. and Simm, G. (1998) Estimation of genetic parameters using health, fertility and production data from a management recording system for dairy cattle. *Animal Science* 66 (3), 577–584.

Reents, R., Jamrozik, J., Schaeffer, L.R. and Dekkers, J.C.M. (1995) Estimation of genetic parameters for test day records of somatic cell score. *Journal of Dairy Science* 78 (12), 2847–2857.

Rogers, G.W., Hargrove, G.L. and Cooper, J.B. (1991) Correlations among somatic-cell scores of milk within and across lactations and linear type traits of Jerseys. *Journal of Dairy Science* 78 (4), 914–920.

Rogers, G.W., Hargrove, G.L. and Cooper, J.B. (1995) Correlations among somatic-cell scores of milk within and across lactations and linear type traits of Jerseys. *Journal of Dairy Science* 78 (4), 914–920.

Schutz, M.M. (1994) Genetic evaluation of somatic cell scores for United-States dairy cattle. *Journal of Dairy Science* 77 (7), 2113–2129.

Schutz, M.M., Vanraden, P.M. and Wiggans, G.R. (1994) Genetic variation in lactation means of somatic cell scores for six breeds of dairy cattle. *Journal of Dairy Science* 77 (1), 284–293.

Sharif, S., Mallard, B.A., Wilkie, B.N., Sargeant, J.M., Scott, H.M., Dekkers, J.C.M. and Leslie, K.E. (1998) Associations of the bovine major histocompatibility complex DRB3 (BoLA-DRB3) alleles with occurrence of disease and milk somatic cell score in Canadian dairy cattle. *Animal Genetics* 29 (3), 185–193.

Shook, G.E. and Schutz, M.M. (1994) Selection on somatic-cell score to improve resistance to mastitis in the United States. *Journal of Dairy Science* 77 (2), 648–658.

VanDorp, T.E., Dekkers, J.C.M., Martin, S.W. and Noordhuizen, J.P.T.M. (1998) Genetic parameters of health disorders, and relationships with 305-day milk yield and conformation traits of registered Holstein cow. *Journal of Dairy Science* 81 (8), 2264–2270.

Weller, J.I., Saran, A. and Zeliger, Y. (1992) Genetic and environmental relationships among somatic cell count, bacterial infection, and clinical mastitis. *Journal of Dairy Science* 75 (9), 2532–2540.

12

Escherichia coli and *Salmonella* Diarrhoea in Pigs

I. Edfors-Lilja[1] and P. Wallgren[2]
[1]*Department of Technology and Natural Sciences, University of Växjö, Växjö, Sweden;* [2]*National Veterinary Institute, Uppsala, Sweden*

Summary

Diarrhoea due to bacterial infection is a problem mainly in young growing animals, including pigs. Among the bacteria that cause diarrhoea are various strains of *Escherichia coli* and *Salmonella*. Considerable genetic variation in resistance and susceptibility has been found for both neonatal diarrhoea caused by *E. coli* carrying K88 fimbriae and post-weaning diarrhoea and oedema disease due to *E. coli* strains with F18 fimbriae, and the loci for both types of 'receptors' have been mapped. In mice, resistance to *Salmonella* infections is associated with the antimicrobial activity of macrophages and is linked with polymorphism in the *Nramp1* gene. The gene has been identified in several species, including the pig, but data are so far lacking concerning the association between polymorphism in the porcine gene and resistance and susceptibility to *Salmonella* infection.

The rapid development in molecular genetics has given us detailed genome maps and the tools to identify and study individual genes. This means that in the near future we may be able to determine the genotype of individual animals and to study the association between 'disease resistance' genes and production traits. This information is needed before we can include 'disease genes' in breeding programmes.

Introduction

Diarrhoea (scours) is a common problem in animal production, affecting mostly the young growing animal. Despite considerable interherd differences, large field studies have reported that a high average frequency (6–7%) of all litters born are affected with diarrhoea pre-weaning (Svensmark *et al.*, 1989a) as well as post-weaning (Svensmark *et al.*, 1989b). In another large study the mortality due to scours pre-weaning has been reported as 2.7% of the piglets born, representing 11.9% of the total mortality during that period (Nielsen *et*

© CAB *International* 2000. *Breeding for Disease Resistance in Farm Animals* (eds R.F.E. Axford, S.C. Bishop, F.W. Nicholas and J.B. Owen)

al., 1974). According to Svensmark *et al.* (1989a), piglets that had experienced pre-weaning diarrhoea reached 25 kg liveweight 2 days later than others, corresponding to a decreased productivity of 3%. Further, the risk for developing diarrhoea post-weaning increased if the piglets had experienced gastrointestinal disorders during the suckling period, and the risk of dying before reaching 25 kg body weight was increased fourfold for piglets having post-weaning diarrhoea (Svensmark *et al.*, 1989b).

The routine use of feed additives (antibiotics) has been prohibited in Swedish pig production since 1986. Initially this led to an increased frequency of post-weaning diarrhoea and to decreased productivity of the piglets (Robertsson and Lundeheim, 1994). Thus, specific management and hygiene demands were required to prevent disease outbreaks in Sweden. The total mortality among piglets was 14.6% pre-weaning and 3.4% post-weaning in Swedish conventional herds during 1992, figures that were reduced to 7.0% and 0.1%, respectively, in specific pathogen free (SPF) herds (Wallgren, 1993). These differences, and the large interherd differences obtained between the conventional herds, demonstrate a large influence of management and environment on health status and productivity of piglets.

During the neonatal period, scours is generally associated with one pathogen, commonly *E. coli* or *Clostridium perfringens*. Among older piglets, various infectious agents may cause diarrhoea, among them bacteria such as *E. coli* and *Salmonella*. However, viruses and protozoa also contribute to the clinical status. Diarrhoea is thus a multifactorial disease where the outcome of an infection is due to many factors and their interactions. The genetic make-up of the bacteria, determining various virulence factors such as fimbriae enabling adherence to intestinal mucosa and enterotoxin production, is essential for the pathogenity of the bacteria. Although management and housing routines influence the frequency and severity of scours, the genotype of the pig also has a large impact on resistance and susceptibility to clinical infection. This chapter will deal mainly with the genetic resistance of pigs to bacterial infections leading to diarrhoea, focusing on *E. coli* and *Salmonella* infections.

E. coli Diarrhoea

The ability of enteropathogenic (EPEC) or enterotoxigenic (ETEC) *E. coli* to adhere to the brush borders of enterocytes is fundamental for the initiation of the infection. Attachment of pathogenic bacteria to the mucosa of the small intestine is mediated by distinct surface antigens, called pili or fimbriae (Duguid and Anderson, 1967). Several fimbrial adhesins have been identified both in animal and human EPEC/ETEC strains. In the pig, strains expressing fimbriae of F4 (K88), F5 (K99), F6 (987P) or F41 types dominate during the neonatal period (Söderlind *et al.*, 1982; Brinton *et al.*, 1983; Gonzáles *et al.*, 1995), while strains expressing other types of fimbriae such as F18ab (F107), F18ac (2134P) and Av24 are found during the post-weaning period (Bertschinger *et al.*, 1990; Nagy *et al.*, 1992, 1996; Hide *et al.*, 1995; Kennan *et al.*, 1995).

Neonatal diarrhoea

Much neonatal diarrhoea is due to infections with *E. coli* strains possessing fimbriae of the F4 (K88) type. The frequency of K88 amongst enterotoxigenic strains isolated in various countries differs somewhat, but these strains mostly dominate (Söderlind and Möllby, 1978; Söderlind *et al.*, 1982, 1988; Brinton *et al.*, 1983).

Three antigenic variants, *ab*, *ac* and *ad*, have been identified for the K88 fimbriae, all containing a common a-type antigen (Ørskov *et al.*, 1964; Guiné and Jansen, 1979). The K88 fimbriae adhere to specific receptors on the intestinal cell brush borders. Early studies showed that the K88ac receptor contained a variety of sugar molecules, such as D-galactoside (Kearns and Gibbons, 1979; Sellwood, 1980), *N*-acetylglucoseamine, *N*-acetylgalactoseamine and D-galactoseamine (Sellwood, 1984). A later study has described the K88ac receptor as a mucin-type sialoglycoprotein (Erickson *et al.*, 1994). Glycoproteins of 210 and 240 kDa binding K88ab and K88ac, but not K88ad, fimbriae have been identified (Seignole *et al.*, 1994; Billey *et al.*, 1998). Another glycoprotein (74 kDa) belonging to the transferrin family that binds *in vitro* to K88ab fimbriae has been detected (Grange and Mouricout, 1996). However, K88ac and K88ad fimbriae did not bind to this intestinal transferrin.

Detection of receptor phenotype

Identification of the receptor phenotype of pigs can be performed by examining the adhesion of *E. coli* K88-positive bacteria to intestinal cell brush borders *in vitro* (Sellwood *et al.*, 1975). A variant of the assay, in which whole enterocytes instead of brush borders are used, has also been described (Rapacz and Hasler-Rapacz, 1986). Mostly, the intestinal specimens have been sampled after slaughter and the enterocytes have been collected by gently rubbing or scraping a segment of the intestine. The technique has also been performed on specimens from intestinal biopsies (Snodgrass *et al.*, 1981). To shorten the assay time, an enzyme immunoassay and an ELISA (enzyme-linked immunosorbent assay) have been developed (Chandler *et al.*, 1986; Valpotic *et al.*, 1989).

Testing potential breeding animals by intestinal biopsies or test matings are costly and cumbersome. Nor can any of the described assays differentiate between pigs carrying one or two copies of the receptor allele, i.e. distinguish between heterozygous and homozygous animals. Identification of the gene coding for the receptor structure will make direct typing of breeding animals possible and the genotype of individual animals can then be determined. Several laboratories are currently performing research towards this goal.

Inheritance of receptor phenotype

A genetic influence on resistance to ETEC was described as long as 30 years ago (Sweeney, 1968) and a Mendelian inheritance with a dominant receptor allele was later found for *E. coli* K88ac (Sellwood *et al.*, 1975; Gibbons *et al.*, 1977). One locus coding for both the K88ab and K88ac receptors was first suggested (Rapacz and Hasler-Rapacz, 1986; Bijlsma and Bouw, 1987), but later studies suggested two closely linked loci (Guérin *et al.*, 1993; Edfors-Lilja *et al.*, 1995).

The inheritance of the receptor for K88ad has been less clear and in several studies a weak adherence phenotype has been identified (Rapacz and Hasler-Rapacz, 1986; Bijlsma and Bouw, 1987; Hu, 1988). Studies by Hu *et al.* (1993) suggest that there are two receptors for K88ad, a high-affinity and a low-affinity receptor, both allelic to the K88ab and K88ac receptor(s).

Strong linkage disequilibrium between the *K88abR* and *K88acR* loci, with very few pigs positive for *K88abR* and negative for *K88acR*, has been found in most breeds studied so far (Bijlsma *et al.*, 1982; Edfors-Lilja *et al.*, 1986, 1995; Rapacz and Hasler-Rapacz, 1986; Hu *et al.*, 1993; Baker *et al.*, 1997). However, in one recent study a somewhat higher frequency of the $K88abR^{+}$–$K88ac^{-}$ phenotype was found in the Hampshire breed, i.e. four of 24 tested pigs (Baker *et al.*, 1997). With a recombination distance of 1–2%, one would expect to find the recombinant haplotypes *K88abR*$^{+}$–*K88acR*$^{-}$ and *K88abR*$^{-}$–*K88acR*$^{+}$ occurring at high frequency in some breeds. As this is not the case, there are two possible explanations for the strong association: (i) that haplotypes either positive for both *K88abR* and *K88acR* or negative for both are favoured by selection; or (ii) that the recombination frequency was overestimated in studies that suggest two loci, by typing errors or incomplete penetrance. The K88ab and K88ac proteins differ only slightly in amino acid composition (Gaastra *et al.*, 1979). The finding that the antibody response is not variant specific (Bijlsma *et al.*, 1987) might explain the linkage disequilibrium, as discussed by Ollivier and Renjifo (1991).

The receptor(s) for K88ab and K88ac have been determined in newborn as well as adult pigs. In contrast, it has been found that the weak adhesion phenotype for K88ad cannot be detected in pigs after the age of approximately 16 weeks (Hu *et al.*, 1993). A similar age influence has also been found for the adhesion of *E. coli* carrying K99 (Runnels *et al.*, 1980) and 987P fimbriae (Dean-Nystrom, 1995). A variation in amount of K88 receptor along the length of the small intestine has been reported (Chandler *et al.*, 1994).

Chromosomal localization and candidate genes

In humans, the P blood group constitutes the adhesion factor for urinary tract infections with pathogenic *E. coli* (Källenius *et al.*, 1980; Svensson *et al.*, 1983). Linkage studies between blood group loci and the K88 receptor have also been performed, and a weak linkage between the L blood group locus and the K88 receptor was found (Vögeli *et al.*, 1992). However, linkage had also been suggested between the transferrin locus and the K88ac receptor (Gibbons *et al.*, 1977; Guérin *et al.*, 1993). The *TF* and *EAL* loci have been assigned conclusively to two different chromosomes, the q31 band on chromosome 13 (Chowdhary *et al.*, 1993) and the q arm of chromosome 4 (Marklund *et al.*, 1993). The establishment of detailed linkage maps in the pig (Rohrer *et al.*, 1994; Archibald *et al.*, 1995; Marklund *et al.*, 1996) has improved the opportunities to map the receptor. The localization of the gene for the K88ac receptor to chromosome 13 has been confirmed in this way, and the locus for the receptor has been localized 7.4 cM proximal to the transferrin locus (Edfors-Lilja *et al.*, 1995). This chromosomal region is homologous to human chromosome 3 and, using comparative mapping, research is in progress to map candidate genes in this region. Ten human chromosome 3 genes have been assigned to porcine chromosome 13 (Van Pouke *et al.*, 1997; Peelman, 1998a). Further mapping indicates that the K88ac receptor locus is localized terminal to the transferrin locus, and one marker showing no recombination with the K88ac receptor locus has been mapped (Peelman, 1998b). Radiation hybrid mapping (Alexander *et al.*, 1998) and intestinal cDNA libraries (Winterø *et al.*, 1996) are other tools currently used to identify markers close to the receptor gene and possibly the causative gene itself.

Selection for the receptor phenotype – performance of sows and fattening pigs

The newborn pig is dependent on the mothering capacity of the sow, which includes the provision of antibodies in the colostrum and milk. Sows lacking the receptor produce low levels of antibodies to K88 after natural exposure or oral vaccination (Sellwood, 1979, 1982; Bijlsma *et al.*, 1987). A small but significantly higher IgG response has been found in receptor-positive pigs 3 weeks after intramuscular immunization, suggesting that the immunization acted as a booster dose in receptor-positive pigs (Edfors-Lilja *et al.*, 1995). This confirms earlier observations where pigs possessing receptors for K88ab and K88ac had a more pronounced IgG response to K88 after subcutaneous immunization than did pigs lacking the receptors (Edfors-Lilja *et al.*, unpublished observations).

Although the receptors mediate increased susceptibility to neonatal *E. coli* diarrhoea, the function and significance of the receptors on a more basic level is not known. A low frequency of pigs possessing the receptor has been identified in breeds not selected for increased growth. No receptor phenotype pigs were identified in the Chinese Meishan breed (Chappuis *et al.*, 1984; Michaels *et al.*, 1994), while a low frequency of the receptor phenotype was found in the Chinese Minzu breed (Michaels *et al.*, 1994). In another study, weak adhesion to intestinal cells, but with no correlation with virulence, was found for Chinese Meishan pigs (Bertin and Duchet-Suchaux, 1991). Both European wild boars that were used as parents in a reference pedigree for gene mapping lacked the receptor (Edfors-Lilja *et al.*, 1995).

Post-weaning diarrhoea

Diarrhoea in the older pig is often associated with strains other than those causing neonatal diarrhoea. The change in diet at weaning and some nutritional components are thought to predispose to diarrhoea and oedema disease. The frequency of these problems differs largely between countries and populations, but breed differences have also been observed. In Switzerland, oedema disease and post-weaning diarrhoea are responsible for considerable economic losses (Bertschinger *et al.*, 1992) and can also be a problem in adult pigs (Sydler *et al.*, 1996). Oedema disease is rarely seen in pigs in Australia, but a majority of strains isolated from pigs with post-weaning diarrhoea were positive for F107 fimbriae (Hide *et al.*, 1995). Similar results have been found in Denmark (Ojeniyi *et al.*, 1992), where several outbreaks of oedema disease have been reported recently (Jorsal *et al.*, 1996). In Sweden, oedema disease has a low frequency, but almost 50% of *E. coli* strains collected from pigs with post-weaning diarrhoea contained the gene for a major subunit of the F107 fimbriae (Kennan *et al.*, 1995).

The F107 fimbriae belong to a group of related adhesins named F18 (Imberechts *et al.*, 1994). Like the K88 fimbriae, the F18 possesses a common antigenic variant *a*, and two variant-specific determinants, *b* and *c* (Rippinger *et al.*, 1995). Another adhesion group is the F17 family (Bertin *et al.*, 1996) which includes *E. coli* strains associated with bovine diarrhoea and human urinary tract infections (Martin *et al.*, 1997).

Detection and inheritance of receptor phenotype

A genetic influence on the frequency of post-weaning diarrhoea and oedema disease was described 30 years ago (Smith and Halls, 1968) and has since been confirmed (Bertschinger *et al.*, 1986). After the development of the adherence assay for identification of K88 receptor phenotype pigs, similar studies were performed to identify pigs resistant to oedema disease and post-weaning diarrhoea. Genetic studies have shown that susceptibility to colonization by F18ab-positive *E. coli* is dominantly inherited (Bertschinger *et al.*, 1993). Further studies have mapped the locus for the F18ab receptor to chromosome 6, close to the genes for blood group system S and the calcium release channel, CRC (Vögeli *et al.*, 1996). Two α (1, 2)-fucosyltransferase genes (*FUT1* and *FUT2*) are closely linked to the S and F18 receptor loci and a polymorphism in *FUT1* co-segregates with *E. coli* F18 adhesion (Meijerink *et al.*, 1997).

Salmonella Diarrhoea

Salmonella infections are an important human health problem in many countries. Swine, poultry, cattle and seafood are important carriers (Wilcock and Schwartz, 1992). There are over 2 million cases of meat and poultry food-borne disease in humans in the USA per year, at a cost approaching US$1.4 billion. Most of this disease is attributed to *Salmonella* and *Campylobacter* infections (Menning, 1988). The most frequently reported *Salmonella* serotype is *S. typhimurium*. In addition to the economic impact of salmonellosis on the human population, it is also an economic disease of swine resulting in lost income to the pork industry. Data by Fedorka-Cray *et al.* (1994) suggest that two types of disease syndromes appear to occur after infection; clinical disease within 48 hours and a subclinical syndrome that may be important in establishing a carrier state.

In Sweden, an official control programme with respect to *Salmonella* spp. has been running since 1961 and was last revised in 1995 (Anonymous, 1995). Infected farms are subject to restrictions including a total ban of movements of animals, with the exception for transportation to sanitary slaughter. Regular controls are performed at the abattoirs, aiming to detect a prevalence of infection at 5% with a confidence level of 95%. The control is based on bacteriological examinations from five ileo-caecal lymph nodes per animal and from surface swabs of approximately 1400 cm^2 per animal. Further, feed plants are controlled with respect to *Salmonella*. Regardless of the source, all isolations of *Salmonella* made have to be reported (Eld *et al.*, 1991; Malmqvist *et al.*, 1995). Together these measurements have accomplished low prevalences of animals positive to *Salmonella* at slaughter, i.e. well below 1% among broilers (Wierup *et al.*, 1992) as well as among ruminants and pigs (Wahlström *et al.*, 1997).

Genetic resistance

In chickens, differences between inbred strains in resistance to various serotypes of *Salmonella*, including *S. typhimurium* and *S. enteritidis*, have been described (Bumstead and Barrow, 1988, 1993). In these studies, birds were challenged orally, intramuscularly and intravenously, suggesting that resistance is not a function of adherence to epithelial cells. It was also shown that the resistance was inherited as a dominant autosomal trait and that it was not linked to the major histocompatibility complex (MHC). Further studies with *S. enteritidis* have confirmed these chicken strain differences in resistance and susceptibility (Guillot *et al.*, 1995; Protais *et al.*, 1996). In a recent linkage study, a region on chicken chromosome 5 has been identified that accounts for more than 50% of the difference in resistance between two chicken lines (Mariani *et al.*, 1998). This region corresponds to part of mouse chromosome 12 and human chromosome 14, regions that so far contain no mapped genes likely to contribute to resistance.

In mice, a high level of resistance to *S. typhimurium* infection and other facultative intracellular bacteria is determined primarily by the *Ity/Lsh/Bcg* gene (Skamene *et al.*, 1982; Lissner *et al.*, 1983). This gene was later identified as *Nramp*, natural resistance-associated macrophage protein gene (Vidal *et al.*, 1993). Nramp is a family of integral membrane proteins that have been identified in several species, including *Drosophila*, plants and yeast (Cellier *et al.*, 1995). The function is not known, but it has been suggested that the Nramp polypeptides are part of a group of transporters or channels spanning the plasma membrane. The Nramp1 protein seems to have many effects in regulating macrophage activation, including respiratory burst activity, synthesis of nitric oxide synthase, antigen processing, MHC class II molecule expression and regulation of production and release of the cytokines TNF-α and IL-1β (reviewed by Blackwell, 1996).

The *Nramp1* gene has been identified and assigned in mice and humans to chromosomes 1 and 2, in chickens to chromosome 7 (Girard-Santosuosso *et al.*, 1997) and in sheep to chromosome 2 (Pitel *et al.*, 1994, 1995). A second gene, *Nramp2*, has been identified in humans and mice and localized to chromosomes 15 and 12, respectively (Gruenheid *et al.*, 1995). In the pig, a full-length cDNA of *Nramp1* was recently sequenced (Tuggle et *al.*, 1997). The gene has been assigned to chromosome 15 and a small population study revealed large allele frequency differences among breeds (Sun *et al.*, 1998). Association between polymorphism and resistance or susceptibility to *Salmonella* infection is rather well documented in mice (Blackwell, 1996). In humans, data suggest an association between polymorphism in the *Nramp1* promoter region and susceptibility to rheumatoid arthritis. Linkage studies to determine whether this polymorphism also contributes to infectious disease susceptibility are in progress (Blackwell, 1996).

Conclusions

Diarrhoea due to bacterial infections is a problem in pig production, both with regard to the loss in productivity and also from a human health view, as the pig can act as a carrier of human infections. Genetic resistance to neonatal diarrhoea caused by *E. coli* carrying K88 fimbriae has been known for many years. Also, genetic resistance to post-weaning diarrhoea or oedema disease due to *E. coli* with F18 fimbriae has been identified, and the loci for both types of 'receptors' have been mapped. The receptor for *E. coli* K88, as well as the receptor for uropathogenic *E. coli*, contains various carbohydrate molecules. The nature of the F18 receptor is not yet known, but a close linkage with two α-fucosyltransferase genes has been found.

Resistance to *Salmonella* infections is not associated with 'receptor' molecules, but with the antimicrobial activity of macrophages. In mice, this resistance is linked with polymorphism in the *Nramp1* gene. The gene has been identified in several species including the pig, but data are so far lacking for any association with resistance or susceptibility to *Salmonella* infection. The rapid development in molecular genetics has given us detailed genome maps and the tools to identify and study individual genes. This means that, in the near future we may be able to select breeding animals of preferred genotype. How to decide which genotype to select? Some data suggest that pig populations not selected for growth have a low frequency of the K88 receptor, but we do not yet have enough results to know whether receptor-phenotype pigs grow faster. Even fewer studies concerning the influence of the F18 receptor on production traits have been reported. As regards *Salmonella* infections, it has been suggested that good resistance might increase the frequency of autoimmune diseases.

In conclusion, we know far more about genetic resistance to bacteria caused diarrhoea than we did some years ago. The development of DNA-based tests will enable us to determine the genotype of individual animals, and hence it will also be possible to study association between 'disease genes' and production traits. Thus, in the near future we will have the knowledge to identify and select the preferred genotypes.

References

Alexander, L.J., Hawken, R., Murtagugh, J., Sun, J., Flickinger, G., Beattie, C.W., Robic, A., Milan, D., Yerle, M., Gellin, J. and Schook, L.B. (1998) Generation of a radiation hybrid map of porcine chromosome 13. *Pig Chromosome 13 Workshop, XXVI ISAG Meeting, 9–14 August, Auckland, New Zealand*, p. 113.

Anonymous (1995) *Swedish Salmonella Control Programmes for Live Animals, Eggs and Meat*. Commision decision of 23 February 1995 (95/50/EC). Issued by the National Veterinary Institute, the Swedish Board of Agriculture and the National Food Administration.

Archibald, A.L., Haley, C.S., Rown, J.F., Coupperwhite, S., McQueen, H.A., Nicholsen, D., Coppieters, W., Van de Weghe, A., Winterø, A.-K., Fredholm, M., Larsen, N.J., Nielsen, V.H., Milan, D., Woloszyn, N., Robic, A., Dalens, M., Riquet, J., Gellin, J., Caritez, J.-C., Hue, D., Burgaud, G., Ollivier, L., Bidanel, J.-P., Vaiman, M., Renard, C., Gelderman, H., Davoli, R., Ruyter, D., Vestege, E.J.M., Groenen, M.A.M., Davies, W., Høyheim, B., Keiserud, A., Andersson, L., Ellegren, H., Johansson, M., Marklund, L., Miller, R.J., Anderson-Dear, D.V., Signer, E. and Jeffreys, A.J. (1995) The PiGMaP consortium linkage map of the pig (*Sus scrofa*). *Mammalian Genome* 6, 157–175.

Baker, D.R., Billey, L.O. and Francis, D.H. (1997) Distribution of K88 *Escherichia coli* adhesive and nonadhesive phenotypes among pigs of four breeds. *Veterinary Microbiology* 54, 123–132.

Bertin, A.M. and Duchet-Suchaux, M.F. (1991) Relationship between virulence and adherence of various enterotoxigenic *Escherichia coli* strains to isolated intestinal cells from Chinese Meishan and European Large White. *American Journal of Veterinary Research* 52, 45–49.

Bertin, Y., Martin, C., Oswald, E. and Girardeau, J.P. (1996) Rapid and specific detection of F17 related pilin and adhesin genes in diarrheic and septicemic *Escherichia coli* strains by multiplex PCR. *Journal of Clinical Microbiology* 34, 2921–2928.

Bertschinger, H.U., Munz-Müller, M., Pfirter, H.P. and Schneider, A. (1986) Vererbte Resistenz gegen Colienterotoxämie beim Schwein. *Zeitschrift für Tierzüchtung und Züchtungsbiologi* 103, 255–264.

Bertschinger, H.U., Bachmann, M., Mettler, C., Pospischil, A., Schraner, E.M., Stamm, M., Sydler, T. and Wild, P. (1990) Adhesive fimbriae produced *in vivo* by *Escherichia coli* O1399:K12(B):H1 associated with enterotoxaemia in pigs. *Veterinary Microbiology* 25, 267–281.

Bertschinger, H.U., Fairbrother, J.M., Nielsen, N.O. and Pohlenz, J.F. (1992) *Escherichia coli* infections. In: Leman, A.D., Straw, B.E., Mengeling, W.L., D'Allaire, S. and Taylor, D.J. (eds) *Diseases of Swine*, 7th edn. Iowa State University Press, Ames, pp. 487–521.

Bertschinger, H.U., Stamm, M. and Vögeli, P. (1993) Inheritance of resistance to oedema disease in the pig: experiments with an *Escherichia coli* strain expressing fimbriae 107. *Veterinary Microbiology* 35, 79–89.

Bijlsma, I.G.W. and Bouw, J. (1987) Inheritance of K88 mediated adhesion of *Escherichia coli* to jejunal brush borders in pigs: a genetic analysis. *Veterinary Research Communication* 11, 509–518.

Bijlsma, I.G.W., de Nijs, A., van der Meer, C. and Frik, J.F. (1982) Different pig phenotypes affect adherence of *Escherichia coli* to jejunal brush borders by K88ab, K88ac or K88ad antigen. *Infection and Immunity* 37, 891–894.

Bijlsma, I.G.W., van Houten, M., Frik, J.F. and Ruitenberg, E.J. (1987) K88 variants K88ab, K88ac, K88ad in oral vaccination of different porcine adhesive phenotypes. Immunological aspects. *Veterinary Immunology and Immunopathology* 16, 235–250.

Billey, L.O., Erickson, A.K. and Francis, D.H. (1998) Multiple receptors on porcine intestinal epithelial cells for the three variants of *Escherichia coli* K88 fimbrial adhesin. *Veterinary Microbiology* 59, 203–212.

Blackwell, J.M. (1996) Structure and function of the natural-resistance-associated macrophage protein (Nramp1), a candidate protein for infectious and autoimmune disease susceptibility. *Molecular Medicine Today* 2, 205–211.

Brinton, C.C., Fusco, P., Wood, S., Jayappa, H.G., Goodnow, R.A. and Strayer, J.G. (1983) A complete vaccine for neonatal swine colibacillosis and the prevalence of *Escherichia coli* pili on swine isolates. *Veterinary Medicine and Small Animal Clinics* 78, 962–966.

Bumstead, N. and Barrow, P.A. (1988) Genetic resistance to *Salmonella typhimurium* in newly hatched chicks. *British Poultry Science* 29, 521–530.

Bumstead, N. and Barrow, P.A. (1993) Resistance to *Salmonella gallinarium, S. pullorum* and *S. enteritidis* in inbred lines of chickens. *Avian Disease* 37, 189–193.

Cellier, M., Privé, G., Belouchi, A., Kwan, T., Rodrigues, V., Chia, W. and Gros, P. (1995) Nramp defines a family of membrane proteins. *Proceedings of the National Academy of Sciences USA* 92, 10089–10093.

Chandler, D.S., Chandler, H.H., Luke, R.K., Tripori, S.R. and Craven, J.A. (1986) Screening of pig intestines for K88 non-adhesive phenotype by enzyme immunoassay. *Veterinary Microbiology* 11, 153–161.

Chandler, D.S., Mynott, T.K., Luke, R.K.J. and Craven, J.A. (1994) The distribution and stability of *Escherichia coli* K88 receptor in the gastrointestinal tract of the pig. *Veterinary Microbiology* 38, 203–215.

Chappuis, J.P., Duval-Iflah, Y., Ollivier, L. and Legault, C. (1984) *Escherichia coli* K88 adhesion: a comparison of Chinese and Large White piglets. *Génétique, Sélection et Evolution* 16, 385–390.

Chowdhary, B.P., Johansson, M., Chaudhary, R., Ellegren, H., Gu, F., Andersson, L. and Gustavsson, I. (1993) *In situ* hybridization mapping and RFLP analysis of the porcine albumin (*ALB*) and transferrin (*TF*) genes. *Animal Genetics* 24, 85–90.

Dean-Nystrom, E. (1995) Identification of intestinal receptors for enterotoxigenic *Escherichia coli. Methods in Enzymology* 253, 315–324.

Duguid, J.P. and Anderson, E.S. (1967) Terminology of bacteria fimbriae, or pili, and their types. *Nature* 215, 89–90.

Edfors-Lilja, I., Petersson, H. and Gahne, B. (1986) Performance of pigs with and without the intestinal receptor for *Escherichia coli* K88. *Animal Production* 42, 381–387.

Edfors-Lilja, I., Gustafsson, U., Duval-Iflah, Y., Ellegren, H., Johansson, M., Juneja, R.K., Marklund, L. and Andersson, L. (1995) The porcine intestinal receptor for *Escherichia coli* K88ab, K88ac: regional localization on chromosome 13 and influence of IgG response to the K88 antigen. *Animal Genetics* 26, 237–242.

Eld, K., Gunnarsson, A., Holmberg, T., Hurvell, B. and Wierup, M. (1991) *Salmonella* isolated from animals and feed stuff in Sweden during 1983–1987. *Acta Veterinaria Scandinavica* 32, 261–277.

Erickson, A.K., Baker, D.R., Bosworth, B.T., Casey, T.A., Benfield, D.A. and Francis, D.H. (1994) Characterisation of porcine intestinal receptors for the K88ac fimbrial adhesin of *Escherichia coli* as mucin-type sialoglycoproteins. *Infection and Immunity* 62, 5404–5410.

Fedorka-Cray, P.J., Whipp, S.C., Isaacson, R.E., Nord, N. and Lager, K. (1994) Transmission of *Salmonella typhimurium* to swine. *Veterinary Microbiology* 41, 333–344.

Gaastra, W., Klemm, P., Walker, J.M. and de Graaf, F.K. (1979) K88 fimbrial proteins: amino- and carboxyl terminal sequences of intact proteins and cyanogen bromide fragments. *FEMS Microbiology Letters* 6, 15–18.

Gibbons, R.A., Sellwood, R., Burrows, M. and Hunter, P.A. (1977) Inheritance of resistance to neonatal diarrhoea in the pig: examination of the genetic system. *Theoretical and Applied Genetics* 81, 65–70.

Girard-Santosuosso, O., Bumstead, N., Lantier, I., Protais, J., Colin, P., Guillot, J.F., Beaumont, C., Malo, D. and Lantier, F. (1997) Partial conservation of the mammalian *NRAMP1* syntenic group on chicken chromosome 7. *Mammalian Genome* 8, 614–616.

Gonzáles, E.A., Vázquez, F., Garabal, J.I. and Blanco, J. (1995) Isolation of K88 antigen variants (ab, ac, ad) from porcine enterotoxigenic *Escherichia coli* belonging to different serotypes. *Microbiology and Immunology* 39, 937–942.

Grange, P.A. and Mouricout, M.A. (1996) Transferrin associated with the porcine intestinal mucosa is a receptor specific for K88ab fimbriae of *Escherichia coli*. *Infection and Immunity* 64, 606–610.

Gruenheid, S., Cellier, M., Vidal, S. and Gros, P. (1995) Identification and characterization of a second mouse *Nramp* gene. *Genomics* 25, 514–525.

Guérin, G., Duval-Iflah, Y., Bonneau, M., Bertaud, M., Guillaume, P. and Ollivier, L. (1993) Evidence for linkage between K88ab and K88ac intestinal receptors to *Escherichia coli* and transferrin loci in pigs. *Animal Genetics* 24, 393–396.

Guillot, J.F., Beaumont, C., Bellatif, F., Mouline, C., Lantier, F., Colin, P. and Protais, J. (1995) Comparison of resistance of various poultry lines to infection by *Salmonella enteritidis*. *Veterinary Research* 26, 81–86.

Guiné, P.A.M. and Jansen, W.H. (1979) Behaviour of *Escherichia coli* K antigens K88ab, K88ac, and K88ad in immunoelectrophoresis, double diffusion and hemagglutination. *Infection and Immunity* 23, 700–705.

Hide, E.J., Connaughton, I.D., Driesen, S.J., Hasse, D., Monckton, R.P. and Sammons, N.G. (1995) The prevalence of F107 fimbriae and their association with Shiga like toxin II in *Escherichia coli* strains from weaned Australian pigs. *Veterinary Microbiology* 47, 235–243.

Hu, Z. (1988) Studies of genetic and expression variations in susceptibility and resistance of swine enterocytes by enteropathogenic K88ad *Escherichia coli*. MSc thesis, University of Wisconsin, Madison, USA.

Hu, Z.L., Hasler-Rapacz, J., Huang, S.C. and Rapacz, J. (1993) Studies in swine on inheritance and variation in expression of small intestinal receptors mediating adhesion of the K88 enteropathogenic *Escherichia coli* variants. *Journal of Heredity* 84, 157–165.

Imberechts, H., Van Pelt, N. De Greve, H. and Lintermans, P. (1994) Sequences related to the major subunit gene *fedA* of F107 fimbriae in porcine *Escherichia coli* strains that express adhesive fimbriae. *FE915 Microbiology Letters* 119, 309–314.

Jorsal, S.E., Aarestrup, F.M., Ahrens, P., Johansen, M. and Baekbo, P. (1996) Oedema disease in Danish pig herds. Transmission by trade of breeding animals. *Proceedings of the 14th International Pig Veterinary Society Congress, Bologna, Italy*, p. 265.

Källenius, G., Möllby, R., Svensson, S.B., Winberg, J., Lundblad, A., Svensson, S. and Cedergren, B. (1980) The P^k antigen as a receptor for the hemagglutinin of pyelonephritic *Escherichia coli*. *FEMS Microbiology Letters* 7, 297–302.

Kearns, M.J. and Gibbons, R.A. (1979) The possible nature of pig intestinal receptor for the K88 antigen of *Escherichia coli*. *FEMS Microbiology Letters* 6, 165–168.

Kennan, R., Söderlind, O. and Conway, P. (1995) Presence of F107, 2134P and Av24 fimbriae on strains of *Escherichia coli* isolated from Swedish piglets with diarrhoea. *Veterinary Microbiology* 43, 123–129.

Lissner, C.R., Swanson, R.N. and O'Brian, A.D. (1983) Genetic control of the innate resistance of mice to *Salmonella typhimurium*: expression of the *Ity* gene in peritoneal and splenic macrophages isolated *in vitro*. *Journal of Immunology* 131, 3006–3013.

Malmqvist, M., Jacobsson, K.G., Häggblom, P., Cerenius, F., Sjöland, L. and Gunnarsson, A. (1995) Salmonella isolated from animals and feedstuffs in Sweden during 1988–1922. *Acta Veterinaria Scandinavica* 36, 21–39.

Mariani, P., Barrow, P.A., Cheng, H.H., Groenen, M.A.M., Negrini, R. and Bumstead, N. (1998) A major quantitative trait locus determining resistance to Salmonellosis is located on chicken chromosome 5. *XXVI ISAG Meeting, 9–14 August, Aukland, New Zealand*, p. 112.

Marklund, L., Winterø, A.K., Thomsen, P.D., Johansson, M., Fredholm, M., Gustafson, U. and Andersson, L. (1993) A linkage group on pig chromosome 4 comprising the loci for blood group L, GBA, ATP1B1 and three microsatellites. *Animal Genetics* 24, 333–338.

Marklund, L., Marklund, L., Davies, W., Ellegren, H., Fredholm, M., Høyheim, B., Johansson Moller, M., Juneja, R.K., Mariani, P., Coppetiers, W., and Andersson, L. (1996) A comprehensive pig linkage map based on a wild pig – Large White intercross. *Animal Genetics* 27, 255–269.

Martin, C., Rousset, E. and De Greve, H. (1997) Human ureopathogenic and bovine septicaemic *Escherichia coli* strains carry an identical F17-related adhesin. *Research in Microbiology* 148, 55–64.

Meijerink, E., Fries, R., Vögeli, P., Masabanda, J., Wigger, G., Stricker, C., Neunschwander, S., Bertschinger, H.U. and Stranzinger, G. (1997) Two alpha (1, 2) fucosyltransferase genes on porcine chromosome 6q11 are closely linked to the blood group inhibitor (S) and *Escherichia coli* F18 receptor (ECF18R) loci. *Mammalian Genome* 8, 736–741.

Menning, E.L. (1988) Danger lurks in your supermarket meat cases. *Journal of the American Veterinary Medical Association* 192, 494–497.

Michaels, R.D., Wipp, S.C. and Rothschild, M.F. (1994) Resistance of Chinese Meishan, Fengjing and Minzu pigs to K88ac$^+$ strain of *Escherichia coli*. *American Journal of Veterinary Research* 55, 333–338.

Nagy, B., Arp, L.H., Moon, H.W. and Casey, T.A. (1992) Colonization of the small intestine of weaned pigs by enterotoxigenic *Escherichia coli* that lack known colonization factors. *Veterinary Pathology* 29, 239–246.

Nagy, B., Awad-Maselmeh, M., Bodoky, T., Munch, P. and Szekrényi, M.T. (1996) Association of shiga like-toxin type II (SLTII) and heat stable enterotoxins with F18ab, F18ac, K88 and F41 fimbriae of *Escherichia coli* from weaned pigs. *Proceedings of the 14th International Pig Veterinary Society Congress, Bologna, Italy*, p. 264.

Nielsen, N.C., Christensen, K., Bille, N. and Larsen, J.L. (1974) Preweaning mortality in pigs. I. Herd investigations. *Nordic Veterinary Medicine* 26, 137–150.

Ojeniyi, B., Ahrens, P., Jorsal, S.E. and Meyling, A. (1992) Detection of enterotoxigenic *Escherichia coli* from pigs with diarrhoea using colony hybridization and ^{35}S labelled probe. *Proceedings of 12th International Pig Veterinary Society Congress, The Hague, The Netherlands*, p. 246.

Ollivier, L. and Rejinfo, X. (1991) Utilisation de la résistance génétique à la colibacillose K88 dans les schémas d'amélioration génétique de porc. *Génétique, Sélection et Evolution* 23, 235–248.

Ørskov, I., Ørskov, F., Sojka, W.J. and Wittig, W. (1964) K antigens K88ab(L) and K88ac in *E. coli*. *Acta Pathologica et Microbiologica Scandinavica* 62, 439–447.

Peelman, L.J. (1998a) Pig chromosome 13 and human chromosome 3: A tale of rearrangements. *Pig Chromosome 13 Workshop, XXVI ISAG Meeting, 9–14 August, Aukland, New Zealand*, p. 113.

Peelman, L.J. (1998b) K88 receptors: How far away? *Pig Chromosome 13 Workshop, XXVI ISAG Meeting, 9–14 August, Aukland, New Zealand*, p. 114.

Pitel, F., Cribui, E.P., Yerle, M., Lahib-Mansais, Y., Lanneluc, I., Lantier, F. and Gellin, J. (1995) Regional localization of the ovine NRAMP gene to chromosome 2q41–q42 by *in situ* hybridization. *Cytogenetics and Cell Genetics* 70, 116–118.

Pitel, F., Lantier, I., Riquet, J., Lanneluc, I., Tabet-Aoul, K., Saïdi-Mentor, N., Lantier, F.

and Gellin, J. (1994) Cloning, sequencing, and localization of an ovine fragment of the *NRAMP* gene, a candidate for the *ITY/LSH/BCG* gene. *Mammalian Genome* 5, 834–835.

Protais, J., Colin, P., Beaumont, C., Guillot, J.F., Lantier, F., Pardon, P. and Bennejean, G. (1996) Line differences in resistance to *Salmonella enteritidis* PT4 infection. *British Poultry Science* 37, 329–339.

Rapacz, J. and Hasler-Rapacz, J. (1986) Polymorphism and inheritance of swine small intestinal receptors mediating adhesion of three serological variants of *Escherichia coli* producing K88 pilus antigen. *Animal Genetics* 17, 305–321.

Rippinger, P., Bertschinger, H.U., Imberechts, H., Nagy, B., Sorg, I., Stamm, M., Wild, P. and Wittig, W. (1995) Comparison of recently described adhesive fimbriae of *Escherichia coli* isolated from porcine postweaning diarrhoea and from oedema disease: proposed designations F18ab and F18ac for the antigenic variants. *Veterinary Microbiology* 45, 281–295.

Robertsson, J.Å. and Lundeheim, N. (1994) Prohibited use of antibiotics as feed additive for growth promotion – effects on piglet health and production parameters. *Proceeding of the 13th International Veterinary Pig Society Congress, Bangkok, Thailand*, p. 282.

Rohrer, G.A., Alexander, L.J., Keele, J.W., Smith, T.P. and Beattie, C.W. (1994) A microsatellite linkage map the porcine genome. *Genetics* 136, 231–245.

Runnels, P.L., Moon, H.W. and Schneider, R.A. (1980) Development of resistance with host age to adhesion of K99$^+$ *Escherichia coli* to isolated intestinal epithelia cells. *Infection and Immunity* 28, 298–300.

Seignole, D., Grange, P., Duval-Iflah, Y. and Mouricout, M. (1994) Characterization of O-glycan moities of the 210 and 240 kDa pig intestinal receptors for *Escherichia coli* K88ac fimbriae. *Microbiology* 140, 2467–2473.

Sellwood, R. (1979) *Escherichia coli* diarrhoea in pigs with and without the K88 receptor. *Veterinary Record* 105, 228–230.

Sellwood, R. (1980) The interaction of the K88 antigen with porcine intestinal epithelial cell brush borders. *Biochimica et Biophysica Acta* 632, 326–335.

Sellwood, R. (1982) *Escherichia coli*-associated porcine diarrhoea: antibacterial activities of colostrum from genetically resistant sows. *Infection and Immunity* 35, 396–401.

Sellwood, R. (1984) An intestinal receptor for the K88 antigen of porcine enterotoxigenic *Escherichia coli*. In: Boedecker, E.C. (ed.) *Attachment of Organisms to the Gut Mucosa*, vol. 2. CRC Press, Boca Raton, Florida, pp. 167–175.

Sellwood, R., Gibbons, R.A., Jones, G.W. and Rutter, J.M. (1975) Adhesion of enteropathogenic *Escherichia coli* to pig intestinal brush borders. The existence of two pig phenotypes. *Journal of Medical Microbiology* 8, 405–411.

Skamene, E., Gros, P., Forget, A., Kongshavn, P.A.L., St-Charles, C. and Taylor, B.A. (1982) Genetic regulation of resistance to intracellular pathogens. *Nature* 297, 506–509.

Smith, H.W. and Halls, S. (1968) The production of oedema disease and diarrhoea in weaned pigs by oral administration of *Escherichia coli*. Factors that influence the course of the experimental disease. *Journal of Medical Microbiology* 4, 467–485.

Snodgrass, D.R., Chandler, D.S. and Makin, T.J. (1981) Inheritance of *Escherichia coli* K88 adhesion in pigs: identification of nonadhesive phenotypes in a commercial herd. *Veterinary Record* 109, 461–463.

Söderlind, O. and Möllby, R. (1978) Studies on *Escherichia coli* in pigs. V. Determination of enterotoxigenity and frequency of O groups and K88 antigen in strains from 200 piglets with neonatal diarrhoea. *Zentralblatt für Veterinärmedizin. Rehie B* 25, 719–728.

Söderlind, O., Olsson, E., Smyth, C.J. and Möllby, R. (1982) Effect of parental vaccination of dams on intestinal *Escherichia coli* in piglets with diarrhoea. *Infection and Immunity* 36, 900–906.

Söderlind, O., Thafvelin, B. and Möllby, R. (1988) Virulence factors in *Escherichia coli* strains isolated from Swedish piglets with diarrhoea. *Journal of Clinical Microbiology* 26, 879–884.

Sun, H.S., Wang, L., Rothschild, M.F. and Tuggle, C.K. (1998) Mapping of the natural resistance-associated macrophage protein 1 (*NRAMP*) gene to pig chromosome 15. *Animal Genetics* 29, 138–140.

Svensmark, B., Jorsal, S.E., Nielsen, K. and Willeberg, P. (1989a) Epidemiological studies of piglet diarrhoea in intensively managed Danish sow herds. I. Pre-weaning diarrhoea. *Acta Veterinaria Scandinavica* 30, 43–53.

Svensmark, B., Jorsal, S.E., Nielsen, K. and Willeberg, P. (1989b) Epidemiological studies of piglet diarrhoea in intensively managed Danish sow herds. II. Post-weaning diarrhoea. *Acta Veterinaria Scandinavica* 30, 55–62.

Svensson, S.B., Hultberg, H., Källenius, G., Korhonen, T.K., Möllby, R. and Winberg, J. (1983) P-fimbriae of pyelo-nephritogenic *Escherichia coli*: identification and chemical characterization of receptors. *Infection* 11, 73–79.

Sweeney, E.J. (1968) *Escherichia coli* enteric disease of swine: observation on herd resistance. *Irish Veterinary Journal* 22, 42–46.

Sydler, T., Buergi, E., Bertschinger, H.U. and Pospischil, A. (1996) Oedema disease in adult swine. *Proceedings of the 14th International Pig Veterinary Congress, Bologna, Italy*, p. 272.

Tuggle, C.K., Schmitz, C.B. and Gingerich-Feil, D. (1997) Rapid communication: cloning of a full-length natural resistance associated macrophage protein (NRAMP1) cDNA. *Journal of Animal Science* 75, 277.

Valpotic, I., Dean, E.A. and Moon, H.W. (1989) Phenotyping of pigs for the presence of intestinal receptors mediating adhesion of enterotoxigenic *Escherichia coli*-bearing K88ac pilus antigen by ELISA. *Veterinarski Arhiv* 59, 161–175.

Van Pouke, M., Sjöberg, A., Mattheeuws, M., Van Zeveren, A., Bouquet, Y., Chowdhary, B.P. and Peelman, L. (1997) Mapping of the *ATP2* and *PCCB* genes on porcine chromosome 13. *Mammalian Genome* 8, 852–853.

Vidal, S.M., Malo, D., Vogan, E., Skamene, E. and Gros, P. (1993) Natural resistance to infection with intracellular parasites: isolation of a candidate for *Bcg*. *Cell* 73, 469–485.

Vögeli, P., Kuhn, B., Kühne, R., Obrist, R., Stranzinger, G., Huang, S.C., Hu, Z.L., Hasler-Rapacz, J. and Rapacz, J. (1992) Evidence for linkage between the swine L blood group and the loci specifying the receptors mediating adhesion of K88 *Escherichia coli* pilus antigens. *Animal Genetics* 23, 19–29.

Vögeli, P., Bertschinger, H.U., Stamm, M., Stricker, C., Hagger, C., Fries, R., Rapacz, J. and Strantzinger, G. (1996) Genes specifying receptors for F18 fimbriated *Escherichia coli*, causing oedema disease and postweaning diarrhoea in pigs, map to chromosome 6. *Animal Genetics* 27, 321–328.

Wahlström, H., Tysén, E., Bergman, T. and Lindqvist, H. (1997) Results of the Swedish *Salmonella* Surveillance programme in Cattle and Pigs during 1996. *8th International Society of Veterinary Epidemiology and Economics*, 8–11 July, Paris, pp. 7.12.1–3.

Wallgren, P. (1993) Infections and immune functions of swine in fattening herds. Dissertation, University of Agricultural Sciences, Uppsala, Sweden.

Wierup, M., Wahlström, H. and Engström, B. (1992) Experience of a 10-year use of competitive exclusion treatment as part of the *Salmonella* control program in Sweden. *International Journal Food Microbiology* 15, 287–291.

Wilcock, B.P. and Schwartz, K.J. (1992) Salmonellosis. In: Leman, A.D., Straw, B., Mengeling, W.L., D'Allaire, S. and Taylor, D.J. (eds) *Diseases of Swine*, 7th edn. Iowa State University Press, Ames, pp. 570–583.

Winterø, A.K., Fredholm, M. and Davies, W. (1996) Evaluation and characterization of a porcine small intestine cDNA library: analysis of 839 clones. *Mammalian Genome* 7, 509–517.

Viruses and Subviruses

IV

13 Viral Diseases

K.A. Schat and C.J. Davies
Department of Microbiology and Immunology, College of Veterinary Medicine, Cornell University, Ithaca, USA

Summary

Genetic variation in resistance to viral diseases is found in all farm animals, thus providing the potential for selection for improved resistance. However, even though progress has been made with chickens, surprisingly limited information is available on the actual application of selection in swine and ruminants. Even in poultry, selection for resistance has been mostly carried out as a research tool with, thus far, limited application at the level of commercial primary breeding lines. The reasons for the lack of application are complex and are discussed in Chapter 17. However, the need to select for improved resistance, especially to Marek's disease (MD) in chickens, is real in view of the increase in virulence of the field viruses and absence of new vaccine strains to be used in the near future. It is doubtful that transgenesis can be used in the near future in chickens to achieve this goal, but improved knowledge of loci associated with resistance may facilitate selection by classical breeding approaches.

Introduction

Viral infections are a fact of life in all organisms, from the most primitive protozoa and bacteria to plants and mammals. Selection for natural resistance to these infections has occurred as part of evolutionary development. Domestication of cattle, sheep, swine and poultry led to selection based on production parameters without necessarily paying attention to disease resistance. Modern agricultural practices are possible, in part, because effective vaccines have been developed to protect these animals against the common viral diseases. However, pathogens can evolve to more virulent pathotypes in vaccinated herds or flocks, especially when animals are held under intensive production systems. For example, Marek's disease (MD) virus (MDV), an alphaherpesvirus causing lymphomas in chickens, has evolved over the past 30 years to much more virulent pathotypes in vaccinated animals, requiring a continuous effort to develop new vaccines (Witter, 1998). This example is also of interest because MD is one of the best examples of selection being used to improve genetic resistance to a specific disease in farm animals (Calnek,

© CAB *International* 2000. *Breeding for Disease Resistance in Farm Animals* (eds R.F.E. Axford, S.C. Bishop, F.W. Nicholas and J.B. Owen)

1985). It is expected that the rapid progress in genomic mapping of farm animals, improved techniques for the development of transgenic animals and increased understanding of viral pathogenesis at the molecular level will lead to improved genetic resistance to viral infections. In this chapter the current knowledge on genetic resistance to viral infections will be reviewed in the context of the pathogenesis of viral infections and mechanisms of genetic resistance to viral infections. In addition, information will be presented on state-of-the-art knowledge for resistance to viral infections in chickens, ruminants and swine. Finally, future developments will be discussed, based on selected examples.

Viral Pathogenesis

Viruses are obligate intracellular parasites. They are incomplete organisms and are dependent on the metabolic and replicative machinery of host cells. Consequently, they can only replicate in host cells that have the necessary cell-surface receptors, and possess the enzymes, cofactors and components that they need for replication. Furthermore, to become established and persist, a virus must evade the innate and adaptive defence mechanisms of the host.

To complete its life cycle a virus must: (i) gain entry to a host; (ii) attach to host tissues; (iii) if the site of replication differs from the site of entry, travel to the target organ(s); (iv) bind to target cells; (v) enter target cells; (vi) uncoat its genome; (vii) replicate its genome; (viii) make structural proteins; (ix) assemble capsids; (x) exit the cells; and (xi) exit the host and transfer to a new host. Because viruses need to be genetically compatible with their hosts, many have restricted host ranges. Since they must get to the target organ, must have an appropriate ligand for binding to the target cells, and, for replication, require host-cell constituents that may not be present in all types of cells or at all times in the cell cycle, viruses have tissue specificity. A minor mutation in a virus can change either its host range or tissue specificity. In addition, viral mutations that affect the replication efficiency, or the ability of a virus to evade the host's defence mechanisms, can have major effects on the virulence, or pathogenic potential, of the virus. As well as being important in terms of the natural evolution of viruses, these mutations are medically important because they are the basis for the development of modified live viral vaccines.

Mechanisms of Genetic Resistance to Viral Infections

Introduction

In principle, genetic resistance to disease can be at two basic levels: resistance to infection and resistance to disease. In the former case, a pathogen cannot establish infection even after exposure by injection, while in the latter case, infection becomes established and the pathogen replicates, but this does not result in the development of disease. Selection for either type of resistance is often based on an empirical approach, in which animals are exposed to a pathogen and survivors are used to produce the next generation. This approach is especially feasible with species such as chickens, which have short generation intervals and produce large numbers of offspring from defined matings. Cole (1968) used this type of selection to obtain MD resistant (N) and susceptible (P) lines of chickens in four generations. It was later learned that the resistance or susceptibility of these lines was linked to the major histocompatibility complex (MHC). Interestingly, Hansen *et al.* (1967) reported at the same time that chickens with the B^{19} genotype were more susceptible to MD than chickens with the B^{21} genotype.

Selection based on resistance to infection or disease has yielded important information from a practical point of view. Unfortunately, and with few exceptions such as MD and lymphoid leucosis caused by avian leucosis virus (ALV), it does not often allow for the identification of the genes responsible for improved resistance. In addition, the use of outbred animals does not facilitate the comparison of data from different sources, even if the main markers (e.g. MHC class I antigens) are apparently similar. Other genes, as well as environmental factors, can influence the outcome of challenge experiments and obscure the results. To minimize the effects of background genes, it is important to use congenic strains, i.e. strains that differ only for the gene of interest. Minimally, 10 backcrosses selecting for the marker gene are needed to obtain congenic lines that are more than 99.9% inbred, which makes this approach impractical for farm animals, except for chickens. The recent development of linkage maps for farm animals allows the search for quantitative trait loci (QTL) (see Chapter 1 for details) associated with disease resistance or susceptibility. Preliminary linkage maps have been published for chickens (Smith *et al.*, 1997 and references cited there in), cattle (Chapter 1), sheep (Chapter 1) and swine (Schook and Alexander, 1997) by several laboratories, and are frequently updated. The use of these linkage maps to search for resistance genes has already yielded some preliminary results for MD (Vallejo *et al.*, 1997; Bumstead, 1998), which are discussed in the section on MD.

Resistance to infection

Resistance to the establishment of infection can be the result of the absence of receptors or the presence of specific gene products interfering with virus replication. Examples can be found in the literature for both. For example, BALB/c mice are fully susceptible to infection with mouse hepatitis virus (MHV), a coronavirus, because they express a 100–110 kDa glycoprotein (gp) serving as a receptor for MHV. In MHV-resistant SJL/J mice, this protein has undergone some mutations, so that the virus cannot bind to it (Williams *et al.*, 1990). Also in mice, *Mx1* is an example of a gene encoding a protein that is capable of blocking virus replication after entrance of MHV into cells. The *Mx1* gene is activated by interferon (α/β-IFN) and the resulting protein blocks virus infection (Staeheli, 1990).

ALV infection in chickens provides the only example of selection for resistance based on the absence of virus receptors in a commercially important species. ALV belongs to the group of retroviridae in which the viral genome is reverse-transcribed from RNA into DNA (Payne and Fadly, 1997). The genome has been completely sequenced and encodes a limited number of genes (*gag*, *pol*, and *env*) (Fig. 13.1A). Important for the infection of susceptible cells is the glycoprotein gp85 expressed on the viral envelope, which is different for all subgroups of ALV (A–E and J in chickens). It was noted early on that some chickens were resistant to infection while others were susceptible (Crittenden *et al.*, 1967). The susceptibility or resistance was linked to the presence or absence of viral receptors. Susceptibility to subgroup A, B and C is associated with dominant alleles at three autosomal loci, namely *TVA**S*, *TVB**S*, and *TVC**S*, respectively. Cells derived from chickens expressing the recessive resistance allele *TVA**R* are resistant to infection with subgroup A (Payne, 1987; Payne and Fadly, 1997). Recently, the genes coding for the receptors for subgroups A, B, D and E have been cloned (Young, 1998). Unfortunately, the frequency of *TVA**R* is low in most commercial lines (Crittenden and Motta, 1969), which makes selection for resistant genotypes unattractive for commercial breeders, because this may result in the loss of commercially important traits. Moreover, the current emphasis has been on eradication of ALV subgroup A from commercial pure lines, thus alleviating the need to select for resistance (Payne and Fadly, 1997).

Resistance to disease

Resistance to disease is often complex, and multiple genes are frequently involved. The MHC complex has probably attracted the most interest as a potential selection tool for increased resistance to viral infections. This is based on the high degree of polymorphism and its importance in generating specific immune responses. MHC class I antigens are essential for the presen-

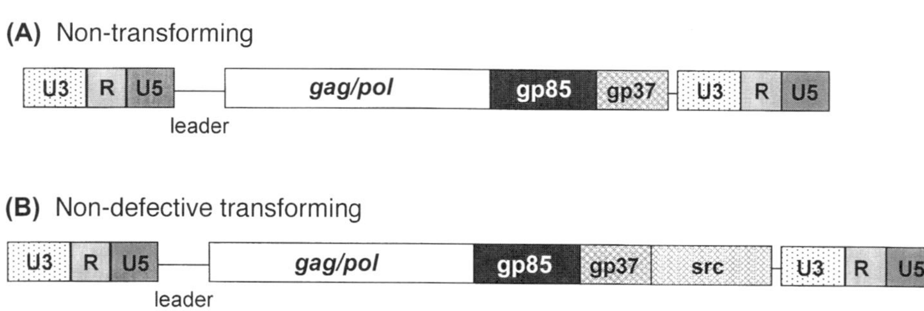

Fig. 13.1. Genome of the non-transforming avian leucosis virus (A) and the non-defective transforming Rous sarcoma virus (B). LTR, Long terminal repeat consisting of U3, R and U5; gag, group-specific antigens; pol, polymerase (reverse transcriptase); env, envelope proteins consisting of gp 85 and gp 37; src, oncogene.

tation of nonapeptides to cytotoxic T cells (CTLs), while class II antigens are important for the presentation of larger peptides, leading to the development of antibodies (reviewed by Kaufman *et al.*, 1995; see Chapter 4). However, there are no convincing examples in which particular mammalian MHC haplotypes are associated with resistance or susceptibility to viral diseases. This is in contrast to chickens, where susceptibility or resistance is strongly linked to specific MHC haplotypes. The basis for the strong association of the $B^{21}B^{21}$ genotype with resistance to MD is a matter of controversy and will be discussed in more detail below. Kaufman *et al.* (1995) and Kaufman and Lamont (1996) described the chicken MHC as a 'Minimal essential MHC' because it is only 30–100 kbp in size, rather than the 4 Mbp in mammals. They speculate that this small size favours selection of the MHC as a 'supergene' allowing selection for stable, specific haplotypes, especially in view of the low frequency of recombination events within the MHC region. Selection would favour the genes in this region staying together. In addition, because many genes have been deleted from the chicken MHC, it may be easier to detect the effects of selection for specific traits such as disease resistance. Other candidate genes for resistance to viral infections have been less well studied and will be discussed in the section on MD.

Genetic Resistance to Viral Diseases in Poultry

Introduction

Genetic resistance to viral diseases in poultry has been studied extensively (Table 13.1), in contrast to studies of other farm animals. In most instances, the mode of resistance has not been determined, but the MHC complex has been implicated in many instances. This was originally facilitated because the MHC complex of chickens contains the B–G family of genes, which are expressed on red blood cells (RBCs). Briles and Briles (1982) characterized many of the B–G alloantigens, and the B-nomenclature used in this chapter is based on their work. Although the function of the B–G antigens remains unknown, their genes are in strong linkage disequilibrium with the B–F and B–L regions of the MHC, which code for class I and class II proteins, respectively. The

Table 13.1. Virus strains for which resistance has been demonstrated in chickens.

Virus family	Virus strain	Genome	Mode of resistance
Herpesviridae	Marek's disease virus	DNA	MHC, others
	Infectious laryngotracheitis virus		Unknown
Retroviridae	Avian leucosis virus	RNA	Virus receptor, MHC
	Rous sarcoma virus		Virus receptor, MHC
Coronaviridae	Infectious bronchitis virus	RNA	Unknown
Birnaviridae	Infectious bursal disease virus	RNA	Unknown
Paramyxoviridae	Newcastle disease virus	RNA	Unknown

strong linkage disequilibrium between the B–G and B–F/B–L regions facilitated typing for class I and II antigens, prior to the advance of molecular techniques, by agglutination of RBCs using alloantisera (Kaufman, 1996; Kaufman and Lamont, 1996; Lamont, 1998). On the same minichromosome, but separated from the MHC by the nucleolar organizer region, is a second region, the *Rfp-Y* region, containing class I and class II genes (Miller *et al.*, 1994, 1996a). The functional importance of the *Rfp-Y* for genetic resistance to disease has not yet been elucidated, but conflicting data have been reported for MD (see below). In addition to the MHC, 11 non-MHC alloantigen systems have been described on red blood cells. At least four of these alloantigens are also expressed on leucocytes (W.E. Briles, DeKalb, Illinois, 1998, personal communication). The potential role of these antigens in disease resistance is currently under investigation. These alloantigens or minor histocompatibility antigens could actually be alloantigen expressions of cluster of differentiation (CD) antigens.

Marek's disease

MD, caused by an alphaherpesvirus, is characterized by the development of immunosuppression and tumours in nerves, visceral organs, skin and muscles. Tumours consist mostly of activated $CD4^+$ T cells, but other T cells can also be transformed. Currently most commercial chickens are vaccinated against MD at hatch or as 17-day-old embryos. Replication of MDV follows the general pattern of all herpesviruses: active replication, causing a lytic infection, is followed by latency and possible reactivation and tumour development. The process of replication and reactivation of latency is strictly regulated: the immediate early (IE) genes are first transcribed and translated. The IE proteins are needed to transactivate transcription of early and late genes. The molecular basis for transformation has not yet been elucidated (Calnek and Witter, 1997). The immune responses for MD have been reviewed in detail by Schat (1996).

Genetic resistance and susceptibility to MD has been well established and can be caused by different genes or combinations of genes. The influence of the MHC region on susceptibility or resistance has been reviewed in detail by Calnek (1985) and Bacon (1987). Especially, the $B^{21}B^{21}$ haplotype has been associated with strong resistance to MD. Bacon and Witter (1992) examined MHC-congenic lines on a $15I_5$ background. Only the line expressing B^{21} was resistant, while all other lines (B^2, B^{12}, B^{13}, B^{15} and B^{19}) were highly susceptible to challenge with MDV. Interestingly, there was no difference in susceptibility of the B^2 lines derived from the susceptible line 7_2 or the resistant line 6_2.

Other investigators have also shown that B^{21}-like haplotypes provide strong protection (e.g. Schat *et al.*, 1981; Abplanalp *et al.*, 1985; Hedemand *et al.*, 1993). These haplotypes are named B^{21}-like, based on a positive haemagglutination assay using an antiserum against B^{21}, but this does not indicate that these haplotypes are identical. Additional haplotypes may also provide resistance, e.g. the B^{11} haplotypes derived from Ancona stocks (Miller *et al.*, 1996b) and the B^{23} haplotypes derived from the New Hampshire 105 line (Schat *et al.*, 1994) are highly resistant to challenge with the very virulent MD strain RB1B. Other haplotypes, such as B^5 are associated with susceptibility (Bacon *et al.*, 1981). The MHC-associated resistance is not absolute; the level of resistance depends on the virulence of the challenge virus. For example the N line ($B^{21}B^{21}$) is resistant to challenge with virulent MDV (e.g. GA-5) but not with very virulent MDV (e.g. RB-1B) strains (Schat *et al.*, 1981). Selection for specific MHC haplotypes may also be important because an influence on vaccine-induced immunity has been noticed. Certain haplotypes respond better to specific vaccines than others under experimental conditions and in commercial chickens (Bacon and Witter, 1994a, b).

These studies only determined that resistance was linked to the MHC, but not if it was associated with the B–G, B–F, or B–L region. Briles *et al.* (1983) reported that the resistance is due to the B–F/B–L region by using specific matings between chickens in which a recombination had occurred between the B–G and the B–F/B–L regions. This was confirmed by Hepkema *et al.* (1993) using commercial, pure lines in which one line had an unusually high frequency of recombinations. Schat *et al.* (1994) examined the resistance of two lines (R2 and R3) which carry independent B^{F2-G23} recombinations. These lines were partly congenic on the UCD-003 ($B^{17}B^{17}$) background after four backcrosses. Chickens with 93% background gene uniformity and homozygous for the recombination were obtained by mating heterozygote $B^{17}B^{F2-G23}$ birds. The recombinant lines were significantly more resistant than the parent UCD-003 line and the NH105 line from which the B^{G23} fragment was obtained. Interestingly, the B^{F2} part was derived from a mating of a susceptible line 7_2 dam with a recombinant $B^{F24-G23}B^2$ sire. Unfortunately, there are no recombinants available between the B–F and B–L loci, preventing the determination of the role of either locus in resistance to MD.

The mechanism of the MHC-associated resistance has not yet been elucidated. It is likely that cell-mediated immune responses are important to curtail the lytic infection and prevent reactivation from latency. The lytic infection occurs mainly in B lymphocytes during the first 3–6 days post-infection, resulting in activation of T cells, which become susceptible to infection as a consequence of activation. Between 6 and 8 days, the infection becomes latent and remains latent in resistant birds, but reactivation from latency and tumour development frequently occur in susceptible chickens (Calnek, 1986; Schat, 1987, 1996). Using several MHC-congenic strains, Abplanalp *et al.* (1985) did not find differences in viral replication, suggesting that MHC-based resistance is not associated with the early lytic infection. This observation is consistent with a potential role for cell-mediated immune responses, as suggested by Omar and Schat (1996, 1997). They reported the development of antigen-specific CTLs against several MD proteins, such as glycoprotein B which appeared 5 days after infection (Omar *et al.*, 1998), phosphoprotein 38 and the IE protein ICP4. The finding that ICP4-specific CTLs were detected in the resistant but not in the susceptible line is of considerable interest, because ICP4 is one of the first proteins to be present in cells when virus is reactivated from latency. It suggests that these CTLs may be able to eliminate cells in which MDV is reactivated in the resistant lines before virus replication is completed and additional cells become infected. More recently, CTLs for a second IE protein were also detected, but these CTLs were detected in the resistant and susceptible lines (Xing and Schat, unpublished data). On the other hand, Kaufman (1996) found that MHC class I expression is lower in $B^{21}B^{21}$ haplotypes than in other haplotypes, which would allow natural killer (NK) cells to be more efficient in eliminating infected cells. However, Chausse *et al.* (1995) reported that MDV infection results in an upregulation of MHC class I antigens, which would negate the potential role of NK cells. Moreover, we have consistently been able to generate reticuloendotheliosis virus-specific CTLs in MD-resistant N2a ($B^{21}B^{21}$) and susceptible P2a ($B^{19}B^{19}$) chickens at comparable levels (Omar and Schat, 1996; Pratt *et al.*, 1992), demonstrating that antigen-specific CTLs can be generated in $B^{21}B^{21}$ chickens.

The importance of the *Rfp-Y* region has been examined by several groups. Wakenell *et al.* (1996) and Miller *et al.* (1996b) reported that the Y^3Y^3 genotype was associated with an increased MD incidence when compared to Y^1Y^1 or Y^1Y^3 genotypes. The influence of the Y^2 haplotype was less clear due to the low number of birds examined. The overall incidence of MD was rather low, because of the strong association of B^{11} with resistance. However, two other groups failed to find an influence of *Rfp-Y*. Lakshmanan and Lamont (1998) found no effect using four lines that had been selected for nine generations for multitrait immunocompetence (Kean *et al.*, 1994). Vallejo *et al.* (1997) examined the potential role of *Rfp-Y* in an intercross of lines 6_3 and 7_2 chickens. These lines have identical B^2B^2 haplotypes but are highly resistant or susceptible to MD, respectively. Line 6_3 contains the Y^{11} haplotype, while line 7_2 carries the Y^{12} haplotype. The lack of influence of these two haplotypes on MD resistance excludes the possibility that the difference in MD resistance between these two lines is caused by the *Rfp-Y* system. Unfortunately, the nomenclature for the *Rfp-Y* locus has not been standardized. Thus, it is not clear whether the same haplotypes were examined by the three groups. Moreover, it is possible that the B-locus exerts a stronger influence than *Rfp-Y*, which could mask the influence of the latter. Clearly, additional studies are needed to further clarify the role of this new system.

In addition to the MHC influence on resistance, a second independent system has been described for the MD-resistant 6_3 and MD susceptible 7_2 lines (Pazderka *et al.*, 1975). Both lines express the same B^2B^2 haplotype, do not respond to lymphocytes from each other in one-way mixed-lymphocyte reactions (Bacon *et al.*, 1986), and the B-LβII genes are identical for the two lines (Pharr *et al.*, 1998). Fredericksen *et al.* (1977, 1982) found two alloantigens Ly-4^a and Th-1^a in line 6_3, that are associated with increased resistance. These antigens are expressed on most peripheral T cells and most thymocytes, respectively. Unfortunately, these antigens have not been characterized by monoclonal antibodies. Studies by Lee *et al.* (1981) and Powell *et al.* (1982), and recently confirmed by Bumstead *et al.* (1997) using a quantitative PCR assay, showed that virus replicates to significantly higher titres in line 7_2 than in line 6_3 chickens, which may contribute to the difference in tumour development. It is not clear which gene(s) are responsible for the restricted virus replication in line 6_3. Clearly, more studies are needed to elucidate the genetic basis for the difference in resistance between these two lines sharing the same MHC.

The development of linkage maps has led to searches for QTL linked to MD resistance. Vallejo *et al.* (1998) reported five regions potentially linked to resistance using F_2 intercross chickens derived from line $6_3 \times 7_2$ matings. These loci, named MD1–5, are located on chromosomes 2, 4, 7, 8 and East-Lansing linkage group 16, respectively. The authors hypothesized that these regions contain genes that may be interactive with each other, suggesting that the susceptibility or resistance in these two lines is a polygenic trait. Recently, several additional QTL linked to MD resistance have been identified in this population (N. Yonash, Storrs, Connecticut, 1998, personal communication). It is tempting to speculate that some of these regions contain the loci that encode the Ly and Th alloantigens discussed above. Bumstead (1998) identified a region (MDV1) on chromosome 1 that is potentially linked to resistance to MD. Based on the location of other markers and in analogy with similar markers on mouse chromosome 6, it is suggested that MDV1 is the chicken homologue of the lectin-like NK cell–antigen complex. The importance of this finding for genetic resistance is not clear at this time, because only limited but conflicting data have been reported on differences in NK cell activity in susceptible versus resistant lines (Sharma and Schat, 1991).

Testing for genetic resistance to MD is a costly process and, as a consequence, several groups have tried to correlate selection for improved immune responsiveness to unrelated antigens with improved resistance to MD. Unfortunately, there is no clear indication that this approach will yield results. Lamont *et al.* (1996) found no differences in MD resistance in lines that had been selected for multiple immunological traits. These lines had been selected for high and low antibody responses against two antigens, as well as wing web responses to phytohaemagglutinin (PHA), and clearance of colloidal carbon (Kean *et al.*, 1994). Pinard *et al.* (1993) were unable to find a clear relationship between resistance or susceptibility to MD and lines selected for high and low response to sheep red blood cells (SRBC). In contrast, Dunnington *et al.* (1986) reported a positive correlation between increased antibody responses to SRBC and MD resistance. Likewise, Steadham *et al.* (1987) reported that selection for high antibody response to glutamic acid–alanine–tyrosine was correlated with increased resistance to MD. Finally, Calnek *et al.* (1989) speculated that high responders to T-cell mitogens (e.g. P-line, $B^{19}B^{19}$) may be more susceptible to MD than low responders (e.g., N-line, $B^{21}B^{21}$) because T-cell activation is important to switch infection from B cells to T cells during the lytic infection. However, no clear relationship could be established between high responses to PHA or concanavalin A and tumour development on an individual bird basis.

Lymphoid leucosis and Rous sarcomas

The cellular resistance to infection with ALV was discussed before. In this section, the resistance to lymphoid leucosis and Rous sarcomas will be examined. In order to do this, it is important to understand the differences between ALV and Rous sarcoma virus (RSV) at the genomic level. The basic genomic structures for ALV and RSV are depicted in Fig. 13.1A and 13.1B. The difference is that RSV, carrying the viral (v)-*src* oncogenic (*onc*) gene, transforms cells rapidly *in vivo*. Transformation by ALV requires activation of cellular *onc* genes by the long terminal repeat (LTR), which generally occurs over a longer period of time (Kung and Maihle, 1987).

Resistance to lymphoid leucosis

Resistance to ALV viraemia and subsequent tumour development has been described in a few instances and seems to be related to the MHC. Bacon *et al.* (1981, 1983, 1985) reported that chickens with B^2B^2 were relatively resistant to LL, while B^5B^5 and $B^{15}B^{15}$ conferred susceptibility. This may depend on other genes, however, because Baba and Humphries (1984) noted that in two unrelated strains, SC (B^2B^2) and TK ($B^{15}B^{21}$), the B^2B^2 haplotype was susceptible to LL. Although the virus spread faster in the SC strain, the authors were unable to explain the difference between the SC and TK strains. In an unrelated study, an association between MHC haplotype and the risk of becoming a virus shedder was noted (Yoo and Sheldon, 1992). A possible explanation for variations in LL incidence was provided by Purchase *et al.* (1977). They transplanted bursal lymphocytes from ALV-infected, susceptible and resistant chicks into chicks treated with cyclophosphamide. LL developed when the cells were transferred from susceptible but not from resistant donors into either host. This finding suggests that the level of resistance may be at the level of the bursal lymphocyte. On the other hand, a possible immunological explanation for differences in susceptibility was provided by Thacker *et al.* (1995). They reported marked differences in generation of ALV-specific CTL between haplotypes.

Resistance to Rous sarcomas

Although RSV does not cause an economically important disease, it has been used extensively to demonstrate a genetic basis for resistance to tumours. The possibility to select for RSV resistance was noted by Gyles and Brown (1971), when they developed lines in which the sarcomas would progress or regress. The ability to regress sarcomas was linked to the MHC complex by several groups (e.g. Collins *et al.*, 1977; Bacon *et al.*, 1981), and more specifically to the B–F/B–L region using MHC recombinant strains (Aeed *et al.*, 1993). The B^2B^2 haplotype is found in regressor lines, while the B^5B^5 haplotype is associated with progressor lines. Based on studies in MHC recombinants, it is generally accepted that the B–F/B–L fragment of the MHC locus is responsible for the rejection (Plachy and Benda, 1981; Aeed *et al.*, 1993; White *et al.*, 1994). White *et al.* (1994) also found that three MHC-recombinant lines BR2, BR3 and BR4, with serologically identical B^{F2-G23} recombination, differed in their ability to reject Rous sarcomas when tested in the fourth backcross generation in the highly inbred UCD-003 ($B^{17}B^{17}$) line. This is interesting because the B-F2 part was derived from a known regressor line. It is not known if mutations in the B-F part are responsible for the differences or if other genes are involved. Recently, LePage *et al.* (1998, personal communication) reported that two minor histocompatibility (D and

L) antigens influenced tumour regression, suggesting the influence of other genes. Because RSV and ALV have the same structural viral proteins (Fig. 13.1), it can be expected that there are similarities between resistance to lymphoid leucosis and Rous sarcomas, which is at least the case for line 6_3 chickens (Bacon *et al.*, 1981). Interestingly, when birds are inoculated with a v-*src* construct, tumours lacking viral proteins develop at the site of inoculation. These tumours were also rejected in accordance with the regressor status, as determined by challenge with RSV. Likewise, the development of metastatic tumours was severely restricted in the B^2B^2 regressor but not in congenic B^5B^5 progressor lines (Taylor *et al.*, 1994). These findings demonstrate that selection for resistance to a tumour produced by the viral src protein is possible using MHC as a marker.

Infectious laryngotracheitis

Infectious laryngotracheitis (ILT) is a respiratory disease caused by a herpes-virus (ILTV). The infection normally remains confined to the upper respiratory tract (Bagust and Guy, 1997). Differences in susceptibility to ILT were reported by Loudovaris *et al.* (1991a) and Poulsen *et al.* (1998), using strains with different MHC haplotypes, but further studies are needed using MHC congenic strains to clarify the role of the MHC locus. Interestingly, *in vitro* infection of macrophages showed that macrophages from resistant lines contained significantly more ILTV antigens than macrophages from the susceptible line. Replication of ILTV in macrophages is rather restrictive, suggesting that resistance may be associated with a more effective uptake and subsequent destruction of ILTV by macrophages in resistant lines compared with susceptible lines (Loudovaris *et al.*, 1991b).

Infectious bursal disease

Infectious bursal disease (IBD) is a highly contagious disease of chickens. IBD virus (IBDV), a birnavirus, replicates in B lymphocytes, primarily in the bursa of Fabricius but also in other lymphoid organs, causing apoptosis of B cells. The disease is characterized by high morbidity, some mortality, and subsequent immunosuppression in survivors. The disease is controlled by vaccination and strict hygiene (Lukert and Saif, 1997). The potential role of the MHC region in resistance to IBDV was examined by two groups using different approaches. Fadly and Bacon (1992) challenged seven congenic lines, developed on the $15I_5$ background, at 1 day and 28 days of age. No differences were found in lesion score or immunosuppression when challenged at 1 day of age. However, B haplotypes did influence IBDV mortality when challenged at 28 days of age. All congenic lines had significantly increased levels of mortality compared to the parent $B^{15}B^{15}$ line. Interestingly, the increase was the lowest in the congenic line expressing B^2B^2 derived from line 7_2 which is highly susceptible to Marek's disease, but not significantly different from the other haplotypes tested, with the exception of $B^{12}B^{12}$. Bumstead *et al.* (1993) compared 11 inbred and partially inbred lines with different MHC haplotypes as well as F_1, F_2 and backcross chickens. They also reported differences in mortality between the lines, but concluded that these were not related to the MHC haplotypes, based on studies using F_2 and backcross chickens. Instead, a single dominant gene seems to be responsible for the partial resistance. It was also found that meat-type strains were more resistant than Leghorn chickens, an observation confirmed independently by Nielsen *et al.* (1998). It is often suggested that the limited gene pool used by commercial producers of light and heavy breeds may have led to the loss of valuable genes for disease resistance. However, Okoye and Aba-Adulugba (1998) reported that indigenous chickens obtained in Nigeria were actually more susceptible to IBDV challenge than commercially available broilers and pullets, although the difference in susceptibility between the latter and indigenous chickens was minimal.

Infectious bronchitis

Infectious bronchitis virus (IBV), a coronavirus, causes respiratory disease and occasionally nephritis. In layer flocks, infection usually causes a drop in egg production and often a decline in eggshell quality. Infectious bronchitis is a serious economic problem and vaccinations are not always successful, due to frequent mutations in the S1 gene coding for the spike protein (Cavanagh and Naqi, 1997). Purchase *et al.* (1966) examined the effect of inoculation of IBV in embryos of line 6, line 7, line 15_I and crosses between these lines. Line 7 embryos died significantly later than embryos of line 6, while embryos of line 15_I were intermediate. The authors suggested that sex-linked genes could be responsible for the differences, because a difference in embryo mortality was detected in reciprocal crosses between chickens of line 6 and line 7. Embryos derived from matings using line 7 males died later than embryos from matings using line 7 females. However, Bumstead *et al.* (1989, 1991) rejected a sex-linked influence, using a model in which birds were challenged with IBV or IBV together with *Escherichia coli*. They reported marked differences in mortality among a number of inbred lines. In some lines, e.g. line 7_2, virus infection alone caused high levels of mortality, and the addition of *E. coli* did not alter the mortality percentage significantly (82.5% versus 86.7%); while in other lines, e.g. line 6_1, the mortality increased drastically (9.9% to 60%) with the addition of *E. coli*. Similar results were reported by Cook *et al.* (1990), after intranasal inoculation with IBV. Additional studies were conducted using lines with high (line 15_I) and low (line C) levels of mortality, to examine the basis for the genetic differences. After challenge with a mild strain of IBV, chicks from line C cleared the virus considerably faster than chicks from line 15_I (Otsuki *et al.*, 1990; Nakamura *et al.*, 1991). Estimation of antibody titres suggested that this is probably caused by increased levels of IBV-specific IgA in the lachrymal fluid and saliva of the line C chicks, compared to line 15_I chicks. Similarly, line 6_1 chicks also had higher levels of IBV-specific IgA in the lachrymal fluid than the susceptible line 7_2 chicks, but no higher levels in the saliva. These data suggest that the genetic differences in resistance may be related to an increased ability to produce IgA. This explanation is probably too simplistic because no differences in IgA levels were found in tracheal secretions (Cook *et al.*, 1992), especially in view of the recent finding that cytotoxic T cells may be important in the clearance of virus-infected cells (Seo and Collisson, 1997).

Newcastle disease

The clinical picture of Newcastle disease (NCD), caused by Newcastle disease virus (NDV), a paramyxovirus, can vary from mild respiratory distress to acute mortality with lesions in the gastrointestinal tract (Alexander, 1997), depending on the virulence of the virus strains. Genetic differences between lines for resistance to NCD have not been reported in chickens. However, it may be possible to select for increased antibody responses to live or killed NCD vaccines. Leitner *et al.* (1994) reported that heavy breeds selected for an early antibody response to killed *E. coli* also had higher titres to either killed or live NDV vaccines than the lines selected for low responses to *E. coli*. On the other hand, selection for high antibody responses to sheep red blood cells resulted in a lower response to a killed NCD vaccine (Dunnington *et al.*, 1992).

Tsai *et al.* (1992) examined the resistance of four lines of turkeys to NDV-induced mortality. Lines RBC1 and 2 were random-bred lines, while E and F were sublines derived from RBC1 and 2, respectively. Line E was selected for improved egg production and line F for increased body weight. Line F had a significantly higher level of mortality than the three other lines, but antibody titres were actually higher in line F than in RBC2. The basis for the genetic difference has not been established, but selection for increased meat production may negatively influence disease resistance.

Conclusions

Although it is difficult to compare results from different research teams, some interesting comparisons can be made on genetic resistance to viral diseases using four lines of birds that are frequently used by different investigators. Line 6 birds are especially of interest because they seem to be resistant to several viral diseases, while line 7, with identical MHC composition, is often susceptible. Perhaps some of the QTL identified by Vallejo *et al.* (1998) are related to specific antiviral activity, such as efficient generation of cytokines or other immune response mechanisms.

Clearly, selection for disease resistance is possible. Yet, the reality is that commercial breeders have not used these opportunities optimally. This is caused, in part, because resistance to disease may not correlate with production traits (see Chapter 17; Bacon, 1987). In addition, selection for resistance for viral diseases may not correlate with resistance to other diseases.

Genetic Resistance to Viral Diseases in Ruminants

Introduction

In the past 15 years, many tools for genetic and immunological characterization of ruminants have become available. Linkage maps and physical maps are now available for both cattle and sheep (see Chapter 1). Many of the genes that encode important immunoregulatory proteins have been cloned and sequenced. The MHC of cattle and sheep have been well characterized, and good methods and reagents for MHC typing have been developed (Davies *et al.*, 1994a, b, 1997; Russell *et al.*, 1997; Maddox, 1998; Chapter 4). Reagents are now available for most of the key bovine and ovine CD antigens and immunoglobulin isotypes (Naessens and Hopkins, 1996; Naessens *et al.*, 1997; Hein, 1998; Naessens, 1998). Finally, antibody reagents, bioassays and molecular methods for analysis of cytokine expression have been developed (Zarlenga *et al.*, 1995; Mertens, 1998; Nash *et al.*, 1998).

Despite all of the advances in ruminant immunology and genetics, there is little specific information about the genetic basis of resistance to viral diseases in ruminants. There are a number of reasons for this lack of progress. First, in contrast to the situation with bacterial, protozoal, helminth and arthropod parasites, vaccination against viral diseases is frequently quite effective. This decreases the impetus for the identification and selection of disease-resistant animals. Secondly, for many host-specific viruses, eradication of the viral agent from individual herds or even whole countries is more practical and cost-effective than selection of disease-resistant ruminant hosts. Examples of success with this approach are eradication of foot-and-mouth disease virus (FMDV) from the USA, and eradication of bovine leukaemia virus (BLV) from a number of European countries. Thirdly, because the transmission and severity of viral diseases are influenced by the genetic make-up of the virus, the prior exposure history of the host population to antigenically cross-reactive naturally occurring and vaccine viruses, environmental factors, and the physiological state of the host, it is difficult to isolate the host genetic component of 'resistance'. Furthermore, since a variety of factors influence the severity of disease, it is often difficult to come up with rational, testable criteria for inherent resistance. A final reason why progress has been slow is that the long generation time and small number of offspring per female, make challenge and selection programmes and the production of congenic lines, prohibitively expensive. This means that the only practical approach is to test for the influence of candidate genes in the course of naturally occurring disease or in challenge experiments. A couple of examples of how this approach can be used are provided below. Host genetic polymorphism is also important in terms of response to vaccination. Consequently, how host MHC polymorphism affects immune responses to component FMDV vaccines will also be discussed.

Bovine leukaemia virus

BLV is a C-type retrovirus and is the cause of enzootic bovine leucosis. BLV can be transmitted vertically by the transfer of infected lymphocytes in milk from a dam to her calf, or horizontally between cattle by biting insects or via contaminated instruments (Pelzer and Sprecher, 1993). The primary target cell for BLV is the B cell, although the virus can also infect macrophages and T cells (Stott *et al.*, 1991). Since retroviruses, such as BLV, become integrated into the host DNA as proviruses, once infected, a host is infected for life. Infected B cells proliferate, resulting in an inversion in the B:T cell ratio from less than one to greater than one in about 70% of infected cattle, and persistent lymphocytosis (PL; a sustained increase in the number of lymphocytes in the peripheral blood) in about 30% of infected cattle (Lewin *et al.*, 1988a; Lewin, 1989). Less than 1% of infected cattle eventually develop lymphosarcoma.

Lewin and Bernoco (1986) provided the first evidence that the bovine MHC, known as the bovine lymphocyte antigen (BoLA) complex, influences the subclinical progression of BLV infection. In this study, an association was shown between resistance or susceptibility to the development of PL and specific BoLA class I alleles. In a subsequent study, it was found that the *BoLA-A14* (formerly *w8.1*) allele was associated with resistance to sero-conversion (infection) as well as resistance to the development of PL (Lewin *et al.*, 1988b). Later studies showed that resistance to PL was more strongly associated with the BoLA class II region (van Eijk *et al.*, 1992), and ultimately with a particular amino acid motif (ER) at positions 70 and 71 of the *DRB3* gene (Xu *et al.*, 1993). A DRB3 motif at amino acids 75–78 was also associated with susceptibility to PL. Furthermore, it was determined that resistance to PL associated with the motif at positions 70–71 was dominant while susceptibility to PL associated with the motif at positions 75–78 was recessive (Xu *et al.*, 1993). The dominant nature of the resistance motif would be consistent with presentation of an immunodominant epitope to class II restricted $CD4^+$ T cells that mediate resistance. The susceptibility motif was found to be identical to a sequence in the viral *pol* gene and, therefore, molecular mimicry may play a role in susceptibility (Xu *et al.*, 1993).

This is a beautiful example of the role of MHC class II genes in resistance to disease. Unfortunately, because different MHC alleles are likely to confer resistance or susceptibility to different pathogens, selection for specific MHC alleles with the concurrent elimination of polymorphism is not a very good idea. The introduction by transgenesis of an extra *DRB3* gene carrying the resistance motif is a scientifically intriguing idea but probably not very practical.

Bovine herpesvirus-1

Bovine herpesvirus type 1 (BHV-1) is the aetiological agent of infectious bovine rhinotracheitis (IBR) and several less common clinical syndromes of cattle (Straub, 1991). Transmission is via nasal secretions and aerosol spread. Following an acute infection, BHV-1 establishes latency in respiratory tract sensory ganglia where the DNA genome of the virus persists as an episome. During periods of stress the virus recrudesces by travelling down the axon and re-establishing an active infection at the original site. This results in renewed shedding and spread to susceptible herdmates. Much of the cytopathology resulting from BHV-1 infection is due to the inflammatory response that the virus induces and secondary bacterial infections (Bielefeldt Ohmann *et al.*, 1991). Cattle are routinely vaccinated with either modified live or killed vaccines. Unfortunately, vaccination does not protect against infection or the development of latency, it only prevents or reduces the severity of the clinical disease (Straub, 1991). Type-1 interferons (α-IFN, β-IFN, ω-IFN and trophoblast IFN) apparently influence the severity of disease, either by inhibiting viral replication or by influencing the severity of the inflammatory response (Ryan *et al.*, 1993; Ryan and Womack, 1997).

Ten α-IFN (IFNA), six β-IFN (IFNB), ten ω-IFN (IFNW) and six trophoblast IFN (IFNT) genes are located on bovine chromosome 8 (Ryan and Womack, 1993). Restriction fragment length polymorphisms were detected in 19 of these 32 genes. In a disease-association study it was found that alleles at three loci (*IFNB1*C*, *IFNW4*2* and *IFNW8*A*) were associated with severe clinical disease while a different allele at one of the same loci (*IFNB1*B*) was associated with mild disease (Ryan *et al.*, 1993; Ryan and Womack, 1997). It was not reported whether the alleles associated with severe disease at the three loci were in linkage disequilibrium.

It is not known which type-1 IFN gene(s) influences the severity of IBR. Furthermore, it is not known what the molecular or functional differences are between different IFN alleles. Consequently, additional research is needed before a rational selection programme can be introduced. Nevertheless, identification of an association between a non-MHC gene and the severity of disease caused by a viral pathogen is significant. Selection for particular interferon alleles, particularly if they infer increased protection against a number of viral pathogens, would be a reasonable way of selecting for improved resistance to viral diseases.

Foot-and-mouth disease virus

Foot-and-mouth disease (FMD) is a highly contagious disease of cattle and other cloven-hoofed animals caused by foot-and-mouth disease virus (FMDV), the only virus in the *Aphthovirus* genus of the *Pirornaviridae* family. FMDV is a small, icosahedral, non-enveloped, single-stranded, positive-sense RNA virus. Viral particles are assembled in the cytoplasm of infected cells and are released by cell lysis. There are seven major FMDV serotypes and numerous subtypes. FMDV infects epithelial cells at the initial site of infection, often the oral mucosa, and then spreads to other sites via the blood. Protective immunity is mediated by antibody and complement (C3b) opsonization, followed by receptor-mediated phagocytosis (McCullough *et al.*, 1992). Immunity can be induced by either natural infection or vaccination. Antibody responses are T-cell, or more specifically T-helper-1, dependent (van Lierop *et al.*, 1995).

Genetic resistance to FMD has not been thoroughly investigated. However, response to vaccination, which is directly related to resistance to disease, is influenced by host genetics. Despite their complex composition, conventional, inactivated FMDV vaccines fail to induce protective immunity in all cattle (Barteling and Vreeswijk, 1991). Conventional whole-virus vaccines are a complex mixture of antigens. It is, therefore, difficult to work out the genetic control of the immune response to a conventional vaccine. Recent work on component vaccines has, however, provided the impetus and opportunity for detailed analysis of the genetic control of FMDV vaccine responses in cattle.

In an effort to create a better FMDV vaccine, Dimarchi *et al.* (1986) created a peptide vaccine composed of two discontinuous segments (amino acids 200–213 and 141–158) of viral-protein-1 (VP1) joined by a proline–proline–serine spacer. The spacer induces a hairpin turn in the peptide and this is meant to make the peptide mimic the three-dimensional structure of the loop region and an adjacent segment of VP1. In the initial trial, vaccination with this peptide induced a high serum neutralization antibody titre in all 12 cattle. However, when challenged with virulent virus, only five out of the 12 cattle were protected against the development of secondary lesions (Dimarchi *et al.*, 1986). Because there was considerable variation in the level of protection, studies were undertaken to evaluate T-cell responses to the Dimarchi peptide (Glass *et al.*, 1991; Glass and Millar, 1994, 1995). The cattle used for these studies were typed serologically for MHC class I antigens and by isoelectric focusing for BoLA-DR, class II antigens. Responses of T-cell lines and clones isolated from peptide-vaccinated cattle were assessed by *in vitro* [^{3}H]thymidine incorporation following stimulation with the Dimarchi peptide or one of several related peptides. These studies revealed that: (i) T-cell responses were MHC class II restricted; (ii) some T-cell clones were restricted by class II molecules other than BoLA-DR, presumably BoLA-DQ; (iii) cattle responded to different T-cell epitopes, depending on their class II haplotype; (iv) MHC class II haplotypes were associated with high and low T-cell responses; and (v) almost half of the cattle recognized a portion of the peptide that included the spacer and that was not part of the native protein. Cattle that preferentially utilize a non-native T-cell epitope when responding to a peptide vaccine could be susceptible to disease because of a lack of virus-specific memory T cells.

Another research group has used four MHC-homozygous and four MHC-heterozygous cattle characterized with the full array of MHC class I and class II typing methods used in the Fifth BoLA Workshop (Davies *et al.*, 1994a, b), to evaluate the specificity of T-cell responses following vaccination with a conventional trivalent FMDV vaccine (van Lierop *et al.*, 1995). Responses to three peptides containing known T-cell epitopes (VP1[35–53], VP2[74–88] and VP4[20–34]) were tested in an *in vitro* proliferation assay. Lymphocytes from all of the cattle responded to purified FMDV, but responses to the peptides varied considerably. Lymphocytes from one of the homozygotes failed to respond to any of the peptides. Lymphocytes from the other three homozygotes responded to one or two of the peptides. The responses of the heterozygotes appeared to be a combination of the responses restricted by each of their haplotypes. This suggests a heterozygous advantage since T-helper cells from a heterozygote would be capable of responding to a wider range of epitopes. All of the cattle, except the one non-responder homozygote, responded to the VP4[20–34] peptide. Consequently, fine mapping of VP4 epitopes was done by testing the responses of lymphocytes from three cattle, two homozygotes and a heterozygote, with one responder and one non-responder haplotype, to 13 overlapping 7- to 15-mer peptides that spanned the VP4 amino acid sequence 17–40. This experiment demonstrated that the three cattle were responding to different VP4 epitopes. Lastly, van Lierop *et al.* (1995) showed that the responding T cells were secreting γ-IFN and suppressing IL-4 transcription. This means that the responding lymphocytes are presumably T-helper-1 cells.

The FMDV MHC restriction data show how important MHC diversity is for assuring resistance to viral diseases at a population level. Most individuals respond to a limited number of immunodominant T-cell epitopes from a given virus. Epitope selection is influenced strongly by the MHC alleles carried by the individual. Furthermore, MHC homozygotes apparently respond to a much more restricted set of epitopes. Consequently, there is probably a significant immunological advantage to being a heterozygote. These factors should be taken into account before undertaking a selective breeding programme that would limit MHC polymorphism. Furthermore, MHC polymorphism needs to be taken into consideration when designing and testing vaccines.

Genetic Resistance to Viral Diseases in Swine

Genetic differences in disease susceptibility have been reported for different breeds of swine. However, most if not all of the research on genetic resistance has been directed toward the characterization of a genetic basis for resistance to bacterial and parasitic diseases (Straw and Rothschild, 1992). Research on genetic resistance to viral diseases in swine has been complicated by a number of factors. Some of the more devastating viral diseases (e.g. classical swine fever (CSF) and foot-and-mouth disease) are controlled by the use of relatively cheap vaccines. In contrast, the old-fashioned approach of selecting for resistance by challenge is a costly and time-consuming procedure, even with the relatively large litters of piglets available for experimental procedures. Moreover, until recently there was a paucity of defined inbred lines of swine. The development of MHC-characterized strains of miniature pigs (Sachs *et al.*, 1976) has not been fully exploited to determine whether differences in MHC phenotype are associated with differences in resistance to viral diseases. Finally, until recently there was a lack of reagents to characterize MHC class I and class II antigens, CD antigens expressed on lymphocytes and cytokines. However, many of these reagents have become available over the past decade (Lunney *et al.*, 1996; Vandenbroeck and Billiau, 1997; Lunney and Butler, 1998). It is expected that selection for genetic resistance to viral diseases will become more important, especially in the EU, where the use of vaccines is being increasingly replaced by management techniques to maintain optimal health.

Limited data suggest that selection for improved resistance to viral disease is possible in swine. Mengeling and Cutlip (1972, 1976) reported marked differences in susceptibility to disease between litters after experimental infection with haemagglutinating encephalomyelitis virus, a coronavirus. The basis for the differences in susceptibility to this virus has not been determined, and breeding experiments have not been performed to confirm that these differences were indeed genetically based. Selection for genetic resistance to African swine fever (ASF), a highly lethal infection caused by an enveloped, icosahedral DNA virus (ASFV), may also be feasible. In pigs, ASFV replicates in macrophages in the spleen, followed by rapid dissemination to other lymphoid organs and massive apoptosis of lymphocytes. The apoptosis is probably caused by the release of factors from virus-infected macrophages. Marked differences in morbidity and mortality have been found between ASFV infections in domesticated swine (*Sus scrofa*) versus bushpigs (*Pothamchoerus porcus*) and warthogs (*Phacochoerus aethiopicus* and *P. africanus*). The latter two species are considered to be the natural reservoir for ASFV and they are resistant to ASF but not to infection: virus replicates in splenic macrophages, but dissemination to other lymphoid organs is limited and apoptosis is not seen at 5 days post-infection (Oura *et al.*, 1998). The reason(s) for the limited infection in macrophages is poorly understood. It may be possible to determine whether there is a genetic basis for the difference in viral replication between *S. scrofa* and *Pothamchoerus porcus*, because crossbreeding between the two species has been reported (Skinner and Smithers, 1990). In the case that the resistance is linked to a single gene, it may become feasible to generate ASF-resistant, transgenic pigs. Recently, it was also suggested that differences in antibody responses to recombinant viral proteins could form a basis for selection between litter mates for improved resistance to challenge with ASFV (Gómez-Puertas *et al.*, 1998). Three-month-old pigs were vaccinated with recombinant viral proteins p54 and/or p30; only pigs vaccinated with the combination were partially protected against disease (Table 13.2). The three pigs that died in this group had a delayed onset and a lower viraemia level than non-vaccinated control levels. Neutralizing antibodies were present in survivors and non-survivors, suggesting that other immune responses may be important. MHC-restricted, ASFV-specific CTLs have indeed been demonstrated (Martins *et al.*, 1993). It would be interesting to determine if breeding from the survivors leads to improved genetic resistance to challenge, with and without vaccination.

Future Developments

Recent technical improvements in the generation of transgenic animals has raised the level of expectations that this approach will produce animals with increased resistance to viral infections (e.g. Crittenden and Salter, 1986; Hawken and Schook, 1998; Hawken *et al.*, 1998). In the short term, however, there is little evidence to support these expectations. The reasons for the lack of progress are the variation in virulence of the infectious agent, environmental

Table 13.2. Protection of pigs vaccinated with recombinant viral proteins p30 and p54 and challenged with African swine fever virus strain E75 (from Gómez-Puertas *et al.*, 1998).

Vaccine[a]	No. of pigs	No. dead	Mean time to death in days
None	5	5	6
Attenuated E75CV1-4	4	0	–
p30	3	3	9
p54	3	3	6
p30 + p54	6	3	26

[a]Pigs received three doses of recombinant protein mixed in complete or incomplete Freund's adjuvant.

influences and the fact that the genetic basis of resistance is still only poorly understood in most cases. Even in the case where resistance is based on a single gene, results have thus far been lacking.

For example, the induction of expression of the mouse *Mx1* (*mMx*) gene by α-IFN or β-IFN prevents the replication of influenza virus in the mouse (Staeheli, 1990). Control of influenza virus in pigs would be beneficial because re-assortment of influenza virus occurs in pigs, leading to new strains (Murphy and Webster, 1995). Müller *et al.* (1992) developed transgenic pigs in which *mMx* was transcribed using the *mMx* promoter. Interestingly, and unexpectedly, these transcripts were not translated into the Mx protein, thus preventing the testing of the original hypothesis that the pigs would be resistant to influenza virus. The reasons for the lack of translation are not understood.

Resistance to infection with ALV is based on the absence of specific cellular receptors that bind to the viral gp85 inserted in the virus envelope (see above). Based on observations by Robinson *et al.* (1981) that chickens carrying *ev3* and *ev6* (defective endogenous virus genes coding for ALV subgroup E glycoproteins) were more resistant to infection with subgroup E ALV, Salter *et al.* (1986, 1987) developed transgenic chickens expressing alv6 subgroup A glycoproteins. These glycoproteins presumably are shed from infected cells and bind to the receptor, thus preventing virus–receptor interactions. These chickens were indeed resistant to infection with subgroup A ALV and did not develop viraemia, antibodies or lymphoid leucosis after challenge (Salter and Crittenden, 1989). However, when the transgenic chickens were challenged at 6 days of incubation with the RPL41 subgroup A field isolate, protection, although significant, was incomplete against viraemia and lymphoma development. After 36 weeks of age, the incidence of lymphomas in the alv6+ (transgenic) birds was 18% and in the alv6– birds 51% ($P < 0.01$). Gavora *et al.* (1995) examined the important question of whether or not there are negative influences of alv6 expression on production parameters. Alv6 heterozygous sires were mated with alv6– dams of two different lines. In both crosses the alv6+ hens reached sexual maturity 4–6 days later, produced fewer eggs until

497 days of age, and had a 4–15% decline in the rate of egg production compared to their alv6– hatchmates, suggesting a 'biological cost' associated with the transgenesis. Although the development of transgenic chickens with increased resistance to ALV certainly demonstrated the feasibility of this approach, the use of transgenic techniques for improved resistance to disease is a long way from becoming applicable in commercial lines. In the meantime, the major breeders of table-egg producing chickens, the group most commonly infected with subgroup A ALV, have eradicated this virus from the pure lines because of the negative effect of ALV on egg production (Payne and Fadly, 1997).

Clearly, the techniques for transgenesis are available, but it remains a costly and time-consuming approach. Moreover, the results are not always positive when biological costs are taken into account.

References

Abplanalp, H., Schat, K.A. and Calnek, B.W. (1985) Resistance to Marek's disease of congenic strains differing in major histocompatibility haplotypes to 3 virus strains. In: Calnek, B.W. and Spencer, J.L. (eds) *International Symposium on Marek's Disease.* American Association of Avian Pathologists, Kennett Square, pp. 347–358.

Aeed, P.A., Briles, W.E., Zsigray, R.M. and Collins, W.M. (1993) Influence of different B-complex recombinants on the outcome of Rous sarcomas in chickens. *Animal Genetics* 24, 177–181.

Alexander, D.J. (1997) Newcastle disease and other *Paramyxoviridae* infection. In: Calnek, B.W., Barnes, H.J., Beard, C.W., McDougald, L.R. and Saif, Y.M. (eds) *Diseases of Poultry*, 10th edn. Iowa State University Press, Ames, pp. 541–569.

Baba, T.W. and Humphries, E.H. (1984) Avian leukosis virus infection: analysis of viremia and DNA integration in susceptible and resistant chicken lines. *Journal of Virology* 51, 123–130.

Bacon, L.D. (1987) Influence of the major histocompatibility complex on disease resistance and productivity. *Poultry Science* 66, 802–811.

Bacon, L.D. and Witter, R.L. (1992) Influence of turkey herpesvirus vaccination on the B-haplotype effect on Marek's disease resistance in 15.B congenic chickens. *Avian Diseases* 36, 378–385.

Bacon, L.D. and Witter, R.L. (1994a) Serotype specificity of B-haplotype influence on the relative efficacy of Marek's disease vaccines. *Avian Diseases* 38, 65–71.

Bacon, L.D. and Witter, R.L. (1994b) B haplotype influence on the relative efficacy of Marek's disease vaccines in commercial chickens. *Poultry Science* 73, 481–487.

Bacon, L.D., Crittenden, L.B., Witter, R.L., Fadly, A. and Motta, J. (1981) *B*-haplotype influence on Marek's disease, Rous sarcoma, and lymphoid leukosis virus-induced tumors in chickens. *Poultry Science* 60, 1132–1139.

Bacon, L.D., Crittenden, L.B., Witter, R.L., Fadly, A. and Motta, J. (1983) B^5 and B^{15} associated with progressive Marek's disease, Rous sarcoma, and avian leukosis virus-induced tumors in inbred $15I_4$ chickens. *Poultry Sciences* 62, 573–578.

Bacon, L.D., Frederickson, T.L., Gilmour, D.L., Fadly, A.M. and Crittenden, L.B. (1985) Tests of association of lymphocyte alloantigen genotypes with resistance to viral oncogenesis in chickens. 2. Rous sarcoma and lymphoid leukosis in progeny derived from $6_3 \times 15I$ and $100 \times 6_3$ crosses. *Poultry Science* 64, 39–47.

Bacon, L.D., Ismail, N. and Motta, J.V. (1986) Allograft and antibody responses of $15I_5$–*B* congenic chickens. In: Weber, W.T. and Ewert D.L. (eds) *Avian Immunology*. Alan R. Liss, New York, pp. 219–233.
Bagust, T.J. and Guy, J.S. (1997) Laryngotracheitis. In: Calnek, B.W., Barnes, H.J., Beard, C.W., McDougald, L.R. and Saif, Y.M. (eds) *Diseases of Poultry*, 10th edn. Iowa State University Press, Ames, pp. 527–539.
Barteling, S.J. and Vreeswijk, J. (1991) Developments in foot-and-mouth disease vaccines. *Vaccine* 9, 75–88.
Bielefeldt Ohmann, H., Babiuk, L.A. and Harland, R. (1991) Cytokine synergy with viral cytopathic effects and bacterial products during the pathogenesis of respiratory tract infection. *Clinical Immunology and Immunopathology* 60, 153–170.
Briles, W.E. and Briles, R.W. (1982) Identification of haplotypes in the chicken major histocompatibility complex (*B*). *Immunogenetics* 15, 449–459.
Briles, W.E., Briles, R.W., Taffs, R.L. and Stone, H.A. (1983) Resistance to a malignant lymphoma in chickens is mapped to subregion of major histocompatibility (B) complex. *Science* 219, 977–979.
Bumstead, N. (1998) Genomic mapping of resistance to Marek's disease. *Avian Pathology* 27, S78–S81.
Bumstead, N., Huggins, M.B. and Cook, J.K.A. (1989) Genetic differences in susceptibility to a mixture of avian infectious bronchitis virus and *Escherichia coli*. *British Poultry Science* 30, 39–48.
Bumstead, N., Millard, B.M., Barrow, P. and Cook, J.K.A. (1991) Genetic basis of disease resistance in chickens. In: Owen, J.B. and Axford R.F.E. (eds) *Breeding for Disease Resistance in Farm Animals*. CAB International, Wallingford, pp. 10–23.
Bumstead, N., Reece, R.L. and Cook, J.K.A. (1993) Genetic differences in susceptibility of chicken lines to infection with infectious bursal disease virus. *Poultry Science* 72, 403–410.
Bumstead, N., Sillibourne, J., Rennie, M., Ross, N. and Davison, F. (1997) Quantification of Marek's disease virus in chicken lymphocytes using the polymerase chain reaction with fluorescence detection. *Journal of Virological Methods* 65, 75–81.
Calnek, B.W. (1985) Genetic resistance. In: Payne, L.N. (ed.) *Marek's Disease. Scientific Basis and Methods of Control*. Martinus Nijhoff, Boston, pp. 293–328.
Calnek, B.W. (1986) Marek's disease – a model for herpesvirus oncology. *CRC Critical Reviews in Microbiology* 12, 293–320.
Calnek, B.W. and Witter, R.L. (1997) Marek's disease. In: Calnek, B.W., Barnes, H.J., Beard, C.W., McDougald, L.R. and Saif, Y.M. (eds) *Diseases of Poultry*, 10th edn. Iowa State University Press, Ames, pp. 367–413.
Calnek, B.W., Adene, D.F., Schat, K.A. and Abplanalp, H. (1989) Immune responses versus susceptibility to Marek's disease. *Poultry Science* 68, 17–26.
Cavanagh, D. and Naqi, S.A. (1997) Infectious bronchitis. In: Calnek, B.W., Barnes, H.J., Beard, C.W., McDougald, L.R. and Saif, Y.M. (eds) *Diseases of Poultry*, 10th edn. Iowa State University Press, Ames, pp. 511–526.
Chausse, A.M., Bernardet, N., Musset, E., Thoraval, P., Vaino, O., Dambrine, G. and Coudert, F. (1995) Expression of MHC class I antigens following Marek's disease infection. In: Davison, T.F., Bumstead, N. and Kaiser, P. (eds) *Advances in Avian Immunology Research*. Carfax Publishing Company, Abingdon, pp. 135–140.
Cole, R.K. (1968) Studies on genetic resistance to Marek's disease. *Avian Diseases* 12, 9–28.
Collins, W.H., Briles, W.E., Zsigray, R.M., Dunlop, W.R., Corbett, A.C., Clark, K.K., Marks, J.L. and McGrail, T.P. (1977) The B locus (MHC) in the chicken: association with the fate of RSV-induced tumors. *Immunogenetics* 5, 333–343.

Cook, J.K.A., Otsuki, K., Huggins, M. and Bumstead, N. (1990) Investigations into resistance of chicken lines to infection with infectious bronchitis virus. In: Cavanagh, D. and Brown, T.D.K. (eds) *Coronaviruses and their Diseases*. Plenum Press, New York, pp. 491–496.

Cook, J.K.A., Otsuki, K., da Silva Martins, N.R., Ellis, M.M. and Huggins, M.B. (1992) The secretory antibody response of inbred lines of chickens to avian bronchitis virus infection. *Avian Pathology* 21, 681–692.

Crittenden, L.B. and Motta, J.V. (1969) A survey of genetic resistance to leukosis sarcoma viruses in commercial stocks of chickens. *Poultry Science* 48, 1751–1757.

Crittenden, L.B. and Salter, D.W. (1986) Gene insertion: current progress and long-term goals. *Avian Diseases* 30, 43–46.

Crittenden, L.B., Stone, H.A., Reamer, R.H. and Okazaki, W. (1967) Two loci controlling genetic cellular resistance to avian leukosis sarcoma viruses. *Journal of Virology* 1, 898–904.

Davies, C.J., Joosten, I., Andersson, L., Arriens, M.A., Bernoco, D., Bissumbhar, B., Byrns, G., van Eijk, M.J.T., Kristensen, B., Lewin, H.A., Mikko, S., Morgan, A.L.G., Muggli-Cockett, N.E., Nilsson, Ph.R., Oliver, R.A., Park, C.A., van der Poel, J.J., Polli, M., Spooner, R.L. and Stewart, J.A. (1994a) Polymorphism of bovine MHC class II genes. Joint report of the Fifth International Bovine Lymphocyte Antigen (BoLA) Workshop, Interlaken, Switzerland, 1 August 1992. *European Journal of Immunogenetics* 21, 259–289.

Davies, C.J., Joosten, I., Bernoco, D., Arriens, M.A., Bester, J., Ceriotti, G., Ellis, S., Hensen, E.J., Hines, H.C., Horin, P., Kristensen, B., Lewin, H.A., Meggiolaro, D., Morgan, A.L.G., Morita, M., Nilsson, Ph.R., Oliver, R.A., Orlova, A., Østergård, H., Park, C.A., Schuberth, H.-J., Simon, M., Spooner, R.L. and Stewart, J.A. (1994b) Polymorphism of bovine MHC class I genes. Joint report of the Fifth International Bovine Lymphocyte Antigen (BoLA) Workshop, Interlaken, Switzerland, 1 August 1992. *European Journal of Immunogenetics* 21, 239–258.

Davies, C.J., Andersson, L., Ellis, S.A., Hensen, E.J., Lewin, H.A., Mikko, S., Muggli-Cockett, N.E., van der Poel, J.J. and Russell, G.C. (1997) Nomenclature for factors of the BoLA system, 1996: report of the ISAG BoLA Nomenclature Committee. *Animal Genetics* 28, 159–168.

Dimarchi, R., Brooke, G., Gale, C., Cracknell, V., Doel, T. and Mowat, N. (1986) Protection of cattle against foot-and-mouth disease by a synthetic peptide. *Science* 232, 639–641.

Dunnington, E.A., Martin, A., Briles, W.E., Briles, R.W. and Siegel, P.B. (1986) Resistance to Marek's disease in chickens selected for high and low antibody responses to sheep red blood cells. *Archiv für Geflügelkunde* 50, 94–96.

Dunnington, E.A., Larsen, C.T., Gross, W.B. and Siegel, P.B. (1992) Antibody responses to combinations of antigens in white leghorn chickens of different background genomes and major histocompatibility complex genotypes. *Poultry Science* 71, 1801–1806.

Fadly, A.M. and Bacon, L.D. (1992) Response of B congenic chickens to infection with infectious bursal disease virus. *Avian Diseases* 36, 871–880.

Fredericksen, T.L., Longenecker, B.M., Pazderka, F., Gilmour, D.G. and Ruth, R.F. (1977) A T-cell antigen system of chickens: Ly-4 and Marek's disease. *Immunogenetics* 5, 535–552.

Fredericksen, T.L., Gilmour, D.G., Bacon, L.D., Witter, R.L. and Motta, J. (1982) Tests of association of lymphocyte alloantigen genoypes with resistance to viral oncogenesis in chickens. 1. Marek's disease in F7 progeny derived from $6_3 \times 15_I$ crosses. *Poultry Science* 61, 2322–2326.

Gavora, J.S., Benkel, B., Spencer, J.L., Gagnon, C. and Crittenden, L.B. (1995) Influence of the alv6 recombinant avian leukosis virus transgene on production traits and infection with avian tumor viruses in chickens. *Poultry Science* 74, 852–863.

Glass, E.J. and Millar, P. (1994) Induction of effective cross-reactive immunity by FMDV peptides is critically dependent upon specific MHC-peptide-T cell interactions. *Immunology* 82, 1–8.

Glass, E.J. and Millar, P. (1995) Bovine T cells preferentially recognize non-viral spacer epitopes in a putative FMDV vaccinal peptide. *Vaccine* 13, 225–229.

Glass, E.J., Oliver, R.A., Collen, T., Doel, T.R., Dimarchi, R. and Spooner, R.L. (1991) MHC class II restricted recognition of FMDV peptides by bovine T cells. *Immunology* 74, 594–599.

Gómez-Puertas, P., Rodriguez, F., Oviedo, J.M., Brun, A., Alonso, C. and Escribano, J.M. (1998) The African swine fever virus proteins p54 and p30 are involved in two distinct steps of virus attachment and both contribute to the antibody-mediated protective immune response. *Virology* 243, 461–471.

Gyles, N.R. and Brown, C.J. (1971) Selection in chickens for retrogression of tumors caused by Rous sarcomas. *Poultry Sciences* 50, 901–905.

Hansen, M.P., Van Zandt, J.N. and Law, G.R.J. (1967) Differences in susceptibility to Marek's disease in chickens carrying two different B blood group allotypes (abstr). *Poultry Science* 46, 1268.

Hawken, R.J. and Schook, L.B. (1998) Classical improvement and molecular genetics in animal production. *Proceedings of the 8th World Congress on Animal Production*, Seoul, Korea, p. 34.

Hawken, R.J., Beattie, C.W. and Schook, L.B. (1998) Resolving the genetics of disease resistance to infectious agents. *Revue Scientifique et technique de l'Office Internationale des Epizooties* 17, 17–25.

Hedemand, J.E., Sørensen, P., Ducro, B.J. and Simonsen, B.J. (1993) Resistance to Marek's disease associated with B^{21}-like haplotypes. *Archiv für Geflügelkunde* 57, 73–76.

Hein, W.R. (1998) Sheep immunology and goat peculiarities: 7. Immunoglobulins. In: Pastoret, P-P., Griebel, P., Bazin, H. and Goverts, A. (eds) *Handbook of Vertebrate Immunology*. Academic Press, San Diego, pp. 503–505.

Hepkema, B.G., Blankert, J.J., Albers, G.A.A., Tilanus, M.G.J., Egberts, E., van der Zijpp, A.J. and Hensen, E.J. (1993) Mapping of susceptibility to Marek's disease within the major histocompatibility (B) complex by refined typing of white leghorn chickens. *Animal Genetics* 24, 283–287.

Kaufman, J. (1996) Structure and function of the major histocompatibility complex of chickens. In: Davison, T.F., Morris, T.R. and Payne, L.N. (eds) *Poultry Immunology*. Carfax Publishing Company, Abingdon, pp. 67–82.

Kaufman, J. and Lamont, S.J. (1996) The chicken major histocompatibility complex. In: Schook, L.B. and Lamont, S.J. (eds) *The Major Histocompatibility Complex in Domestic Animal Species*. CRC Press, Boca Raton, Florida, pp. 35–64.

Kaufman, J., Völk, H. and Wallny, H.-J. (1995) A minimal essential Mhc and an unrecognized Mhc: two extremes in selection for polymorphism. *Immunological Reviews* 143, 63–88.

Kean, R.P., Cahaner, A., Freeman, A.E. and Lamont, S.J. (1994) Direct and correlated responses to multitrait, divergent selection for immunocompetence. *Poultry Science* 73, 18–32.

Kung, H.-J. and Maihle, N.J. (1987) Molecular basis of oncogenesis by non-acute avian retroviruses. In: de Boer, G.F. (ed.) *Avian Leukosis*. Martinus Nijhoff, Boston, pp. 77–99.

Laksmanan, N. and Lamont, S.J. (1998) Rfp-Y region polymorphism and Marek's disease resistance in multitrait immunocompetence-selected chicken lines. *Poultry Science* 77, 538–541.

Lamont, S.J. (1998) Impact of genetics on disease resistance. *Poultry Science* 77, 1111–1118.

Lamont, S.J., Lakshmanan, N. and Kaiser, M.G. (1996) Effect of selection for multitrait immune response on innate and vaccinal resistance to Marek's disease. In: Silva, R.F., Cheng, H.H., Coussens, P.M., Lee, L.F. and Velicer, L.F. (eds) *Current Research on Marek's Disease.* American Association of Avian Pathologists, Kennett Square, pp. 1–7.

Lee, L.F., Powell, P.C., Rennie, M., Ross, L.J.N. and Payne, L.N. (1981) Nature of genetic resistance to Marek's disease in chickens. *Journal National Cancer Institute* 66, 789–796.

Leitner, G., Gutman, M., Heller, E.D., Yonash, N. and Cahaner, A. (1994) Parental effect on the humoral immune response to *Escherichia coli* and Newcastle disease virus in young broiler chicks. *Poultry Science* 73, 1534–1541.

Lewin, H.A. (1989) Disease resistance and immune response genes in cattle: strategies for their detection and evidence of their existence. *Journal of Dairy Science* 72, 1334–1348.

Lewin, H.A. and Bernoco, D. (1986) Evidence for BoLA-linked resistance and susceptibility to subclinical progression of bovine leukaemia virus infection. *Animal Genetics* 17, 197–207.

Lewin, H.A., Wu, M.-C., Nolan, T.J. and Stewart, J.A. (1988a) Peripheral B lymphocyte percentage as an indicator of subclinical progression of bovine leukemia virus infection. *Journal of Dairy Science* 71, 2526–2534.

Lewin, H.A., Wu, M., Stewart, J.A. and Nolan, T.J. (1988b) Association between BoLA and subclinical bovine leukemia virus infection in a herd of Holstein–Friesian cows. *Immunogenetics* 27, 338–344.

Loudovaris, T., Yoo, B.H. and Fahey, K.J. (1991a) Genetic resistance to infectious laryngotracheitis in inbred lines of White Leghorn chickens. *Avian Pathology* 20, 357–361.

Loudovaris, T., Calnek, B.W., Yoo, B.H. and Fahey, K.J. (1991b) Genetic susceptibility of chicken macrophages to *in vitro* infection with infectious laryngotracheitis virus. *Avian Pathology* 20, 29–302.

Lukert, P.D. and Saif, Y.M. (1997) Infectious bursal disease. In: Calnek, B.W., Barnes, H.J., Beard, C.W., McDougald, L.R. and Saif, Y.M. (eds) *Diseases of Poultry*, 10th edn. Iowa State University Press, Ames, pp. 721–738.

Lunney, J., Salmueller, A., Pauly, T., Boyd, P., Hyatt, S., Strom, D., Martin, S. and Zuckermann, F. (1996) Cellular immune responses controlling infectious diseases. In: Tumbleson, M.E. and Schook, L.B. (eds) *Advances in Swine in Biomedical Research.* Plenum Press, New York, pp. 307–316.

Lunney, J.K. and Butler, J.E. (1998) Immunogenetics. In: Rothschild, M.F. and Ruvinsky, A. (eds) *The Genetics of Pigs.* CAB International, Wallingford, pp. 163–197.

McCullough, K.C., De Simone, F., Brocchi, E., Capucci, L., Crowther, J.R. and Kihm, U. (1992) Protective immune response against foot-and mouth disease. *Journal of Virology* 66, 1835–1840.

Maddox, J.F. (1998) Sheep immunology and goat peculiarities: 8. Ovine major histocompatibility complex antigens (Ovar). In: Pastoret, P.-P., Griebel, P., Bazin, H. and Goverts, A. (eds) *Handbook of Vertebrate Immunology.* Academic Press, San Diego, pp. 505–509.

Martins, C.L.V., Lawman, M.J.P., Scholl, T., Mebus, C.A. and Lunney, J.K. (1993) African swine fever virus specific porcine cytotoxic T cell activity. *Archives of Virology* 129, 211–225.

Mengeling, W.L. and Cutlip, R.C. (1972) Experimentally induced infection of newborn pigs with hemagglutinating encephalomyelitis virus strain 67N. *American Journal Veterinary Research* 33, 953–956.

Mengeling, W.L. and Cutlip, R.C. (1976) Pathogenicity of field isolants of hemagglutinating encephalomyelitis virus for neonatal pigs. *Journal American Veterinary Medical Association* 168, 236–239.

Mertens, B. (1998) Immunology of Cattle: 4. Cytokines. In: Pastoret, P.-P., Griebel, P., Bazin, H. and Goverts, A. (eds) *Handbook of Vertebrate Immunology*. Academic Press, San Diego, pp. 447–454.

Miller, M.M., Goto, R., Bernot, A., Zoorob, R., Auffray, C., Bumstead, N. and Briles, W.E. (1994) Two MHC class I and MHC class II genes map to the *Rfp-Y* system outside the B-complex. *Proceedings of the National Academy of Sciences USA* 91, 4397–4401.

Miller, M.M., Goto, R., Taylor, R.L., Jr, Zoorob, R., Auffray, C., Bumstead, N., Briles, R.W., Briles, W.E. and Bloom, S.E. (1996a) Assignment of *Rfp-Y* to the chicken microchromosomes and evidence for high frequency recombination associated with nucleolar organizer region. *Proceedings of the National Academy of Sciences USA* 93, 3958–3962.

Miller, M.M., Afanassieff, M., Wakenell, P.S., Briles, W.E., Chase, W.B. and Schwartz, R.D. (1996b) *Rfp-Y* genotypes in populations under challenge with Marek's disease virus. In: Silva, R.F., Cheng, H.H., Coussens, P.M., Lee, L.F. and Velicer, L.F. (eds) *Current Research on Marek's Disease.* American Association of Avian Pathologists, Kennett Square, pp. 47–51.

Müller, M., Brenig, B., Winnacker, E.-L., and Brem, G. (1992) Transgenic pigs carrying cDNA copies encoding the murine *Mx1* protein which confers resistance to influenza virus infection. *Gene* 121, 263–270.

Murphy, B.R. and Webster, R.G. (1995) Orthomyxoviruses. In: Fields, B.N., Knipe, D.M., Howley, P.M., Chanock, R.M., Melnick, J.L., Monath, T.P., Roizman, B. and Straus, S.E. (eds) *Field's Virology.* Lippincott Raven, New York, Vol. I, pp. 1397–1447.

Naessens, J. (1998) Immunology of Cattle: 6. Immunoglobulins. In: Pastoret, P.-P., Griebel, P., Bazin, H. and Goverts, A. (eds) *Handbook of Vertebrate Immunology.* Academic Press, San Diego, pp. 456–459.

Naessens, J. and Hopkins, J. (1996) Introduction and summary of workshop findings. *Veterinary Immunology and Immunopathology* 52, 213–235.

Naessens, J., Howard, C.J. and Hopkins, J. (1997) Nomenclature and characterization of leukocyte differentiation antigens in ruminants. *Immunology Today* 18, 365–368.

Nakamura, K., Cook, J.K.A., Otsuki, K., Huggins, M.B. and Frazier, J.A. (1991) Comparative study of respiratory lesions in two chicken lines of different susceptibility infected with infectious bronchitis virus: histology, ultrastructure and immunohistochemistry. *Avian Pathology* 20, 241–257.

Nash, A.D., Hawken, R.J., Andrews, A.E., Maddox, J.F. and Martin, H.M. (1998) Sheep immunology and goat peculiarities: 5. Sheep cytokines. In: Pastoret, P.-P., Griebel, P., Bazin, H. and Goverts, A. (eds) *Handbook of Vertebrate Immunology.* Academic Press, San Diego, pp. 496–500.

Nielsen, O.L., Sørensen, P., Hedemond, J.E., Laurssen, S.B. and Jørgensen, P.H. (1998) Inflammatory response of different chicken lines and B haplotypes to infection with infectious bursal disease virus. *Avian Pathology* 27, 181–189.

Okoye, J.O.A. and Aba-Adulugba, E.P. (1998) Comparative study of the resistance or susceptibility of local Nigerian and exotic chickens to infectious bursal disease. *Avian Pathology* 27, 168–177.

Omar, A.R. and Schat, K.A. (1996) Syngeneic Marek's disease virus (MDV)-specific cell-mediated immune responses against immediate early, late, and unique MDV proteins. *Virology* 222, 87–99.

Omar, A.R. and Schat, K.A. (1997) Characterization of Marek's disease herpesvirus-specific cytotoxic T lymphocytes in chickens inoculated with a nononcogenic vaccine strain of MDV. *Immunology* 90, 579–585.

Omar, A.R., Schat, K.A., Lee, L.F. and Hunt, H.D. (1998) Cytotoxic T lymphocyte response in chickens immunized with a recombinant fowlpox virus expressing Marek's disease herpesvirus glycoprotein B. *Veterinary Immunology and Immunopathology* 62, 73–82.

Otsuki, K., Huggins, M.B. and Cook, J.K.A. (1990) Comparison of the susceptibility to avian infectious bronchitis virus infection of two inbred lines of white leghorn chickens. *Avian Pathology* 19, 467–475.

Oura, C.A.L., Powell, P.P., Anderson, E. and Parkhouse, R.M.E. (1998) The pathogenesis of African swine fever in the resistant bushpig. *Journal of General Virology* 79, 1439–1443.

Payne, L.N. (1987) Epizootiology of avian leukosis virus infections. In: de Boer, G.F. (ed.) *Avian Leukosis.* Martinus Nijhoff, Boston, pp. 47–75.

Payne, L.N. and Fadly, A.M. (1997) Leukosis/sarcoma group. In: Calnek, B.W., Barnes, H.J., Beard, C.W., McDougald, L.R. and Saif, Y.M. (eds) *Diseases of Poultry*, 10th edn. Iowa State University Press, Ames, pp. 414–466.

Pazderka, F., Longenecker, B.M., Law, G.R.J., Stone, H.A. and Ruth, R.F. (1975) Histocompatibility of chicken populations selected for resistance to Marek's disease. *Immunogenetics* 2, 93–105.

Pelzer, K.D. and Sprecher, D.J. (1993) Controlling BLV infection on dairy operations. *Veterinary Medicine* 88, 275–281.

Pharr, G.T., Dodgson, J.B., Hunt, H.D. and Bacon, L.D. (1998) Class II MHC cDNAs in 15I5 B-congenic chickens. *Immunogenetics* 26, 350–354.

Pinard, M.-H., Janss, L.L.G., Maatman, R. and Noordhuizen, J.P.T.M. (1993) Effect of divergent selection for immune responsiveness and of major histocompatibility complex on resistance to Marek's disease in chickens. *Poultry Sciences* 72, 391–402.

Plachy, J. and Benda, V. (1981) Location of the gene responsible for Rous sarcoma regression in the B-F region of the B complex (MHC) in the chicken. *Folia Biologica (Praque)* 27, 363–368.

Poulsen, D.J., Thureen, D.R. and Keeler, C.L. Jr (1998) Comparison of disease susceptibility and resistance in three lines of chickens experimentally infected with infectious laryngotracheitis virus. *Poultry Science* 77, 17–21.

Powell, P.C., Lee, L.F., Mustill, B.M. and Rennie, M. (1982) The mechanism of genetic resistance to Marek's disease in chickens. *International Journal of Cancer* 29, 69–76.

Pratt, W.D., Morgan, R.W. and Schat, K.A. (1992) Cell-mediated cytolysis of lymphoblastoid cells expressing Marek's disease-specific polypeptides. *Veterinary Microbiology* 33, 93–99.

Purchase, H.G., Cunningham, C.H. and Burmester, B.R. (1966) Genetic differences among chicken embryos in response to inoculation with an isolate of infectious bronchitis virus. *Avian Diseases* 10, 162–172.

Purchase, H.G., Gilmour, D.G., Romero, C.H. and Okazaki, W. (1977) Post infection genetic resistance to avian lymphoid leukosis resides in a B target cell. *Nature* 270, 61–62.

Robinson, H.L., Astrin, S.M., Senior, A.M. and Salazar, F.H. (1981) Host susceptibility to endogenous viruses: defective, glycoprotein-expressing proviruses interfere with infection. *Journal of Virology* 40, 745–751.

Russell, G.C., Davies, C.J., Andersson, L., Ellis, S.A., Hensen, E.J., Lewin, H.A., Mikko, S., Muggli-Cockett, N.E. and van der Poel, J.J. (1997) BoLA class II nucleotide sequences, 1996: report of the ISAG BoLA Nomenclature Committee. *Animal Genetics* 28, 169–180.

Ryan, A.M. and Womack, J.E. (1993) Type I interferon genes in cattle: restriction fragment length polymorphisms, gene numbers and physical organization of bovine chromosome 8. *Animal Genetics* 24, 9–16.

Ryan, A.M. and Womack, J.E. (1997) A molecular genetic approach to improved animal health: the effect of interferon genotype on the severity of experimental bovine herpesvirus-1 infection. *Veterinary Clinics of North America: Food Animal Practice* 13, 401–409.

Ryan, A.M., Hutcheson, D.P. and Womack, J.E. (1993) Type-I interferon genotypes and severity of clinical disease in cattle inoculated with bovine herpesvirus 1. *American Journal of Veterinary Research* 54, 73–79.

Sachs, D.H., Leight, G., Cone, J., Schwartz, S., Stuart, L. and Rosenberg, S. (1976) Transplantation in miniature swine. I. Fixation of the major histocompatibility complex. *Transplantation* 22, 559–567.

Salter, D.W. and Crittenden, L.B. (1989) Artificial insertion of a dominant gene for resistance to avian leukosis into the germ line of the chicken. *Theoretical and Applied Genetics* 77, 457–461.

Salter, D.W., Smith, E.J., Hughes, S.H., Wright, S.E., Fadly, A.M., Witter, R.L. and Crittenden, L.B. (1986) Gene insertion into the chicken germ line by retroviruses. *Poultry Science* 65, 1445–1458.

Salter, D.W., Smith, E.J., Hughes, S.H., Wright, S.E. and Crittenden, L.B. (1987) Transgenic chickens: insertion of retroviral genes into the chicken germ line. *Virology* 157, 236–240.

Schat, K.A. (1987) Marek's disease: a model for protection against herpesvirus-induced tumours. *Cancer Surveys* 6, 1–37.

Schat, K.A. (1996) Immunity to Marek's disease, lymphoid leukosis and reticuloendotheliosis. In: Davison, T.F., Morris, T.R. and Payne, L.N. (eds) *Poultry Immunology*. Carfax Publishing Company, Abingdon, pp. 209–233.

Schat, K.A., Calnek, B.W. and Fabricant, J. (1981) Influence of oncogenicity of Marek's disease on evaluation of genetic resistance. *Poultry Science* 60, 2259–2566.

Schat, K.A., Taylor, R.L. Jr and Briles, W.E. (1994) Resistance to Marek's disease in chickens with recombinant haplotypes of the major histocompatibility (B) complex. *Poultry Science* 73, 502–508.

Schook, L.B. and Alexander, L. (1997) Mapping the porcine genome. *Pig News and Information* 18, 253N–256N.

Seo, S.H. and Collisson, E.W. (1997) Specific cytotoxic T lymphocytes are involved in *in vivo* clearance of infectious bronchitis virus. *Journal of Virology* 71, 5173–5177.

Sharma, J.M. and Schat, K.A. (1991) Natural immune functions. In: Sharma, J.M. (ed.) *Avian Cellular Immunology*. CRC Press, Boca Raton, Florida, pp. 51–70.

Skinner, J.D. and Smithers, R.H.N. (1990) *The Mammals of the Southern African Subregion*. University of Pretoria, Pretoria.

Smith, E.J., Lyons, L.A., Cheng, H.H. and Suchyta, S.P. (1997) Comparative mapping of the chicken genome using the East Lansing reference population. *Poultry Science* 76, 743–747.

Staeheli, P. (1990) Interferon-induced proteins and the antiviral state. *Advances in Virus Research* 38, 147–200.

Steadham, E.M., Lamont, S.J., Kujdych, I. and Nordskog, A.W. (1987) Association of Marek's disease with Ea-B and immune response genes in subline and F_2 populations of the Iowa sate S1 leghorn line. *Poultry Science* 66, 571–575.

Stott, M.L., Thurmond, M.C., Dunn, S.J., Osburn, B.I. and Stott, J.L. (1991) Integrated bovine leukosis proviral DNA in T helper and T cytotoxic/suppressor lymphocytes. *Journal of General Virology* 72, 307–315.

Straub, O.C. (1991) BHV1 infections: relevance and spread in Europe. *Comparative Immunology, Microbiology and Infectious Diseases* 14, 175–186.

Straw, B.E. and Rothschild, M.F. (1992) Genetic influences on liability to acquired disease. In: Straw, B.E., Leman, A.D., Mengeling, W.L., D'Allaire, S. and Taylor, D.J. (eds) *Diseases of Swine.* Iowa State University Press, Ames, pp. 709–717.

Thacker, E.I., Fulton, J.E. and Hunt, H.D. (1995) *In vitro* analysis of a primary, MHC-restricted, cytotoxic T lymphocyte response to avian leukosis virus (ALV) using target cells expressing MHC class I cDNA inserted into a recombinant ALV vector. *Journal of Virology* 69, 6439–6444.

Taylor, R.L. Jr, England, J.M., Kopen, G.C., Christou, A.E. and Halpern, M.S. (1994) Major histocompatibility (*B*) complex control of formation of v-*src*-induced metastases. *Virology* 205, 569–573.

Tsai, H.J., Saif, Y.M., Nestor, K.E., Emmerson, D.A. and Patterson, R.A. (1992) Genetic variation in resistance of turkeys to experimental infection with Newcastle disease virus. *Avian Diseases* 36, 491–820.

Vallejo, R.L., Pharr, G.T., Liu, H.C., Cheng, H.H., Witter, R.L. and Bacon, L.D. (1997) Non-association between *Rfp-Y* major histocompatibility complex-like genes and susceptibility to Marek's disease virus-induced tumours in $6_3 \times 7_2$ F_2 intercross chickens. *Animal Genetics* 28, 331–337.

Vallejo, R.L., Bacon, L.D., Liu, H.C., Witter, R.L., Groenen, M.A.M., Hillel, J. and Cheng, H.H. (1998) Genetic mapping of quantitative trait loci affecting susceptibility to Marek's disease virus induced tumors in F-2 intercross chickens. *Genetics* 148, 349–360.

Vandenbroeck, K. and Billiau, A. (1997) Recent progress in the molecular characterization of porcine cytokines. In: Schijns, V.E.C.J. and Horzinek, M.C. (eds) *Cytokines in Veterinary Medicine.* CAB International, Wallingford, pp. 139–154.

van Eijk, M.J.T., Stewart-Haynes, J.A., Beever, J.E., Fernando, R.L. and Lewin, H.A. (1992) Development of persistent lymphocytosis in cattle is closely associated with DRB2. *Immunogenetics* 37, 64–68.

van Lierop, M.-J.C., Nilsson, P.R., Wagenaar, J.P.A., van Noort, J.M., Campbell, J.D.M., Glass, E.J., Joosten, I. and Hensen, E.J. (1995) The influence of MHC polymorphism on the selection of T-cell determinants of FMDV in cattle. *Immunology* 84, 79–85.

Wakenell, P.S., Miller, M.M., Goto, R.M., Gauderman, W.J. and Briles, W.E. (1996) Association between *Rfp-Y* haplotype and the incidence of Marek's disease in chickens. *Immunogenetics* 44, 242–245.

White, E.C., Briles, W.E., Briles, R.W. and Taylor, R.L. Jr (1994) Response of six major histocompatibility complex recombinant haplotypes to Rous sarcomas. *Poultry Science* 73, 836–842.

Williams, R.K., Snyder, S.W. and Holmes, K.V. (1990) MHV-resistant SJL/J mice express a non-functional homolog to the MHV receptor protein. In: Cavanagh, D. and Brown, T.D.K. (eds) *Coronaviruses and their Diseases.* Plenum Press, New York, pp. 45–50.

Witter, R.L. (1998) The changing landscape of Marek's disease. *Avian Pathology* 27, S46–S53.

Xu, A., van Eijk, M.J.T., Park, C. and Lewin, H.A. (1993) Polymorphism in BoLA-DRB3 exon 2 correlates with resistance to persistant lymphocytosis caused by bowine leukemia virus. *The Journal of Immunology* 151, 6977–6985.
Yoo, B.H. and Sheldon, B.L. (1992) Association of the major histocompatibility complex with avian leukosis virus infection in chickens. *British Poultry Science* 33, 613–620.
Young, J.A.T. (1998) Avian leukosis virus-receptor interactions. *Avian Pathology* 27, S21–S25.
Zarlenga, D.S., Canals, A. and Gasbarre, L. (1995) Method for constructing internal standards for use in competitive PCR. *BioTechniques* 19, 324–326.

14

Diseases Caused by Maedi–Visna and Other Ovine Lentiviruses

J.C. DeMartini[1], A. de la Concha-Bermejillo[2], J.O. Carlson[3] and R.A. Bowen[4]

Departments of [1]Pathology, [3]Microbiology and [4]Physiology, Colorado State University, Fort Collins, USA; [2]Department of Pathobiology, Texas A&M University, Texas Agricultural Experiment Station, San Angelo, USA

Summary

Maedi–visna and other ovine lentiviruses (OvLV) are worldwide causes of slowly progressive diseases of lungs, brains, joints and mammary glands of infected sheep. Infection by these retroviruses is characterized by lifelong persistence, horizontal and vertical transmission, and insidious onset of disease in a proportion of infected adults. In the absence of effective vaccines, OvLV infection is controlled by using test and cull schemes or by prevention of vertical (milk) transmission of the virus. Developing breeding stock that are resistant to OvLV infection or OvLV-associated diseases is an alternative approach. Traditionally, this would involve selection and propagation of individuals that exhibit resistance to OvLV infection or disease. However, identification of such animals is so problematic that a selection programme has never been undertaken. For the same reason, no searches for DNA markers have been undertaken. Based on rapidly developing information on molecular characteristics of a related lentivirus, HIV, and new technologies for introducing genes into the mammalian genome, it is now possible to produce sheep that are transgenic for viral or other genes that may induce resistance to OvLV infection or disease. Candidate genes for introduction include viral structural protein genes such as env, mutated viral regulatory protein genes such as *tat* and *rev*, genes encoding viral antisense RNA, and genes encoding ribozymes capable of cleaving viral RNA within infected cells. Success of these approaches will depend upon optimizing expression of the desired gene in appropriate target cells without adversely affecting cell or host physiological functions. After constructing transgenic sheep that stably express the desired gene, their ability to resist OvLV infection or disease must be assessed by natural or experimental challenge. Currently available tools of biotechnology and knowledge of lentivirus molecular virology and pathogenesis provide a sense of optimism that progress in controlling animal retroviral diseases through these means is imminent.

© CAB *International* 2000. *Breeding for Disease Resistance in Farm Animals* (eds R.F.E. Axford, S.C. Bishop, F.W. Nicholas and J.B. Owen)

Introduction

Maedi and visna (maedi synonyms: ovine progressive pneumonia in the USA, la bouhite in France, Graaff-Reinet disease in South Africa, and zwoegerziekte in the Netherlands) are chronic multisystemic diseases of sheep caused by ovine lentivirus (OvLV; Petursson *et al.*, 1992; de la Concha-Bermejillo, 1997). Maedi, an Icelandic word meaning dyspnoea, is used to describe a chronic interstitial pneumonia, whereas the term visna (meaning wasting) refers to a slow, progressive disease of the central nervous system resulting in paralysis; chronic mastitis and arthritis are also caused by OvLV. Ovine lentivirus infection has been reported in most sheep-raising countries of the world, with the notable exception of Australia and New Zealand (Dawson, 1980).

Ovine lentivirus, the cause of these diverse disease syndromes, is a prototype virus of the genus *Lentivirus* within the family *Retroviridae*. Ovine lentiviruses share morphological, genetic and pathogenic characteristics with other lentiviruses, including caprine arthritis encephalitis virus (CAEV), bovine immunodeficiency virus (BIV), equine infectious anaemia virus (EIAV), feline immunodeficiency virus (FIV), simian immunodeficiency virus (SIV) and human immunodeficiency virus (HIV) (Narayan and Clements, 1989). Because of very similar genomic organization, pathogenicity and epidemiology, OvLV and CAEV are often grouped as small ruminant lentiviruses; these viruses can be transmitted between their respective hosts (Banks *et al.*, 1983).

Retrovirus infections have proven difficult to control in domestic animals because of their capacity to integrate into the genome of the host, to induce a persistent infection and to vary the antigenicity of envelope glycoproteins (Pearson *et al.*, 1989; DeMartini *et al.*, 1991; Brodie *et al.*, 1992a). Since effective vaccines have not been developed, small ruminant lentivirus diseases are primarily controlled using test and culling schemes to eliminate infected animals. In this chapter, we integrate new knowledge concerning the biology of lentiviruses and their interaction with the host with approaches based on selective breeding, genetics, embryo manipulation and molecular biology to describe novel strategies for the control of lentivirus-associated diseases of sheep.

Characteristics of Ovine Lentiviruses and their Replication

The OvLV genome, as in other lentiviruses, consists of two identical positive-sense single-stranded RNA subunits of 9.4 kb containing structural and regulatory genes (Fig. 14.1; Sonigo *et al.*, 1985; Narayan and Clements, 1990; Clements and Zink, 1996). The structural genes of the virus are *gag*, *pol* and *env*. The *gag* gene of OvLV is highly conserved and encodes information for one major and two minor core proteins that are proteolytically processed from a 53 kDa gag precursor protein by the virion-encoded protease. The major core protein, referred to as capsid protein (CA, p24–p27), elicits a strong antibody response and has been used to develop sensitive diagnostic assays. The two small core proteins, the matrix protein (MA, p16) and the nucleocapsid

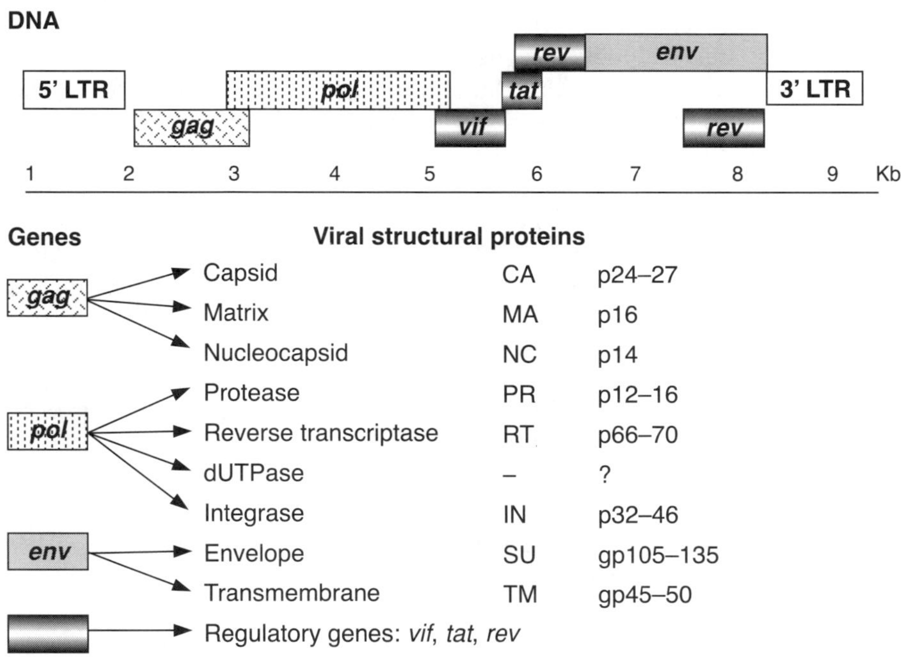

Fig. 14.1. Ovine lentivirus proviral genomic organization, regulatory genes, and structural genes and their protein products.

protein (NC, p14) are poorly immunogenic. The *pol* gene of OvLV encodes information for reverse transcriptase (RT), a protease (PR), an endonuclease/integrase (IN) and a dUTPase. These proteins play important roles in viral nucleic acid transcription, protein synthesis and integration into host-cell DNA, during replication. The dUTPase is the only OvLV gene that is not essential for replication, and visna virus lacking a functional dUTPase gene can replicate well in sheep macrophages *in vitro* (Turelli *et al.*, 1996). The *env* gene encodes the information for the envelope surface glycoprotein (SU, gp105) and the transmembrane protein (TM), both of which are used in serological assays along with CA (Juste *et al.*, 1995). The envelope proteins are synthesized as a large precursor protein that is cleaved by a cellular protease into the SU and the TM glycoproteins. The SU protein carries the neutralization and fusion epitopes. Antigenic variation within the SU protein confers biological and serological properties of different isolates (Braun *et al.*, 1987; Mwaengo *et al.*, 1997; Cheevers *et al.*, 1999). As with other lentiviruses, the proviral DNA of OvLV is flanked on both ends by long terminal repeats (LTRs) that contain enhancer–promoter elements for the initiation of DNA transcription and play a role in tissue tropisms (Small *et al.*, 1989). A unique genetic feature of all lentiviruses is the presence of small regulatory genes located between the *pol* and *env* genes and the 3′ terminus, *vif*, *tat* and *rev*, that encode proteins that regulate viral replication.

An important characteristic of lentiviruses is that their replication cycle includes the integration of a DNA intermediate (the provirus) into the host cell's chromosomes. After virus infection, RT transcribes the genomic viral RNA into a double-stranded DNA copy. After circularization, the enzyme IN inserts proviral DNA into the host-cell DNA. The viral genome thus becomes part of the cellular DNA and is duplicated during cell division. As a result, OvLV infection of sheep is potentially lifelong and the virus can be isolated from seropositive sheep years after the original infection (Haase, 1986).

Ovine lentivirus replicates primarily in monocytes and macrophages of the blood, lung, spleen, bone marrow and brain of infected animals (Brodie *et al.*, 1992a; Narayan *et al.*, 1993; Clements and Zink, 1996). *In vitro*, macrophages and primary choroid plexus, lung, trachea, cornea and synovial membrane cells are permissive. Cytopathic effects include the formation of multinucleated syncytial cells and cell lysis, and genetic variants of these have been described (Woodward *et al.*, 1995).

Clinical Features, Pathogenesis and Immunology of Ovine Lentivirus Infections

The most common clinical manifestations of OvLV infection are progressive respiratory failure and cachexia in mature sheep (Bulgin, 1990). Some animals may develop indurative mastitis, swollen joints, or paresis or paralysis. Affected ewes often give birth to small, weak lambs. Once clinical disease becomes apparent, sheep die within 6–8 months, usually due to anoxia or secondary bacterial infections.

The lungs of sheep with OvLV-associated lymphoid interstitial pneumonia (LIP) are non-collapsing and heavy, and there may be 1–2 mm diameter grey foci scattered throughout the parenchyma; red–grey consolidation may involve the cranioventral lobes, particularly if secondary bronchopneumonia is present (de la Concha-Bermejillo, 1997; Brodie *et al.*, 1998). The microscopic features of LIP include lymphocyte hyperplasia surrounding airways and blood vessels, infiltration of the interalveolar septa and alveoli by mononuclear cells, bronchiolar smooth muscle hyperplasia and fibrosis. $CD8^+$ and $CD4^+$ T cells are increased in lung compartments of infected sheep (Watt *et al.*, 1992). Pulmonary lymph nodes are markedly hyperplastic due to increased numbers of B cells and T cells in lymphoid follicles and germinal centres and paracortical zones (Ellis and DeMartini, 1985).

More than 60% of the ewes in OvLV-affected flocks have chronic lymphocytic mastitis and resultant reduction in milk production (Dawson, 1987). Lambs from ewes with OvLV-induced mastitis have reduced weaning weights (Pekelder *et al.*, 1994). Lentivirus-associated arthritis typically begins insidiously 2–3 years after infection, with weight loss and swelling of the carpal and tarsal joints (Kennedy-Stoskopf, 1989). Microscopically, there is infiltration of mononuclear cells and synovial villous hyperplasia, which may progress to chronic osteoarthritis. Sheep affected by OvLV-associated neurological disease (visna) show aberrations of gait followed by progressive paraplegia of the rear limbs, quadriplegia and wasting (Petursson *et al.*, 1990).

Spinal taps reveal mononuclear pleocytosis in the cerebrospinal fluid. Microscopically there is periventricular encephalitis characterized by ependymal necrosis, widespread demyelination and prominent perivascular lymphocytic cuffing (Petursson *et al.*, 1992; Georgsson, 1994). Although CNS manifestations of OvLV infection are uncommon in the USA, a study showed that 18% of naturally infected sheep had subclinical mononuclear cell infiltration in the leptomeninges, cerebral white matter, choroid plexus or cervical spinal cord. Histiocytes in these lesions contained lentiviral CA protein or RNA (Brodie *et al.*, 1995).

The pathogenesis of OvLV-induced disease depends on host, viral and environmental factors. Among the viral factors, the virus load in infected animals, which may be influenced by host genetics, seems to be the most important (Brodie *et al.*, 1992b; de la Concha-Bermejillo *et al.*, 1995). Infected macrophages in the lungs and lymphoid tissues express viral proteins on their surface in close association with MHC antigens. This dual signal is recognized by T cells, leading to the production of inflammatory cytokines that recruit other inflammatory cells, leading to chronic inflammation (Lairmore *et al.*, 1988a; Narayan, 1990; Luján *et al.*, 1994).

During the incubation period, there is an acute viraemia followed by an immune response that restricts virus replication to low levels but fails to eliminate the virus completely (Juste *et al.*, 1998). Seroconversion, detectable by ELISA (DeMartini *et al.*, 1999) or Western blotting assays, usually occurs between 2 and 8 weeks post-infection (Kajikawa *et al.*, 1990; Brodie *et al.*, 1993; Juste *et al.*, 1998, DeMartini *et al.*, 1999). The ensuing decline in viraemia coincident with seroconversion suggests that viraemia is under immune control. However, despite the presence of neutralizing antibodies (Cheevers *et al.*, 1999) and cell-mediated immunity (CMI), OvLV persists in cells of the macrophage lineage of infected animals, often as integrated provirus. Since viral gene expression is activated only when the monocytes mature into macrophages, monocytes form a reservoir of latently infected cells which escape immune surveillance and perpetuate infection (Haase, 1986). This state of cellular latency appears to be controlled by an interaction of cellular and viral transcription factors that regulate viral RNA expression (Gendelman *et al.*, 1985; Staskus *et al.*, 1991; Clements and Zink, 1996).

Cell-mediated immunity is thought to play a more important role in protection against lentivirus infections than humoral immunity. OvLV-specific T-cell proliferation responses to both purified virions and recombinant p25 have been demonstrated (Reyburn *et al.*, 1992; Bird *et al.*, 1993). Using T cells from OvLV-infected animals, it has been shown that OvLV-infected macrophages can induce cytotoxic T-cell (CTL) activity; in turn, these CTLs can kill infected macrophages (Lee *et al.*, 1994; McConnell *et al.*, 1996). This cytotoxic activity is mediated by $CD8^+$ T cells (Blacklaws *et al.*, 1994). However, because a pool of latently infected cells that is inaccessible to T-cell-mediated killing remains in lymphoid organs, CMI fails to eradicate the virus from infected animals. This pool of persistently infected cells gives rise to bursts of virus replication, leading to chronic inflammation.

Role of Virus Strains, Host Genetics and Cofactors in OvLV Pathogenesis

The outcome of lentivirus infections is the result of complex interactions among genetically diverse virus strains, the host genome and the environment. Initial studies showed that OvLV strains that grow slowly and produce little cytopathic effect *in vitro* caused little disease in lambs but more lytic virus strains often produced severe disease (Lairmore *et al.*, 1987, 1988b). Although OvLV strains show a predominant phenotype *in vitro* (lytic or non-lytic), more recent experiments indicate that in an infected individual, lentiviruses are present as a 'quasispecies', a genetically heterogeneous viral population (Woodward *et al.*, 1994; Pieniazek *et al.*, 1995; Clements and Zink, 1996).

The influence of host genetics on the outcome of lentivirus infections has attracted a great deal of attention in recent years. Several major histocompatibility complex (MHC) genes or haplotypes appear to influence disease progression, although the effects are complex and may depend on interactions with other host genes. For example, in humans infected with HIV, the haplotype A1B8DR3 appears to be associated with faster progression to AIDS. In one study, eight of 11 haemophiliacs infected in 1984 who had developed AIDS displayed this haplotype, whereas none of seven HIV+ asymptomatic people carried it (Peixinho and Mendes, 1994). In HIV-infected children, the *DR3* allele is associated with accelerated progression to AIDS while *DPB1* is associated with survival (Just *et al.*, 1995). Similarly, disease progression in simian immunodeficiency virus (SIV)-infected monkeys appears to be influenced by MHC genotype. In SIV-infected monkeys, the lack of the *Mamu-A26* allele correlates with a more rapid progression to SAIDS (Bontrop *et al.*, 1996). In the case of CAEV, a lentivirus closely related to OvLV, susceptibility to arthritis has been associated with differences in frequencies of certain caprine leucocyte antigens (CLAs). Thus, animals of the Saanen breed carrying the CLA Be7 specificity are less prone to develop CAEV-induced arthritis than are goats lacking this specificity (Ruff and Lazary, 1988).

Breed differences in susceptibility to OvLV infection and disease supports the notion that host genetic factors influence the outcome of OvLV infection. Reports suggest that Finnish breeds have a greater tendency to become infected by OvLV than the Ile de France, Rambouillet or Columbia breeds (Gates *et al.*, 1978; Houwers *et al.*, 1989). Breed-related resistance to OvLV-induced disease was suggested by studies in Iceland indicating that progression of lung lesions was relatively delayed in crosses between Icelandic sheep and Border Leicester rams (Palsson, 1976). In contrast to these findings, Border Leicester sheep were found to be more likely than Columbia sheep to develop multisystemic lesions in response to experimental or natural OvLV infection (Cutlip *et al.*, 1986). Furthermore, the Awassi breed of sheep is highly susceptible to infection with OvLV, but in a 20-year study, no pure Awassi sheep has developed disease (Perk *et al.*, 1996). Since interpretation of breed-susceptibility studies is complicated by variation in viral strains and host genetics, further rigorously controlled research would be required to confirm a particular breed or group of sheep as resistant to OvLV infection or disease.

More recent experiments using artificially created isogenic twin lambs have confirmed that host genetic factors may play an important role in determining the extent and severity of OvLV-induced pulmonary lesions (de la Concha-Bermejillo *et al.*, 1995). In this study, the degree of pulmonary LIP at necropsy was independent of the virus strain used for inoculation, and the amount of OvLV proviral DNA in alveolar macrophages correlated with the degree of LIP. This genetic basis for susceptibility or resistance of the host to lentivirus-induced disease could explain many of the differences in disease progression in OvLV-infected sheep.

It is well known that a proportion of sheep in an OvLV-positive flock remain seronegative for life. Although such animals may have never been exposed to OvLV or may not have responded immunologically to infection, it is also possible that they possess innate resistance to OvLV infection. In the case of HIV, recent experimental evidence indicates that in addition to the primary CD4 receptor found on both monocytes/macrophages and lymphocytes, chemokine receptors play a crucial role in virus entry and susceptibility to infection. Macrophage-trophic strains use the CCR5 chemokine receptor while T-cell-trophic strains utilize the CXCR4 chemokine receptor (Bjorndal *et al.*, 1997). A 32 bp deletion has been found in the CCR5 gene, and being homozygous for this deletion appears to confer resistance to infection (Fauci, 1996; Samson *et al.*, 1996; Rana *et al.*, 1997). In one study, 4.5% of highly exposed seronegative individuals were homozygous for this deletion, but none of the HIV-infected people were homozygous for the deletion. Moreover, although heterozygosity did not appear to confer resistance to infection, it did appear to delay the progression to AIDS (Zimmerman *et al.*, 1997). Since only 4.5% of the highly exposed seronegative people were homozygous for the CCR5 mutation, it is probable that other genetic factors influencing infection remain to be discovered. Furthermore, polymorphisms in other chemokine receptor genes may affect HIV disease progression. The role of other cofactors in lentivirus pathogenesis cannot be ignored. In one study, pulmonary abscesses due to *Corynebacterium pseudotuberculosis* infection were present in 53% of OvLV-seropositive sheep, all of which had some degree of LIP, but were only present in 30% of OvLV-seronegative animals, none of which had LIP (Ellis *et al.*, 1990). A synergistic basis for this was suggested by data indicating that *C. pseudotuberculosis* induced the secretion of tumour necrosis factor-α (TNF-α) by ovine macrophages which, in turn, enhanced OvLV replication (Ellis *et al.*, 1991). Mycoplasma agents also have been associated with LIP in OvLV-infected sheep (DeMartini *et al.*, 1993). These findings suggest a role for intercurrent bacterial infections in the pathogenesis of OvLV-induced disease.

Transmission of Ovine Lentivirus and Traditional Methods of Control

Transmission of OvLV commonly occurs via lactogenic or aerosol routes, but the importance of each route in the epizootiology of the disease is debatable.

In flocks where the infection is enzootic, the primary route of infection seems to be from ewe to lamb through infected colostrum (Petursson *et al*., 1992); cell-associated OvLV can be detected in the milk of infected ewes (Ouzrout and Lerondelle, 1990; Brodie *et al*., 1994). On the other hand, serological surveys throughout different sheep-producing areas of the USA show a wide range of OvLV prevalence, indicating that other factors and routes of transmission may play a role in the epidemiology of the disease. Because a positive correlation has been found between the proportion of sheep seropositive to OvLV and age of infected animals, it has been speculated that OvLV spreads laterally, probably by the respiratory route (Cutlip *et al*., 1977, 1992). However, in a recent study in Texas in which eight OvLV-free lambs were placed in close contact in open pens with 32 experimentally inoculated lambs for 8 months, none of the contact controls became infected (de la Concha-Bermejillo, unpublished observation). In addition, OvLV prevalence in Texas, where 20% of all US sheep are raised, is less than 1% compared to a 26% national average (Cutlip *et al*., 1992; de la Concha-Bermejillo *et al*., 1998). In Texas, most sheep are raised under extensive conditions in a mostly dry and hot climate, which suggests that housing and close contact during lambing in cold weather may play an important role in OvLV transmission. In a European study of flocks with mixed infected and uninfected ewes, approximately 37% of the offspring of infected ewes and about 20% of the offspring of uninfected ewes became OvLV-infected within 1 year (Houwers *et al*., 1989). Close confinement in winter housing, perhaps combined with genetic susceptibility of Icelandic sheep breeds, was suspected to have been a factor contributing to the explosive OvLV outbreak in Iceland (Palsson, 1976). Although less common, vertical transmission from mother to fetus has also been reported (Cutlip *et al*., 1981; Brodie *et al*., 1994). Venereal transmission of OvLV has not been reported; however, recent information indicates that OvLV-inoculated animals co-infected with *Brucella ovis* can shed the virus in semen (de la Concha-Bermejillo *et al*., 1996).

To date, the only effective means for controlling OvLV-induced disease has been through removal of infected sheep or prevention of spread of the virus. This approach was undertaken on a large scale in Iceland in the 1950s when over 600,000 sheep in infected flocks were slaughtered and the farms were repopulated with unexposed sheep (Palsson, 1976). More recently, OvLV infection has been eliminated from flocks by removing lambs from seropositive ewes before nursing and rearing them in isolation, or by repeated testing and culling of seropositive ewes and their progeny (Houwers *et al*., 1983, 1984, 1987; Cutlip and Lehmkuhl, 1986). Voluntary eradication programmes based on these methods have been established in The Netherlands, Great Britain and Belgium, and have achieved considerable success. However, since such methods are expensive and are not readily adopted by farmers, they may not be as successfully employed in countries with larger sheep populations where the seroprevalence of OvLV infection is high. This necessitates the development of alternative strategies for control of OvLV (Blacklaws *et al*., 1995).

Strategies for the Control of Ovine Lentivirus Infections in Individual Animals

Presently, there are no effective vaccines or treatments against OvLV infections. Developing a vaccine to protect against OvLV infection or disease has been a challenging problem because of antigenic variation, latency and the complex interactions that lentiviruses have evolved with their host. Studies of unsuccessful vaccination against OvLV in sheep have involved heat-, formalin- or ethyleneimine-inactivated vaccines, with or without Freund's incomplete or aluminium hydroxide adjuvant (Cutlip *et al.*, 1987; Pearson *et al.*, 1989). In one study, sheep immunized with purified virions homogenized in Freund's complete adjuvant developed more severe CNS lesions than the controls, suggesting a role for cell-mediated immunity in the development of CNS lesions (Nathanson *et al.*, 1981). Alternative novel methods of vaccination against OvLV, such as the use of naked proviral DNA, are presently being studied (Perk *et al.*, 1996).

Several compounds, including phosphonoformate, 2′,3′-dideoxynuclosides and α-interferon, have been shown to inhibit OvLV replication *in vitro* (Frank *et al.*, 1987). In addition, recombinant ovine τ-interferon has been shown to have antiviral activity against OvLV (Juste *et al.*, 1996). However, due to the irreversible nature of the infection, secondary to the integration of the virus genome into the host cell's chromosomes, drug-based approaches for the treatment of OvLV infection would appear to have no practical application.

Cytokine-expressing viruses have been envisioned as a novel approach for the development of safe and efficacious live-attenuated vaccines against retroviruses (Giavedoni *et al.*, 1992, 1997). Retroviral vectors efficiently transfer gene sequences into cells and promote their stable expression (Naldini *et al.*, 1996). Some simple retroviruses that have been engineered are replication-competent; others need helper packaging cell lines. A replication-competent SIV with a deletion in the nef gene ($SIV_{\Delta nef}$) that expresses high levels of human γ-interferon (γ-IFN) (SIV_{HyIFN}) has been constructed (Giavedoni and Yilma, 1996). Rhesus macaques vaccinated with SIV_{HyIFN} had a lower viral load than macaques similarly immunized with $SIV_{\Delta nef}$. Viral loads remained low in the SIV_{HyIFN}-vaccinated group and the animals remained healthy for more than 32 weeks after challenged with virulent SIV_{mac251}, suggesting that the modified virus expressing γ-IFN provided effective protection from subsequent infection by virulent strains of virus (Giavedoni *et al.*, 1997).

Gene therapeutics, using replication-defective lentiviruses and packaging cell lines, is the latest addition to a multitude of approaches that could be used to combat lentivirus infections. Replication-defective lentiviruses are seen as a safer approach for gene delivery than replication-competent retroviruses. Multiplasmid expression systems can be used to generate lentivirus-derived vector particles by transient transfection that result in stable *in vivo* gene transfer into terminally differentiated cells (Naldini *et al.*, 1996).

Selection and Breeding of Ovine Lentivirus-resistant Animals

During the past few decades, animal breeders have been very successful in breeding animals that are superior in the production of milk, eggs, meat and wool. However, selection for disease resistance has been largely ignored (Gavora and Spencer, 1983). Selection for resistance to infectious disease requires the identification of specific genes or genetic markers linked to resistance either to infection or to disease.

Establishing a selection programme for resistance to ovine lentivirus infection is difficult because the OvLV viral receptor is unknown. However, one way to identify animals that may be resistant to OvLV infection would be to study the small percentage of animals in a highly seropositive flock that remain seronegative for life. Although such animals may never have been exposed to OvLV or may have become infected early in life with development of tolerance to the virus, they also may possess innate resistance to OvLV infection, perhaps because of a lack of specific viral receptors. If this can be documented, the progeny of animals that remain seronegative in an infected flock should be selected for breeding stock. Of course, production characteristics of the foundation flock must also be taken into account.

Another selection strategy would be to propagate progeny of sheep that become infected by OvLV (and seroconvert) but do not develop disease. In flocks naturally infected with OvLV, a proportion of infected animals do not develop disease, suggesting that there are factors that make individuals resistant to OvLV-induced disease (DeMartini *et al.*, 1991). Previous research indicates that resistance or susceptibility to OvLV is the result of host genetic factors (de la Concha-Bermejillo *et al.*, 1995). Therefore, the identification of genetic markers that could be used to recognize disease-resistant animals would be a powerful tool for the development of resistant breeding stock. To identify genetic markers that correlate with disease resistance, families of sheep that have segregating alleles for resistance or susceptibility to OvLV-induced disease would need to be created. One approach would be to inoculate sheep that have different genetic backgrounds with OvLV and to evaluate their response, to estimate variation within these animals. Because signs of OvLV-induced disease are non-specific, usually develop late in life, and because animals would have to be killed to evaluate severity of disease, a marker for prediction of OvLV-induced disease would need to be used. Presently, virus load is considered the best predictor of the severity of disease (Brodie *et al.*, 1992b; de la Concha-Bermejillo *et al.*, 1995). In this scenario, the male with the highest response (high virus load) could be mated to the females with the lowest response (low virus load) and the male with the lowest response could be mated to the females with the highest response. The resulting F_1 progeny of these matings would each have alleles for resistance, from the low-response parent, and susceptibility, from the high-response parent. F_1 males then could be mated to unrelated females. The resulting paternal half-sib families of progeny would then be a useful resource for linkage analysis because there will be segregation of putative alleles. Genes that could be targeted for analysis in the created families include tumour necrosis factor-α (TNF-α), interleukin-1

(IL-1), interleukin-2 (IL-2), interleukin-6 (IL-6) and MHC genes. All these genes have been shown to play an important role in the immune response against lentiviruses. A similar approach could be used to select animals that are resistant to infection.

Strategies for Developing Ovine Lentivirus-resistant Transgenic Sheep

Controlling an infectious disease by genetic modification of the host requires identification or development of a 'resistance gene' and methods for introducing that DNA into the germ line of animals such that it will be appropriately expressed in the required target cells. Ultimately, hemizygous founder animals bearing such transgenes must be amplified to generate populations of resistant animals by production of homozygotes, possibly through the use of embryo transfer and artificial insemination, or by deriving offspring from fetal or adult cells.

Methods for producing transgenic sheep

Production of transgenic mice has become a widely used and powerful technique for studying normal and abnormal gene expression (Camper, 1987; Hanahan, 1989). Most commonly, a sequence of DNA containing both regulatory and coding regions is cut from a plasmid and purified, then microinjected into one pronucleus of a recently fertilized, one-cell embryo. In a fraction of surviving embryos, the transgene has been integrated into the genome and is carried in both somatic and germ cells. Assuming the transgene integrates into a site that does not silence transcription, expression is dependent largely on the regulatory region of the transgene. For example, the visna LTR has been shown to direct expression of a reporter gene in macrophages, lymphocytes and the central nervous system of transgenic mice (Small *et al.*, 1989).

Microinjection of DNA has been used by several groups to produce transgenic sheep (Pursel and Rexroad, 1993; Clements *et al.*, 1994; Janne *et al.*, 1994). Success, as measured by the number of transgenic lambs obtained per injected ova, has been much lower than that obtained with mice, generally less than 1%. As a result of this low efficiency and the expense of animal maintenance, production of transgenic sheep is very much more costly than production of transgenic mice.

Another approach to production of transgenic animals is infection of the embryo with a recombinant retrovirus. Infection is followed at very high frequency by integration of the corresponding provirus, which typically contains a transgene flanked by retroviral LTRs. This technique has been used to produce transgenic mice (Van der Putten *et al.*, 1985; Soriano *et al.*, 1986), and infection of early bovine embryos with such viruses resulted in a high percentage of transgenic embryos and fetuses (Haskell and Bowen, 1995). Retroviral infection of early embryos is clearly an efficient means of producing transgenic animals, but several important limitations exist, among which are the high incidence of mosaicism in transgenic offspring and the limited capacity of the virus for insertion of foreign sequences. None the less, this technique may be of value for production of transgenic sheep, particularly for research purposes.

A final strategy for production of transgenic sheep is to introduce the transgene into cells in culture, then use those cells to reconstitute a whole animal. Remarkable progress has recently been attained in cloning sheep by fusion of enucleated oocytes with fibroblasts. Use of fetal fibroblasts appears most useful, but one cloned sheep appears to have derived from fusion of an oocyte with a cell taken from an adult ewe (Wilmut *et al.*, 1997). If a transgene is introduced into the fibroblasts prior to fusion with oocytes, the resulting animals will be transgenic. Already this technique has been applied to production of sheep transgenic for human factor IX (Schnieke *et al.*, 1997), and initial studies suggest that it may be significantly more efficient than DNA microinjection for production of transgenic ruminants.

Several additional constraints must be considered with regard to producing populations of any type of transgenic sheep. First, even with the use of modern reproduction technology, such projects will take considerable time, particularly if homozygous breeding animals are desired. Secondly, one must expect that a fraction of transgenic animals produced will not express the transgene at levels sufficient to alter phenotype, and an additional number will have some type of defect, due to problems such as insertional mutagenesis. Finally, a disease-resistant sheep will be of marginal value if production characteristics are sacrificed or neglected during the years required to generate a population of transgenic animals.

Approaches to engineering lentivirus resistance

A number of gene therapy approaches have been developed against HIV. Most of these involve expression of genes that interfere with some aspect of the infectious process. This has sometimes been called intracellular immunization. Although it seems unlikely that gene therapy will ever be economically feasible in sheep, some of the intracellular immunization strategies might be the basis for transgenic strategies to modify the germ line to develop sheep that are resistant to ovine lentiviral infections.

Virus structural protein genes

One of the prominent strategies for developing virus resistance has been to develop transgenic organisms that express a critical structural protein of the virus. The rationale for this approach is that many viruses show interference to superinfection. That is, infection with one virus will prevent subsequent infection with related viruses. For example, expression of viral envelope glycoproteins has been used to protect chickens against avian leucosis retroviruses. Robinson *et al.* (1981) showed that some strains of chickens that carried defective endogenous proviruses with a type E envelope were resistant to infection with other type E retroviruses. This resistance correlated with expression of the envelope glycoprotein of the endogenous virus. Transgenic chickens have been developed that express the type A envelope from a defective provirus (Crittenden *et al.*, 1989; Salter and Crittenden, 1989). These chickens are resistant to infection by subgroup A but not subgroup B avian leucosis virus. The molecular basis of this is thought to be due to interaction of the envelope protein with receptor proteins either on the surface of the cell or in the endoplasmic reticulum (Federspiel *et al.*, 1989; Delwart *et al.*, 1992). This interaction blocks the binding of the virus to receptor protein and therefore interferes with entry of the virus into the cell.

Replication interference between OvLV subtypes has been described (Jolly and Narayan, 1989). OvLV strains related by phenotype and genotype cross-interfered whereas different subtypes did not, thus permitting superinfection of cultured cells or animals by different lentivirus subtypes; the interference seemed to occur at the level of binding of the virus to the cell. By analogy with the avian leucosis virus system, transgenic sheep expressing the gene for OvLV envelope glycoprotein could be protected from OvLV infection. Three transgenic sheep containing the visna virus envelope gene under the control of the visna LTR promoter have been constructed (Clements *et al.*, 1994). The transgenic sheep were healthy, and no deleterious effects or clinical abnormalities from the transgene were observed. Expression of Env glycoprotein was observed in *in vitro* cultured differentiated macrophages and lymphocytes from the animals. Expression was also observed in a number of tissues including lung, spleen, brain and several others (Clements *et al.*, 1996). Two of the animals developed antibodies against the Env glycoprotein. Although lentivirus glycoproteins tend to cause cell fusion and syncytium formation *in vitro*, this did not seem to be a problem *in vivo*. To date, challenge studies to determine the susceptibility of these animals or their offspring to OvLV have not been reported.

Virus regulatory protein genes

A second strategy might be to express a viral non-structural protein in the transgenic animal. As discussed above, lentiviruses have regulatory proteins that modulate viral gene expression during an infection. Generally these are transacting factors that control the temporal order and tissue specificity of expression of the viral genes at either a transcriptional or post-transcriptional level. Cell culture studies suggest that expression of certain types of mutant regulatory proteins can interfere with the normal course of a viral infection. These mutants must be transdominant and thus able to disrupt the function of normal regulatory proteins in an infection. In both OvLV and HIV, the regulatory proteins of the *tat* and *rev* genes are necessary for a successful infection (Neuveut *et al.*, 1993; Toohey and Haase, 1994). The Tat protein is a transacting transcriptional regulatory protein required for efficient transcription from the LTR promoter. Rev protein seems to be necessary for transport of singly spliced and unspliced viral RNAs from the nucleus to the cytoplasm where they can act as mRNAs. During the course of an infection, the Rev protein must accumulate in order for the synthesis of viral structural proteins to proceed. It seems to function by binding to a site, the Rev responsive element (RRE), in these RNAs to mediate their transport into the cytoplasm for translation. Certain HIV *tat* and *rev* mutants are deficient in their functions and are transdominant and thus prevent either synthesis or transport of RNA from the nucleus (Green and Ishina, 1989; Malim *et al.*, 1989). Thus cells and perhaps transgenic animals that express these mutant proteins would not allow the normal growth of the virus because of aberrant regulation at either the transcriptional or post-transcriptional level.

The visna virus Tat protein influences transcription from the viral LTR promoter and cellular promoters (Neuveut *et al.*, 1993). Thus the effect of dominant mutants on the expression of both viral and cellular genes must be considered. Transgenic mice that express the wild-type visna Tat protein have been constructed (Velutini *et al.*, 1994). These mice show lymphoproliferative disorders in several tissues, including lung, spleen, lymph nodes and skin. Whether transdominant *tat* mutants would cause these lesions is unknown. Domains of the Tat protein associated with transcriptional and pathogenic effects have been defined (Carruth *et al.*, 1994; Philippon *et al.*, 1994), and careful attention to these properties may allow the design and construction of mutant proteins with the desired properties.

The visna virus Rev protein may be an even more attractive target for intracellular immunization strategies. The Rev protein shares its amino-terminal 48 amino acids with the Env protein precursor, and the exon encoding the remaining 119 amino acids is contained within the *env* gene, though in a different reading frame from the Env protein. The transgenic sheep with the visna *env* gene described above (Clements *et al.*, 1994), therefore, carried a functional *rev* gene, and Rev protein expression was detected in the sheep. Indeed Rev protein is necessary for expression of Env protein (Schoborg and Clements, 1994; Toohey and Haase, 1994). Since the transgenic sheep did not show any obvious pathological effects, it does not appear that *rev* expression is deleterious to the sheep, and it seems unlikely that the expression of transdominant *rev* mutants would cause problems. Two other anti-Rev strategies have been devised for intracellular immunization against HIV (Inouye *et al.*, 1997). Single-chain antibodies have been constructed that bind Rev protein and interfere with its function in the cell. Also an RNA decoy consisting of the minimal Rev-binding domain of the RRE was developed. When either of these elements was expressed in cultured cells, HIV replication was significantly reduced, and when both were expressed together, HIV production was almost eliminated. Transgenic sheep expressing similar constructs targeting the visna Rev protein might be resistant to OvLV infection.

Viral antisense RNA

Another strategy for interfering with viral infections is to express antisense RNA representing portions of viral RNA. Antisense RNA presumably forms a double-stranded RNA with the target RNA which can disrupt normal function of the target in the cell. Several naturally occurring antisense mechanisms have been described.

1. Antisense RNA can prevent the normal folding of an RNA molecule. The frequency of initiation of plasmid Col E1 DNA replication is controlled by an antisense mechanism (Polisky, 1988). In this case, the RNA primer for replication must be folded into a certain conformation to act as a primer. An antisense RNA to a portion of the primer prevents this folding and thus regulates the frequency of initiation.
2. Antisense RNA can interfere with translation of an mRNA. The frequency of translation of the mRNA for the Tn10 transposase is controlled by an antisense RNA that binds to the ribosome-binding site of the mRNA and prevents initiation of translation (Simons and Kleckner, 1983).
3. Antisense RNA can target an RNA molecule for modification and perhaps degradation. In *Xenopus* oocytes, a portion of the gene for basic fibroblast growth factor is transcribed in both directions. This allows association of the two transcripts in this region, which is then a substrate for a modifying enzyme that converts adenine residues in double-stranded RNA to inosines. This disrupts the double strands and results in inactivation of the mRNA (Kimelman and Kirschner, 1989).

The design of artificial antisense mechanisms for altering the course of virus infections has been attempted for several viruses, including HIV. Vectors that express antisense RNA to various portions of the HIV genome have been designed and used to stably transform cultured cells (Liu *et al.*, 1997; Veres *et al.*, 1998). Although these cells are resistant to challenge by HIV to various degrees, there is still considerable room for optimization and refinement of the expression constructs.

Ribozymes

A variation on the antisense strategy is to design and express ribozymes that cleave viral RNA within the infected cell. Certain RNA molecules have been shown to be capable of performing enzymic reactions. Among these are several plant satellite virus and virusoid and viroid RNAs that cleave their own replicative form RNAs during replication to generate progeny genomes. Based upon studies to locate and define the active sites of these self-cleaving RNAs (Forster and Symons, 1987; Hampel and Tritz, 1989), it has been possible to design ribozymes that bind RNA targets by base pairing and carry the active site for cleaving the RNA at the binding site (Uhlenbeck, 1987; Haseloff and Gerlach, 1988; Hampel *et al.*, 1990). Since these ribozymes are RNA molecules themselves, they could be expressed in cells and would cleave any substrate RNA that they encounter. Cleavage of HIV-1 genomes by ribozymes has been reported in HIV-infected cell cultures (e.g. Sarver *et al.*, 1990; Gervaix *et al.*, 1997; Smith *et al.*, 1997), and it seems likely that ribozymes could be designed against ovine lentiviruses as well. Once again there is considerable opportunity for refinement of ribozyme strategies. For example, targeting the ribozyme to the appropriate intracellular compartment is critical to successful knockout of viruses by ribozymes (Sullenger and Cech, 1993).

Other strategies

An additional strategy to develop resistant transgenic animals would be to design a toxin gene which would be expressed only in virus-infected cells (e.g. under control of virus-specific transactivating factors). Such cells would commit suicide by expressing the toxin before allowing the virus to multiply. This would prevent the spread of the virus to other cells. Another possibility would be to express a soluble form of the virus receptor which would circulate in the bloodstream and bind virus before it could infect cells, and thus protect the animal. This receptor could also be coupled to antibody Fc regions to assist in destruction of virus or infected cells by the immune system (Byrn *et al.*, 1990). These types of experiments are highly speculative and little has been done to test these concepts at the organism level.

Acknowledgements

This work was supported in part by Public Health Service grant R01 HL53244 from the National Institutes of Health and by USDA grant USDA-CSRS 90-37266-5566.

References

Banks, K.L., Adams, D.S., McGuire, T.C. and Carlson, J. (1983) Experimental infection of sheep by caprine arthritis-encephalitis virus and goats by progressive pneumonia virus. *American Journal of Veterinary Research* 44, 2307–2311.

Bird, P., Blacklaws, B., Reyburn, H.T., Allen, D., Hopkins, J., Sargan, D. and McConnell, I. (1993) Early events in immune evasion by the lentivirus maedi–visna occurring within infected lymphoid tissue. *Journal of Virology* 67, 5187–5197.

Bjorndal, A., Deng, H., Jansson, M., Fiore, J.R., Colognesi, C., Karlsson, A., Albert, J., Scarlatti, G., Littman, D.R. and Fenyo, E.M. (1997) Coreceptor usage of primary human immunodeficiency virus type 1 isolates varies according to biological phenotype. *Journal of Virology* 71, 7478–7487.

Blacklaws, B., Bird, P. and McConnell, I. (1994) Pathogenesis and immunity in lentivirus infections of small ruminants. In: Goddeeris, B.M.L. and Morrison, W.I. (eds) *Cell-mediated Immunity in Ruminants.* CRC Press, Boca Raton, pp. 199–212.

Blacklaws, B.A., Bird, P., Allen, D., Roy, D.J., MacLennan, C.M., Hopkins, J., Sargan, D.R. and McConnell, I. (1995) Initial lentivirus–host interactions within lymph nodes: a study of maedi–visna virus infection in sheep. *Journal of Virology* 69, 1400–1407.

Bontrop, R.E., Otting, N., Niphuis, H., Noort, R., Teeuwsen, V. and Heeney, J.L. (1996) The role of major histocompatibility complex polymorphisms on SIV infection in rhesus macaques. *Immunology Letters* 51, 35–38.

Braun, M.J., Clements, J.E. and Gonda, M.A. (1987) The visna virus genome: evidence for a hypervariable site in the *env* gene and sequence homology among lentivirus envelope proteins. *Journal of Virology* 61, 4046–4054.

Brodie, S.J., Snowder, G.D. and DeMartini, J.C. (1992a) Ovine progressive pneumonia: advances and prospects for control. *Sheep Research Journal* 8, 116–126.

Brodie, S.J., Marcom, K.A., Pearson, L.D., Anderson, B.C., de la Concha-Bermejillo, A., Ellis, J.A. and DeMartini, J.C. (1992b) Effects of virus load in the pathogenesis of lentivirus-induced lymphoid interstitial pneumonia. *Journal of Infectious Diseases* 166, 531–541.

Brodie, S.J., Pearson, L.D., Snowder, G.D. and DeMartini, J.C. (1993) Host–virus interaction as defined by amplification of viral DNA and serology in lentivirus-infected sheep. *Archives of Virology* 130, 413–428.

Brodie, S.J., de la Concha-Bermejillo, A., Koenig, G., Snowder, G.D. and DeMartini, J.C. (1994) Maternal factors associated with prenatal transmission of ovine lentiviruses. *Journal of Infectious Diseases* 169, 531–541.

Brodie, S.J., Bickle, H.M. and DeMartini, J.C. (1995) Virological markers in cerebrospinal fluid are predictive of ovine lentivirus-associated subclinical encephalomyelitis. *Clinical Immunology and Immunopathology* 77, 14–18.

Brodie, S.J., de la Concha-Bermejillo, A., Snowder, G.D. and DeMartini, J.C. (1998) Current concepts in the epizootiology, diagnosis, and economic importance of ovine progressive pneumonia in North America: a review. *Small Ruminant Research* 27, 1–17.

Bulgin, M.S. (1990) Ovine progressive pneumonia, caprine arthritis-encephalitis, and related lentiviral diseases of sheep and goats. *Veterinary Clinics of North America. Food Animal Practice* 6, 691–704.

Byrn, R.A., Mordenti, J., Lucas, C., Smith, D., Maesters, S.A., Johnson, J.S., Cossum, P., Chemow, S.M., Wurm, F.M., Gregory, T., Groopman, J.E. and Capon, D.J. (1990) Biological properties of a CD4 immunoadhesin. *Nature* 344, 667–670.

Camper, S.A. (1987) Research applications of transgenic mice. *Biotechniques* 7, 638–650.

Carruth, L.M., Hardwick, J.M., Morse, B.A. and Clements, J.E. (1994) Visna virus tat protein: a potent transcription factor with both activator and suppressor domains. *Journal of Virology* 68, 6137–6146.

Cheevers, W.P., Cordery-Cotter, R., McGuire, T.C. and DeMartini, J.C. (1999) Neutralizing antibody responses and evolution of antigenic variants in monozygotic twin lambs infected with phenotypically distinct ovine lentiviruses. *Virology* 258, 382–388.

Clements, J.E. and Zink, M.C. (1996) Molecular biology and pathogenesis of animal lentivirus infections. *Clinical Microbiology Reviews* 9, 100–117.

Clements, J.E., Wall, R.J., Narayan, O., Hauer, D., Schoborg, R., Sheffer, D., Powell, A., Carruth, L.M., Zink, M.C. and Rexroad, C.E. (1994) Development of transgenic sheep that express the visna virus envelope gene. *Virology* 200, 370–380.

Clements, J.E., Hu, L., Lindstrom, L., Powell, A., Rexroad, C. and Zink, M.C. (1996) Molecular studies of visna virus gene expression: analysis of envelope gene expression in transgenic mice. *AIDS Research and Human Retroviruses* 12, 421–423.

Crittenden, L.B., Salter, D.W. and Federspiel, M.J. (1989) Segregation, viral phenotype, and proviral structure of 23 avian leukosis virus inserts in the germline of chickens. *Theoretical and Applied Genetics* 77, 505–515.

Cutlip, R.C. and Lehmkuhl, H.D. (1986) Eradication of ovine progressive pneumonia from sheep flocks. *Journal of the American Veterinary Medical Association* 188, 1026–1027.

Cutlip, R.C., Jackson, T.A. and Laird, G.A. (1977) Prevalence of ovine progressive pneumonia in a sampling of cull sheep from Western and Midwestern United States. *American Journal of Veterinary Research* 38, 2091–2093.

Cutlip, R.C., Lehmkuhl, H.D. and Jackson, T.A. (1981) Intrauterine transmission of ovine progressive pneumonia virus. *American Journal of Veterinary Research* 42, 1795–1797.

Cutlip, R.C., Lehmkuhl, H.D., Brogden, K.A. and Sacks, J.M. (1986) Breed susceptibility to ovine progressive pneumonia (maedi/visna) virus. *Veterinary Microbiology* 12, 283–288.

Cutlip, R.C., Lehmkuhl, H.D., Brogden, K.A. and Schmerr, M.J.F. (1987) Failure of experimental vaccines to protect against infection with ovine progressive pneumonia (maedi–visna) virus. *Veterinary Microbiology* 13, 201–204.

Cutlip, R.C., Lehmkuhl, H.D., Sacks, J.M. and Weaver, A.L. (1992) Seroprevalence of ovine progressive pneumonia virus in sheep in the United States as assessed by analyses of voluntarily submitted samples. *American Journal of Veterinary Research* 53, 976–979.

Dawson, M. (1980) Maedi/visna: a review. *Veterinary Record* 106, 212–216.

Dawson, M. (1987) Pathogenesis of maedi-visna. *Veterinary Record* 120, 451–454.

de la Concha-Bermejillo, A. (1997) Maedi–visna and ovine progressive pneumona. *Veterinary Clinics of North America. Food Animal Practice* 13, 13–33.

de la Concha-Bermejillo, A., Brodie, S.J., Magnus-Corral, S., Bowen, R.A. and DeMartini, J.C. (1995) Pathologic and serologic responses of isogeneic twin lambs to phenotypically distinct lentiviruses. *Journal of Acquired Immune Deficiency Syndromes* 8, 116–123.

de la Concha-Bermejillo, A., Magnus-Corral, S., Brodie, S.J. and DeMartini, J.C. (1996) Venereal shedding of ovine lentivirus in infected rams. *American Journal of Veterinary Research* 57, 684–687.

de la Concha-Bermejillo, A., Shelton, M. and DeMartini, J.C. (1998) Seroprevalence of ovine progressive pneumonia in Texas. *Sheep and Goat Research Journal* 14, 127–132.

Delwart, E.L., Buchschacher, G.L.J., Freed, E.O. and Panganiban, A.T. (1992) Analysis of HIV-1 envelope mutants and pseudotyping of replication-defective HIV-1 vectors by genetic complementation. *AIDS Research and Human Retroviruses* 8, 1669–1677.

DeMartini, J.C., Bowen, R.A., Carlson, J.O. and de la Concha-Bermejillo, A. (1991) Strategies for the genetic control of ovine lentivirus infections. In: Owen, J.B. and Axford, R.F.E. (eds) *Breeding for Disease Resistance in Farm Animals.* CAB International, Wallingford, UK, pp. 293–314.

DeMartini, J.C., Brodie, S.J., de la Concha-Bermejillo, A., Ellis, J.A. and Lairmore, M.D. (1993) Pathogenesis of lymphoid interstitial pneumonia in natural and experimental ovine lentivirus infection. *Clinical Infectious Diseases* 17, 236–242.

DeMartini, J.C., Halsey, W., Boshoff, C., York, D. and Howell, M. (1999) Comparison of a maedi-visna virus CA-TM fusion protein ELISA with other assays for detecting sheep infected with North American ovine lentivirus strains. *Veterinary Immunology and Immunopathology*, (in press).

Ellis, J.A. and DeMartini, J.C. (1985) Immunomorphologic and morphometric changes in pulmonary lymph nodes of sheep with progressive pneumonia. *Veterinary Pathology* 22, 32–41.

Ellis, J.A., Hawk, D.A., Holler, L.D., Mills, K.W. and Pratt, D.L. (1990) Differential antibody responses to *Corynebacterium pseudotuberculosis* in sheep with naturally acquired caseous lymphadenitis. *Journal of the American Veterinary Medical Association* 196, 1609–1613.

Ellis, J.A., Lairmore, M.D., O'Toole, D.T. and Campos, M. (1991) Differential induction of tumor necrosis factor alpha in ovine pulmonary alveolar marcophages following infection with *Corynebacterium pseudotuberculosis*, *Pasteurella haemolytica*, or lentivirus. *Infection and Immunology* 59, 3254–3260.

Fauci, A.S. (1996) Resistance to HIV-1 infection: it's in the genes. *Nature Medicine* 2, 966–967.

Federspiel, M.J., Crittenden, L.B. and Hughes, S.H. (1989) Expression of avian reticuloendotheliosis virus envelope confers host resistance. *Virology* 173, 167–177.

Forster, A.C. and Symons, R.H. (1987) Self cleavage of plus and minus RNAs of a virusoid and a structural model for the active sites. *Cell* 49, 211–220.

Frank, K.B., McKernan, P.A., Smith, R.A. and Smee, D.F. (1987) Visna virus as an *in vitro* model for human immunodeficiency virus and inhibition by ribavirin, phosphonoformate, and 2′,3′-dideoxynucleosides. *Antimicrobial Agents and Chemotherapy* 31, 1369–1374.

Gates, N.L., Winward, L.D., Gorham, J.R. and Shen, D.T. (1978) Serologic survey of prevalence of ovine progressive pneumonia in Idaho range sheep. *Journal of the American Veterinary Medical Association* 173, 1575–1577.

Gavora, J.S. and Spencer, J.L. (1983) Breeding for immune responsiveness and disease resistance. *Animal Blood Groups Biochemistry Genetics* 14, 159.

Gendelman, H.E., Narayan, O., Molineaux, S., Clements, J.E. and Ghotbi, Z. (1985) Slow, persistent replication of lentiviruses: role of tissue macrophages and macrophage precursors in bone marrow. *Proceedings of the National Academy of Sciences USA* 82, 7086–7090.

Georgsson, G. (1994) Neuropathological aspects of lentiviral infections. *Annals of the New York Academy of Sciences* 724, 50–67.

Gervaix, A., Kraus, G. and Wong-Staal, F. (1997) Multigene antiviral vectors inhibit diverse human immunodeficiency virus type 1 clades. *Journal of Virology* 71, 3048–3053.

Giavedoni, L.D. and Yilma, T. (1996) Construction and characterization of replication-competent simian immunodeficiency virus vectors that express gamma interferon. *Journal of Virology* 70, 2247–2251.

Giavedoni, L.D., Jones, L., Gardner, M.B., Gibson, H.L., Ng, C.T., Barr, P.J. and Yilma, T. (1992) Vaccinia virus recombinants expressing chimeric proteins of human immunodeficiency virus and gamma interferon are attenuated for nude mice. *Proceedings of the National Academy of Sciences USA* 89, 3409–3413.

Giavedoni, L., Ahmad, S., Jones, L. and Yilma, T. (1997) Expression of gamma interferon by simian immunodeficiency virus increases attenuation and reduces postchallenge virus load in vaccinated rhesus macaques. *Journal of Virology* 71, 866–872.

Green, M.R. and Ishina, M.L.P.M. (1989) Mutational analysis of HIV-1 tat minimal domain peptides: identification of trans-dominant mutants that suppress HIV-LT R-driven gene expression. *Cell* 58, 215–223.

Haase, A.T. (1986) Pathogenesis of lentivirus infections. *Nature* 322, 130–136.

Hampel, A. and Tritz, R. (1989) RNA catalytic properties of the minimum (-)sTRSV sequence. *Biochemistry* 28, 4929–4933.

Hampel, A., Tritz, R., Hicks, M. and Cruz, P. (1990) 'Hairpin' catalytic RNA model: evidence for helices and sequence requirement for substrate RNA. *Nucleic Acids Research* 18, 299–304.

Hanahan, D. (1989) Transgenic mice as probes into complex systems. *Science* 246, 1265–1275.

Haseloff, J. and Gerlach, W.L. (1988) Simple RNA enzymes with new and highly specific endoribonuclease activities. *Nature* 334, 585–591.

Haskell, R.E. and Bowen, R.A. (1995) Efficient production of transgenic cattle by retroviral infection of early embryos. *Molecular Reproduction and Development* 40, 386–390.

Houwers, D.J., Konig, C.K., De Boer, G.F. and Schaake, J.J. (1983) Maedi–visna control in sheep. I. Artificial rearing of colostrum-deprived lambs. *Veterinary Microbiology* 8, 179–185.

Houwers, D.J., Schaake, J.J. and De Boer, G.F. (1984) Maedi–visna control in sheep II.

Half-yearly serological testing with culling of positive ewes and progeny. *Veterinary Microbiology* 9, 445–451.

Houwers, D.J., König, C.D.W., Bakker, J., de Boer, M.J., Pekelder, J.J., Sol, J., Vellema, P. and de Vries, G. (1987) Maedi–visna control in sheep III: results and evaluation of a voluntary control program in the Netherlands over a period of four years. *Veterinary Quarterly* 9, 29S–36S.

Houwers, D.J., Visscher, A.H. and Defize, P.R. (1989) Importance of ewe/lamb relationship and breed in the epidemiology of maedi–visna virus infections. *Research in Veterinary Science* 46, 5–8.

Inouye, R.T., Du, B., Boldt-Houle, D., Ferrante, A., Park, I.W., Hammer, S.M., Duan, L., Groopman, J.E., Pomerantz, R.J. and Terwilliger, E.F. (1997) Potent inhibition of human immunodeficiency virus type 1 in primary T cells and alveolar macrophages by a combination anti-rev strategy delivered in an adeno-associated virus vector. *Journal of Virology* 71, 4071–4078.

Janne, J., Hyttinen, J.M., Peura, T., Tolvanen, M., Alhonen, L., Sinervirta, R. and Halmekyto, M. (1994) Transgenic bioreactors. *International Journal of Biochemistry* 26, 859–870.

Jolly, P.E. and Narayan, O. (1989) Evidence for interference, coinfections, and intertypic virus enhancement of infection by ovine–caprine lentiviruses. *Journal of Virology* 63, 4682–4688.

Just, J.J., Abrams, E., Louie, L.G., Urbano, R., Wara, D., Nicholas, S.W., Stein, Z. and King, M.C. (1995) Influence of host genotype on progression to acquired immunodeficiency syndrome among children infected with human immunodeficiency virus type 1. *Journal of Pediatrics* 127, 544–549.

Juste, R.A., Kwang, J. and de la Concha-Bermejillo, A. (1995) Comparative evaluation of the agar gel immunodiffusion test and recombinant ELISA for the diagnosis of ovine progressive pneumonia. *Proceedings of the 99th Annual Meeting of the United States Animal Health Association*, pp. 536–545.

Juste, R.A., Ott, T.L., Kwang, J., Bazer, F.W. and de la Concha-Bermejillo, A. (1996) Effects of recombinant interferon-τ on ovine lentivirus replication. *Journal of International Cytokine Research* 16, 989–994.

Juste, R.A., Kwang, J. and de la Concha-Bermejillo, A. (1998) Dynamics of cell-associated viremia and antibody response during the early phase of lentivirus infection in sheep. *American Journal of Veterinary Research* 59, 563–568.

Kajikawa, O., Lairmore, M.D. and DeMartini, J.C. (1990) Analysis of antibody responses to phenotypically distinct lentiviruses. *Journal of Clinical Microbiology* 28, 764–770.

Kennedy-Stoskopf, S. (1989) Pathogenesis of lentivirus-induced arthritis. A review. *Rheumatology International* 9, 129–136.

Kimelman, D. and Kirschner, M.W. (1989) An antisense in RNA directs the covalent modification of the transcript encoding fibroblast growth factor in *Xenopus* oocytes. *Cell* 59, 687–696.

Lairmore, M.D., Akita, G.Y., Russell, H.R. and DeMartini, J.C. (1987) Replication and cytopathic effects of ovine lentivirus strains in alveolar macrophages correlate with *in vivo* pathogenicity. *Journal of Virology* 61, 4038–4042.

Lairmore, M.D., Butera, S.T., Callahan, G.N. and DeMartini, J.C. (1988a) Spontaneous interferon production by pulmonary leukocytes is associated with lentivirus-induced lymphoid interstitial pneumonia. *Journal of Immunology* 140, 779–785.

Lairmore, M.D., Poulson, J.M., Adducci, T.A. and DeMartini, J.C. (1988b) Lentivirus-induced lymphoproliferative disease. Comparative pathogenicity distinct ovine lentivirus strains. *American Journal of Pathology* 130, 80–89.

Lee, W.C., McConnell, I. and Blacklaws, B.A. (1994) Cytotoxic activity against maedi–

visna virus-infected macrophages. *Journal of Virology* 68, 8331–8338.

Liu, D.L., Donegan, J., Nuovo, G., Mitra, D. and Laurence, J. (1997) Stable human immunodeficiency virus type 1 HIV-1 resistance in transformed CD4+ monocyte cells treated with mulitargeting HIV-1 antisense sequences incorporated into U1 snRNA. *Journal of Virology* 71, 4079–4085.

Luján, L., Begara, I., Collie, D.D.S. and Watt, N.J. (1994) Ovine lentivirus (maedi-visna virus) protein expression in sheep alveolar macrophages. *Veterinary Pathology* 31, 695–703.

McConnell, I., Blacklaws, B.A., Bird, P., Lee, W.C., Roy, D.J. and Sargan, D. (1996) Lentivirus replication in lymphoid tissue: use of lymphatic cannulation to study the initial stages of infection and immunity. *AIDS Research and Human Retroviruses* 12, 417–420.

Malim, M.H., Bohnlein, S., Hauber, J. and Cullen, B.R. (1989) Functional dissection of the HIV-1 *rev* transactivator – derivation of a *trans*-dominant repressor of *rev* function. *Cell* 47, 537–540.

Mwaengo, D.M., Grant, R.F., DeMartini, J.C. and Carlson, J.O. (1997) Envelope glycoprotein nucleotide sequences and genetic characterization of North American ovine lentiviruses. *Virology* 238, 135–144.

Naldini, L., Blomer, U., Gallay, P., Ory, D., Mulligan, R., Gage, F.H., Verma, I.M. and Trono, D. (1996) *In vivo* gene delivery and stable transduction of nondividing cells by a lentiviral vector. *Science* 272, 263–267.

Narayan, O. (1990) Immunopathology of lentiviral infections in ungulate animals. *Current Opinions in Immunology* 2, 399–402.

Narayan, O. and Clements, J.E. (1989) Biology and pathogenesis of lentiviruses. *Journal of General Virology* 70, 1617–1639.

Narayan, O. and Clements, J.E. (1990) Biology and pathogenesis of lentiviruses of ruminant animals. In: Gallo, R.C. and Wong-Staal, F. (eds) *Retrovirus Biology and Human Disease.* Marcel Dekker, New York, pp. 117–146.

Narayan, O., Zink, M.C., Gorrell, M., Crane, S., Huso, D., Jolly, P., Saltarelli, M., Adams, R.B. and Clements, J.E. (1993) The lentiviruses of sheep and goats. In: Levy, J.A. (ed.) *The Retroviridae.* New York, Plenum Press, pp. 229–255.

Nathanson, N., Martin, J.R., Georgsson, G., Pálsson, P.A., Lutley, R.E. and Petursson, G. (1981) The effect of post-infection immunization on the severity of experimental visna. *Journal of Comparative Pathology* 91, 185–191.

Neuveut, C., Vigne, R., Clements, J.E. and Sire, J. (1993) The visna transcriptional activator tat: effects of the viral LTR and on cellular genes. *Virology* 197, 236–243.

Ouzrout, R. and Lerondelle, C. (1990) Expression of visna–maedi virus in mammary secretions of a seropositive ewe during gestation and an artificial induction of lactation. *Annales de Recherches Vétérinaires* 21, 69–73.

Palsson, P.A. (1976) Maedi and visna in sheep. In: Kimberlin, R.H. (ed.) *Slow Virus Diseases of Animals and Man.* North-Holland, Amsterdam, pp. 17–43.

Pearson, L.D., Poss, M.L. and DeMartini, J.C. (1989) Animal lentivirus vaccines: problems and prospects. *Veterinary Immunology and Immnunopathology* 20, 183–212.

Peixinho, Z.F. and Mendes, N.F. (1994) HLA antigens and resistance to HIV. *Journal of Clinical and Laboratory Immunology* 8, 456–458.

Pekelder, J.J., Veenink, G.K., Akkermans, J.P., Van Eldik, P., Elving, L. and Houwers, D.J. (1994) Ovine lentivirus induced indurative lymphocytic mastitis and its effect on the growth of lambs. *Veterinary Record* 134, 348–350.

Perk, K., Yaniv, A., Gazit, A. and DeMartini, J.C. (1996) Evaluation of vaccines for ovine lentivirus infection. *AIDS Research and Human Retroviruses* 12, 425–426.

Petursson, G., Georgsson, G. and Palsson, P. (1990) Maedi–visna virus. In: Dinker, Z. and

Morein, B. (eds) *Virus Infections of Ruminants.* Elsevier Science Publishers, Amsterdam, pp. 431–440.

Petursson, G., Andresdottir, V. and Andresson, O.S. (1992) Lentivirus diseases of sheep and goats: maedi–visna and caprine arthritis encephalitis. In: Speedy, A.W. (ed.) *Progress in Sheep and Goat Research.* CAB International, Wallingford, UK, pp. 107–129.

Philippon, V., Vellutini, C., Gambarelli, D., Harkiss, G., Arbuthnott, G., Metzger, G., Roubin, R. and Filippi, P. (1994) The basic domain of the lentiviral tat protein is responsible for damages in mouse brain: involvement of cytokines. *Virology* 205, 519–529.

Pieniazek, D., Janini, L.M., Ramos, A., Tanuri, A., Schechter, M., Peralta, J.M., Vicente, A.C., Pieniazek, N.K., Schochetman, G. and Rayfield, M.A. (1995) HIV-1 patients may harbor viruses of different phylogenetic subtypes: implications for the evolution of the HIV/AIDS pandemic. *Emerging Infectious Diseases* 1, 86–88.

Polisky, B. (1988) Col E1 replication control circuitry: sense from antisense. *Cell* 55, 929–932.

Pursel, V.G. and Rexroad, C.E.J. (1993) Recent progress in the transgenic modification of swine and sheep. *Molecular Reproduction and Development* 36, 251–254.

Rana, S., Besson, G., Cook, D.G., Rucker, J., Smyth, R.J., Yi, Y., Turner, J.D., Guo, H.H., Du, J.G., Peiper, S.C., Lavi, E., Samson, M., Libert, F., Liesnard, C., Vassart, G., Doms, R.W., Parmentier, M. and Collman, R.G. (1997) Role of CCR5 in infection of primary macrophages and lymphocytes by macrophage-tropic strains of human immunodeficiency virus: resistance to patient-derived and prototype isolates resulting from the delta ccr5 mutation. *Journal of Virology* 71, 3219–3227.

Reyburn, H.T., Roy, D.J., Blacklaws, B.A., Sargan, D.R., Watt, N.J. and McConnell, I. (1992) Characteristics of the T cell-mediated immune response to maedi–visna virus. *Virology* 191, 1009–1012.

Robinson, H.L., Astrin, S.M., Senior, A.M. and Salazar, F.H. (1981) Host susceptibility to endogenous viruses: defective, glycoprotein-expressing proviruses interfere with infection. *Journal of Virology* 40, 745–751.

Ruff, G. and Lazary, S. (1988) Evidence for linkage between the caprine leucocyte antigen (CLA) system and susceptibility to CAE virus-induced arthritis in goats. *Immunogenetics* 28, 303–309.

Salter, D.W. and Crittenden, L.B. (1989) Artificial insertion of a dominant gene for resistance to avian leukosis virus into the germ line of the chicken. *Theoretical and Applied Genetics* 77, 457–461.

Samson, M., Libert, F., Doranz, B.J., Rucker, J., Liesnard, C., Farber, C.M., Saragosti, S., Lapoumeroulie, C., Cognaux, J., Forceille, C., Muyldermans, G., Verhofstede, C., Burtonboy, G., Georges, M., Imai, T., Rana, S., Yi, Y., Smyth, R.J., Collman, R.G., Doms, R.W., Vassart, G. and Parmentier, M. (1996) Resistance to HIV-1 infection in caucasian individuals bearing mutant alleles of the CCR-5 chemokine receptor gene. *Nature* 382, 722–725.

Sarver, N., Cantin, E.M., Chang, P.S., Zola, J.A., Ladne, P.A., Stephens, D.A. and Rossi, J.J. (1990) Ribozymes as potential anti-HIV-1 therapeutic agents. *Science* 247, 1222–1225.

Schnieke, A.E., Kind, A.J., Ritchie, W.A., Mycock, K., Scott, A.R., Ritchie, M., Wilmut, I., Colman, A. and Campbell, K.H. (1997) Human factor IX transgenic sheep produced by transfer of nuclei from transfected fetal fibroblasts. *Science* 278, 2130–2133.

Schoborg, R. and Clements, J.E. (1994) The rev protein of visna virus is localized to the nucleus of infected cells. *Virology* 202, 485–490.

Simons, R.W. and Kleckner, N. (1983) Translational control of ISIO transposition. *Cell* 34, 683–691.

Small, J.A., Bieberich, C., Ghotbi, Z., Hess, J., Scangos, G.A. and Clements, J.E. (1989) The visna virus long terminal repeat directs expression of a reporter gene in activated macrophages, lymphocytes, and the central nervous systems of transgenic mice. *Journal of Virology* 63, 1891–1896.

Smith, S.M., Maldarelli, F. and Jeang, K.-T. (1997) Efficient expression by an alphavirus replicon of a functional ribozyme targeted to human immunodeficiency virus. *Journal of Virology* 71, 9713–9721.

Sonigo, P., Alizon, M., Staskus, K., Klatzmann, D., Cole, S., Danos, O., Retzel, E., Tiollsais, P., Haase, A. and Wain-Hobson, S. (1985) Nucleotide sequence of the visna lentivirus: relationship to the AIDS virus. *Cell* 42, 369–382.

Soriano, P., Cone, R.D., Mulligan, R.C. and Jaenisch, R. (1986) Tissue-specific and ectopic expression of genes introduced into transgenic mice by retroviruses. *Science* 234, 1409–1413.

Staskus, K.A., Couch, L., Bitterman, P., Retzel, E.F., Zupancic, M., List, J. and Haase, A.T. (1991) *In situ* amplification of visna virus DNA in tissue sections reveals a reservoir of latently infected cells. *Microbial Pathogenesis* 11, 67–76.

Sullenger, B.A. and Cech, T.R. (1993) Tethering ribozymes to a retroviral packaging signal for destruction of viral RNA. *Science* 262, 1566–1569.

Toohey, K.L. and Haase, A.T. (1994) The rev gene of visna virus is required for productive infection. *Virology* 200, 276–280.

Turelli, P., Petursson, G., Guiguen, F., Mornex, J.F., Vigne, R. and Querat, G. (1996) Replication properties of dUTPase-deficient mutants of caprine and ovine lentiviruses. *Journal of Virology* 70, 1213–1217.

Uhlenbeck, O.C. (1987) A small catalytic oligoribonucleotide. *Nature* 328, 596–600.

Van der Putten, H., Botteri, F.M., Miller, A.D., Rosenfeld, M.G., Fan, H., Evans, R.M. and Verma, I.M. (1985) Efficient insertion of genes into the mouse germ line via retroviral vectors. *Proceedings of the National Academy of Sciences USA* 82, 6148–6152.

Velutini, C., Philippon, V., Gambarelli, D.G., Horschowski, N., Nave, K.-A., Navarro, J.M., Auphan, M., Courcoul, M.-A. and Filippi, P. (1994) The maedi–visna tat protein induces multiorgan lymphoid hyperplasia in transgenic mice. *Virology* 68, 4955–4962.

Veres, G., Junker, U., Baker, J., Barske, C., Kalfoglou, C., Ilves, H., Escaich, S., Kaneshima, H. and Bohnlein, E. (1998) Comparative analysis of intracellularly expressed antisense RNAs as inhibitors of human immunodeficiency virus type 1 replication. *Virology* 72, 1894–1901.

Watt, N.J., MacIntyre, N., Collie, D., Sargan, D. and McConnell, I. (1992) Phenotypic analysis of lymphocyte populations in the lungs and regional lymphoid tissue of sheep naturally infected with maedi visna virus. *Clinical and Experimental Immunology* 90, 204–208.

Wilmut, I., Schnieke, A.E., McWhir, J., Kind, A.J. and Campbell, K.H. (1997) Viable offspring derived from fetal and adult mammalian cells. *Nature* 385, 810–813.

Woodward, T.M., Carlson, J. and DeMartini, J.C. (1994) Analysis of ovine lentiviral genomic variability by denaturing gradient gel electrophoresis. *Biotechniques* 17, 366–371.

Woodward, T.M., Carlson, J.O., de la Concha-Bermejillo, A. and DeMartini, J.C. (1995) Biological and genetic changes in ovine lentivirus strains following passage in isogeneic twin lambs. *Journal of Acquired Immune Deficiency Syndromes* 8, 124–133.

Zimmerman, P.A., Buckler-White, A., Alkhatib, G., Spalding, T., Kubofcik, J., Combadiere, C., Weissman, D., Cohen, O., Rubbert, A., Lam, G., Vaccarezza, M., Kennedy, P.E., Kumaraswami, V., Giorgi, J.V., Detels, R., Hunter, J., Chopek, M., Berger, E.A., Fauci, A.S., Nutman, T.B. and Murphy, P.M. (1997) Inherited resistance to HIV-1 conferred by an inactivating mutation in CC chemokine receptor 5: studies in populations with contrasting clinical phenotypes, defined racial background, and quantified risk. *Molecular Medicine* 3, 23–36.

15

Transmissible Spongiform Encephalopathies

N. Hunter
Institute for Animal Health, Neuropathogenesis Unit, Edinburgh, UK

Summary

Transmissible spongiform encephalopathies (TSEs) are neurodegenerative diseases of disputed aetiology. Hallmarks of TSEs include distinctive pathology, the involvement of a host-encoded protein (PrP) in disease development and genetic control of susceptibility. Scrapie in sheep can be controlled by breeding for resistant genotypes of the PrP gene, and this may also be true for scrapie in goats, but cattle have so far shown no linkage of genetic markers with BSE incidence. Breeding for resistant sheep will reduce the number of clinical scrapie cases but attention should also be paid to the possible emergence of new scrapie strains and the potential importance of hidden infection in carrier animals.

Introduction

Transmissible spongiform encephalopathies (TSEs) are degenerative diseases which affect the central nervous system of many mammals including scrapie in sheep and goats, bovine spongiform encephalopathy (BSE) of cattle and Creutzfeldt–Jakob disease (CJD) of humans. TSEs are fatal and are characterized by long asymptomatic incubation periods which, in ruminants and humans, may last for years. Scrapie has been endemic in sheep in Europe for at least 250 years and, like other TSEs, can be transmitted from affected to healthy animals by experimental injection or feeding of diseased tissues. BSE, on the other hand, is a relatively new disease, first recorded in the 1980s, but much of the understanding of this cattle disease has come from the background of research on sheep scrapie in its laboratory-mouse adapted forms.

Diagnosis of TSE is based on clinical signs and post-mortem brain pathology. Common features in fixed brain sections include the widespread formation of membrane-bound vesicles or vacuoles (the 'sponge' in

© CAB *International* 2000. *Breeding for Disease Resistance in Farm Animals* (eds R.F.E. Axford, S.C. Bishop, F.W. Nicholas and J.B. Owen)

spongiform) in cell bodies of nerve cells, vacuolation of the extracellular space and a proliferation of astroglial cells in the absence of demyelination or other overt inflammatory responses. These features are most pronounced in the terminal stages of disease and it is very difficult to diagnose TSE by histopathology in mammals which are not exhibiting symptoms.

More promising as a pre-clinical diagnostic test is the presence in tissue sections and extracts taken from affected individuals of aggregated abnormal forms of a host protein called PrP, or prion protein. During the pre-clinical phase, the normal protein, PrP^{C}, is changed in conformation by an unknown mechanism, to become the aggregated, partially protease resistant form, PrP^{SC}. This disease-specific marker has been found in peripheral tissues suitable for biopsy, over a year before scrapie signs have developed in sheep.

The agent responsible for causing TSEs is unusual. This pathogen has some of the properties of a conventional virus but can survive normal virucidal procedures, such as prolonged exposure to formalin, dry heat and some regimes of autoclaving. The molecular structure of the pathogen is not known for certain and it can only be detected by transmission to other animals. Biochemical studies have indicated that the PrP^{SC} protein is constantly associated with infectivity and this has given rise to the prion hypothesis, which suggests that this protein may be the sole constituent of the infectious particle. Others, while still recognizing the importance of PrP in these diseases, propose that there are other factors involved.

Breeding for disease resistance is now the most powerful means of controlling scrapie in sheep. The *PrP* gene in many polymorphic variants has been shown to be genetically linked to disease incidence, both of experimental TSE in sheep and goats and of natural scrapie in a wide range of sheep breeds. This chapter reviews the current knowledge of the genetics and biochemistry of TSE susceptibility in sheep and goats and describes the reasons why such advances have not proved to be so useful with cattle and BSE.

Clinical Signs and Pathology

The best studied of the transmissible spongiform encephalopathies (TSEs) is scrapie, which occurs naturally in sheep and goats. Written descriptions of scrapie (and complaints about lack of government intervention to control the disease) have been made for over 250 years in various parts of Europe, including England and Germany (Parry, 1984). Scrapie is a notifiable disease in EU countries but the exact number of cases occurring each year are not known for certain – there is undoubtedly some under-reporting. Several countries are regarded as being free of the disease, notably Australia and New Zealand which have stringent procedures controlling the import of new sheep bloodlines, including many years of quarantine, in order to keep scrapie out.

Clinical signs of scrapie in sheep (Dickinson, 1976) can last from 2 weeks to 6 months and begin with mildly impaired social behaviour such as unusual restlessness and signs of nervousness. In later stages of the clinical course, the general condition of the animal begins to deteriorate, sometimes accompanied by a change in fleece colour. Pruritus (itching with resultant inflammation) can result from the animal scratching its body against fence posts or by biting the affected area (Parry, 1984), for example around the base of the tail and, occasionally, the whole of the side of the body. In the final stages of scrapie, although the appetite may appear normal, the animals lose the ability to feed themselves and the condition really starts to degenerate. Scrapie does not seem to alter reproductive ability until muscle wasting interferes with the ability to move. Lambs can be born successfully to mothers in the clinical phase of the disease and rams remain fertile and active even when showing scrapie symptoms (Parry, 1984, Foster and MacKenzie, unpublished).

Reported clinical descriptions often vary, however. In a group of scrapie-affected sheep from Shetland between 1985 and 1991 (Clark and Moar, 1992), most animals showed signs of pruritus and emaciation, others had pruritus, emaciation and hyperaesthesia (oversensitivity) and others showed all these signs plus ataxia. In a report of clinical signs in some Japanese sheep, (Onodera and Hayashi, 1994) some animals (Suffolks and Corriedales) showed signs of pruritus but others (Corriedales) had died for no obvious reason – scrapie was diagnosed after histopathological examination. These differences both in symptoms and clinical phase may simply be due to breed characteristics or may indicate the presence in the field of different strains of scrapie.

It is very difficult to detect, by histopathology, those animals with scrapie which are not visibly affected by the disease. At terminal stages, however, common neurological lesions in the brain include neuronal degeneration with the formation of vacuoles and proliferation of astroglial cells, but no demyelination or other overt inflammatory responses. Vacuolation is not present in the same parts of the brain in all scrapie cases, for example, one study of scrapie-affected sheep in Britain described seven different patterns of vacuolation (Wood *et al.*, 1997) across ten brain regions.

In another study of both naturally affected and experimentally challenged sheep, vacuolation occurred in areas such as the dorsal vagus nucleus, cerebellum and thalamic nuclei (Foster *et al.*, 1996b). Vacuolation was described as 'seldom apparent' without detection of the presence of PrP^{SC} in the vicinity; however, PrP^{SC} was sometimes present in areas with no vacuolation. The presence of the disease-associated PrP^{SC} can be detected in the pre-clinical phase (Schreuder *et al.*, 1996) and is therefore of greater potential interest as a diagnostic test than is vacuolation.

BSE-affected cattle become very difficult to handle and show increasing signs of ataxia, altered behaviour with fear or aggression and sensitivity to noises and to touch. Affected animals spend less time ruminating than healthy cattle (Austin and Pollin, 1993) although their physiological drive to eat appears to remain normal. Several studies have noted that BSE cattle have low heart rates (brachycardia) which may be related to the low food intake associated with reduced rumination, or which may indicate that there is some damage to the vagus during disease development (Austin *et al.*, 1997). BSE-affected cattle also show significant neuronal loss in the brain (Jeffrey and Halliday, 1994) and the appearance of vacuolar lesions in brain sections is very similar to that seen in sheep scrapie. BSE was confirmed as a TSE by the demonstration of the diagnostic TSE-related PrP fibrils in brain extracts (Hope *et al.*, 1988) and by transmission of the disease to mice (Bruce *et al.*, 1994).

The TSE Agent

TSEs are caused by an unusual pathogen. It has some of the biological properties of a conventional virus, for example heritable strain characteristics and host-range limitations, but can survive normal virucidal procedures such as exposure to formalin, dry heat and autoclaving. The molecular structure of the scrapie pathogen is not known for certain and it can only be detected by transmission to other animals, hence it is very difficult to monitor and control. Biochemical and physical studies of scrapie infectivity have shown that protein is an essential part of the infectious particle and many believe that protein is the sole factor necessary to cause disease. An abnormal form (PrP^{SC}) of a host protein (PrP^{C}) has been found to be closely associated with infectivity, and it is PrP^{SC} which is regarded by proponents of the prion hypothesis as forming part, or all, of the infectious entity, either as an infectious protein (Prusiner, 1982) or as simply a genetic disease (for review, see Prusiner and Scott, 1997). Others have proposed that PrP^{SC} protects the pathogen and that the heritable scrapie strain-specific information is carried on another molecule, probably a nucleic acid (Bruce and Dickinson, 1987) – this is the virino hypothesis – and there are still those who valiantly continue to point out that there is evidence for the involvement of a conventional virus (Manuelidis *et al.*, 1987; Diringer, 1995; Dron and Manuelidis, 1996). At the very least, PrP^{SC} is a disease-specific marker and is useful in post-mortem diagnosis and in the development of pre-clinical biopsy procedures (Schreuder *et al.*, 1996).

TSE Genetics and Breeding for Resistance

In this section the other feature of PrP will be described in some detail – that is the linkage of various polymorphic forms of the *PrP* gene with disease incidence and the ability to make use of this linkage in breeding sheep for resistance to scrapie. The *PrP* gene has recently been mapped to sheep chromosome 13q15, to 13q17 in cattle, 13q15 in goats and 14q15 in river buffalo (Iannuzzi *et al.*, 1998). Although PrP genetics has greatly improved the prospects for eradication of scrapie, breeding for BSE resistance in cattle has not, so far, been possible.

TSE genetics in sheep

Studies of natural scrapie in sheep have confirmed the importance of three amino acid codons in the sheep *PrP* gene (136, 154 and 171) (Belt *et al.*, 1995;

Clouscard *et al.*, 1995; Hunter *et al.*, 1996), originally shown to be associated with differing incubation periods following experimental challenge of sheep with different sources of scrapie and BSE (Goldmann *et al.*, 1991a, 1994). A diagram of the sheep *PrP* gene structure (similar in all species) is shown in Fig. 15.1 and the polymorphic codons are described in Table 15.1. There are breed differences in *PrP* allele frequencies and in disease-associated alleles, however, some clear genetic rules have emerged. The most resistant sheep *PrP* genotype is $AA_{136}RR_{154}RR_{171}$. Out of hundreds of scrapie-affected sheep worldwide, only one animal of this genotype has been reported with scrapie – a Japanese Suffolk sheep (Ikeda *et al.*, 1995). Sheep of $AA_{136}RR_{154}RR_{171}$ genotype are also resistant to experimental challenge with both scrapie and BSE (Goldmann *et al.*, 1994). At the other end of the scale, sheep of genotypes which encode QQ_{171} are more susceptible to scrapie. For example, in Suffolk sheep,

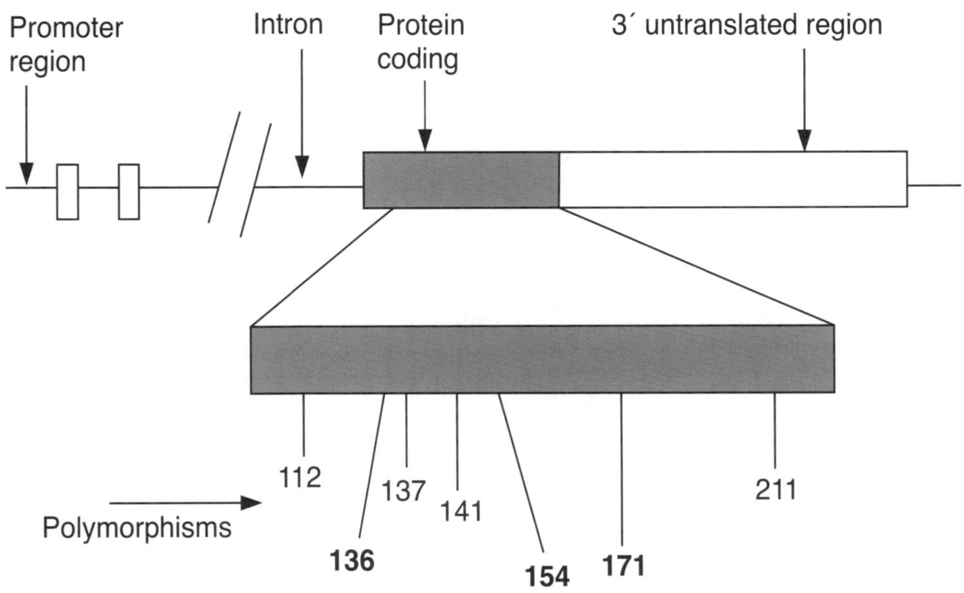

Fig. 15.1. Sheep *PrP* gene structure.

Table 15.1. Sheep *PrP* gene polymorphic amino acid codons.

Codon	Amino acid	Abbreviation
136	Valine	V_{136}
	Alanine	A_{136}
154	Histidine	H_{154}
	Arginine	R_{154}
171	Glutamine	Q_{171}
	Arginine	R_{171}

$AA_{136}RR_{154}QQ_{171}$ animals are most susceptible (Table 15.2A), although not all sheep of this genotype develop scrapie (Westaway *et al.*, 1994; Hunter *et al.*, 1997a). There is much less PrP genetic variation in Suffolk sheep than in some other breeds, the so-called 'valine breeds'. Breeds such as Cheviots, Swaledales and Shetlands encode *PrP* gene alleles with valine at codon 136, and sheep with the genotype $VV_{136}RR_{154}QQ_{171}$ appear to be extremely susceptible to scrapie (Table 15.2B) (Hunter *et al.*, 1994a, 1996). $VV_{136}RR_{154}QQ_{171}$ is a rare genotype and when it does occur, is almost always in scrapie-affected sheep and so it has been suggested that scrapie may be simply a genetic disease (Ridley and Baker, 1995). However, healthy animals of this genotype can live up to 8 years of age, well past the usual age-at-death from scrapie (2–4 years) (Hunter *et al.*, 1996) and susceptible sheep genotypes are easily found in countries which are free of scrapie disease (Hunter *et al.*, 1997b). The genetic disease hypothesis seems less likely than an aetiology that involves host genetic control of susceptibility to an infecting agent.

In some 'valine breed' sheep flocks affected by scrapie, there is a survival advantage if genotypes encode certain *PrP* alleles, such as $A_{136}H_{154}Q_{171}$ and $A_{136}R_{154}R_{171}$, so that despite having a high-risk allele such as $V_{136}R_{154}Q_{171}$, animals are unlikely to develop scrapie if their genotypes are $VA_{136}HR_{154}QQ_{171}$ or $VA_{136}RR_{154}RQ_{171}$ (Hunter *et al.*, 1996). However, this has not been found to be the case in all outbreaks, and there are many breed-related differences that give slightly more risk of disease in some genotypes than in others. PrP genetic testing is now available commercially to sheep breeders, being offered first to Swaledale and to Suffolk breeders (Dawson, 1994; Hosie and Dawson, 1996). As more genotyping information has become available, a general scheme, applicable across many breeds, is about to be put into operation (Dawson *et al.*, 1998), using a shortened system of nomenclature and giving a 'risk assessment' according to breed and to genotype, ranging from R1 (lowest risk) to R5 (highest risk) in individual sheep and their progeny.

Despite the steadily accumulating data on PrP genetic linkage and scrapie incidence, there remain some differences between flocks and between breeds

Table 15.2. Suffolk and Cheviot sheep PrP genetics and scrapie.

Risk of scrapie	Genotype	Abbreviated form[a]
(A) Suffolks		
Lowest	$AA_{136}RR_{154}RR_{171}$	RR_{171}
Low	$AA_{136}RR_{154}RQ_{171}$	RQ_{171}
Highest	$AA_{136}RR_{154}QQ_{171}$	QQ_{171}
(B) Cheviots		
Lowest	$AA_{136}RR_{154}RR_{171}$	–
Low	$VA_{136}RR_{154}RQ_{171}$	–
Low	$VA_{136}HR_{154}QQ_{171}$	–
High	$VA_{136}RR_{154}QQ_{171}$	–
Highest	$VA_{136}RR_{154}QQ_{171}$	–

[a]Variation only occurs at codon 171 in Suffolks, so the genotype is often abbreviated.

which are still to be explained. For example, the genotype that is most susceptible in Suffolks ($AA_{136}RR_{154}QQ_{171}$) is usually found to be resistant to natural scrapie in Cheviots, a 'valine breed'. In addition, some flocks of 'valine breeds' show scrapie only in animals encoding the $V_{136}R_{154}Q_{171}$ allele (homozygotes and heterozygotes) whereas in other flocks, despite the existence of $V_{136}R_{154}Q_{171}$-encoding sheep, scrapie appears to target only QQ_{171} genotypes and occurs in both $VA_{136}RR_{154}QQ_{171}$ and $AA_{136}RR_{154}QQ_{171}$ sheep (Clouscard *et al.*, 1995; Hunter *et al.*, 1997c) (Table 15.3). The answer could be that natural scrapie is similar to experimental TSEs in sheep where different sources of scrapie and BSE apparently target sheep according to their genotype at either codon 136 or codon 171. Following challenge by subcutaneous injection, the scrapie source SSBP/1 affects Cheviot sheep encoding the $V_{136}R_{154}Q_{171}$ allele whereas the scrapie source CH1641 and experimental BSE target sheep primarily according to codon 171 genotype producing disease with shortest incubation period in sheep that are QQ_{171} (Goldmann *et al.*, 1994). Extending these findings to the naturally affected sheep, it is possible that there are also various types or strains of natural scrapie which target either particular sheep breeds or different *PrP* codons. The usual way to investigate this at the moment is by passage of natural scrapie sources into a panel of mouse strains in which incubation periods and damage to particular brain regions give characteristic patterns (Bruce *et al.*, 1994). However, it is also possible that scrapie strains may produce PrP^{SC} protein with distinct patterns on immunoblots of SDS-PAGE gels (Hill *et al.*, 1997) and this is currently under investigation as a strain-typing method in several laboratories throughout the world.

Genetics of TSE in goats

Goats have been said always to succumb to experimental challenge with any source of TSE, for example, the transmission of SSBP/1 scrapie to all of a group of crossbred goats by intracerebral (ic) injection (Pattison *et al.*, 1959) was followed by its passage from goats to goats again with 100% incidence. In

Table 15.3. Unusual scrapie genetics in a flock of Poll Dorset sheep: frequencies (%) of PrP genotype in scrapie-affected and healthy sheep (data taken from Hunter *et al.*, 1997c).

PrP genotype	Scrapie sheep (%)	Healthy sheep (%)
$VA_{136}RR_{154}QQ_{171}$	50	0
$VA_{136}RR_{154}RQ_{171}$	0	19
$AA_{136}RR_{154}QQ_{171}$	50	13
$AA_{136}HR_{154}QQ_{171}$	0	6
$AA_{136}RR_{154}RQ_{171}$	0	31
$AA_{136}HR_{154}RQ_{171}$	0	13
$AA_{136}RR_{154}RR_{171}$	0	19

U.W.B. LIBRARY P.C.B. LLYFRGELL

a review (Pattison and Millson, 1962) 24 experiments carried out between 1956 and 1962 were described. In total 170 goats were ic injected with scrapie brain homogenates and all animals succumbed, with incubation periods ranging from 7 to 22 months, with a mean of 12 months. This finding of 100% susceptibility with goats of various types has been reported several times, but experiments have usually been carried out with goats of uncontrolled genetic background; for example, the crossbreds in Pattison's studies had a mixture of Anglo-Nubian, Toggenburg, Saanen and British Alpine ancestry (Dickinson, 1976), so it might have been thought that the variation in incubation period was just the result of background breed differences.

However, recent studies of the goat *PrP* gene have found that this may not be the case. The goat *PrP* gene has very high homology with both the sheep and the bovine *PrP* genes and analysis of the goat *PrP* gene has revealed several different alleles. Four *PrP* alleles have been found, the result of variation at codons 142 (isoleucine or methionine), 143 (histidine or arginine) and 240 (serine or proline) (Obermaier *et al.*, 1995; Goldmann *et al.*, 1996). One *PrP* allele was identical to the most common $A_{136}R_{154}Q_{171}$ sheep *PrP* allele. The polymorphism at codon 142 appeared to be associated with differing disease incubation period in goats experimentally infected with BSE, with CH1641 sheep scrapie and with a source of mouse-passaged scrapie, ME7. Isoleucine at codon 142 was associated with longer incubation periods. However, these experiments were conducted using the ic route of infection, not one expected to bear any resemblance to a 'natural' route. In order to be useful, these results must be extended to natural scrapie-affected goats, however, preliminary studies indicated that none of four natural scrapie-affected goats encoded methionine at codon 142 (Goldmann *et al.*, 1996).

Recently a new caprine *PrP* gene allele has been reported, containing only three instead of the usual five copies of a short peptide repeat and an additional tryptophan to glycine change at codon 102. The three-repeat allele is not pathogenic in the heterozygous state; however, it may result in a longer incubation period following challenge with SSBP/1 scrapie (Goldmann *et al.*, 1998).

Genetics of BSE in cattle

When BSE was found in cattle, the cattle *PrP* gene was searched for resistance or susceptibility markers similar to those that had been found in sheep. The bovine *PrP* gene coding region was first sequenced in 1991 (Goldmann *et al.*, 1991b) and this sequence was subsequently confirmed by two other groups (Yoshimoto *et al.*, 1992; Prusiner *et al.*, 1993). Allowing for various polymorphic forms of the gene in each species, there is very little difference (>90% identity) between the cattle and sheep *PrP* gene. The cattle gene, however, turned out to be remarkably invariant compared both to the sheep and to the human *PrP* genes. There has been one major polymorphism described – that of a difference in the numbers of an octapeptide repeat within the PrP protein (Goldmann *et al.*, 1991b). In humans many variations in the octapeptide repeat number have been described, ranging from 4 to 14, some of which have strong linkage to the incidence of human TSE (Poulter *et al.*, 1992). In cattle the most common number of octapeptide repeats is six and three genotypes have been described as 5:5, 6:6 with the heterozogote 6:5. In a study that compared the frequencies of these genotypes in BSE-affected and healthy cattle, there was no difference in frequency between the two groups, with around 90% of animals being 6:6 and 10% 6:5. The genotype 5:5 was rare (<1%) and was not found in the BSE-affected animals (Hunter *et al.*, 1994b) (Table 15.4). It is possible that the 5:5 genotype is associated with resistance because all BSE cattle had at least one copy of the 6 repeat allele but this remains to be established by, for example, direct experimental challenge of 5:5 cattle.

Cattle appear to be unusual in not (so far) demonstrating a PrP-related link with TSE incidence. BSE challenge does show such differences in other species – mice (Bruce *et al.*, 1994), sheep (Goldmann *et al.*, 1994) and goats (Goldmann *et al.*, 1996). It may be that all cattle would be susceptible to BSE if they received a high-enough dose of infection, or it may be that a polymorphism in the gene region controlling expression of the gene could provide such a link.

There is some evidence of heightened risk of BSE in offspring of BSE-affected cows (Wilesmith and Ryan, 1997) and several studies attempting to find evidence of inheritance of genetic control of susceptibility, rather than maternal transmission of disease itself, have not conclusively ruled out some element of genetic control of susceptibility in cattle (Curnow and Hau, 1996; Donnelly *et al.*, 1997; Ferguson *et al.*, 1997; Wilesmith *et al.*, 1997). It is normally the case that family studies are carried out in order to get an indication of whether there is genetic association with disease incidence. In the case of TSEs, however, the gene that controls disease incidence is known (the *PrP* gene) and it might be thought to be unnecessary to carry out family studies of cattle affected by BSE. However, in the light of the findings that the bovine PrP coding region does not show disease-linked variation, one such family study is of interest (Neibergs *et al.*, 1994). The analysis revealed three possible alleles in the *PrP* gene region generating the genotypes AA, BB and AB. The source of the change in DNA that resulted in the A type as opposed to the B

Table 15.4. PrP octarepeat genotype frequencies in healthy and BSE-affected cattle (data taken from Hunter *et al.*, 1994b).

		Frequency (%)		
Cattle group	*n*[a]	6:6	6:5	5:5
BSE-affected cattle	172	91	9	0
Single-herd study				
Healthy cattle	85	89	11	0
BSE-affected cattle	5	100	0	0
Unrelated healthy cattle	108	82	17	1

[a]Number of cattle.

type is unknown, however, BSE-affected animals and their relatives were found to be more likely to have the AA genotype than the other animals analysed, with BSE-affected animals giving AA frequency of 48%, their relatives 58% and unrelated healthy animals 29%. Although the AA genotype cannot be regarded as a marker for BSE susceptibility in these cattle, it is suggestive that there may be some genetic linkage with disease incidence outwith the *PrP* gene coding region itself. It is interesting that in this study, non-UK cattle (Boran and N'Dama from Kenya, Friesian Sahiwal, Brahman/ Brahman crosses from Australia and Brangus from the USA) had extremely low frequencies of AA (5%), suggesting that something is indeed genetically different about UK cattle, however, this has never been followed up.

Potential and Pitfalls

The advantages in using sheep PrP genotyping to breed against scrapie susceptibility is that within a number of years (depending on frequencies of genotypes) a flock can be rendered free of scrapie symptoms appearing in its sheep. On the basis of present knowledge, it is believed that this approach will work in getting rid of clinical scrapie and, as such, is the most powerful measure of scrapie disease control that has ever been available to the sheep breeder. In previous years, the standard advice has been simply to cull out the maternal line of affected animals, as it had been noted that scrapie 'runs in families'. This approach is unlikely to result in disease eradication as epidemiological studies have suggested that most cases do not arise through maternal (vertical) transmission but instead result from contact with infectious animals (horizontal transmission) (Hoinville, 1996). Computer modelling of slaughter of affected lines suggests that even if 100% of vertical transmission was eliminated, only very small reduction in the number of cases per year would be achieved (Woolhouse *et al.*, 1998). An additional problem would be the unnecessary loss of bloodlines. In, for example, a Suffolk flock that is harbouring scrapie infection but in which genotypes are unknown, it is perfectly possible for a scrapie-affected ewe to produce resistant offspring if she was QQ_{171} and was mated with an RR_{171} ram. The progeny (RQ_{171}) would be at low risk of disease and would be needlessly slaughtered. On the other hand two healthy, low-risk, RQ_{171} animals could produce some QQ_{171} progeny, highly susceptible to scrapie. The extremely drastic alternative option is to kill every animal in an affected flock and to repopulate with sheep from elsewhere. However, this approach, when adopted in Iceland, led to the re-emergence some years later of scrapie in the newly introduced animals (Palsson, 1979).

The potential disadvantages in changing *PrP* genotype frequencies by the use of $AA_{136}RR_{154}RR_{171}$ rams include, first, that there might, by chance, be loss of desirable breed characteristics. This problem is avoidable by means of selection on the basis both of *PrP* genetics and of breed conformation. The second problem with selection of only one genotype is that there may be very rare scrapie strains that could emerge to cause scrapie in $AA_{136}RR_{154}RR_{171}$ sheep. There is one documented case of scrapie occurring in a sheep of this genotype, in Japan (Ikeda *et al.*, 1995). Mixed genotypes, avoiding only the most susceptible animals, offer the best protection against the emergence of new natural scrapie strains.

The third problem relates to the possibility that 'carrier' sheep may exist and, if so, that they could pass on infection to other sheep. In this scenario, $AA_{136}RR_{154}RR_{171}$ sheep would not show signs of clinical scrapie but would harbour the infectious agent, posing a threat by maintaining the infectious agent and creating a potential situation in which a new strain, capable of causing clinical scrapie, could be selected. Even if a new strain did not appear, computer modelling predicts that infection could remain hidden in the flock for many decades (Woolhouse *et al.*, 1998), threatening the health of any susceptible animals introduced into the flock. There is some limited evidence for the existence of hidden infectivity. One of the earliest studies of scrapie in mice (Chandler, 1963) showed that incubation period was dose-dependent. In other words, incubation period lengthened as the amount of infectivity in the inoculum was reduced. At very low levels of scrapie infection, replication may take such a long time to build up, that disease does not develop within the animal's normal life span (Dickinson *et al.*, 1975). Such a situation, if shown to exist in sheep (and as yet there is no proof of this) could lead to the maintenance of a low level of infection in a flock without any signs of symptoms. Such sheep could be shedding scrapie infectivity for many years and be a source of infection for their flockmates. More recent evidence supporting such a phenomenon comes from another mouse model (scrapie strain 87V injected intraperitoneally (ip) as a 1% brain homogenate into IM mice) where there is long-term persistence of high scrapie titres in the mouse spleen without either accompanying replication in the brain or development of scrapie symptoms. This could be a mechanism by which carrier-status animals can be produced (Collis and Kimberlin, 1985). Cross-species persistence may also occur, hamster scrapie injected into mice does not produce disease but the hamster scrapie remains in brain and spleen of the mice and can be recovered in a form still able to infect hamsters (Race and Chesebro, 1998).

In mathematical models of scrapie outbreaks, the most effective means of controlling scrapie depend on the availability of good pre-clinical diagnostic tests, which are capable of detecting infected animals very early in their symptomless incubation period (Woolhouse *et al.*, 1998). Recognition and slaughter of pre-clinical animals less than half way to development of scrapie could have a significant effect on disease prevalence. As bioassay is an expensive and time-consuming option for study of hidden scrapie infection and does not even always work (Fraser, 1983), very sensitive methods are being developed to reveal the presence of the infection-associated protein PrP^{SC} and these will, in the future, be used to study resistant sheep, to give an indication of whether they are truly resistant or whether they harbour infection at subclinical levels.

Breeding for disease resistance in goats, although potentially possible, would be subject to the same comments as above. However, in cattle the option to control TSE disease by breeding for resistance is not available – there are no genetic markers linked to BSE. BSE in the UK is in decline as a result of the physical measures taken to control cattle food, along with the slaughter of any animal considered at risk of disease (Anderson *et al.*, 1996). Because of this, it may be thought that there is no point trying to understand the genetics of BSE. However, BSE has apparently infected other species, including humans (Bruce *et al.*, 1997), and it has the potential to spread into sheep (Goldmann *et al.*, 1994) where its pathogenesis may be different from that in cattle (Foster *et al.*, 1996a). Knowledge and understanding of BSE may therefore protect us from similar new diseases in the future.

Conclusions

In conclusion, the use of genotyping in breeding for TSE resistance is currently possible only in sheep. It should be encouraged, but not to the exclusion of all other desirable bloodline characteristics. In the future, large-scale testing of PrP genetic selection will take place and there should also be careful monitoring of genotypes of any scrapie cases that occur, in order to reveal any developing problems, such as changes in the genotypes succumbing to scrapie. In addressing the problems and concerns of the sheep breeder, science continues to address the remaining questions about the aetiology of TSEs – the unusual pathogens.

References

Anderson, R.M., Donnelly, C.A., Ferguson, N.M., Woolhouse, M.E.J., Watt, C.J., Udy, H.J., Mawhinney, S., Dunstan, S.P., Southwood, T.R.E., Wilesmith, J.W., Ryan, J.B.M., Hoinville, L J., Hillerton, J.E., Austin, A.R. and Wells, G.A.H. (1996) Transmission dynamics and epidemiology of BSE in British cattle. *Nature* 382, 779–788.

Austin, A. and Pollin, M. (1993) Reduced rumination in bovine spongiform encephalopathy and scrapie. *Veterinary Record* 132, 324–325.

Austin, A.R., Pawson, L., Meek, S. and Webster, S. (1997) Abnormalities of heart rate and rhythm in bovine spongiform encephalopathy. *Veterinary Record* 141, 352–357.

Belt, P.B.G.M., Muileman, I.H., Schreuder, B.E.C., Bos-de Ruijter, J., Gielkens, A.L.J. and Smits, M.A. (1995) Identification of five allelic variants of the sheep PrP gene and their association with natural scrapie. *Journal of General Virology* 76, 509–517.

Bruce, M.E. and Dickinson, A.G. (1987) Biological evidence that scrapie agent has an independent genome. *Journal of General Virology* 68, 79–89.

Bruce, M., Chree, A., McConnell, I., Foster, J., Pearson, G. and Fraser, H. (1994) Transmission of bovine spongiform encephalopathy and scrapie to mice – strain variation and the species barrier. *Philosophical Transactions of the Royal Society of London Series B – Biological Sciences* 343, 405–411.

Bruce, M.E., Will, R.G., Ironside, J.W., McConnell, I., Drummond, D., Suttie, A., McCardle, L., Chree, A., Hope, J., Birkett, C., Cousens, S., Fraser, H. and Bostock, C.J. (1997) Transmissions to mice indicate that 'new variant' CJD is caused by the BSE agent. *Nature* 389, 488–501.

Chandler, R.L. (1963) Experimental scrapie in the mouse. *Research in Veterinary Science* 4, 276–285.

Clark, A.M. and Moar, J.A. (1992). Scrapie: a clinical assessment. *Veterinary Record* 130, 377–378.

Clouscard, C., Beaudry, P., Elsen, J.M., Milan, D., Dussaucy, M., Bounneau, C., Schelcher, F., Chatelain, J., Launay, J.M. and Laplanche, J.L. (1995). Different allelic effects of the codons 136 and 171 of the prion protein gene in sheep with natural scrapie. *Journal of General Virology* 76, 2097–2101.

Collis, S.C. and Kimberlin, R.H. (1985). Long-term persistence of scrapie infection in mouse spleens in the absence of clinical-disease. *FEMs Microbiology Letters* 29, 111–114.

Curnow, R.N. and Hau, C.M. (1996) The incidence of bovine spongiform encephalopathy in the progeny of affected sires and dams. *Veterinary Record* 138, 407–408.

Dawson, M.M. (1994) Genotyping service for Swaledale sheep. *Veterinary Record* 134, 583–584.

Dawson, M., Hoinville, L.J., Hosie, B.D. and Hunter, N. (1998) Guidance on the use of PrP genotyping as an aid to the control of clinical scrapie. *Veterinary Record* 142, 623–625.

Dickinson, A.G. (1976) Scrapie in sheep and goats. In: Kimberlin, R. (ed.) *Slow Virus Diseases of Animals and Man.* North-Holland, Amsterdam, pp. 209–241.

Dickinson, A.G., Fraser, H. and Outram, G.W. (1975) Scrapie incubation time can exceed natural lifespan. *Nature* 256, 732–733.

Diringer, H. (1995) Proposed link between transmissible spongiform encephalopathies of man and animals. *Lancet* 346, 1208–1210.

Donnelly, C.A., Ferguson, N.M., Ghani, A.C., Wilesmith, J.W. and Anderson, R.M. (1997) Analysis of dam-calf pairs of BSE cases: confirmation of a maternal risk enhancement. *Proceedings of the Royal Society of London Series B – Biological Sciences* 264, 1647–1656.

Dron, M. and Manuelidis, L. (1996) Visualisation of viral candidate cDNAs in infectious brain fractions from Creutzfeldt–Jacob disease by representational difference analysis. *Journal of Neurovirology* 2, 240–248.

Ferguson, N.M., Donnelly, C.A., Woolhouse, M.E.J. and Anderson, R.M. (1997) Genetic interpretation of heightened risk of BSE in offspring of affected dams. *Proceedings of the Royal Society of London Series B – Biological Sciences* 264, 1445–1455.

Foster, J.D., Bruce, M., McConnell, I., Chree, A. and Fraser, H. (1996a) Detection of BSE infectivity in brain and spleen of experimentally infected sheep. *Veterinary Record* 138, 546–548.

Foster, J.D., Wilson, M. and Hunter, N. (1996b) Immunolocalisation of the prion protein (PrP) in the brains of sheep with scrapie. *The Veterinary Record* 139, 512–515.

Fraser, H. (1983) A survey of primary transmission of Icelandic scrapie (rida) to mice. In: Court, L. and Cathala, F. (eds) *Virus Non Conventionnels et Affection du System Nerveux Central.* Masson, Paris, pp. 34–46.

Goldmann, W., Hunter, N., Benson, G., Foster, J.D. and Hope, J. (1991a) Different scrapie-associated fibril proteins (PrP) are encoded by lines of sheep selected for different alleles of the Sip gene. *Journal of General Virology* 72, 2411–2417.

Goldmann, W., Hunter, N., Martin, T., Dawson, M. and Hope, J. (1991b) Different forms of the bovine PrP gene have five or six copies of a short, G–C-rich element within the protein coding exon. *Journal of General Virology* 72, 201–204.

Goldmann, W., Hunter, N., Smith, G., Foster, J. and Hope, J. (1994) PrP genotype and agent effects in scrapie – change in allelic interaction with different isolates of agent in sheep, a natural host of scrapie. *Journal of General Virology* 75, 989–995.

Goldmann, W., Martin, T., Foster, J., Hughes, S., Smith, G., Hughes, K., Dawson, M. and Hunter, N. (1996) Novel polymorphisms in the caprine PrP gene: a codon 142 mutation associated with scrapie incubation period. *Journal of General Virology* 77, 2885–2891.

Goldmann, W., Chong, A., Foster, J., Hope, J. and Hunter, N. (1998) The shortest known prion gene allele occurs in goats, has only three octapeptide repeats and is non-pathogenic. *Journal of General Virology* 79, 3173–3176.

Hill, A.F., Desbruslais, M., Joiner, S., Sidle, K.C.L., Gowland, I., Collinge, J., Doey, L.J. and Lantos, P. (1997) The same prion strain causes vCJD and BSE. *Nature* 389, 448–450.

Hoinville, L.J. (1996) A review of the epidemiology of scrapie in sheep. *Reviews in Science and Technology of the Office International des Epizooties* 15, 827–852.

Hope, J., Reekie, L.J.D., Hunter, N., Multhaup, G., Beyreuther, K., White, H., Scott, A.C., Stack, M.J., Dawson, M. and Wells, G.A.H. (1988) Fibrils from brains of cows with new cattle disease contain scrapie-associated protein. *Nature* 336, 390–392.

Hosie, B. and Dawson, M. (1996) Scrapie genotyping for Suffolk sheep. *The Veterinary Record* 138, 215–216.

Hunter, N., Goldmann, W., Smith, G. and Hope, J. (1994a) The association of a codon 136 PrP gene variant with the occurrence of natural scrapie. *Archives of Virology* 137, 171–177.

Hunter, N., Goldmann, W., Smith, G. and Hope, J. (1994b) Frequencies of PrP gene variants in healthy cattle and cattle with BSE in Scotland. *The Veterinary Record* 135, 400–403.

Hunter, N., Foster, J., Goldmann, W., Stear, M., Hope, J. and Bostock, C. (1996) Natural scrapie in a closed flock of Cheviot sheep occurs only in specific PrP genotypes. *Archives of Virology* 141, 809–824.

Hunter, N., Moore, L., Hosie, B., Dingwall, W. and Greig, A. (1997a) Natural scrapie in a flock of Suffolk sheep in Scotland is associated with PrP genotype. *Veterinary Record* 140, 59–63.

Hunter, N., Cairns, D., Foster, J., Smith, G., Goldmann, W. and Donnelly, K. (1997b) Is scrapie a genetic disease? Evidence from scrapie-free countries. *Nature* 386, 137.

Hunter, N., Goldmann, W., Foster, J., Cairns, D. and Smith, G. (1997c) Natural scrapie and PrP genotype: case-control studies in British sheep. *The Veterinary Record* 141, 137–140.

Iannuzzi, L., Palomba, R., Di Meo, G.P., Perucatti, A. and Ferrara, L. (1998) Comparative FISH-mapping of the prion protein gene (PRNP) on cattle, river buffalo, sheep and goat chromosomes. *Cytogenetics and Cell Genetics* 81, 202–204.

Ikeda, T., Horiuchi, M., Ishiguro, N., Muramatsu, Y., Kai-Uwe, G. and Shinagawa, M. (1995) Amino acid polymorphisms of PrP with reference to onset of scrapie in Suffolk and Corriedale sheep in Japan. *Journal of General Virology* 76, 2577–2581.

Jeffrey, M. and Halliday, W.G. (1994) Numbers of neurons in vacuolated and non-vacuolated neuroanatomical nuclei in bovine spongiform encephalopathy-affected brains. *Journal of Comparative Pathology* 110, 287-293.

Manuelidis, L., Sklaviadis, T. and Manuelidis, E.E. (1987) Evidence suggesting that PrP is not the infectious agent in Creutzfeldt–Jakob disease. *EMBO Journal* 6, 341–347.

Neibergs, H.L., Ryan, A.M., Womack, J.E., Spooner, R.L. and Williams, J.L. (1994) Polymorphism analysis of the prion gene in BSE-affected and unaffected cattle. *Animal Genetics* 25, 313–317.

Obermaier, G., Kretzschmar, H.A., Hafner, A., Heubeck, D. and Dahme, E. (1995) Spongiform central nervous system myelinopathy in African dwarf goats. *Journal of Comparative Pathology* 113, 357–372.

Onodera, T. and Hayashi, T. (1994) Diversity of clinical signs in natural scrapie cases occurring in Japan. *Japan Agricultural Research Quarterly* 28, 59–61.

Palsson, P.A. (1979) Rida (scrapie) in Iceland and its epidemiology. In: Prusiner, S.B. and Hadlow, W.J. (eds) *Slow Transmissible Diseases of the Nervous System.* Academic

Press, New York, 1, 357–366.

Parry, H. (1984) *Scrapie.* Academic Press, London.

Pattison, I.H., Gordon, W.S. and Millson, G.C. (1959) Experimental production of scrapie in goats. *Journal of Comparative Pathology* 69, 300–312.

Pattison, I.H. and Millson, G.C. (1962) Distribution of the scrapie agent in the tissues of experimentally inoculated goats. *Journal of Comparative Pathology* 72, 233–244.

Poulter, M., Baker, H.F., Frith, C.D., Leach, M., Lofthouse, R., Ridley, R.M., Shah, T., Owen, F., Collinge, J. and Brown, J. (1992) Inherited prion disease with 144 base pair gene insertion. 1. Genealogical and molecular studies. *Brain* 115, 675–685.

Prusiner, S.B. (1982) Novel proteinaceous infectious particles cause scrapie. *Science* 216, 136–144.

Prusiner, S.B. and Scott, M.R. (1997) Genetics of prions. *Annual Review of Genetics* 31, 139–175.

Prusiner, S.B., Miklos, F., Scott, M., Serban, H., Taraboulos, A., Gabriel, J.-M., Wells, G.A.H., Wilesmith, J.W., Bradley, R., DeArmond, S.J. and Kristensson, K. (1993) Immunologic and molecular biologic studies of prion proteins in bovine spongiform encephalopathy. *Journal of Infectious Diseases* 136, 602–613.

Race, R. and Chesebro, B. (1998) Scrapie infectivity found in resistant species. *Nature* 392, 770.

Ridley, R.M. and Baker, H.F. (1995) The myth of maternal transmission of spongiform encephalopathy. *British Medical Journal* 311, 1071–1075.

Schreuder, B.E.C., van Keulen, L.J.M., Vromans, M.E.W., Langeveld, J.P.M. and Smits, M.A. (1996) Pre clinical test for prion diseases. *Nature* 381, 563.

Westaway, D., Zuliani, V., Cooper, C.M., Dacosta, M., Neuman, S., Jenny, A.L., Detwiler, L. and Prusiner, S.B. (1994) Homozygosity for prion protein alleles encoding glutamine-171 renders sheep susceptible to natural scrapie. *Genes and Development* 8, 959–969.

Wilesmith, J. and Ryan, J. (1997) Absence of BSE in the offspring of pedigree suckler cows affected by BSE in Great Britain. *Veterinary Record* 141, 250–251.

Wilesmith, J., Wells, G., Ryan, J., Gavier-Widen, D. and Simmons, M. (1997) A cohort study to examine maternally-associated risk factors for bovine spongiform encephalopathy. *Veterinary Record* 141, 239–243.

Wood, J., McGill, I., Done, S. and Bradley, R. (1997) Neuropathology of scrapie: a study of the distribution patterns of brain lesions in 222 cases of natural scrapie in sheep, 1982–1991. *The Veterinary Record* 140, 167–174.

Woolhouse, M.E.J., Stringer, S.M., Matthews, L., Hunter, N. and Anderson, R. (1998) Epidemiology and control of scrapie within a sheep flock. *Proceedings of the Royal Society of London Series B* 265, 1205–1210.

Yoshimoto, J., Iinumo, T., Ishiguro, N., Imamura, M. and Shinagawa, M. (1992) Comparative sequence analysis and expression of bovine PrP gene in mouse-L929 cells. *Virus Genes* 6, 343–356.

Production Diseases

V

16

Genetics of Susceptibility in Cattle and Sheep

C.A. Morris
Ruakura Agricultural Research Centre, Hamilton, New Zealand

Summary

Studies of host resistance to production disease in cattle and sheep have shown paternal half-sib heritabilities averaging 0.31 and realized heritabilities averaging 0.28. These values are similar to those for production traits. A summary of 16 selection experiments in cattle and sheep in New Zealand and Australia was undertaken. Four of these were single-generation studies. For the other 12, the average length of time under selection was 14.9 years (four generations) and the average rate at which the resistant or susceptible herds/flocks diverged from the control was 0.065 σ_p year^{-1}; this again was very similar to the rates in experimental beef herds selected for growth.

Industry herds and flocks are taking up opportunities to select for improved disease resistance using natural or artificial testing procedures identified under research conditions. As genetic markers are identified in future years, extra opportunities will be provided for herds and flocks to improve disease resistance by way of genetic marker traits.

Introduction

On cattle and sheep farms, production diseases are part of the usual farm environment. Farmers in many countries help their animals to fight these challenges by the use of vaccination, antibiotics, anthelmintic drenches, fungicides and insecticides. Alternatively, animals have been saved from the challenge by various management options, such as removal to another pasture, to another type of feed, to yards or to indoor housing. With increasing consumer resistance to residues in animal products, farmers will eventually be left with the choices of continuing to remove animals from the various disease challenges or of breeding or buying stock that is genetically resistant to these

© CAB *International* 2000. *Breeding for Disease Resistance in Farm Animals* (eds R.F.E. Axford, S.C. Bishop, F.W. Nicholas and J.B. Owen)

challenges. Alternatively, other biological controls may be used to limit the challenge to susceptible cattle and sheep.

In some cases the degree of natural immunity depends on the age or physiological status of the animal. Examples are internal parasites in sheep and cattle, where suckling animals are only gradually challenged as they consume more solid food and those over about a year of age have good resistance. Periparturient animals lose their immunity temporarily, which can have major management implications for the spread of diseases such as internal parasites in grazing sheep and mastitis in dairy cattle. Potentially, one of the forces with biggest impact for changing the host–environment relationships in the long term is selection for host resistance. In a previous review, Morris (1991) summarized heritability estimates for host resistance to various diseases in cattle and sheep. The present chapter seeks to extend that by summarizing results of studies involving screening and genetic selection for increased resistance or susceptibility to disease, and discusses how this variation may be harnessed within industry herds or flocks.

Underlying Philosophy

In practice, the incentives to apply genetic selection principles to disease traits in sire-breeding herds or flocks derive from the following.

1. A realization that the breeding objective is often incomplete without a disease trait(s).
2. The increasing cost or the decreasing availability of effective management strategies and drug therapies.
3. Ethical concerns about continuing to treat animals with drugs, although with perhaps equal ethical concerns to minimize the suffering experienced by diseased animals.
4. Increasing consumer preferences for residue-free animal products.

Two philosophies on genetic selection for resistance are prevalent in New Zealand. In the first (type 1), the practice is to select mainly for production traits, in the confidence that resistant animals will be among those selected. This is the philosophy used by the New Zealand Dairy Board (Harris *et al.*, 1996), with about 20 proven bulls each year being selected from a young-bull team of 250. Dairy cows are expected to produce high yields when challenged routinely with production diseases such as mastitis, facial eczema (a fungal toxin on autumn pasture) and (to a lesser extent, because of routine precautions taken by farmers) pasture bloat and hypomagnesaemia.

In the second (type 2), the practice is to treat the animal's response to a disease as another recorded trait and as part of a multitrait breeding objective. This is the approach being adopted by some New Zealand sheep breeders, with selection objectives combining disease trait(s) and the usual output traits (more lambs, meat and wool).

In the Morris (1991) review, it was suggested that it was too early to distinguish between these two types of selection procedures on merit. The structure of the industry has locked some groups of bull or ram breeders into their present systems of sire evaluation and sire selection, and they find it difficult to change the structure to include selection for disease traits. However, even dairy-industry sire-evaluation schemes have been modified in some countries in the past 5–10 years to include some disease traits, e.g. somatic cell score in the USA (Schutz and Wiggans, 1998). In the sheep industries of New Zealand and Australia, some ram breeders are now incorporating selection for disease traits, e.g. against facial eczema (Morris *et al.*, 1994) or nematode parasite infection (McEwan *et al.*, 1997), even though economic values for these traits are difficult to calculate (Woolaston and Baker, 1996).

Type of Challenge

One of the important design details in selecting for disease resistance is the testing procedure, including the choice between a natural or artificial challenge. Using a natural challenge on pasture may expose animals (especially the susceptible ones) to an intolerable assault. However, to obtain a uniform natural challenge, animals need to have the same daily food intake and uniform diet composition. In the absence of this, it is necessary at least to show that a natural challenge leads to repeatable and heritable variation.

If a natural challenge is used for selection, as at Ruakura with selection for high or low internal parasite infection, one method of checking the effects of variable food intake is to determine if the selection flocks also show corresponding differences under a standard artificial challenge (Baker *et al.*, 1990). Conversely, if a standard artificial challenge is used for selection, such as with the facial eczema resistant, control and susceptible selection flocks at Ruakura, then the flock responses to a natural challenge must be confirmed (Morris *et al.*, 1995b). This is to say, the real-life commercial conditions must be cross-checked against more ideal controlled conditions.

Performance versus Progeny Testing

In choosing between performance and progeny testing, the usual compromise is between: (i) a short generation interval and lower accuracy with performance testing, and (ii) a longer generation interval but greater accuracy (and generally fewer sires to choose from) with progeny testing. In the case of disease traits, the candidate's performance may pose difficulties of measurement. The candidate can be tested if the disease is, say, bloat in cattle, where the condition is reversible provided that intensive precautions are taken. However, for a trait like resistance to facial eczema, there may be long-term irreversible effects of ingesting toxic spores.

In the early generations of a project to select for resistance, some degree of protection from the disease will be necessary, because the candidate's self-protection may be limited. One alternative (e.g. Baker *et al.*, 1990) is to employ a milder challenge, and then to use best linear unbiased prediction (BLUP) (a statistical technique that allows the incorporation of data from all ancestors and half-sibs or other relatives) to supplement the performance test, and thus provide a more accurate breeding value estimate for the candidate. If an indicator trait has to be used, this may compromise accuracy for the goal itself (see below).

Experimental Selection within Herds and Flocks

A number of experimental herds and flocks are now being selected for disease traits, and these resources have been monitored over long time periods. Recently Morris (1998) reviewed those single-trait selection experiments for disease resistance or susceptibility, which are completed or still under way in cattle and sheep. The review was restricted to New Zealand and Australia, where most of the single-trait selection studies in cattle and sheep have been carried out. This generality does not apply to pigs or chickens, which will not be described here.

Experiments included

Estimates of direct responses to selection for disease resistance or susceptibility are given in Table 16.1, with results standardized in phenotypic standard deviation units (σ_p). For 12 multigeneration studies, selection was applied for an average of 14.9 years (approximately four generations), and all but two are still continuing. It is notable how little work is being done in dairy cattle. Flock or herd sizes were most commonly around 100 females per line. Four other experiments (selection for resistance to facial eczema in dairy cattle, to dermatophilosis in sheep, to reduced faecal egg count (FEC) in beef cattle (not shown in Table 16.1) and for resistance to ticks in beef cattle) were run for one generation each. One further experiment (not included here) is selection for resistance to footrot, which has been under way since 1993 in Australia, but where no results have yet been published (H.W. Raadsma, personal communication 1997). The cattle FEC study was a single-year experiment with F_2 *et sequen* Africander × Hereford cattle in Queensland (Esdale *et al.*, 1986), with a single-sample repeatability of 0.20 ± 0.05, a significant progeny group difference in FEC ($P < 0.005$) for three high and three low FEC bulls, and a realized heritability for the mean of four FECs of 0.52, giving a single-record heritability of about 0.2.

Heritabilities and responses

The heritability estimates for single records ranged from 0.13 to 0.45, with an unweighted mean of 0.28. Some of the traits and animal resources were the same as those cited by Morris (1991), where the average value from paternal half-sib estimates was 0.31. Generally the estimates in Table 16.1 were animal

Table 16.1. Summary of direct selection responses in New Zealand (NZ) and Australian (AUS) sheep and cattle studies[a].

Trait	Species	Breed	Years	Lines	Size	Difference	h^2	Reference
Faecal egg count	Sheep, NZ	Romney	1979–P	R, C, S	100	1.95(18)	0.28±0.05	Morris *et al.* (1997b)
		Romney	1985–92	R, C, S	100	0.86(8)	0.34±0.05	Morris *et al.* (1997b)
		Perendale	1986–P	R, S	90	1.27(10)	0.28±0.05	Morris *et al.* (1997b)
	Sheep, AUS	Merino H[b]	1977–P	R, C, S	110	1.83(15)	0.29±0.03	Woolaston and Piper (1996)
		Merino T[b]	1975–P	R, C, S	100	2.19(22)	0.38±0.04	Woolaston and Eady (1995)
		Merino	1988–92	R, S	100	0.82(5)	0.42±0.14	Cummins *et al.* (1991)
		Merino	1988–P	R, C	700, 100	0.40(8)	0.28±0.05	Greeff *et al.* (1995)
Resilience[c]	Sheep, NZ	Romney	1994–P	R, C	70	0.43(2)	0.14±0.03	Morris and Bisset (1996)
Facial eczema	Sheep, NZ	Romney	1975–P	R, C, S	120	2.34(22)	0.45±0.05	Morris *et al.* (1995b)
	Cattle, NZ	Jersey	1990	R, S	–	0.41	0.31±0.10	Morris *et al.* (1991a)
Ryegrass staggers	Sheep, NZ	Romney	1993–P	R, S	100	0.86(4)	0.13±0.05	Morris *et al.* (1995a)
Bloat	Cattle, NZ	FJ[d]	1973–P	R, S	30	1.20(23)	0.19±0.04	Morris *et al.* (1997a)
Body strike[e]	Sheep, AUS	Merino	1974–P	R, S	175–100	0.26(15)	0.26±0.12	Raadsma (1991)
Dermatophilosis	Sheep, AUS	Merino	1985–86	R, S	–	0.25	0.13±0.05	Lewer *et al.* (1987)
Cattle ticks	Cattle, AUS	AIS[d]	1967–70	R, S	–	0.83	0.39	Utech *et al.* (1978)

[a] P, present year (1997); R, resistant; C, control; S, susceptible; difference in selection response (on the transformed scale, if used) is between the extreme lines over the years shown in parentheses, in phenotypic SD units; size = average flock/herd numbers per line (size is not given in single year or single generation tests); h^2 = single-record heritability, from the base population or from a REML analysis of all years and all flocks/herds in the experiment.

[b] H, *Haemonchus* lines; where R, increased resistance to *Haemonchus* (IRH); C, control *Haemonchus* (CH); S, decreased resistance to *Haemonchus* (DRH); T, *Trichostrongylus* lines (data: R.R. Woolaston and R.G. Windon, personal communication 1997).

[c] High resilience = high productivity (post-weaning growth) and low dags, during extended period of parasitic challenge.

[d] FJ = Friesian–Jersey synthetic; AIS = Australian Illawarra Shorthorn.

[e] Body strike and fleece rot: either or both traits in various years (artificial and/or natural challenge); h^2 here was from a random-bred flock, and was on the observed scale, whereas h^2 for liability on the underlying scale was 0.54±0.25.

model values, whereas in the earlier review they were sire model estimates. Average divergences in selection response achieved between the resistant (R) and susceptible (S) lines, or between the R and control (C) lines are also given in Table 16.1, with the number of years of data analysed shown in brackets. For the New Zealand experiments, this involved including the 1995 or 1996 birth year, even though these results may have been later than the most recent published trial design and heritability estimates. Results for 12 experiments (excluding single year or generation experiments) showed an average annual divergence of selection line from control line of 0.065 σ_p year^{-1}, assuming symmetry. Alternatively, restricting the summary to the seven experiments that were evaluated after 10 or more years, the annual divergence estimate was 0.049 σ_p year^{-1}.

For comparison, Mrode's (1988) review of 29 growth-rate, weight-for-age and efficiency selection experiments in cattle found a mid-parent selection *differential* averaging 0.213 σ_p year^{-1}. Combining this with realized heritability estimates in cattle averaging 0.304 (Koch *et al.*, 1982) also leads to a mean realized response of 0.065 σ_p year^{-1}. These estimates for both disease and growth traits do not necessarily indicate maximal achievable rates of response because the selection lines have generally also been used as resource flocks for studies of the underlying biology. Parallel studies can affect selection intenities (especially among males) and also the numbers of females available for the next mating year. In many cases, flock or herd sizes were small, and mating restrictions were applied to reduce early inbreeding rates, which also meant reduced rates of progress. However, even in an 'ideal' selection programme with selection intensity averaging 1.0 (about 2 in males and close to zero in females) and assuming an average heritability of 0.28, the expected annual response would be only 0.08 σ_p if the generation interval was, say, 3.5 years, which is not much greater than the average of 0.065 σ_p from data in Table 16.1.

The range of divergences across experiments also reflects the wide initial screening used in some experiments, whereas others used only a single flock or herd as the base. Absolute rates of direct response and σ_p may both depend on the mean incidence of disease in the control. The ultimate rate of response in a resistance line may diminish as the line reaches fixation or as σ_p declines, or both. Nevertheless, responses across experiments appear to have been encouragingly high. For the traits in Table 16.1, major genes have been implicated in three cases (out of eight 'diseases'), i.e. faecal egg count (FEC), bloat and cattle ticks.

Symmetry

In cases where there were R, C and S lines, the symmetry of response could be tested. For the New Zealand FEC lines measured on the log transformed scale, the S flocks diverged at 0.71 times the rate achieved in the R flocks (Morris *et al.*, 1997b), and similar results have been observed in the Australian FEC (*Haemonchus*) lines (S.J. Eady, personal communication 1997). Asymmetry in the New Zealand Romney FEC lines was explained by a 9.8% higher net reproductive rate and by a 22% higher selection differential in the R than the S line (Morris *et al.*, 1997c). Asymmetry, with the same probable explanations, also occurred in the facial eczema lines of sheep (Morris *et al.*, 1991b, 1995b). Additionally, the heritability estimates tended to be higher in the R than S facial eczema flocks. For the bloat herds, breeding value estimates in the absence of a control line showed that the responses were greater in the S herd than in the R herd; there was no difference in net reproductive rate (Morris *et al.*, 1995c), but the presence of a putative major gene (recessive for susceptibility) and a lower heritability in the R line led to the asymmetry (Morris *et al.*, 1997a).

Screening

Generally, authors have not reported the proportion of the response contributed by screening from the foundation generation. However, this proportion depends on how much genetic variation is to be found between flocks rather than within one particular flock, and also on the relative size of the recorded population across all candidate flocks or within one flock. The use of progeny testing across flocks for disease traits solves the problem of different levels of disease challenge being experienced from flock to flock.

With most screening exercises, the potential genetic lift offered by screening for a polygenic trait declines after the foundation matings as the élite herd or flock improves in genetic merit. However, for a single gene trait, the screening opportunities do not diminish with time; identifying the outliers from a large population remains important.

In the case of some transmissible diseases, it could be undesirable to introduce screened animals to the élite flock or herd; in this case, the use of artificial insemination or embryo transfer may have to be considered.

Indicator Traits

When performance testing is used, some disease traits require the use of an indicator trait, for example resistance of sheep to internal parasites. The indicator most commonly used so far for resistance to internal parasite infection is FEC (Woolaston, 1990). It is assumed that egg numbers and subsequent worm numbers closely reflect the level of parasitic challenge to the host. However, sheep that are tolerant of (rather than resistant to) worms will not be accurately identified by egg counting, if the real objective is resistance to the effects of parasitism rather than resistance to parasitic infection. Other points are that it can be reasonably assumed that the rate of egg production by each parasite does not vary to an important degree (or it is controlled by the host) and, with some exceptions, worm species cannot be identified separately by egg counting.

Marker-assisted Selection and Candidate Genes

Now that genetic linkage maps are available for the common farm animal species, work has intensified to find quantitative trait loci (QTL) and candidate genes for each disease condition. A linked marker may often be identified before a QTL or a candidate gene is identified. In New Zealand and Australia the most common approach has been to generate first crosses and backcrosses between selection lines, and then to test for linkage within the backcross population. This is in spite of the fact that the average selection period in cattle and sheep (14.9 years, Table 16.1) is only about four generations, so not all alleles at relevant loci will have become fixed by directional selection. Alternatively, in other countries or other species, breed crosses and intercrosses or backcrosses have been used. The objective, however, with selection-line crosses or breed crosses is the same, namely to identify associations between a gene or a marker and an animal's phenotype.

Examples of the markers found in pigs have been reviewed recently by Rothschild (1998) for eight disease traits. Their relevance to cattle and sheep is because of the homologous nature of many gene sequences across mammalian species (Womack and Moll, 1986). In dairy cattle, three markers for somatic-cell count have been found in preliminary German studies (Reinsch *et al.*, 1998). In sheep, a marker for susceptibility to copper deficiency has been identified in Britain (Neary *et al.*, 1998). Candidate genes include a salivary protein in cattle for bloat susceptibility (Rajan *et al.*, 1996; Wheeler *et al.*, 1998), and different proteins in two British sheep breeds for susceptibility to nematode infection (Gulland *et al.*, 1993; Schwaiger *et al.*, 1995).

Another approach to identifying the presence of genes for disease resistance is by way of a mixed-inheritance segregation analysis (McEwan and Kerr, 1998). This procedure attempts to separate out from the data the effects of a segregating major gene and the effects of multifactorial (i.e. polygenic) inheritance. Examples of applying this technique include identifying a gene for resistance to cattle ticks under northern Australian conditions (Kerr *et al.*, 1994), identifying a gene for resistance to nematode infection in sheep (McEwan and Kerr, 1998), and a gene for susceptibility to bloat in cattle (Morris *et al.*, 1997a). Among other things, this type of analysis estimates the extent and direction of dominance, the size of the genotype effects and the contribution of the segregating locus to all genetic variation observed in a particular trait and population.

Opportunities for marker-assisted selection are greatest for traits that are sex-limited or expressed at a late age, e.g. carcass traits, some disease traits. Opportunities also depend on how the particular trait in the particular population is affected by genetic variation (single gene or polygenic), and by the relative costs of phenotyping and genotyping (Muir and Stick, 1998). However, the great potential of marker-assisted selection for disease traits is that it bypasses the need for challenging animals with the disease organism, and the associated ethical concerns.

Discussion

Responses

This chapter has demonstrated that selection for increased disease resistance in sheep and cattle is feasible, and that responses have been achieved under experimental conditions. In dairy cattle at an industry level, genetic response for reduced incidence of mastitis is being achieved in the countries that pioneered this type of selection, e.g. Norway and Sweden (Solbu and Lie, 1990).

Most diseases under investigation show variation that is in part heritable, e.g. tuberculosis in cattle (Petukhov *et al.*, 1998); leucosis in cattle (Lewin and Bernoco, 1986; Kulikova and Petukhov, 1994) with a positive genetic correlation between tuberculosis and leucosis also reported by Kulikova and Petukhov (1994); brucellosis in cattle (Templeton *et al.*, 1990); scrapie in sheep (Goldmann *et al.*, 1990); and BSE in cattle (Neibergs *et al.*, 1994). There are also circumstantial data from France (1938–1952) on foot-and-mouth disease in a dairy herd, suggesting genetic differences in host susceptibility (Prat, 1952). Two diseases apparently not yet subjected to genetics study are *Brucella ovis* in sheep and Johnes disease in cattle or sheep.

In some of the longer-term disease-selection studies, attention is now turning to the genetic correlations between disease and production traits. Piper and Barger (1988) have argued that, where genetic correlations are estimated in the field, their magnitude can be affected by the size of the disease challenge. Despite this imperfection, the correlation under field conditions is still likely to be relevant to the commercial situation.

Another question of experimental design arises from the need to compare selection lines. Should they be run together in the same environment, leading to cross-contamination? Or should they be run in separate farmlets and replicated? Our experience with high- and low-FEC selection lines confirms that some different correlated responses are observed, e.g. for fleece weight in parasitized animals (Morris *et al.*, 1997c) for Romney lines run together, compared with data from D.M. Leathwick (personal communication 1997), for Perendale lines run separately and replicated.

Ethical limitations

In spite of ethical concerns about the need to challenge and record affected animals, there seems to be no alternative when initial genetics studies for a new disease are undertaken. This is before any phenotypic selection is undertaken commercially, or while searching for a QTL, or while testing a possible candidate gene. Later, when a DNA marker test has been developed, then susceptible animals can be identified, treated and saved from the disease challenge.

Research directions

It is interesting to note that, only 10 years ago, the possibility or feasibility of selection for resistance to internal parasite infection was being seriously questioned (Piper and Barger, 1988). It is now generally accepted amongst scientists and industry that genetic selection for resistance to internal parasites in sheep is a feasible option, although not all ram breeders choose to take up the option. A secondary advantage from improving disease resistance is the reduction in feed and capital stock which is possible as a result of lower levels of disease mortality and higher productivity, to produce the same output from lower breeding stock numbers.

It seems to be easier to obtain research funding for genetic studies of disease traits than for production traits at present, presumably for welfare reasons.

Conclusions

Selection for disease resistance under field conditions is a realistic option, although there will be less selection opportunity for production traits at the same time.

References

Baker, R.L., Watson, T.G., Bisset, S.A. and Vlassoff, A. (1990) Breeding Romney sheep which are resistant to gastro-intestinal parasites. *Proceedings of the Australian Association of Animal Breeding and Genetics* 8, 173–178.

Cummins, L.J., Thompson, R.L., Yong, W.K., Riffkin, G.G., Goddard, M.E., Callinan, A.P.L. and Saunders, M.J. (1991) Genetics of *Ostertagia* selection lines. In: Gray, G.D. and Woolaston, R.R. (eds) *Breeding for Disease Resistance in Sheep.* Australian Wool Corporation, Melbourne, pp. 11–18.

Esdale, C.R., Leutton, R.D., O'Rourke, P.K. and Rudder, T.H. (1986) The effect of sire selection for helminth egg counts on progeny helminth egg counts and live weight. *Proceedings of the Australian Society of Animal Production* 16, 199–202.

Goldmann, W., Hunter, N., Foster, J.D., Salbaum, J.M., Beyreuther, K. and Hope, J. (1990) Two alleles of a neural protein gene linked to scrapie in sheep. *Proceedings of the National Academy of Sciences USA* 87, 2476–2480.

Greeff, J.C., Karlsson, L.J.E. and Harris, J.F. (1995) Heritability of faecal worm egg count at different times of the year in a mediterranean environment. *Proceedings of the Australian Association of Animal Breeding and Genetics* 11, 117–121.

Gulland, F.M.D., Albon, S.D., Pemberton, J.M., Moorcroft, P.R. and Clutton-Brock, T.H. (1993) Parasite-associated polymorphism in a cyclic ungulate population. *Proceedings of the Royal Society, London, B,* 254, 7–13.

Harris, B.L., Clark, J.M. and Jackson, R.G. (1996) Across breed evaluation of dairy cattle. *Proceedings of the New Zealand Society of Animal Production* 56, 12–15.

Kerr, R.J., Frisch, J.E. and Kinghorn, B.P. (1994) Evidence for a major gene for tick resistance in cattle. *Proceedings of the 5th World Congress on Genetics Applied to Livestock Production* 20, 265–268.

Koch, R.M., Gregory, K.E. and Cundiff, L.V. (1982) Critical analysis of selection methods and experiments in beef cattle and consequences upon selection programs applied. *Proceedings of the 2nd World Congress on Genetics Applied to Livestock Production* 5, 514–526.

Kulikova, S.G. and Petukhov, V.L. (1994) Genetic correlation of cattle resistance to tuberculosis and leucosis. *Proceedings of the 5th World Congress on Genetics Applied to Livestock Production* 20, 300–301.

Lewer, R.P., Gherardi, S.G. and Sutherland, S.S. (1987) Realised heritability estimates for resistance to dermatophilosis in Merino sheep. In: McGuirk, B.J. (ed.) *Merino Improvement Programs in Australia.* Australian Wool Corp, Melbourne, pp. 347–350.

Lewin, H.A. and Bernoco, D. (1986) Evidence of BoLA-linked resistance and susceptibility to subclinical progression of bovine leukaemia virus infection. *Animal Genetics* 17, 197–207.

McEwan, J.C. and Kerr, R.J. (1998) Further evidence that major genes affect host resistance to nematode parasites in Coopworth sheep. *Proceedings of the 6th World Congress on Genetics Applied to Livestock Production* 27, 335–338.

McEwan, J.C., Dodds, K.G., Greer, G.J., Bain, W.E., Wright, C.S., Green, R.S. and Watson, T.G. (1997) Genotype rankings for host resistance to internal parasites for sheep grazed in contrasting environments. *Proceedings of the Association for the Advancement of Animal Breeding and Genetics* 12, 40–44.

Morris, C.A. (1991) Screening and selection for disease resistance – repercussions for genetic improvement. In: Owen, J.B. and Axford, R.F.E. (eds) *Breeding for Disease Resistance in Farm Animals.* CAB International, Wallingford, UK, pp. 123–135.

Morris, C.A. (1998) Responses to selection for disease resistance in sheep and cattle in New Zealand and Australia. *Proceedings of the 6th World Congress on Genetics Applied to Livestock Production* 27, 295–302.

Morris, C.A. and Bisset, S.A. (1996) Progress on genetic studies of resilience to nematode parasites in sheep. *Proceedings of the New Zealand Society of Animal Production* 56, 91–93.

Morris, C.A., Towers, N.R., Smith, B.L. and Southey, B.R. (1991a) Progeny testing bulls for susceptibility to facial eczema. *New Zealand Journal of Agricultural Research* 34, 413–417.

Morris, C.A., Towers, N.R., Wesselink, C. and Southey, B.R. (1991b) Effects of facial eczema on ewe reproduction and postnatal lamb survival in Romney sheep. *New Zealand Journal of Agricultural Research* 34, 407–412.

Morris, C.A., Towers, N.R., Wesselink, C. and Wheeler, M. (1994) Selection for or against facial eczema susceptibility in sheep. *Proceedings of the New Zealand Society of Animal Production* 54, 263–266.

Morris, C.A., Towers, N.R., Wheeler, M. and Amyes, N.C. (1995a) A note on the genetics of resistance or susceptibility to ryegrass staggers in sheep. *New Zealand Journal of Agricultural Research* 38, 367–371.

Morris, C.A., Towers, N.R., Wheeler, M. and Wesselink, C. (1995b) Selection for or against facial eczema susceptibility in Romney sheep, as monitored by serum concentrations of a liver enzyme. *New Zealand Journal of Agricultural Research* 38, 211–219.

Morris, C.A., Cullen, N.G. and Carruthers, V.R. (1995c) A note on the pregnancy statistics of cows, and perinatal survival and birth weight of calves, in two herds selected for high or low susceptibility to bloat. *New Zealand Journal of Agricultural Research* 38, 205–209.

Morris, C.A., Cullen, N.G. and Geertsema, H.G. (1997a) Genetic studies of bloat susceptibility in cattle. *Proceedings of the New Zealand Society of Animal Production* 57, 19–21.

Morris, C.A., Vlassoff, A., Bisset, S.A., Baker, R.L. and Watson, T.G. (1997b) Direct responses to selection for divergence in faecal nematode egg count in young Romney and Perendale sheep. *Proceedings of the Association for the Advancement of Animal Breeding and Genetics* 12, 413–416.

Morris, C.A., Vlassoff, A., Bisset, S.A., Baker, R.L., West, C.J. and Hurford, A.P. (1997c) Responses of Romney sheep to selection for resistance or susceptibility to nematode infection. *Animal Science* 64, 319–329.

Mrode, R.A. (1988) Selection experiments in beef cattle. Part 2. A review of responses and correlated responses. *Animal Breeding Abstracts* 56, 155–167.

Muir, W.M. and Stick, D.A. (1998) Relative advantage of combining genes with major effects in breeding programs: simulation results. *Proceedings of the 6th World Congress on Genetics Applied to Livestock Production* 26, 357–360.

Neary, D.M., Sutcliffe, E., Haley, C.S. and Woolliams, J. (1998) Single marker QTL mapping for copper in the plasma of sheep. *Proceedings of the 6th World Congress on Genetics Applied to Livestock Production* 26, 437–440.

Neibergs, H.L., Ryan, A.M., Womack, J.E., Spooner, R.L. and Williams, J.L. (1994) Polymorphism analysis of the prion gene in BSE-affected and unaffected cattle. *Animal Genetics* 25, 313–317.

Petukhov, V.L., Kochnev, N.N., Panov, B.L., Korotkevich, O.S., Kulikova, S.G. and Marenkov, V.G. (1998) Genetics of cattle resistance to tuberculosis. *Proceedings of the 6th World Congress on Genetics Applied to Livestock Production* 27, 365–366.

Piper, L.R. and Barger, I.A. (1988) Resistance to gastro-intestinal strongyles: feasibility of a breeding programme. *Proceedings of the 2nd World Congress on Sheep and Beef Cattle Breeding* 1, 593–611.

Prat, J. (1952) Sur la transmission héréditaire naturalle contre la fièvre aphteuse chez certains bovins. *Bulletin Société des Sciences Vétérinaires de Lyon*, 297–302.

Raadsma, H.W. (1991) Fleece rot and body strike in Merino sheep. V. Heritability of liability to body strike in weaner sheep under flywave conditions. *Australian Journal of Agricultural Research* 42, 279–293.

Rajan, G.H., Morris, C.A., Carruthers, V.R., Wilkins, R.J. and Wheeler, T.T. (1996) The relative abundance of a salivary protein, bSP30, is correlated with susceptibility to bloat in cattle herds selected for high or low bloat susceptibility. *Animal Genetics* 27, 407–414.

Reinsch, N., Xu, N., Thomsen, H., Looft, C., Kalm, E., Grupe, S., Kuhn, C., Schwerin, M., Leyhe, B., Hiendleder, S., Erhard, G., Medjugorac, I., Russ, I., Forster, M., Brenig, B., Reents, R. and Averdunk, G. (1998) First results on somatic cell count loci from the ADR bovine mapping project. *Proceedings of the 6th World Congress on Genetics Applied to Livestock Production* 26, 426–428.

Rothschild, M.F. (1998) Identification of quantitative trait loci and interesting candidate genes in the pig: progress and prospects. *Proceedings of the 6th World Congress on Genetics Applied to Livestock Production* 26, 403–409.

Schutz, M.M. and Wiggans, G.R. (1998) Current status of genetic evaluation of somatic cell scores for US dairy cattle. *Proceedings of the 6th World Congress on Genetics Applied to Livestock Production* 25, 15–18.

Schwaiger, F.W., Gostomski, D., Stear, M.J., Duncan, J.L., McKellar, Q.A., Epplen, J.L. and Buitkamp, J. (1995) An ovine major histocompatibility complex DRB1 allele is associated with low faecal egg counts following natural, predominantly *Ostertagia circumcincta* infection. *International Journal for Parasitology* 25, 815–822.

Solbu, H. and Lie, O. (1990) Selection for disease resistance in dairy cattle. *Proceedings of the 4th World Congress on Genetics Applied to Livestock Production* 16, 445–448.

Templeton, J.W., Estes, D.M., Price, R.E., Smith, R. and Adams, L.G. (1990) Immunogenetics of natural resistance to bovine brucellosis. *Proceedings of the 4th World Congress on Genetics Applied to Livestock Production* 16, 396–399.

Utech, K.B.W., Seifert, G.W. and Wharton, R.H. (1978) Breeding Australian Illawarra Shorthorn cattle for resistance to *Boophilus microplus*. 1. Factors affecting resistance. *Australian Journal of Agricultural Research* 29, 411–422.

Wheeler, T.T., Haigh, B.J., Wilkins, R.J., McCracken, J.Y. and Morris, C.A. (1998) A candidate gene marker for bloat susceptibility in cattle? *Proceedings of the New Zealand Society of Animal Production* 58, 10–12.

Womack, J.E. and Moll, Y.D. (1986) Gene map of the cow: conservation of linkage with mouse and man. *Journal of Heredity* 77, 2–7.

Woolaston, R.R. (1990) Genetic improvement of resistance to internal parasites in sheep. *Proceedings of the Australian Association of Animal Breeding and Genetics* 8, 163–171.

Woolaston, R.R. and Baker, R.L. (1996) Prospects of breeding small ruminants for resistance to internal parasites. *International Journal for Parasitology* 26, 845–855.

Woolaston, R.R. and Eady, S.J. (1995) Australian Research on genetic resistance to nematode parasites. In: Gray, G.D., Woolaston, R.R. and Eaton, B.T. (eds) *Breeding for Resistance to Infectious Diseases in Small Ruminants.* Australian Centre for International Agricultural Research, Canberra, pp. 53–75.

Woolaston, R.R. and Piper, L.R. (1996) Selection of Merino sheep for resistance to *Haemonchus contortus*: genetic variation. *Animal Science* 62, 451–460.

17

Breeding for Resistance to Production Diseases in Poultry

B.H. Thorp[1] and E. Luiting[2]

[1]Ross Breeders Ltd, Newbridge, Midlothian, UK; [2]Roslin Institute (Edinburgh), Roslin, UK

Summary

This chapter considers production diseases in poultry and opportunities to address these problems by means of breeding. These diseases comprise both metabolic conditions and infectious diseases. There is now considerable evidence for genetic variation in susceptibility, resistance or tolerance to many of these conditions or diseases, including ascites, sudden death syndrome, spondylolisthesis (kinky back), crooked toes, dyschondroplasia, rickets, osteoporosis, infectious bursal disease, Marek's disease, lymphoid leucosis, salmonellosis and coccidiosis. Additionally, genetic variation is also seen for many components of the immune response. The resource allocation theory of Beilharz *et al.* (1993) predicts that selection under environmentally limiting conditions, as currently practised in many commercial breeding programmes, will result in unfavourable genetic correlations between fitness and disease-resistance traits. This will lead to an increase in many of the observed disease and metabolic conditions. Challenges remain to quantify the actual importance of resistance to specific diseases, learn more of the genes involved in disease resistance and incorporate this knowledge into breeding programmes, using the concepts of resource allocation with regard to the relationship between disease resistance and production traits.

Introduction

The first section of this chapter considers reasons for breeding for disease resistance. Before dealing with the various selection strategies for effectively improving resistance to diseases, some terminologies and philosophies are defined and discussed. The next section examines the progress and success of strategies of breeding for resistance to production diseases. The diseases are divided into the categories of metabolic disorders and infectious diseases. The methodologies used and progress made for the common diseases in each category are discussed.

© CAB *International* 2000. *Breeding for Disease Resistance in Farm Animals* (eds R.F.E. Axford, S.C. Bishop, F.W. Nicholas and J.B. Owen)

Reasons for Breeding for Disease Resistance

Disease problems are controlled by a variety of means, including the barriers of biosecurity, biological methods such as vaccination and competitive exclusion, the use of pharmaceuticals (antibiotics and anticoccidials) and genetic selection. In reality a combination of these is used, with the combination varying with the prevalent pathogens. The application of prophylactic controls is increasingly constrained by the introduction of greater restrictions on the use of medications, in particular antibiotics.

Costs of production diseases can be estimated at between 10% and 20% of total production costs (OIE, 1998). Production diseases in poultry are those diseases that result in a drop in output either through death or dysfunction of the birds. In addition zoonoses, i.e. salmonellae, should be considered. This is not because of their direct effect on production, which may be minimal, but because of their potential effect on reduced profitability through a reduced value of the product. Poultry are the most cost-driven of all the animal production systems and the least expensive methods of disease control are used. Small variations in liveability have a minor effect on economic flock performance relative to similar variations in other performance traits. As an example, Table 17.1 shows the effects of the same relative improvement in various performance traits of broiler parent stock on ultimate production efficiency (Albers, 1993). The economic impact of liveability in broiler breeders turns out to be less than 20% of that of reproduction-related traits. In layer stock this percentage is even lower.

This means that disease resistance will not be selected for if the cost in a loss of genetic improvements in other traits is too great and there are other effective methods of disease control. For example, to select for effective resistance to *Salmonella* would currently cost so much in performance that it would be totally infeasible. However, the cost of feed decontamination and site biosecurity is sufficiently low that the UK industry is prepared to adopt these alternate methods of *Salmonella* control. Alternatively, Marek's susceptibility can be identified effectively by a simple blood test at the primary breeder level and the cost of genetic selection against the major histocompatibility complex (MHC) *B19* allele to enhance resistance may be acceptable.

Although decisions in the poultry industry are largely and increasingly driven by economic considerations, the psychological impact of flock morbid-

Table 17.1. Effect of 1% improvement of various performance traits on cost of day-old chicks produced by broiler parent stock (from Albers, 1993).

Trait	Effect of 1% improvement of trait on day-old chick cost (cents)[a]
Rearing mortality	–0.01
Mortality during lay	–0.11
Hatching egg production	–0.38
Egg hatchability	–0.45

[a]Calculated for average Dutch conditions.

ity and mortality on the farmer and society cannot be ignored. Mortality rates above a certain threshold are considered unacceptable. Because of the psychological impact of mortality and morbidity on customers and the public, most breeding companies probably pay more attention to this trait than is warranted by its economic value to the buyers of breeding stock (Albers, 1993). Furthermore, some degree of risk management by the breeding company in connection to internationalization (markets without vaccination programmes, markets with special environments and new, modified or emerging poultry pathogens) will play a role too, as will community restrictions if present.

The principal commercial breeds of the poultry industry are in the hands of a few primary breeding companies. Approximately four turkey primary breeders, five commercial egg-laying primary breeders and five broiler primary breeders supply over 95% of the world's commercial poultry. Over the next 5–10 years the number of primary breeders is likely to halve. Selection is therefore carried out in small, closed populations. A single cockerel within an élite pedigree population can potentially give rise to 2 million broilers. Selection decisions can have a huge impact on the health and welfare of stock globally. An improvement in, say, resistance to Marek's disease could be distributed throughout the world in 3–4 years, conversely, adverse genetic changes can be distributed just as effectively (McKay, 1998).

Disease Resistance as a Composite Trait

Disease resistance may be divided into resistance to infections and resistance to disease development. Resistance to infections ('true' resistance) reduces or prevents infections. Complete resistance is rare, is usually specific to an individual pathogens and is usually receptor-related and inherited through a single gene. Partial resistance to specific infections is a consequence of mechanisms on and within the surfaces of the body, respiratory and alimentary systems. Resistance to disease development (disease tolerance) is less specific with regard to pathogen type and is usually polygenically inherited. As an example, Table 17.2 shows the resistance mechanisms of lymphoid leucosis virus in chickens. The lymphoid leucosis virus attaches and penetrates cells of the host using a low-density lipoprotein as a surface receptor. Disease development

Table 17.2. Resistance mechanisms of lymphoid leucosis virus in chickens (from Gavora, 1997).

Type	Description	Inheritance
Resistance to infection	Absence of cellular receptors for the virus	Single gene
Resistance to tumour development	Natural killer cells and immune mechanisms involving MHC	Polygene

into tumours involves an immune response influenced by the MHC (Gavora, 1997).

The immune system can be considered as functioning at three interactive levels. Non-specific phagocytic activity with no memory ability, plus two forms of immunity with memory – cell-mediated immunity and antibody production. It has been demonstrated experimentally that these three levels of function are under separate polygenic control in chickens (Van der Zijpp, 1983). Hence, it is better to differentiate between general and specific disease resistance instead of considering general and specific disease resistance as a single entity.

Genetic selection against production disease must focus on improving the ability of the bird to produce an effective response to disease challenges and on maintaining production performance, be that egg output, growth or food conversion ratio (FCR). The poultry breeder would like to identify a characteristic of general disease resistance effective against a wide range of pathogens (Gavora, 1990), but such a characteristic has not yet been identified. There are good reasons why a general resistance against a wide range of pathogens is not readily selected for. This is shown by looking at a subset of pathogens, the major bacterial diseases. The pathogenesis of infections with different bacterial agents is very different, and different components of the immune system are involved. *Pasteurella* and *Escherichia coli* in most species of poultry cause septicaemia, whereas *Mycobacterium* and *Salmonella* are primarily intracellular pathogens. Also, some organisms, such as *Salmonella* and *Campylobacter*, can exist in the gut of poultry in a disease-free state. This can have a marked effect on disease control; for example, *Salmonella* vaccine may protect against systemic infection but is less effective at preventing gut colonization.

Simple tests, such as the antibody response to sheep red blood cells, can be considered as a means of selecting for improved response to an antigen. However, this approach can result in far too simple a view being taken. Some straightforward experimental work with Marek's disease challenge on lines selected for high and low antibody responses to sheep red blood cells indicated a greater Marek's mortality in the high antibody-responding lines (Martin *et al.*, 1989). Further studies indicate that the selection for antibody response to sheep red blood cells is not an effective indirect way of improving resistance to Marek's disease (Pinard, 1992).

Moreover, in an undisturbed system containing a host and pathogen the virulence of viruses and bacteria tends to decline over generations. This is a simple consequence of group selection for survival of the pathogen and the host. The balance of this system is disturbed when new threats are used to challenge the pathogen, such as pharmaceuticals or vaccines. This may lead to the evolution of more virulent strains of pathogens, for example the recent emergence of a highly virulent Marek's disease (Witter, 1988). The fixation of an allele in a population that confers resistance to Marek is therefore no guarantee that the population will still be resistant to a potentially different Marek challenge in the future.

Disease Resistance as a Fitness Trait: Laying Hens versus Broilers

Poultry selection has primarily been for production traits and therefore, historically, many diseases that would be detrimental to production are self-limiting or controlled. For example some of the forms of avian leucosis virus infection in egg layers will result in mortality during lay. When selection pressure is for maximal egg numbers then birds affected with tumours caused by leucosis will be selected against as susceptibility to viral infection has a negative correlation on egg production (Hartmann, 1989). Egg production per hen housed may be seen as the ultimate fitness trait in laying hens. At the same time this trait has been the primary selection criterion in laying hen breeding over the past few decades. Hence artificial selection has done something similar to natural selection: selection for maximum overall fitness. The consequence of direct selection for overall fitness is an increase in all fitness component traits, such as survival and number of eggs. In general this type of response in the component traits will continue until environmental resources become limiting. This is true both for natural and artificial selection as long as artificial selection is for high overall fitness. When environmental resources become limiting, natural selection will have no effect whereas artificial selection will lead to another response in hen housed egg production at the expense of some other trait(s). These concepts are captured in the environmental resource allocation theory by Beilharz *et al.* (1993), summarized in Fig. 17.1.

Conversely, there is for some diseases a negative correlation between production traits and disease resistance, for example in commercial layers there is an unfavourable correlation between egg weight and mortality to Marek's disease (Gavora, 1990). Overall fitness is a multiplicative trait: it results from its

$$F = A \times B \times C$$

F = overall fitness
A, B, C = fitness components

$$R = (k_{PROD} \times \text{PROD}) + (k_F \times F) + \ldots$$

or

$$R = (k_{PROD}) \times \text{PROD}) + (k_A \times A) + (k_B \times B) + (k_C \times C) + \ldots$$

R = total environmental resources
(e.g. energy, protein, calcium intakes and body reserves)

PROD, F, A, B, C, . . . =
resource-demanding processes
(production, overall fitness, fitness components, . . .)

k_{PROD}, k_F, k_A, k_B, k_C, . . . = resource conversion factors.

Under some environmental conditions,
R may be limiting.

Fig. 17.1. A summary of the environmental resources allocation theory by Beilharz *et al.* (1993).

component traits by multiplying them rather than adding them up (for example: percentage survival × number of eggs per survived hen × average egg weight). As an algebraic consequence, selection for maximum overall fitness leads to intermediate optimum (not maximum) levels of each of the component traits, which in turn leads to negative genetic correlations among these when environmental resources become limiting. In such a situation direct selection for a single fitness component trait (e.g. average egg weight) will lead to a negative correlated response in another fitness component trait (e.g. survival). Environmental resource-limiting conditions may have been present in laying hens since breeders started to add feed efficiency to egg production as a selection criterion.

Selection for commercially important traits such as growth and feed efficiency in broilers can result in undesirable correlated responses in disease resistance. For example, the selection for increased muscle mass has been associated with deep pectoral myopathy (Siller *et al.*, 1978) and reduced immunological competence (Dunnington *et al.*, 1987). More recently, selection for rapid growth and increased feed efficiency is associated with an increased incidence of skeletal disorders and cardiovascular disorders, including ascites and sudden death syndrome (Julian, 1993). In an environmental resource-limiting situation there is competition for resources among production- and fitness-related traits. It follows that in such conditions, a direct response to selection for production traits (e.g. growth, feed efficiency) will be at the expense of fitness component traits (e.g. survival).

Modern primary breeders function in a biosecure, minimal disease environment. The absence of exposure to disease at this level of the breeding pyramid has the potential to reduce the ability to select for enhanced disease resistance. Indeed, there is the distinct possibility of losing disease resistance while selecting for production traits. This may result in an increased incidence and severity of disease problems in the less biosecure commercial environment (i.e. an environmental resource-limiting situation). To surmount this problem the primary breeders have had to use various methods of challenging the birds and selecting the best stock. These methods tend to rely on indirect selection or assessing the response to the use of commercially available, registered live vaccines. Another way forward may be the analysis of genetic markers, such as microsatellites, restriction fragment length polymorphisms (RFLPs) or single nucleotide polymorphisms (SNPs), enabling allele frequencies in populations to be established. RFLP technology was applied to determine the allele frequencies for growth hormone gene in 12 non-inbred strains of White Leghorn evolved from three different genetic bases (Kuhnlein *et al.*, 1997). The strains had been selected for egg production, Marek's resistance or avian leucosis resistance. Selection for disease resistance was consistently correlated with an increased frequency of one allele, this allele was also associated with delayed onset of ovulation and higher persistency of ovulation. With this knowledge it may be possible to select for the disease resistance allele and either co-select for a normal onset of ovulation or, because onset of ovulation is so susceptible to environment, to modify management of the selected line to encourage earlier onset of ovulation. Recent data from White Leghorn strains suggest that MHC class II genes are likely candidates for the investigation of quantitative trait loci in egg production and disease resistance traits such as Marek's resistance (Lakshmanan *et al.*, 1997). Interestingly, this same study also suggests that the traits associated with MHC class II bands may be unique to each genetic background.

Methods of Breeding for Disease Resistance

There are three main strategies for the genetic improvement of disease-resistance traits in domestic animals. These are: (i) conventional breeding – direct selection and indirect selection based on phenotypic traits; (ii) indirect selection via markers utilizing associated DNA polymorphisms; and (iii) transgenic approaches. Figure 17.2 gives a schematic view of the various selection strategies.

Direct selection is the simplest and has been applied to Marek's disease (Friars *et al.*, 1972) and Newcastle Disease (Gordon *et al.*, 1971). This approach required large numbers of animals, facilities for disease challenge, possibly progeny testing and an increased generation time. The alternative is to select indirectly for a correlated marker trait that is non-pathological and measurable in the live animal or to select for a marker gene.

Marker-assisted selection (MAS) has been employed effectively in other species, for example the halothane locus in the pig. There is the potential for this to be of value in specific and non-specific disease resistance in poultry. The genes in the chicken MHC exert major genetic control over host resistance to autoimmune, viral, bacterial and parasitic disease. MHC typing has already demonstrated its value in the control of Marek's disease. Birds,

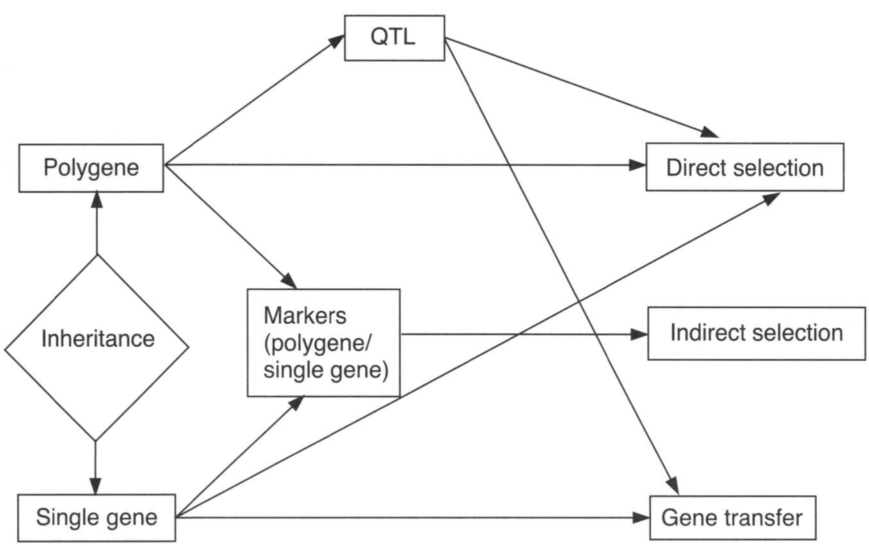

Fig. 17.2. A strategy for selection for disease resistance.

including poultry, have a much smaller, more compact and simpler MHC than mammals. Also, in contrast to mammals, resistance and susceptibility to some viral pathogens are determined by particular MHC haplotypes. Some of the peptide motifs for dominant class I molecules from chicken MHC haplotypes have been determined and may offer simple explanations of some disease associations, though other disease associations may be due to polymorphisms in the level of expression of MHC class II molecules (Kaufman and Wallney, 1996). Indeed, some common chicken MHC haplotypes express only one class I molecule at high levels and selection on a single MHC gene should be strong, in contrast to the situation in mammals (Kaufman and Salomonsen, 1997). The finding that MHC class II bands are associated with production traits or with Marek's disease resistance suggests that MHC class II genes are likely candidates for the investigation of quantitative trait loci (Lakshmanan *et al.*, 1997). Endogenous viral (*ev*) genes have also been associated with quantitative traits such as egg production, egg weight and resistance and susceptibility to Marek's disease (Aggrey *et al.*, 1998).

Transgenic approaches are still in the experimental phase but may become available tools for poultry breeders. The commercial use of this methodology is likely to be strongly influenced by public attitudes. One tested use was the introduction of resistance to lymphoid leucosis. Such a gene transfer was performed by the insertion of a recombinant avian leucosis retroviral genome (expressing a viral envelope protein) into the chicken genome (Salter *et al.*, 1987). Such recombinant viruses are candidate vectors for the insertion of foreign genes. Chickens transgenic for a defective avian leucosis virus (ALV) provirus that expresses envelope glycoprotein, but not infectious virus, were resistant to infection with subgroup A ALV (Crittenden and Salter, 1992); however, egg production performance seemed to be unfavourably influenced (Gavora, 1997).

These various selection methods can differ greatly in effectiveness and costs, both the direct measurement costs as well as the indirect costs of decreasing production. These will greatly influence the decision-making of the breeder. Depending on the existing breeding programme and the breeding goal, costs will influence which and how many animals, which, when and how many measurements have to be taken. Table 17.3 gives a rough summary of the various selection methods with their cost factors and effectiveness.

Methodologies and Progress of Breeding Strategies for Common Diseases

Metabolic disorders

Under normal non-epidemic situations there is more loss from metabolic than from infectious diseases, especially in broilers but also in laying hens. The main disorders to consider are cardiovascular and musculoskeletal. Within growing broilers the most prevalent metabolic diseases are ascites and, historically,

Table 17.3. Methods of selection for disease resistance (modified after Rothschild, 1991; Gavora, 1997).

Type of selection	Method	Consequences for production of breeders	Cost	Effectiveness
Direct	Observe breeding stock	None	Near zero	Questionable
	Challenge breeding stock	Negative	Low to high	Good
	Challenge sibs or progeny of breeding stock	None	High	Good
	Challenge clones	None	Very high	Good
Indirect	Use markers for disease resistance	None	Low to high	Low to good
Gene transfer	Construct resistant genotypes	None?	Very high	Excellent

dyschondroplasia (Leenstra, 1993). The genetic background of metabolic disorders is primarily polygenic and the incidence and severity are strongly influenced by nutrition and environment.

The rise in the incidence of these conditions is a consequence of selection for production traits under environmental resource-limiting conditions (Dunnington, 1990; Beilharz *et al.*, 1993), leaving less resources for fitness traits, with a failure to incorporate selection against metabolic disorders. As a consequence, the typical pathologies seen result from the production pressure applied to the bird. For example, the commercial egg-layer has been selected for a low body weight with a consequentially low skeletal mass. She has also been selected for high egg production. She is more prone to an inability to meet the demands of egg production for calcium from her skeleton, resulting in osteoporosis. As another example, the commercial broiler was, for a substantial period, selected for rapid growth and feed efficiency, there was no selection for cardiac function. This resulted in increased heart failure and ascites through a mismatch in cardiac output and the physiological requirements of a larger carcass.

Direct selection for a low incidence of the metabolic disorders is the most obvious way forward, especially when nutritionally or environmentally challenged in order to create an environmentally resource-limiting situation (i.e. suboptimum conditions). Another option would be indirect selection for a certain desired resource allocation pattern, if it can be defined and measured.

The primary breeding companies have, over recent years, evolved strategies to deal with these problems, with some success. It must be remembered that benefits of the introduction of new methods of selection against metabolic and other diseases take at least 5 years to be seen at the level of the broiler chicken, although it may be slightly shorter for the commercial layer. This is a consequence of the number of generations between the pedigree or élite birds and the commercial end product.

Cardiovascular

ASCITES. The main condition of concern is ascites, the accumulation of fluid within the abdominal cavity as a consequence of heart failure. Historically, ascites was seen in broilers and broiler breeders at high altitude, where there is an increased cardiovascular and respiratory load. Since the early 1980s ascites has been seen with variable frequency in most flocks at sea level. It is considered, and has been shown experimentally, that selection in broiler populations for high feed efficiency will result in a reduction in oxygen consumption; exposing these birds to a changing environment, a higher metabolic rate and increased oxygen requirements may induce cardiopulmonary disturbances, heart failure and ascites (Scheele, 1996). Experimental models of ascites have been established and lines have been selected as ascites-susceptible and ascites-resistant. These lines and experimental models have been used to attempt to find a phenotypic marker of ascites susceptibility that can be used commercially to identify individuals. The most promising progress in this direction has been the use of an assay of plasma troponin T. Troponin T is a product of the myocardium that is increased when this tissue is stressed. Sera from ascitic broiler chickens reveal significant concentrations of troponin T, in contrast to sera from healthy birds (Maxwell *et al.*, 1995). Selection over one generation for low or high levels of cardiac-derived troponin T in plasma shows a moderately high heritability of 0.38, with no difference in body weight of the parents of the two lines (Maxwell *et al.*, 1998).

SUDDEN DEATH SYNDROME (SDS). This condition is seen in broilers and is characterized by death of apparently healthy birds with a full digestive tract, but often some indication of a cardiovascular disorder. It is seen as lung congestion, possibly oedema and possible enlarged heart. There is little information on the heritability of the syndrome but indications are that it is low (Chambers, 1986).

Skeletal

The genetics of skeletal disorders are described in some detail by Sørensen (1992). It was stated that genetic selection for fast growth caused an increase in the incidence and severity of skeletal disorders (for example, spondylolisthesis: Riddell, 1973; Leach and Gay, 1987). Indeed some researchers have stated that leg disorders are a direct consequence of selection for increasing growth rate (Reiland, 1978). This conclusion is disputed by Riddell (1973), who identifies feral forms of avian species that grow faster than commercial broilers and turkeys; but of course the question is to what extent the environmental resource allocation has been changed by selection, rather than what the absolute growth rate is.

The highly heritable skeletal deformities (for example spondylolisthesis or 'kinky back', a deformity of the thoracic spine causing compression of the spinal cord) will usually produce sufficient changes within the individual that they will 'in practice' be culled within a commercial breeding programme (Wise, 1973). A greater challenge is the eradication of lowly heritable skeletal disorders with a significant environmental influence. The inheritance of skeletal disorders is demonstrated by a selection experiment of Leenstra *et al.* (1984). From a base population they selected over three generations: (i) for high 6-week body weight (R); (ii) against twisted legs (AD); and (iii) against twisted legs followed by selection for a high 6-week body weight (K). This study also shows the effect of genotype–environment interactions, indicating that for selection against twisted legs a cage environment is more effective (Table 17.4).

It has also been demonstrated that crooked toes in poultry are an inherited condition. Selection over three generations using a leg disorder index reduced the incidence of crooked toes from 30% to 15% (Sørensen *et al.*, 1980), although Sørensen (1992) notes that in a less-affected line the selection response was less marked.

Datasets from three lines in which crooked toes, bow-legs and valgus deformities are analysed indicate genetic correlations with body weight of 0.22, 0.26 and 0.10 (Mercer and Hill, 1984). These correlations are sufficiently low to allow for combined selection for increased body weight and leg health. Indeed, recent studies of lameness in broilers (Thorp, 1997) and turkeys (Wilson, personal communication) within the UK indicate a low incidence of skeletal disorders and those commonly seen were of infectious or nutritional nature. This suggests that the selection programmes used by the primary breeding companies are effective against the bone deformities and other inherited disorders commonly seen (Reiland *et al.*, 1978; Leach and Gay, 1987).

DYSCHONDROPLASIA. Dyschondroplasia results in a thickening of the bone growth plate due to the failure of chondrocytes to differentiate. The condition is most commonly seen in the proximal tibiotarsus, though it is also seen in other long bone extremities. The Lixiscope is a hand-held X-ray-like device that can be used to detect tibial dyschondroplasia (TD) (Bartels *et al.*, 1989) at an accuracy of 92% (Thorp *et al.*, 1997). This phenotypic measure of the expression of TD has been used to establish divergent lines with a high and low incidence of TD (Wong-Valle *et al.*, 1993). An increased incidence and severity of tibial dyschondroplasia is seen when diets are marginal in calcium (Rennie *et al.*, 1993). In a selected high TD line there is effective prevention of

Table 17.4. The effect on frequency of twisted legs and body weight in broilers from three generations of selection in cages and on litter (from Leenstra *et al.*, 1984).

	Frequency of twisted legs (%)					
	In cages		On litter		Body weight at 6 weeks	
Line[a]	Male	Female	Male	Female	Male	Female
R	49	28	18	12	1969	1708
AD	12	7	5	2	1808	1555
K	27	12	8	5	1941	1677

[a]Lines selected over three generations for high 6-week body weight (R), against twisted legs (AD), or both (K).

tibial dyschondroplasia by treatment with a vitamin D_3 metabolite (Thorp *et al.*, 1993) and in another high TD selection line (Parkinson *et al.*, 1996) there is an increased incidence of TD associated with depression of a circulating vitamin D_3 metabolite. These studies, when combined, suggest that tibial dyschondroplasia seen in broilers may be an inherited form of rickets. It is unknown whether the aetiology corresponds to the effect of a single gene, corresponding to the major gene suggested by Sheridan *et al.* (1978).

RICKETS. Rickets is precipitated by a calcium, vitamin D_3 or phosphorus deficiency; susceptible strains have been selected (Austic *et al.*, 1977).

OSTEOPOROSIS. Osteoporosis of egg-layers contributes to bone fractures during the laying period and especially on depletion of flocks. Studies (Fleming *et al.*, 1997) indicate a high heritability for some skeletal characteristics at the end of the laying period and indicate that selection may be a useful tool for improving resistance to osteoporosis. Humeral and tibial bone strength and keel radiographic density were positively correlated with each other (Fleming *et al.*, 1997). A 'bone index' comprising these three bone traits was found to have a heritability of 0.48. Preliminary genetic correlations with egg production were neutral (Bishop, personal communication) and those with body weight were positive. These findings obviously give the option of a combined selection for high egg production and reduced severity of osteoporosis at the end of lay (for example by a restricted selection index designed to improve egg production and bone characteristics but not body weight). Moreover, increased bone density in the humerus at 25 weeks of age is associated with greater bone strength at the end of lay (Fleming *et al.*, 1998), which could shift the moment of osteoporosis selection to the beginning of the laying period. However, initial evidence indicates that there seems to be an unfavourable genetic correlation between shell strength and humeral and tibial strength measurements (Bishop, personal communication).

Infectious diseases

General/non-specific disease resistance

Non-specific disease resistance can be considered as the general ability of the bird to produce a rapid and effective T-cell response or antibodies to a wide range of pathogens. To investigate immune response, chickens have been divergently selected for high and low antibody response to sheep red blood cells 5 days after intramuscular immunization at 37 days of age (Van der Zijpp and Nieuwland, 1986). These divergent lines vary in their response to various T-cell dependent antigens (Parmentier *et al.*, 1994), vaccines (Parmentier *et al.*, 1996) and bovine serum albumin (BSA) (Parmentier *et al.*, 1994). Interestingly, the more effective responders have a lower body weight. This may support the assumption that modern, fast-growing commercial strains of chickens which are selected in the 'sterile' environment of biosecure breeding programmes are more susceptible to disease than the village chicken that is exposed to a wide range of pathogens. Some studies show that this is not true for some specific pathogens. For example, infectious bursal disease (IBD) challenge experiments have shown a greater susceptibility in local Nigerian chickens compared to broilers (Okoye and Aba-Adulugba, 1998), and IBD has also been shown to cause greater mortality in laying strain compared to meat-type lines (Nielsen *et al.*, 1998) with their higher growth rates. But, of course, the question is to what extent the environmental resource allocation has been changed by selection, rather than what the absolute growth rate is. Furthermore, the improved liveability of local strains may be due to tolerance of climatic extremes and should not be confused with inherent resistance to infectious disease.

Age is a key factor in non-specific resistance. For example, there is increased resistance to infectious laryngotracheitis virus (ILTV) in 4-week-old compared to 2-week-old birds (Poulsen *et al.*, 1998). This is attributed to lower numbers of circulating CD8 lymphocytes in the younger birds, although the immaturity of other components of the immune system is likely to contribute too. The onset of immune competence in chickens can be modified by selection. Pitcovski *et al.* (1987) showed that, in chickens divergently selected for a high or low early response to *E. coli* and Newcastle disease virus vaccine, the response was 68% higher in the high line. The heritability in the high line was approximately 0.7 and mortality after *E. coli* challenge was reduced in the high line to one-third of that in the low line.

The use of antibody responses, phagocytosis and T-cell responses may enable an index of total immunocompetence to be produced. This index could be used to improve general disease resistance. However, this is complicated by interpretation of the differing effects of challenge agents, mode of challenge and age at challenge, timing of response and interactions with maternal antibodies and nutrition. Gross *et al.* (1980) reported that selection for persistence of antibody response resulted in a line that was more resistant to all infectious agents tested than the line selected for non-persistence. Another option would be indirect selection for a certain desired resource allocation pattern, provided it can be defined and measured.

Specific disease resistance

Complete resistance to a pathogen is rare (an exception is lymphoid leucosis); resistance is usually specific for an individual pathogen and is inherited as a single gene (Gavora, 1990). The greatest area of study has been the influence of MHC genes on susceptibility and resistance to avian oncogenic viral infections. Located within the MHC region are the *B-F*, *B-L* and *B-G* genes, the first two of these have been associated with disease resistance, although confusingly some alleles may not cause improved resistance across a range of poultry strains whereas they do in other strains (Cole, 1985). The role of most recently identified, genetically independent, MHC-like regions such as *Rfp-Y* is unclear. In this case it has been shown that the frequency of one polymorphism was significantly different between divergently selected multitrait immunocompetence lines in one replicate only; therefore, the impact of multitrait immunocompetence selection on *Rfp-Y* polymorphisms is inconclusive (Pharr *et al.*, 1996). No association has been found between the PvuII-defined

Rfp-Y region polymorphisms with either innate or vaccine-induced MD resistance (Lakshmanan and Lamont, 1998). It is clear that over time our increasing understanding of the MHC regions and their alleles will enable more precise selection for specific disease resistance.

MAREK'S DISEASE. The association between genetic resistance to Marek's disease and the chicken MHC is widely known (Longenecker *et al.*, 1977). The B blood groups provide convenient genetic markers of the MHC complex with associated resistance to Marek's disease (Briles *et al.*, 1977). There are three classes of cell-membrane antigens clearly defined by serology, histogenetic, biochemical and molecular methods. Two of these are homologous to classes I and II of mammals, *B-F* and *B-L*, the third class, *B-G*, has not been detected in mammals (Plachy *et al.*, 1992). Genetic resistance to Marek's disease is associated with the *B-F* region of the MHC. Across 40 lines of chickens the strongest MHC association is the resistance of the chicken B21 haplotype to classical Marek's disease virus and the susceptibility of the B19 haplotype (Kaufman and Salomonsen, 1997).

Large differences in susceptibility are also seen between lines with a common MHC haplotype. Crosses suggest a small number of genes may have a large effect and the current development of genetic maps makes the search for genes responsible for non-MHC associated Marek's resistance possible (Bumstead, 1998). Methods of investigation include the identification of antigenic loci expressed on bursal cells, lymphocytes and thymocytes. Resistance could occur at different points during the potential progression of Marek's disease, such as the replication or spread of the virus, the likelihood of an infected cell becoming transformed, or differences in the immune control of infection or tumour development (Ruth, 1977). Traditionally, mortality or tumour development has been used to assess resistance. These compound measures of the infectious process are not ideal for mapping studies and have led to the development of quantitative PCR to assess viral levels throughout infection (Bumstead, 1998). This method indicates, in segregating birds of crosses between two lines with differences in resistance, a significant correlation between development of tumours and viral load in lymphocytes. Mapping studies (Bumstead, 1998; Vallejo *et al.*, 1998) of lines that share the same MHC haplotype indicate a number of chromosomal regions possibly containing resistance alleles. The use of representational difference analysis (RDA) identified eight clones, four of which lie in one region on chicken chromosome 1. This region shows a significant ($P < 0.0001$) level of association with resistance in the mapped birds and in another independent F_2 population. This locus is now designated *MDV1* and is mapped to a position in chickens that is equivalent to regions in mammals containing the lectin-like natural killer (NK) cell antigen complex (Bumstead, 1998).

It is likely that the use of biotechnological methods such as hybridization and DNA studies will be used more and more frequently to identify birds with resistance and susceptibility to Marek's disease rapidly and effectively and to ensure that undesirable genetic material is eradicated from a breeding programme.

LYMPHOID LEUCOSIS. Lymphoid leucosis is a consequence of exogenous leucosis virus infection of subgroup A, B, C, D or J. Early infection, either vertical or horizontal, may lead to tumour formation as a consequence of cell transformation. Resistance is at a number of levels. 'True' resistance to the viral infection is the first method of defence and is by specific cell receptors for the specific leucosis viruses. If the receptor is not present, infection of the cell cannot take place. Immune response is the second method of defence, and stress and concurrent immunosuppressive disease may cause increased susceptibility. Resistance to cell transformation and tumour formation is the third method of defence. The interaction of these defence mechanisms determines the pattern of disease and is genetically influenced. For example, infection of day-old Brown Leghorn chicks by injection or contact with J-strain leucosis virus (HPRS 103) results in antibody-positive, but virus-negative (Ab+, V–) chicks, which do not develop tumours or secrete the virus. In contrast, infection of day-old meat chicks results in most chicks becoming antibody positive and virus negative, but some chicks may become tolerant viraemics (Ab–, V+). These individuals may spread the infection horizontally and vertically and are more prone to tumour development (Payne, 1998).

The endogenous viruses belong to subgroup E. Chickens expressing subgroup E envelope proteins or complete virus have been shown to be less active in producing antibody to exogenous avian leucosis viruses and develop higher rates of lymphoid leucosis (Crittenden, 1991). Such genotypes are likely to be selected against in an environment with a high exogenous leucosis virus challenge, but not in an environment free of exogenous leucosis virus. This has been shown in White Leghorns, where the influence of selection for egg production traits in four strains resulted in significant changes of the frequency of endogenous viral genes related to avian leucosis (Kuhnlein *et al.*, 1989).

More recent studies have focused on the identification of potential markers of resistance and susceptibility to leucosis virus infections. Investigations of the cellular receptors for the viral envelope look promising and have been identified for subgroups A, B, D and E (Young, 1998). The Tva receptor appears specific for subgroup A and the Tvb receptors for viral subgroups B, D and E. RFLP analysis of F_2 progeny from a cross between susceptible ($tv\text{-}a^s/tv\text{-}a^s$) and resistant ($tv\text{-}a^r/tv\text{-}a^r$) genotypes indicates that the cloned gene maps to *tv-a* (Bates *et al.*, 1998). Different *tv-b* alleles cause three different patterns of susceptibility to the B, D and E subgroups, varying from permitting infection by all three subgroups to not allowing leucosis virus entry (Weiss, 1993). Novel selection strategies have been suggested for leucosis viruses. An example is the selection, *in vitro*, for genetic resistance to an acutely transforming virus with tropism similar to J-subgroup leucosis virus (Payne, 1998).

SALMONELLAE. Some early work demonstrated that selection for an early increase in body temperature of chicks reduced susceptibility to *Salmonella pullorum* (Hutt, 1935; Hutt and Crawford, 1960). Inbred lines have been shown to have differences in resistance to *Salmonella typhimurium* and crosses suggest an autosomally dominant resistance gene, with the absence of maternal effects or an influence of the MHC (Bumstead and Barrow, 1988).

Recent studies have demonstrated that the genes *Nramp1* and *TNC* are linked with resistance to salmonellosis (Hu *et al.*, 1997).

In selecting for *Salmonella* resistance the goal has to be clear. The object is to select for the ability of the bird to rid itself of *Salmonella* organisms rather than resistance to clinical infection, as it is the risk of zoonosis from carrier birds that is the greatest concern.

COCCIDIOSIS. There is large variation in the susceptibility of different lines to *Eimeria*. Challenge experiments first demonstrated strain and sex differences in the susceptibility to coccidia (Rosenberg, 1941). Recently, inbred lines of chickens have also been found to differ in resistance and susceptibility to infection with various *Eimeria* species (Bumstead and Millard, 1992). Comparison of the numbers of oocysts produced by the different lines indicated that there may be common genetic factors affecting susceptibility to six of the seven *Eimeria* species. Surprisingly, there appeared to be an inverse relationship between susceptibility to *E. tenella* and susceptibility to the other *Eimeria* species: lines that produced most oocysts of *E. tenella* produced least oocysts of the other *Eimeria* species and vice versa.

Some differences in resistance to coccidiosis have been associated in part with the MHC haplotype (Gavora and Spencer, 1978). But non-MHC genes may, perhaps, predominate (Lillehoj *et al.*, 1989). Studies using F_2 progeny from a cross between a resistant and a susceptible line to search for a genetic marker for *E. tenella* resistance and susceptibility failed to find any effect of MHC and did not generate any useful markers (Pinard-van der Laan *et al.*, 1998). These findings confirm our poor understanding of resistance mechanisms to parasitic infections and lend support to the view that complex interactions of MHC genes and non-MHC genes influence the host's response to *Eimeria* infection.

In addition to variation in the resistance or susceptibility of birds to 'natural' infection there is variation in the responses to vaccines (Lillehoj, 1991). There is little documented information available from poultry. In mice, vaccinating with a crude antigenic preparation of coccidia, while protecting phenotypically susceptible animals, failed to protect phenotypically resistant mice (Rose *et al.*, 1994). This sort of result encourages the primary breeder to examine the type of infections, natural or vaccinal, that are used to 'enhance' genetic resistance within a breeding programme. Also it emphasizes the importance of utilizing data from vaccinal responses and not relying solely on performance data.

Concluding Remarks

Primary poultry breeders are concerned with increased disease resistance in their breeding programmes, although commercial considerations do not necessarily give it a high priority. The addition of biotechnology methods to the conventional breeding methods is not going to change this commercial fact. Conventional breeding methods will remain the most important approach to improve disease resistance and will increasingly be combined with information coming primarily from immunology and gene mapping. Introduction of new genetically engineered improvements to resistance mechanisms will only be used if economically and psychologically justified. The results described above for common diseases show the poor state of our knowledge of the genes involved in disease resistance and the mechanisms by which they operate. The same can be said about the role of resource allocation with regard to the relationship between disease resistance and production traits.

References

Aggrey, S.E., Kuhnlein, U., Gavora, J.S. and Zadworny, D. (1998) Association of endogenous viral genes with quantitative traits in chickens selected for high egg production and susceptibility or resistance to Mareks disease. *British Poultry Science* 39, 39–41.

Albers, G.A.A. (1993) Breeding for disease resistance: fact and fiction. *Archiv für Geflügelkunde* 57, 56–58.

Austic, R., Baker, D. and Cole, R. (1977) Susceptibility of a dwarf strain of chicken to rickets. *Poultry Science* 56, 285–291.

Bartels, J.E., McDaniel, G.R. and Hoer, F.J. (1989) Radiographic diagnosis of tibial dyschondroplasia in broilers: a field selection technique. *Avian Diseases* 33, 254–257.

Bates, P., Rong, L., Varmus, H.E., Young, J.A.T. and Crittenden, L.B. (1998) Genetic mapping of the cloned subgroup A avian sarcoma and leukosis virus receptor gene to the TVA locus. *Journal of Virology* 72, 2505–2508.

Beilharz, R.G., Luxford, B.G. and Wilkinson, J.L. (1993) Quantitative genetics and evolution: is our understanding of genetics sufficient to explain evolution? *Journal of Animal Breeding and Genetics* 110, 161–170.

Briles, W.E., Stone, H.A. and Cole, R.K. (1977) Marek's disease: effects of B histocompatibility alloalleles in resistant and susceptible chicken lines. *Science* 195, 193–195.

Bumstead, N. (1998) Genomic mapping of resistance to Marek's disease. *Avian Pathology* 27, S78–S81.

Bumstead, N. and Barrow, P. (1988) Genetics of resistance to *Salmonella typhimurium* in newly hatched chicks. *British Poultry Science* 29, 521–529.

Bumstead, N. and Millard, B.J. (1992) Variation in susceptibility of inbred lines of chickens to seven species of *Eimeria. Parasitology* 104 (3), 407–413.

Chambers, J.R. (1986) Heritability of crippling and acute death syndrome in sire and dam strains of broiler chickens. *Poultry Science* 65 (Suppl.), 23.

Cole, R.K. (1985) Natural resistance to Marek's disease. In: Calnek, B.W. and Spencer, J.L. (eds) *International Symposium on Marek's Disease.* American Association of Avian Pathology, Pennsylvania.

Crittenden, L.B. (1991) Retroviral elements in the genome of the chicken: implications for poultry genetics and breeding. *Critical Reviews in Poultry Biology* 3, 73–109.

Crittenden, L.B. and Salter, D.W. (1992) A transgene, *alv6*, that expresses the envelope of subgroup A avian leukosis virus reduces the rate of congenital transmission of a field strain of avian leucosis virus. *Poultry Science* 71, 799–806.

Dunnington, E.A. (1990) Selection and homeostasis. *Proceedings of the 4th World Congress on Genetics Applied to Livestock Production, Edinburgh UK* XVI, 5–12.

Dunnington, E.A., Martin, A. and Siegel, P.B. (1987) Antibody responses to sheep erythrocytes in early and late feathering chicks in a broiler line. *Poultry Science* 66, 2060–2062.

Fleming, R., Bishop, S.C., McCormack, H.A., Flock, D.K. and Whitehead, C.C. (1997) Heritability of bone characteristics affecting osteoporosis in laying hens. *British Poultry Science* 38, S21.

Fleming, R., McCormack, H.A., McTeir, L. and Whitehead, C.C. (1998) Medullary bone and humeral breaking strength in laying hens. *Research in Veterinary Science* 64, 63–67.

Friars, G.W., Chambers, J.R., Kennedy, A. and Smith, A.D. (1972) Selection for resistance to Marek's Disease in conjunction with other economic traits in chickens. *Avian Diseases* 16, 2–10.

Gavora, J.S. (1990) New directions in poultry genetics. Disease genetics. In: Crawford, R.D. (ed.) *Poultry Breeding and Genetics.* Elsevier, Amsterdam, pp. 805–846.

Gavora, J.S. (1997) Genetic resistance in control of disease in poultry. *Zootecnica International* 20 (8), 48–53.

Gavora, J.S. and Spencer, J.L. (1978) Breeding for genetic resistance to disease: specific or general. *Worlds Poultry Science* 34, 137–148.

Gordon, C.D., Beard, C.W., Hopkins, S.R. and Siegel, H.S. (1971) Chick mortality as a criterion of selection towards resistance or susceptibility to Newcastle disease. *Poultry Science* 50, 783–789.

Gross, W.G., Siegel, P.B., Hall, R.W., Domermuth, C.H. and DuBoise, R.T. (1980) Production and persistence of antibodies in chickens to sheep erythrocytes. 2. Resistance to infectious diseases. *Poultry Science* 59, 205–210.

Hartmann, W. (1989) Evaluation of 'major genes' affecting disease resistance in poultry in respect of their potential for commercial breeding. *Progress in Clinical Biology Research* 307, 221–231.

Hu, J., Bumstead, N., Barrow, P.A., Sebastiani, G., Olien, L., Morgan, K. and Malo, D. (1997) Resistance to Salmonellosis in the chicken is linked to *NRAMP1* and *TNC. Genome Research* 7, 693–704.

Hutt, F.B. (1935) On the physiological basis of genetic resistance to *Salmonella pullorum* in the fowl. *American Nature* 69, 66–67.

Hutt, F.B. and Crawford, R.D. (1960) On breeding chickens resistant to pullorum disease without exposure there to. *Canadian Journal of Genetics and Cytology* 2, 357–370.

Julian, R.J. (1993) Ascites in poultry. *Avian Pathology* 22, 419–454.

Kaufman, J. and Salomonsen, J. (1997) The 'minimal essential MHC' revisited: both peptide-binding and cell surface expression level of MHC molecules are polymorphisms selected by pathogens in chickens. *Hereditas* 127, 67–73.

Kaufman, J. and Wallney, H.J. (1996) Chicken MHC molecules, disease resistance and the evolutionary origin of birds. *Current Topics in Microbiology and Immunology* 212 129–141.

Kuhnlein, U., Sabour, M., Gavora, J.S., Fairfull, R.W. and Bernon, D.E. (1989) Influence of selection for egg production and Marek's disease resistance on the incidence of endogenous viral genes in White Leghorns. *Poultry Science* 68, 1161–1167.

Kuhnlein, U., Ni, L., Weingend, S., Gavora, J.S., Fairfull, W. and Zadworny, D. (1997) DNA polymorphisms in the chicken growth hormone gene: response to selection for disease resistance and association with egg production. *Animal Genetics* 28, 116–123.

Lakshmanan N. and Lamont S.J. (1998) Rfp-Y region polymorphism and Marek's disease resistance in multitrait immunocompetence-selected chicken lines. *Poultry Science* 77, 538–541.

Lakshmanan N, Gavora, J.S. and Lamont S.J. (1997) Major histocompatibility complex class II DNA polymorphisms in chicken strains selected for Marek's disease resistance and egg production or for egg production alone. *Poultry Science* 76, 1517–1523.

Leach, R.M. and Gay, C.V. (1987) Role of epiphyseal cartilage in endochondral bone formation. *Journal of Nutrition* 117, 784–790.

Leenstra, F.R. (1993) Selection for disease resistance in poultry: general or specific? *Archiv für Geflügelkunde* 57, 69–73.

Leenstra, F.R., van Voorst, A. and Haye, U. (1984) Genetic aspects of twisted legs in broiler sire strain. *Annales Agriculturae Fenniae* 23, 261–270.

Lillehoj, H.S. (1991) Cell mediated immunity in parasitic and bacterial diseases. In: Sharma J.M. (ed.) *Avian Cellular Immunology*. CRC Press, Boca Raton, Florida.

Lillehoj, H.S., Ruff, M.D., Bacon, L.D., Lamont, S.J. and Jeffers, T.K. (1989) Genetic control of immunity to *Eimeria tenella*. Interaction of MHC genes and non-MHC linked genes influences levels of disease susceptibility in chickens. *Veterinary Immunology and Immunopathology* 20 (2), 135–148.

Longenecker, B.M., Pazderka, F., Gavora, J.S., Spencer, J.L., Stephens, E.A., Witter, R.L. and Ruth, R.F. (1977) Role of the major histocompatibility complex in resistance to Marek's disease: restriction of the growth of JMW-MD tumours in genetically resistant birds. *Poultry Science* 77, 538–541.

McKay, J.C. (1998) A breeder's approach to avian neoplasia. *Avian Pathology* 27, S74–S77.

Martin, A., Dunnington, E.A., Briles, W.E. and Siegel, P.B. (1989) Marek's disease and major histocompatibility complex haplotypes in chickens selected for high and low antibody response. *Animal Genetics* 20, 407–414.

Maxwell, M.H., Robertson, G.W. and Moseley, D. (1995) Serum troponin T values in 7-day-old hypoxia- and hyperoxia-treated, and 10-day-old ascitic and debilitated, commercial broiler chicks. *Avian Pathology* 24, 333–346.

Maxwell, M.H., Robertson, G.W., Baustra-Ortega, J. and Hocking, P.M. (1998) A preliminary estimate of the heritability of plasma troponin T in broiler chickens. *British Poultry Science* 39, 16–19.

Mercer, J.T. and Hill, W.G. (1984) Estimation of genetic parameters for skeletal defects in broiler chickens. *Heredity* 53, 193–204.

Nielsen, O.L., Sorensen, P., Hedemand, J.E., Laursen, S.B. and Jorgensen, P.H. (1998) Inflammatory response of different chicken lines and B haplotypes to infection with infectious bursal disease. *Avian Pathology* 27, 181–189.

OIE (1998) Genetic resistance to animal diseases. *Scientific and Technical Review of the International Office of Epizooties* 17, 12–26.

Okoye, J.O.A. and Aba-Adulugba, E.P. (1998) Comparative study of the resistance or susceptibility of local Nigerian and exotic chickens to infectious bursal disease. *Avian Pathology* 27, 168–177.

Parkinson, G., Thorp, B.H., Azuolas, J. and Vianio, A.J. (1996) Sequential studies of endochondral ossification and serum 1,25-dihydroxycholecalciferol in broiler chickens between day old and 21 days of age. *Research in Veterinary Science* 60, 173–178.

Parmentier, H.K., Siemonsma, R. and Nieuwland, M.G.B. (1994) Immune responses to bovine serum albumin in chicken lines divergently selected for antibody responses to sheep red blood cells. *Poultry Science* 733, 825–835.

Parmentier, H.K., Nieuwland, M.G.B., Rijke, E., De Vries, G. and Schrama, J.W. (1996) Divergent antibody responses to vaccines and divergent body weights of chicken lines selected for high and low humoral responses to sheep red blood cells. *Avian Diseases* 40, 634–644.

Payne, L.N. (1998) HPRS-103: a retrovirus strikes back. The emergence of subgroup J avian leucosis virus. *Avian Pathology* 27, S36–S45.

Pharr, G.T., Gwynn, A.V. and Bacon, L.D. (1996) Histocompatibility antigen(s) linked to Rfp-Y (Mhc-like) genes in the chicken. *Immunogenetics* 45, 52–58.

Pinard, M.L. (1992) Selection for immunoresponsiveness in chickens: effects of the major histocompatibility complex. PhD thesis, University of Wageningen, The Netherlands.

Pinard-van der Laan, M.H., Monvoisin, J.L., Pery, P., Hamet, N. and Thomas, M. (1997) Comparison of outbred lines of chickens for resistance to experimental infection with coccidiosis (*Eimeria tenella*). *Poultry Science* 77, 185–191.

Pitcovski, J., Heller, D.E., Cahaner, A. and Peleg, B.A. (1987) Selection for early responsiveness of chicks to *E. coli* and Newcastle Disease virus. *Poultry Science* 66, 1276–1282.

Plachy, J., Pink, J.R. and Hala, K. (1992) Biology of the chicken MHC (B complex). *Critical Reviews of Immunology* 12, 47–79.

Poulsen, D.J., Thureen, D.R. and Keeler, J.R. (1998) Comparison of the susceptibility and resistance in three lines of chickens experimentally infected with infectious laryngotracheitis virus. *Poultry Science* 77, 17–21.

Reiland, S., Ollson, S.E., Poulos, P.W. Jr and Elwinger, K. (1978) Normal and pathologic skeletal development in broiler and leghorn chickens. *Acta Radiologica* 358 (Suppl.), 277–298.

Rennie, J.S., Whitehead, C.C. and Thorp, B.H. (1993) The effect of dietary 1,25-dihydroxycholecalciferol in preventing tibial dyschondroplasia in broiler fed diets imbalanced in calcium and phosphorous. *Journal of Nutrition* 69, 809–816.

Riddell, C. (1973) Studies on spondylolisthesis (Kinky back) in broiler chickens. *Avian Pathology* 2, 295–304.

Rose, M.E., Hesketh, P. and Wakelin, D. (1994) Immunisation against experimental coccidiosis produces contrasting results in inbred mice of differing susceptibility to infection. *Infection and Immunity* 62, 733–737.

Rosenberg, M.M. (1941) A study of the inheritance of resistance to *E. tenella* in the domestic fowl. *Poultry Science* 20, 472.

Rothschild, M.F. (1991) Selection under challenging environments. In: Owen, J.B. and Axford, R.F.E. (eds) *Breeding for Disease Resistance in Farm Animals.* CAB International, Wallingford, UK, pp. 73–85.

Ruth, R.F. (1977) Role of the major histocompatibility complex in resistance to Marek's disease: restriction of the growth of JMW-MD tumours in genetically resistant birds. *Poultry Science* 77, 538–541.

Salter, D.W., Smith, E.J., Hughes, S.H., Wright, S.E. and Crittenden, L.B. (1987) Transgenic chickens: insertion of retroviral genes into the chicken germ line. *Virology* 157, 236–240.

Scheele, C.W. (1996) Ascites in chickens: oxygen consumption and requirements related to its occurrence. PhD Thesis from Landbouwwuninersiteit Wageningen, ID-DLO Box 65, 8200 AD Lelystad, The Netherlands.

Sheridan, A.K., Howlett, C.R. and Burton, R.W. (1978) The inheritance of tibial dyschondroplasia in broilers. *British Poultry Science* 19, 491–499.

Siller, W.G., Wright, P.A.L. and Martindale, L. (1978) Exercise induced deep pectoral myopathy in broiler fowls and turkeys. *Veterinary Science Communications* 2, 331–336.

Sørensen, P. (1992) Genetics of leg disorders. In: Whitehead, C.C. (ed.) *Bone Biology and Skeletal Disorders in Poultry.* Carfax Publishing Company, Abingdon.

Sørensen, P., Nielsen, V. and Kold, N. (1980) The breeding work with slaughter chickens in the breeding station Stryno. *National Institute of Animal Science, Copenhagen,*

Report 505, 1–115.

Thorp, B.H., Ducro, B., Whitehead, C.C., Farquharson, C. and Sorensen, P. (1993) Avian tibial dyschondroplasia: the interaction of genetic selection and dietary 1,25-dihydroxycholecalciferol. *Avian Pathology* 22, 311–324.

Thorp, B.H., Dick, L., Jeffreries, D., Houston, B. and Wilson, J. (1997) An assessment of the efficacy of the Lixiscope for the detection of tibial dyschondroplasia. *Avian Pathology* 26, 97–104.

Vallejo, R.L., Bacon, L.D., Liu, H.-C., Witter, R.L., Groenen, M.A.M., Hillel, J. and Cheng, H.H. (1988) Genetic mapping of quantitative trait loci affecting susceptibility to Marek's disease virus induced tumours in F2 intercross chickens. *Genetics* 148, 349–360.

Van der Zijpp, A.J. (1983) The effect of genetic origin, source of antigen and dose of antigen on the immune response of cockerels. *Poultry Science* 62, 205–211.

Van der Zijpp, A.J. and Nieuwland, M.G.B. (1986) Immunological characterisation of lines selected for high and low antibody production. *Proceedings of the 7th European Poultry Conference* 1, 211–215.

Weiss, R.A. (1993) Cellular receptors and viral glycoproteins involved in retrovirus entry. In: Levy, J.A. (ed.) *The Retroviruses*. Plenum Press, New York, pp. 1–108.

Wise, D.R. (1973) The incidence and aetiology of avian spondylolisthesis (Kinky back). *Research in Veterinary Science* 14, 1–10.

Witter, R.L. (1988) Very virulent Marek's disease viruses: importance and control. *Proceedings of the 18th Worlds Poultry Congress*, pp. 92–97.

Wong-Valle, J., McDaniel, G.R., Kuhlers, D.L. and Bartels, J. (1993) Correlated responses to selection for high or low incidence of tibial dyschondroplasia in broilers. *Poultry Science* 72, 1621–1639.

Young, J.A. (1998) Avian leucosis virus-receptor interactions. *Avian Pathology* 27, S21–S25.

18

Genetic Aspects of Health and Disease Resistance in Pigs

B.N. Wilkie and B.A. Mallard
Department of Pathobiology, University of Guelph, Guelph, Canada

Summary

Evidence for genetic control of susceptibility to diseases due to single-allele-controlled defects in structure and physiological functions in pigs is substantial and such diseases can be controlled effectively by detecting the unfavourable alleles. A more complex situation prevails in the pursuit of genetic resistance to infectious disease, which, like most production traits, is under polygenic control. The objective of directly enhancing resistance to specific diseases by observing and quantifying variation in resistant and susceptible phenotypes would seem to be impractical. This is because of the improbability of making objective evaluations on sufficient breeding individuals for all of the likely target diseases. Also, there exists a clear risk of inducing enhanced susceptibility to one set of diseases as a result of breeding for resistance to the opposite set, given the polarized pathogenesis and host resistance mechanisms broadly relating to extracellular and intracellular parasitism. It may also be unlikely that manipulable major genes will be identified for resistance to many, or any, of the economically important infectious diseases. Alternative approaches aimed at achieving enhanced general resistance to infectious disease by breeding for improved antibody and cell-mediated immune response appear to be feasible, given the heritability of selected traits and ability to detect and breed for high- and low-immune-response phenotypes. High immune response has several apparently advantageous correlates, including rate of gain, and is apparently associated with stress-coping phenotype and social rank. Although the major histocompatablity genes have been associated with immune response, it does not appear likely, given available information, that any particular haplotype might afford superior health in pigs.

© CAB *International* 2000. *Breeding for Disease Resistance in Farm Animals* (eds R.F.E. Axford, S.C. Bishop, F.W. Nicholas and J.B. Owen)

Introduction

Genetic approaches to enhanced health in livestock have two principal objectives and related rationales: specific and general disease resistance. The former hypothesizes major genes controlling key resistance and susceptibility functions for each possible disease, while the latter acknowledges the polygenic nature of resistance-related traits involving host response to infection. In each case, the difficulty of objective measurement of disease and health phenotypes in breeding animals is an impediment. Variation in health should be assessed to reflect resistance to disease. More realistically, signs of disease are observed to measure susceptibility. This is most feasible for diseases involving defective alleles controlling physiological functions or structure that have well-defined aetiology and pathogenesis reflected in clear signs, such as those associated with porcine stress syndrome (PSS). However, infectious diseases are more complex since they involve variation in pathogen, host, host response and environment, making quantitative assessment difficult. Opportunity to observe infectious disease in breeding animals is very limited since management practices are designed to reduce disease incidence, and occurrence of any given disease is unpredictable. Inducing infection in enough individuals to provide meaningful data is not feasible. To adequately observe susceptibility and resistance phenotypes, the average disease incidence should be at least 50%, although with leptospirosis occurring at 20–30% over a 4-year period it was possible to select for resistant pigs (Pryztulski and Porzeczkowska, 1980). The feasibility of meaningfully observing and quantifying signs of all diseases contributing to losses in livestock production would seem to be low. However, the Swedish national system of disease recording and procedures for genetic evaluation of disease and fertility traits in cattle may provide for meaningful progress to be made when disease prevalence is less than 50% and heritability is low (0.02–0.10) (Lindhé and Philipsson, 1998). In systems such as this, selection may be based upon progeny group means when the progeny group is between 50 and 150.

Alternative strategies for genetically enhancing disease resistance might involve indirect, multitrait selection for candidate phenotypes and genotypes reflecting variation in host resistance-mediating functions, such as innate and immune response (Gavora and Spencer, 1983; Mallard *et al.*, 1998; Wilkie *et al.*, 1998). By this or other approaches, general improvement in resistance to infectious diseases may be achieved in high-producing livestock to derive substrate populations in which to delete or add disease-specific susceptibility or resistance alleles. In pigs, PSS has been controlled using an allele-specific polymerase chain reaction to identify and remove carriers (Rempel *et al.*, 1993). Major genes potentially conferring direct resistance advantage (Vogeli *et al.*, 1996), or indirectly enhancing mediation of resistance (Muller *et al.*, 1992), could be introduced, possibly using transgenic and marker-assisted selection methods (Georges and Massey, 1991; Soller, 1994), to populations already improved for general disease resistance. It is likely that strategies will build upon existing methods of environmental and host management, including vaccination, to enhance resistance to specific infectious diseases.

Vaccination and other prophylactic and therapeutic measures may be more efficient in animals with enhanced inherent disease resistance and immune response (Fig. 18.1).

Genetic Control of Infectious Disease

Genetic effects on resistance and susceptibility to disease in animals have been recognized for some time (Hutt, 1958). In a study of children adopted before 1 year of age, the cause of premature death was correlated with cause of death in the natural or adoptive parents to estimate genetic and environmental influences (Sørensen *et al.*, 1988). The relative risks (RR) of adoptees and biological parents dying before 50 years of age from infectious disease (RR 5.813, $P \leq 0.001$) or vascular disease (RR 4.52, $P \leq 0.001$) were higher than for cancer (RR 1.19, $P \geq 0.05$), for which the adoptive parents and adoptees were at higher relative risk (RR 5.16, $P \leq 0.05$). Similar effects were observed in

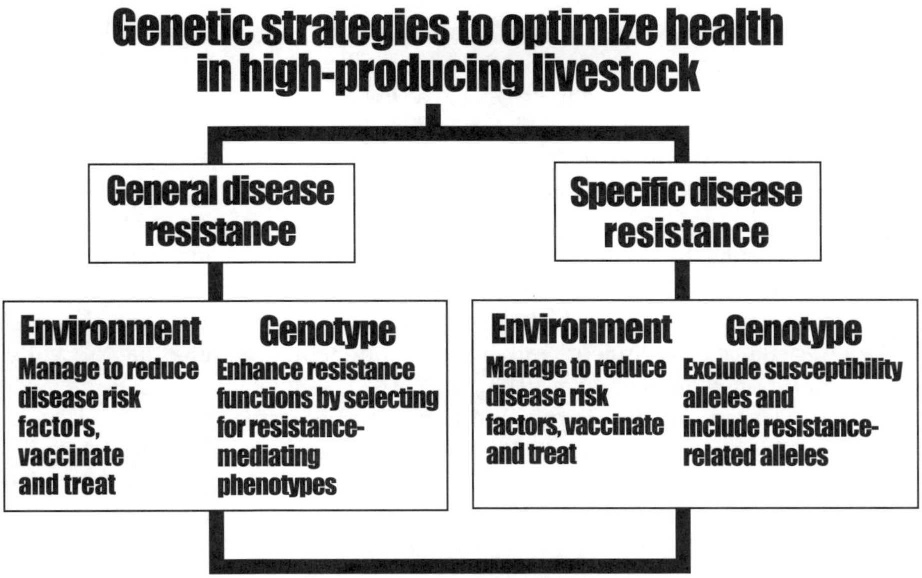

Fig. 18.1. Genetic strategies for enhancing health in livestock. Alternative strategies for optimizing health in livestock may involve methods for improving general disease resistance, most likely by selecting for phenotypes associated with high performance of traits mediating host innate and immunologically specific effector functions. Disease-specific approaches are most likely to be relevant to monogenic traits for which the susceptibility phenotype can be controlled by avoiding the related alleles. Both approaches require optimal management of the environment, and animals improved for general disease resistance are expected to be ideally suited for additional enhancement by deletion or addition of genes. In this scheme the polygenic traits associated with health are analogous to those controlling production, which may also be enhanced by use of major genes if and when these are identified.

relationship to death occurring prior to 70 years of age. These data indicate a strong genetic component in determining death from infectious disease, while death from cancer has a stronger influence from family environment.

The accuracy of traditional genetic selection schemes are dependent on heritability (h^2) of the trait in question as well as the source and quantity of phenotypic information (candidate individual, parents, siblings or progeny). The h^2 indicates the relationship between additive genetic variance and the phenotypic variance and, together with selection intensity and generation interval, dictate the genetic improvement that may be anticipated. Generally, h^2 for disease resistance is low, hence genetic progress based on selection theory would be expected to be slow.

Heritability of respiratory disease in Landrace and Yorkshire pigs was estimated, from sire components of variance and covariance analysis, at 0.14 within the Swedish pig progeny testing programme, in which incidence of pneumonia of all types diagnosed post-mortem was 22.7%. Landrace had a lower frequency of pneumonia than Yorkshire ($P \leq 0.001$) while the opposite was true for incidence of atrophic rhinitis ($P \leq 0.05$), for which h^2 was 0.16 (Lundheim, 1979). Genetic correlations between respiratory diseases and production traits were not significant but sick pigs grew more slowly, were leaner and had smaller area of the longissimus dorsi muscles. Heritability for enteric disease was estimated to be 0.59 in a similar study (Lundheim, 1988). Resistance to porcine brucellosis was shown to be heritable and controlled by relatively few genes (Cameron *et al.*, 1940, 1942). Mating of resistant × resistant phenotypes produced 128 progeny of which 76.6% were resistant, 22.6% were of uncertain status and 0.8% were susceptible after one generation of mass selection. Resistance to intracellular bacterial infection, including brucellosis, in mice, cattle and other species has been attributed to a gene, *Nramp1*, within the *Ity/Lsh/Bcg* locus (Qureshi *et al.*, 1996). However, the importance of this allele in resistance to virulent rather than attenuated strains has recently been questioned (Medina *et al.*, 1996).

Based upon microagglutinating serum antibody induced by natural exposure to *Leptospira icterohaemorrhagiae*, *L. grippotyphosa*, *L. serjoe*, *L. tarassovi* and *L. pomona* in Large White boars and sows and their progeny, h^2 for resistance to leptospirosis was estimated by half-sib correlation to be 0.20–021 (Pryztulski and Porzeczkowska, 1980). Pigs having the highest, intermediate and lowest titres of antibody were categorized as susceptible (highest) or resistant (lowest), and progeny of matings within the highest and lowest categories differed significantly ($P \leq 0.001$) in serum antibody. There was no significant difference in production traits between groups. Since actual infection with *Leptospira* was not quantified, the assumption that high or low serum antibody titres reflect susceptibility and resistance, respectively, may be questioned. However, given conditions of natural exposure, it is likely that the categories do indicate variation in host response to infection.

The relative paucity of information on genetic aspects of resistance and susceptibility to complex infectious disease likely reflects the difficulties associated with objectively observing and quantifying disease in individuals on a scale sufficient to permit evaluation of genetic and environmental effects. Health records from Norway suggested that risk of disease increased in sows that had previously been ill and in litters from sows that had previously had diseased litters (Lingass, 1991).

Susceptibility and Resistance Controlled by Single Alleles

Disease caused by defective alleles can be resolved in populations by testing for the phenotype of the allele itself and reducing the frequency of the unfavourable allele by culling or by avoiding use of carriers in breeding. For the complex traits of health or disease resistance, like production traits, inheritance is polygenic and controlled by multiple quantitative trait loci (QTL) which may be occupied by alleles favourable or unfavourable to the trait. The phenotype is due to the sum of favourable and unfavourable alleles (Soller, 1994). Identification of QTLs is facilitated by progress in genome mapping. Genes having a major effect on health-related traits could, in theory, be introgressed to enhance average resistance. However, QTL mapping requires comprehensive genome maps and a large number of within-family phenotyped individuals; identifying these for each of the many target infectious diseases is a challenge (Soller, 1994).

The PSS phenotype depends upon inheritance of an allele (*ryr1*) of the skeletal muscle ryanodine receptor, a calcium release channel, which has a mutation involving C to T substitution at base pair 1843 (Rempel *et al.*, 1993). The mutation predisposes to stress-induced malignant hyperthermia (MHS) due to abnormal release of Ca^{2+} from the sarcoplasmic reticulum of skeletal muscle, possibly as a result of altered control of Mg^{2+} in the myoplasm (Owen *et al.*, 1997). Pigs that are susceptible to PSS and MHS are lean and have relatively low fat content in muscle and adipose tissue. The lean Pietrain breed had the highest frequency of the *ryr1* allele (Archibald, 1991; Hartemann *et al.*, 1997). The *ryr1* mutation is associated with halothane-induced tremors and halothane challenge identifies PSS; however, detection of the *ryr1* mutation at the genetic level more accurately identifies homozygous or heterozygous pigs for implementation of control procedures (Rempel *et al.*, 1993) and has been used successfully in the control of PSS.

Deficiency of the complement regulatory protein, factor H, in Norwegian Yorkshire pigs is inherited as an autosomal recessive trait with complete penetrance and is associated with early piglet death due to membranoproliferative glomerulonephritis induced by deposition of complement component C3 (Jansen *et al.*, 1995). Plasma factor H measurements by enzyme-linked immunosorbent assay (ELISA) identify homozygous deficient and heterozygous healthy carriers which represented 13.5% of the population (Hogasen *et al.*, 1997). To eradicate the disease carriers were excluded from breeding.

Several other pig diseases have been described that involve defective alleles, or are suggestive of allelic control of physiological functions or susceptibility to neoplasia. Inherited vitamin D-deficiency rickets occurs in homozygotes that are deficient in renal 25-hydroxycholecalciferol-1-hydroxylase (Winkler *et al.*, 1986). The bleeding disorder, porcine von Willebrand disease, is associated with a restriction fragment length polymorphism in or near the locus controlling the von Willebrand factor (vWF) and likely represents a mutation in the vWF gene (Bahou *et al.*, 1988). Mutations associated with the apolipoprotein B gene are associated with hypercholesterolaemia and coronary artery disease due to atherosclerosis (Rapacz *et al.*, 1986; Maeda *et al.*, 1988). Heritable myasthenia and tremor of leg muscles in Pietran pigs ('campus syndrome') (Richter *et al.*, 1995) and hypertrophic cardiac myopathy occurring at a frequency of 5.26% in Durocs, 22.98% in Landrace and 5.56% in Yorkshires in Taiwan (Huang *et al.*, 1996) have also been described, but the genetic basis for these is not known. Cutaneous malignant melanoma of Sinclair miniature pigs is inherited and controlled by two loci, one within the major histocompatability gene complex (MHC), which produces a phenotype in which a mutant allele of the second non-MHC locus can initiate tumour development (Tissot *et al.*, 1987). An inherited form of lymphosarcoma has been described in Large White pigs (McTaggart *et al.*, 1982). Although the prevalence of these conditions is low in commercial pigs, they may be useful models of human diseases and illustrate the potential for unfavourable mutant alleles to arise in pigs and to cause economically important disease.

Intestinal brush border cell receptors for the three known variants (K88ac, K88ab and K88ad) of the K88 fimbrial adhesin of enteropathogenic *Escherichia coli* are expressed in susceptible pigs (Edfors-Lilja, 1991). In a sample of 24 pigs in each of the Chester White, Duroc, Hampshire and Yorkshire breeds, overall frequency of the receptor-negative and hence resistant phenotype, was 28%, although breeds differed in receptor phenotype (Baker *et al.*, 1997). Inheritance of the K88ac non-receptor phenotype was thought to be recessive, although this is debated (Edfors-Lilja, 1991). Similarly, binding of K99-positive enteropathogenic *E. coli* to pig intestine requires expression of a ganglioside receptor, the active form of which is most prevalent in piglets, the susceptible age group, rather than adults (Teneberg *et al.*, 1990). Receptors for the F18 fimbriated *E. coli* that cause post-weaning diarrhoea and oedema disease of pigs are the dominantly expressed *B* allele in susceptible pigs at the ECF18R locus while resistant animals carry the recessive *b* allele (Vogeli *et al.*, 1996). While it might be feasible to breed selectively for the resistant host phenotype in each of these cases, it is unlikely that the relevant bacterial phenotype will remain predominant if ability to adhere to intestinal brush border receptors is a requirement for successful colonization. Given the high mutation rate of bacteria, it would seem unwise to entertain genetic modification of pigs to counter any given bacterial genotype, since novel strains unaffected by the resistance phenotype are likely to arise and to be selected. Selection to eliminate expression of these receptors may also compromise other, as yet undescribed, physiological function(s) of the intestinal brush border cells.

Candidate Phenotypes for Indirect Selection for Disease Resistance

Difficulties associated with measuring disease in breeding populations as a basis for genetic selection for resistance are such that indirect selection for more readily quantified traits correlating with resistance to infectious disease is an attractive alternative. It is probable that variation in the principal resistance-mediating functions, innate and immunologically specific responses, correlate with variation in resistance to infection and infectious disease. If favourable phenotypes can be quantified and heritability is sufficiently high, it may be possible to enhance resistance by selective breeding (Gavora and Spencer, 1983; Mallard *et al.*, 1992, 1998; Wilkie *et al.*, 1998).

In effect, it may be more feasible to measure phenotypes that reflect response to infection rather than disease itself. Variation in phenotype of resistance-mediating functions is controlled by QTLs, the additive effect of which results in the observed phenotype. The greatest influence upon variation in QTLs may come from regulatory genes rather than from structural genes and, in immune system performance, the cytokine genes may have an important influence (Mitchison, 1997). The polarized antibody and cell-mediated immune (CMI) responses of inbred mouse strains, which are associated with resistance and susceptibility to extracellular and intracellular pathogens respectively, are controlled by cytokines produced in characteristic patterns, Th1 and Th2, which steer the immune response towards CMI or antibody production respectively (Reed and Scott, 1993; Wilkie *et al.*, 1998). Early in infection, interaction between pathogens and cells of the innate resistance-mediating system, such as natural killer (NK) cells, can induce cytokines that direct development of a response that is characteristic of the host–parasite interaction and leads to resistance or susceptibility as a function of the host genotype, at least in inbred mice (Reed and Scott, 1993).

Because of the dichotomous CMI/Th1 and antibody/Th2 responses appropriate to resistance to intracellular and extracellular pathogens, strategies involving selection for resistance-mediating traits must avoid bias to one or the other phenotype, which would likely result in enhanced resistance to one set of pathogens at the expense of increased susceptibility to the opposite set. In pigs and other animals, the response to infection, or artificial immunization, is due to integrated functions of the neuroendocrine and immune systems, such that effectively limiting infection reduces the 'sickness response' attributable to action of cytokines produced by the activated immune system cells (Husband, 1995). The cytokines (IL-1, IL-6 and TNF-α) involved in the acute-phase response immediately after infection or immunization induce fever, somnolence, lethargy, depression and metabolic changes that decrease muscle growth (Husband, 1995). Prolonged release of these cytokines, as in chronic or repeated infections, reduces fitness and growth. Through effects on the pituitary, they also increase adrenocorticotrophic hormone (ACTH) release and elevate adrenal corticosteroid production. Corticosteroids modulate immune response and inflammation, including potentially altering the bias of CMI and antigen responses towards antibody by promoting the

Th2 cytokine IL-4 (Daynes and Araneo, 1989). Stress-coping strategies vary in pigs and other animals such that active and passive responses are associated with lower and higher relative basal plasma corticosteroid concentrations and correspondingly greater ability to produce CMI and antibody responses, respectively (Hessing *et al.*, 1995). The active responders to stress occupy a higher social rank than the passive responders and are more resistant to Aujesky's disease induced by experimental pseudorabies virus infection (Hessing *et al.*, 1994). Social rank, stress-coping behaviour and ability to respond immunologically are stable characteristics of individual pigs which are measurable by 21 days of age and correlate positively with birth weight (Hessing *et al.*, 1994). These correlated traits are likely under genetic control and measurements of ability to respond immunologically describe a phenotype that reflects overall capacity to cope with environmental stimuli. Given the risks inherent in the extreme phenotypes for immune response, strategies to enhance resistance by altering immune response phenotype and related genotype should seek to improve both CMI and antibody producing ability (Mallard *et al.*, 1992). While it is not clear what the murine Th1 and Th2 cytokine pattern equivalents are in pigs, it appears likely that porcine cytokines do correlate with differentiated immune responses and hence reflect bias to CMI or antibody responses as well as resulting resistance and inflammation (Reddy, 1998).

Selection for Enhanced Immune Response

Provided suitable phenotypes can be identified, objectively measured and have sufficiently high heritability, selection for immune response may be a feasible indirect approach to enhanced disease resistance and productivity in pigs (Mallard *et al.*, 1992, 1998; Wilkie *et al.*, 1998). The objective of selection should be to increase the average individual and population ability to produce CMI and antibody responses rather than to enhance either response at the expense of the other. This strategy assumes that in pigs, polarized response phenotypes, as represented by the inbred mouse strains C57BL/6, C3H/HeN (Th1 strains) and BALB/c (Th2 strain) (Fig. 18.2), would be polarized in susceptibility and resistance to pathogens adapted to extracellular and intracellular survival within infected hosts. Hence, to optimize general resistance to infectious disease, selection of individuals capable of producing either response would be desirable (Fig. 18.2).

Biozzi and co-workers selected mice for high antibody or CMI responses with correlated variation in response to infection such that enhanced CMI favoured resistance to intracellular pathogens and vice versa (Biozzi *et al.*, 1984). Selected lines of mice also varied in longevity and in incidence of spontaneous tumours (Covelli *et al.*, 1989) and had high or low immune response to antigens which were not included in the selection. Variance analysis of genes controlling high antibody response suggested ten independently segregating loci with an MHC-linked gene and a gene associated with the *Igh* locus contributing 10% and 15% to the phenotype. Other loci were identified on

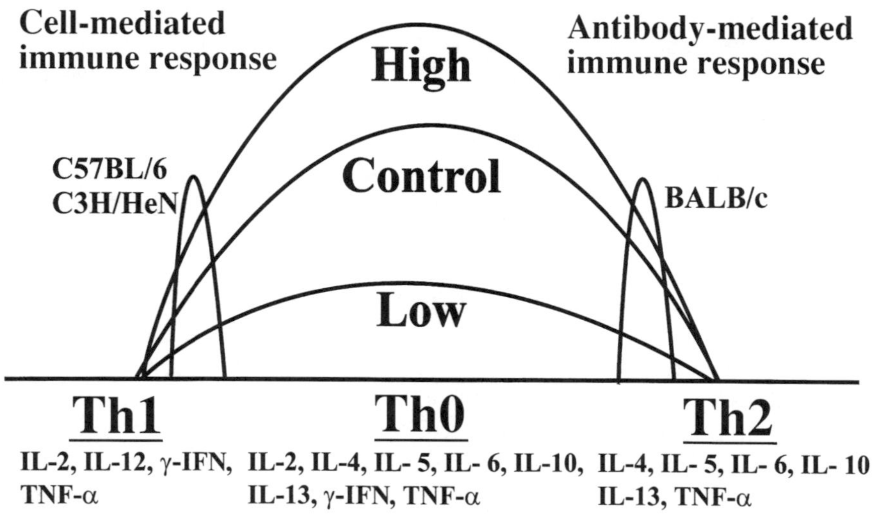

Fig. 18.2. Immune response phenotype. The polarity of cell-mediated immunity (CMI) and antibody response is correlated in mice with regulatory cytokine patterns classified as Th1 and Th2, which are derived under antigenic stimulus from the Th0 pattern. Inbred mouse strains (C57BL/6, C3H/HeN, BALB/c) are genetically predisposed to one or the other phenotype and hence vary in resistance and susceptibility to pathogens for which antibody or CMI response are critical. To derive pigs that have improved general resistance to infectious disease, selection based upon combined estimated breeding values for antibody and CMI have resulted in high and low lines that differ in several immune response and other traits. This approach assumes that each line can develop CMI or antibody response with approximately equal probability as appropriate to the stimulus, and that the quantity and quality of each response varies under the influence of quantitative trait loci which have assorted differently by line.

chromosomes 4, 6 and 8 using polymorphic microsatellites (Puel *et al.*, 1995), confirming the polygenic control of antibody response in mice.

Genetic variation has been described for putative immune response-related traits in Yorkshire pigs (Edfors-Lillja *et al.*, 1994). For neutrophil phagocytosis, concanavalin A (Con-A)-induced lymphocyte blastogenesis and IL-2 production, h^2 was moderate (0.3–0.4), while for serum immunoglobulin concentration and α-IFN production it was low (0.0–0.08). The Con-A, IL-2 and α-IFN production were correlated and it was concluded that if these traits indicated general immune competence in swine it should be possible to alter this characteristic by selection.

Yorkshire pigs were tested for antibody and CMI phenotype and bred to derive high (HIR) and low (LIR) responding lines, as well as a control (C) line (Mallard *et al.*, 1992, 1998; Wilkie *et al.*, 1998). To enhance both CMI and antibody production in individuals, selection was based upon summed estimated breeding values (EBV) for four traits reflecting specific and non-specific indicators of cellular and humoral resonses. The traits were: (i) cutaneous double-fold skin thickness response 24 h after injection of purified protein derivative (PPD) of tuberculin (delayed-type hypersensitivity, DTH) in pigs immunized 14 days previously with bacillus Calmette–Guérin; (ii) serum antibody to hen egg-white lysozyme (HEWL) at day 21 after primary and secondary immunizations on days 0 and 14; (iii) *in vitro* blastogenesis of blood mononuclear cells to Con-A; and (iv) serum concentration of IgG. Initial selection involved testing at 60 days of age, approximately 100 piglets from 34 litters by 15 sires and, at subsequent generations (G1–G8), randomly selecting for testing two females and one male per litter from approximately 20 litters per line. At each generation, five boars were used for breeding in each line.

Estimates of heritability varied at each generation within narrow limits and for G8 were: antibody, 0.27; DTH, 0.16; IgG, 0.07; and Con-A, 0.16. Lines diverged to G3, when the HIR and LIR were separated by 2 EBV standard deviations. Inbreeding increased progressively in each line, with average for all lines 0.013 in G2 and 0.196 in G8. Genetic correlations between DTH and antibody were low at G8 ($r = 0.09$) and there were little or no line differences in IgG, which was in negative correlation ($r = -0.6$) to antibody response. Unlike mice selected for antibody response, in which there was an inverse correlation with macrophage function (Biozzi *et al.*, 1984), uptake and killing of *Salmonella typhimurium* by blood mononuclear cells had extremely low heritability in these experiments (Mallard *et al.*, 1992) and there were no line-related differences in expression of swine leucocyte antigen (SLA) II or superoxide anion production by cultured blood monocytes (Groves *et al.*, 1993). Antibody response was superior in HIR to a variety of antigens, including the synthetic peptide (T,G)-A--L, sheep red blood cells (Mallard *et al.*, 1992), carbohydrate and lipopolysaccharide antigens of *Actinobacillus pleuropneumoniae* post-vaccination (Magnusson *et al.*, 1997), influenza virus post-vaccination (Wilkie and Mallard, 1998) and *Mycoplasma hyorrhinis* post-infection (Magnusson *et al.*, 1998). Frequency of non-responders to vaccination with inactivated influenza virus and *A. pleuropneumoniae* was significantly greater in LIR versus HIR pigs (Wilkie and Mallard, 1998). Mean avidity of antibody to HEWL was significantly higher in HIR versus LIR pigs (Appleyard *et al.*, 1992a). While HIR pigs developed less polyserositis after infection with *M. hyorrhinis*, arthritis was more severe (Magnusson *et al.*, 1998). In arthritic joint-associated cells, mRNA for IL-6 and α-IFN was greater than in LIR (Reddy, 1998). The HIR pigs also produced more plasma-binding proteins for the chemokines RANTES, NIP-1B and IL-8 as well as more immunoreactive haptoglobulin in response to infection with *A. pleuropneumoniae* (Banga, 1997). Natural killer cells were less frequent in LIR than in C or HIR pigs and response in frequency and lytic function for K562 target cells was significant for C and HIR but not LIR after exposure to modified live transmissible gastroenteritis virus (Raymond and Wilkie, 1998). Rate of gain (days to 100 kg) was significantly greater in HIR by comparison with C and LIR, such that HIR pigs reached market weight approximately 10 days before C animals regardless of the premises on which they were grown (Mallard *et al.*, 1998).

By simultaneously selecting for CMI and antibody-related traits, it has been possible to derive lines of pigs that have high or low immune response when compared with control, unselected pigs. Antibody response is increased to several tested antigens and the selection generally enhanced the immune response. There are advantages for HIR pigs in response to vaccination and rate of gain. Our unpublished results indicate significantly fewer mummified fetuses in HIR versus C and LIR after natural exposure to porcine parvovirus and, as noted above, HIR develop less polyserositis but more arthritis in response to *M. hyorrhinis* infection. It is apparent that there are line differences in cytokine gene expression as well as in production of inflammation-modulating proteins. Natural killer cells, which are important producers of immune response-steering cytokines in mice (Reed and Scott, 1993), differ in frequency and number by line. Taken together, these observations suggest advantages for HIR but they also indicate that for some stimuli HIR pigs may be more prone to generate inflammation than are the LIR animals. However, it should be possible to adjust the selection index to minimize negative effects and optimize breeding criteria. Enhanced ability of HIR pigs to produce antibody of high avidity and in high amount is likely to improve protection following infection with viruses and other pathogens since binding strength of antibody to antigen (avidity) may be a key quality of effective antibody (Salmi, 1991). The reason for advantages in growth rate are not known but this may reflect efficient response to clinical and subclinical infection with reduced duration of unfavourable 'endocrine-immune gradient' conditions which are associated with illness, including reduced muscle growth (Husband, 1995).

Response to Vaccination

Vaccination artificially induces immune response in order to enhance resistance to challenge with virulent strains of the homologous infectious organism. To be efficacious, the vaccine must be immunogenic in a large proportion of recipients. Immunogenicity is a function of antigenicity and host ability to respond to the antigenic stimulus. Efficacy of current vaccines therefore varies as a function of variable host immune response, which is in part genetically controlled. Significant differences were observed between breeds of pigs and between dams in response to modified live pseudorabies virus vaccine (Rothschild *et al.*, 1984a) and response to both the pseudorabies vaccine and a *Bordetella bronchiseptica* bacterin were significantly and positively correlated to rate of gain (Meeker *et al.*, 1987a). Heritability of response to pseudorabies vaccine was 0.18 ± 0.09 and for response to *B. bronchiseptica* bacterin 0.15 ± 0.07 and 0.52 ± 0.15 at 56 and 119 days post-vaccination, respectively (Meeker *et al.*, 1987b). These values are similar to those reported for antibody response to HEWL (Mallard *et al.*, 1998) as is the observation that immune responsiveness correlates positively with rate of growth.

In response to vaccination with *B. bronchiseptica* bacterin, Chester White pigs produced more antibody than did Yorkshire and Landrace, while Duroc and Hampshire had the lowest response (Rothschild *et al.*, 1984b). The Duroc breed was also found to be inferior to Yorkshires in antibody response to the test antigen HEWL (Wilkie and Mallard, 1998).

In the HIR and LIR lines of Yorkshire pigs, frequency of non-response to *A. pleuropneumoniae* carbohydrate antigen type 1 was 7 and 29%, while for the lipopolysaccharide antigen type 1 it was 11 and 27%, respectively (Magnusson *et al.*, 1997). For each antigen, the HIR pigs made significantly more antibody than did the LIR. Non-responders occurred only in the LIR group (38% non-responders) after vaccination with inactivated influenza virus, and the HIR were significantly better responders than C or LIR pigs (Wilkie and Mallard, 1998).

Opportunity exists to enhance vaccine efficacy in pigs not only by vaccine development, but also by altering the genetic ability of the target recipient populations to respond. Enhanced genetic resistance to disease need not therefore rely only on inherent ability to resist infection but it may also take advantage of improved response to vaccines. Poor immunogenicity and failure to induce antibody of high avidity are recognized problems associated with vaccines based on highly purified recombinant peptides (Mulchany *et al.*, 1992). These problems may be overcome in animals bred for enhanced immune response (Wilkie and Mallard, 1998).

Candidate Genes for Health and Immune Response

Since the association of murine major histocompatability genes with regulation of immune response to defined antigens in inbred mice (Benacerraf and McDevitt, 1972), these so-called 'immune response genes' have been the subject of numerous investigations to test the hypothesis that polymorphisms in the MHC loci are associated with a variety of disease resistance, innate immunity and immune response traits, as well as production traits. The pig MHC, or SLA, has been tested for influence on many traits with varying results.

Correlations between SLA haplotypes or alleles and immune response have been reported. In Large White pigs immunized with bovine immunoglobulin, dinitrophenyl, human serum albumin, sheep red blood cells, poly-L-lysine, hog cholera virus, polydextran or HEWL, only HEWL induced responses correlated with the SLA haplotypes studied, and the correlations were most obvious at low immunizing doses (Vaiman *et al.*, 1978a). Both SLA class I and II alleles differed in frequency by social rank in offspring of Yorkshire × Dutch Landrace pigs and the distribution also correlated with varying response to pseudorabies virus infection (Hessing *et al.*, 1994) and to both antibody and CMI (Hessing *et al.*, 1995).

Miniature pigs bred for fixed SLA haplotypes (Sachs *et al.*, 1976) have been used to estimate SLA effects on several traits. One of the miniature pig haplotypes, $SLA^{a/a}$ was associated with higher response by individuals of several breeds to *B. bronchiseptica* bacterin (Rothschild *et al.*, 1984b). *The* $SLA^{a/a}$ haplotype was also associated with significant reduction in muscle larvae of *Trichinella spiralis* after primary infection (Madden *et al.*, 1990). In contrast, the haplotypes $SLA^{d/d}$, $SLA^{d/g}$ and $SLA^{g/g}$ which share MHC class II genes, were better than $SLA^{a/a}$, $SLA^{a/c}$, $SLA^{a/d}$, $SLA^{c/c}$ and $SLA^{c/d}$ in producing antibody to HEWL, sheep red blood cells and (T,G-A--L) and in developing DTH to PPD (Mallard *et al.*, 1989a). In this study, frequency of non-responders to immunization was not influenced by SLA haplotype, and while SLA effects were related to differences in immune response much greater effects were associated with dam, sire and litter. Similarly, the $SLA^{d/d}$, $SLA^{d/g}$ and $SLA^{g/g}$ haplotypes had significantly more serum IgG than the others (Mallard *et al.*, 1989b) and $SLA^{d/d}$ pigs produced antibody of greater avidity (Appleyard *et al.*, 1992b). The SLA haplotypes did not differ significantly in lytic complement activity (CH_{50}), however, the $SLA^{d/d}$, $SLA^{d/g}$ and $SLA^{g/g}$ haplotypes did not increase CH_{50} after vaccination. Significant differences in serum CH_{50} were observed in comparisons between two other SLA haplotypes (Vaiman *et al.*, 1978b). In response to vaccination with an *aro*A mutant *Salmonella typhimurium*, litter had a significant effect on antibody response to the *O*-polysaccharide (*O*-ps) but SLA haplotypes varied significantly in lymphocyte blastogenesis induced by *O*-ps, in that $SLA^{d/d}$, $SLA^{d/g}$ and $SLA^{g/g}$ behaved as a response group having lower early and higher late responses than the other haplotypes (Lumsden *et al.*, 1993). Haplotype had a significant effect on *in vitro* uptake and killing of *Staphylococcus aureus* and *S. typhimurium*, again with $SLA^{d/g}$ and $SLA^{g/g}$ acting as a response group (Lacey *et al.*, 1989).

Classically, the MHC gene products behave as response/non-response regulators in inbred strains (Benacerraf and McDevitt, 1972), while associations reported for pigs do not suggest control of non-response but rather of the quantity and quality of immune response. This may reflect contribution of linked genes for complement (Lie *et al.*, 1987), tumour necrosis factor, the transporters of antigenic peptides (TAP1 and TAP2) (Vaske *et al.*, 1994 a, b) and the low molecular weight polypeptide complex (LMP) that proteolytically processes endogenous antigen for binding to MHC I. It has recently been reported that binding affinity of peptides to murine MHC class II gene products varies between strains and for a single gene product peptides vary in binding strength. High-affinity binding induced Th1 (γ-IFN) while low affinity steered towards the Th2 cytokine IL-4 (Kumar *et al.*, 1995). This suggests that the MHC gene products themselves, together with linked genes, may behave as QTLs with additive effects in determining quantity and quality of immune response and host resistance phenotype.

Conclusions

Genetic predisposition to disease due to monogenic traits has been described for several conditions in pigs and the use of specific phenotypic and or genotypic tests for detection and control of this type of disease is well established. Evidence for genetic control of host resistance-mediating traits, especially innate and specific response to infection, is accumulating and it appears likely that health and productivity may be improved indirectly by altering these polygenic traits. With increasing knowledge of the pig genome and of methods for exploitation of useful individual genes, it may become feasible to introgress useful health-related alleles into populations already improved by multitrait selection. At present there do not appear to be sufficiently compelling candidate genes for resistance to infectious disease to warrant attempts to use them.

References

Appleyard, G., Wilkie, B.N., Kennedy, B.W. and Mallard, B.A. (1992a) Antibody avidity in Yorkshire pigs of high and low immune response groups. *Veterinary Immunology and Immunopathology* 31, 229–240.

Appleyard, G., Mallard, B.A., Kennedy, B.W. and Wilkie, B.N. (1992b) Antibody avidity in swine lymphocyte antigen-defined miniature pigs. *Canadian Journal of Veterinary Research* 56, 303–307.

Archibald, A.L. (1991) Inherited halothane-induced malignant hyperthermia in pigs. In: Owen, J.B. and Axford, R.F.E. (eds) *Breeding for Disease Resistance in Farm Animals.* CAB International, Wallingford, pp. 449–466.

Bahou, W.F., Bowie, E.J., Fass, D.N. and Ginsburg, D. (1988) Molecular analysis of porcine von Willebrand disease: tight linkage to the von Willebrand factor locus. *Blood* 72, 308–313.

Baker, D.R., Billey, L.O. and Francis, D.H. (1997) Distribution of K88 *Escherichia coli*-adhesive and nonadhesive phenotypes among pigs of four breeds. *Veterinary Microbiology* 54, 123–132.

Banga, H.S.. (1997) Cytokine-binding and acute-phase proteins in pigs. PhD thesis, The University of Guelph, Guelph, Canada.

Benacerraf, B. and McDevitt, H. (1972) Histocompatability linked immune response genes. A new class of genes that controls the formation of specific immune response has been identified. *Science* 175, 273–279.

Biozzi, G., Mouton, D., Stiffel, C. and Bouthiller, Y. (1984) A major role of the macrophage in quantitative genetic regulation of immunoresponsiveness and antiinfectious immunity. *Advances in Immunology* 36, 189–234.

Cameron, H.S., Gregory, P.W. and Hughes, E.H. (1940) Studies on genetic resistance in swine to *Brucella* infection. II. A bacteriological examination of resistant stock *Cornell Veterinarian* 31, 21–24.

Cameron, H.S., Hughes, E.H. and Gregory, P.W. (1942) Genetic resistance to brucellosis in swine. *Journal of Animal Science* 1, 106–110.

Covelli, V., Mouton, D., di Majo, V., Bouthillier, Y., Bangrazi, C., Mevel, J.-C., Rebessi, S., Doria, G. and Biozzi, G. (1989) Inheritance of immune responsiveness, life span and disease resistance in interline crosses of mice selected for high or low multispecific antibody production. *Journal of Immunology* 142, 1224–1234.

Daynes, R.A. and Araneo, B.A. (1989) Contrasting effects of glucocorticoid on the capacity of T-cells to produce growth factors interleukin 2 and interleukin 4. *European Journal of Immunology* 19, 2319–2325.

Edfors-Lilja, I. (1991) *Escherichia coli* resistance in pigs. In: Owen, J.B. and Axford, R.F.E. (eds) *Breeding for Disease Resistance in Farm Animals.* CAB International, Wallingford, pp. 424–435.

Edfors-Lilja, I., Wattrang, E., Magnusson, U. and Fossum, C. (1994) Genetic variation in parameters reflecting immune competence in swine. *Veterinary Immunology and Immunopathology* 40, 1–16.

Gavora, J.S. and Spencer, J.L. (1983) Breeding for immune responsiveness and disease resistance. *Animal Blood Groups and Biochemical Genetics* 14, 159–180.

Georges, M. and Massey, J.M. (1991) Velogenetics or the synergistic use of marker assisted selection and germ-line manipulation. *Theriogenology* 35, 151–159.

Groves, T.C., Wilkie, B.N., Kennedy, B.W. and Mallard, B.A. (1993) The effect of selection of swine for high and low immune-responsiveness on monocyte superoxide anion production and class II MHC antigen expression. *Veterinary Immunology and*

Immunopathology 36, 347–358.

Hartmann, S., Otten, W., Kratzmair, M., Seewald, M.J., Iaizzo, P.A. and Eichinger, H.M. (1997) Influences of breed, sex, and susceptibility to malignant hyperthermia on lipid composition of skeletal muscle and adipose tissue in swine. *American Journal of Veterinary Research* 58, 738–743.

Hessing, M.J.C., Scheepens, C.J.M., Schouten, W.G.P., Tielen, M.J.M. and Wiepkema, P.R. (1994) Social rank and disease susceptibility in pigs. *Veterinary Immunology and Immunopathology* 43, 373–387.

Hessing, M.J.C., Coenen, G.J., Vaiman, M. and Renard, C. (1995) Individual differences in cell-mediated and humoral immunity in pigs. *Veterinary Immunology and Immunopathology* 45, 97–113.

Hogasen, K., Jansen, J.H. and Harboe, M. (1997) Eradication of porcine factor H deficiency in Norway. *Veterinary Record* 140, 392–395.

Huang, S.Y., Tsou, H.L., Chiu, Y.T., Shyu, J.J., Lin, J.H. and Liu, S.K. (1996) Heritability estimate of hypertrophic cardiomyopathy in pigs (*Sus scrofa domestica*). *Laboratory Animal Science* 46, 310–314.

Husband, A.J. (1995) The immune system and integrated homeostasis. *Immunology and Cell Biology* 73, 377–382.

Hutt, F.B. (1958) *Genetic Resistance to Disease in Domestic Animals.* Comstock, Ithaca, New York.

Jansen, J.H., Hogasen, K. and Grondahl, A.M. (1995) Porcine membranoproliferative glomerulonephritis type II: and autosomal recessive deficiency of factor H. *Veterinary Record* 137, 240–244.

Kumar, V., Bhardwaj, V., Soares, L., Alexander, J., Sette, A. and Sercarz, E. (1995) Major histocompatability complex binding affinity of an antigenic determinant is crucial for the differential secretion of interleukin 4/5 or interferon γ by T cells. *Proceedings of the National Academy of Sciences USA* 92, 9510–9514.

Lacey, C., Wilkie, B.N., Kennedy, B.W. and Mallard, B.A. (1989) Genetic and other effects on bacterial phagocytosis and killing by cultured peripheral blood monocytes of SLA-defined miniature pigs. *Animal Genetics* 20, 371–382.

Lie, W.R., Rothschild, M.F. and Warner, C.M. (1987) Mapping of C2, Bf, and C4 genes to the swine major histocompatability comples (swine leucocyte antigen). *Journal of Immunology* 139, 3388–3395.

Lindhé, B. and Philipsson, J. (1998) Conventional breeding programmes and genetic resistance to animal diseases. *Review Scientifique de Technique de l'Office Internationale des Epizootiologie* 17, 291–301.

Lingass, F. (1991) Epidemiological and genetical studies in Norwegian pig herds. IV. Breed effects, recurrence of disease, and relationship between disease and some performance traits. *Acta Veterinaria Scandinavica* 32, 107–114.

Lumsden, J.S., Kennedy, B.W., Mallard, B.A. and Wilkie, B.N. (1993) The influence of swine major histocompatability genes on antibody and cell-mediated immune response to immunization with an aromatic-dependent mutant of *Salmonella typhimurium*. *Canadian Journal of Veterinary Research* 57, 14–18.

Lundheim, N. (1979) Genetic analysis of respiratory diseases in pigs. *Acta Agricultura Scandinavica* 29, 209–215.

Lundheim, N. (1988) Health disorders and growth performance at a Swedish pig progeny testing station. *Acta Agricultura Scandinavica* 38, 77–88.

McTaggart, H.S., Imlah, P. and Head, K.W. (1982) Causes of death and sex differences in survival times of pigs with untreated hereditory lymphsarcoma (leukaemia). *Journal of the National Cancer Institute* 68, 239–248.

Madden, K.B., Murrell, K.D. and Lunney, J.K. (1990) *Trichinella spiralis*: major histocompatability complex-associated elimination of encysted muscle larvae in swine. *Experimental Parasitology* 70, 443–451.

Maeda, N., Ebert, D.L., Doers, T.M., Newman, M., Hasler-Rapacz, J. and Smithies, O. (1988) Molecular genetics of the apolipoprotein B gene in pigs in relation to atherosclerosis. *Gene* 70, 213–229.

Magnusson, U., Bossé, J., Mallard, B.A., Rosendal, S. and Wilkie, B.N. (1997) Antibody response to *Actinobacillus pleuropneumoniae* antigens after vaccination of pigs bred for high and low immune response. *Vaccine* 15, 997–1000.

Magnusson, U., Wilkie, B.N., Mallard, B.A., Rosendal, S. and Kennedy, B. (1998) *Mycoplasma hyorrhinis* infection of pigs selectively bred for high and low immune response. *Veterinary Immunology and Immunopathology* 61, 83–96.

Mallard, B.A., Wilkie, B.N. and Kennedy, B.W. (1989a) Genetic and other effects on antibody and cell mediated immune response in swine leucocyte antigen (SLA)-defined miniature pigs. *Animal Genetics* 20, 167–178.

Mallard, B.A., Wilkie, B.N. and Kennedy, B.W. (1989b) The influence of the swine major histocompatability genes (SLA) on variation in serum immunoglobulin (Ig) concentration. *Veterinary Immunology and Immunopathology* 21, 139–151.

Mallard, B.A., Wilkie, B.N., Kennedy, B.W. and Quinton, M. (1992) Use of estimated breeding values in a selection index to breed Yorkshire pigs for high and low immune and innate resistance factors. *Animal Biotechnology* 3, 257–280.

Mallard, B.A., Wilkie, B.N., Kennedy, B.W., Gibson, J. and Quinton, M. (1998) Immune responsiveness in swine: eight generations of selection for high and low immune response in Yorkshire pigs. *Proceedings of the 6th World Congress on Genetics Applied to Livestock Production*, Armidale, pp. 1–8.

Medina, E., Rogerson, B.J. and North, R.J. (1996) The *Nramp1* antimicrobial resistance gene segregates independently of resistance to virulent *Mycobacterium tuberculosis*. *Immunology* 88, 479–481.

Meeker, D.L., Rothschild, M.F., Cristian, L.L., Warner, C.M. and Hill, H.T. (1987a) Genetic control of immune response to pseudorabies and atrophic rhinitis vaccines: I. Heterosis, general combining ability and relationship to growth and backfat. *Journal of Animal Science* 64, 407–413.

Meeker, D.L., Rothschild, M.F., Cristian, L.L., Warner, C.M. and Hill, H.T. (1987b) Genetic control of immune response to pseudorabies and atrophic rhinitis vaccines: I. Comparison of additive direct and maternal genetic effects. *Journal of Animal Science* 64, 414–419.

Mitchison, A. (1997) Partitioning of genetic variation between regulatory and coding gene segments: the predominance of software variation in genes encoding introvert proteins. *Immunogenetics* 46, 46–52.

Mulchany, G., Reid, E., Dimarchi, R.D. and Noel, T.R. (1992) Maturation of functional antibody affinity in animals immunized with synthetic foot-and-mouth disease virus. *Research in Veterinary Science* 52, 133–140.

Muller, M., Brenig, B., Winnacker, E.L. and Brem, G. (1992) Transgenic pigs carrying cDNA copies encoding the murine Mx1 protein which confers resistance to influenza virus infection. *Gene* 121, 263–270.

Owen, V.J., Taske, N.L. and Lamb, G.D. (1997) Reduced Mg^{2+} inhibition of Ca^{2+} release in muscle fibers of pigs susceptible to malignant hyperthermia. *American Journal of Physiology* 272, 203–211.

Przytulski, T. and Porzeczkowska, D. (1980) Studies on genetic resistance to leptospirosis in pigs. *British Veterinary Journal* 136, 25–32.

Puel, A., Groot, P.C., Lathrop, M.G., Demant, P. and Mouton, D. (1995) Mapping genes controlling quantitative antibody production in Biozzi mice. *The Journal of*

Immunology 154, 5799–5805.

Qureshi, T., Templeton, J.W. and Adams, L.G. (1996) Intracellular survival of *Brucella abortus, Mycobacterium bovis* BGG, *Salmonella dublin*, and *Salmonella typhimurium* in macrophages from cattle genetically resistant to *Brucella abortus. Veterinary Immunology and Innunopathology* 50, 55–65.

Rapacz, J., Hasler-Rapacz, J., Taylor, K.M., Checovich, W.J. and Attie, A.D. (1986) Lipoprotein mutations in pigs are associated with elevated plasma cholesterol and atherosclerosis. *Science* 234, 1573–1577.

Raymond, C. and Wilkie, B.N. (1998) Natural killer cell frequency in pigs selectively bred for high and low antibody and cell-mediated immune response: response to vaccination with modified live transmissible gastroenteritis virus. *Natural Immunity* 374, 1–10.

Reddy, J. (1998) Cytokines in pigs bred selectively for high and low immune response. PhD thesis, The University of Guelph, Guelph, Canada.

Reed, S.G. and Scott, P. (1993) T-cell and cytokine responses in leishmaniasis. *Current Opinion in Immunology* 5, 524–531.

Rempel, W.E., Lu, M., el Kandelgy, S., Kennedy, C.F., Irvin, L.R., Mickelson, J.R. and Louis, C.F. (1993) Relative accuracy of the halothane challenge test and a molecular genetic test in detecting the gene for porcine stress syndrome. *Journal of Animal Science* 71, 1395–1399.

Richter, A., Wissel, J., Harlizius, B., Simon, D., Schelosky, L., Scholz, U., Poewe, W. and Loscher, W. (1995) The 'campus syndrome' in pigs: neurological, neurophysiological, and neuropharmacological characterization of a new genetic animal model of high-frequency tremor. *Experimental Neurology* 134, 205–213.

Rothschild, M.F., Hill, H.T., Christian, L.L. and Warner, C.M. (1984a) Genetic differences in serum-neutralizing titers of pigs after vaccination with pseudorabies modified live vaccine. *American Journal of Veterinary Research* 45, 1216–1218.

Rothschild, M.F., Chen, H.L., Christian, L.L., Lie, W.R., Venier, L., Cooper, M., Briggs, C. and Warner, C.M. (1984b) Breed and swine lymphocyte antigen haplotype differences in agglutination titers following vaccination with *B. bronchiseptica. Journal of Animal Science* 59, 643–649.

Sachs, D.H., Leight, G., Cone, J., Schwarz, S., Stuart, L. and Rosenberg, S.A. (1976) Transplantation in miniature swine. I. Fixation of the major histocompatibility complex. *Transplantation* 22, 559–567.

Salmi, A.A. (1991) Antibody affinity and protection in virus infections. *Current Opinion in Immunology* 3, 503–506.

Soller, M. (1994) Marker assisted selection – an overview. *Animal Biotechnology* 5, 193–207.

Sørensen, T.I.A., Nielsen, G.G., Andersen, P.K. and Teasdale, M.A. (1988) Genetic and environmental influences on premature death in adult adoptees. *The New England Journal of Medicine* 318, 727–732.

Teneberg, S., Willemsen, P., de Graaf, F.K. and Karlsson, K.A. (1990) Receptor-active glycolipids of the epithelial cells of the small intestine of young and adult pigs in relation to susceptibility to infection with *Escherichia coli* K99. *FEBS Letters* 263, 10–14.

Tissot, R.G., Beattie, C.W. and Amoss, M.S. Jr (1987) Inheritance of Sinclair swine cutaneous malignant melanoma. *Cancer Research* 47, 5542–5545.

Vaiman, M., Metzger, J.-J., Renard, C. and Vila, J.-P. (1978a) Immune response gene(s) controlling humoral anti-lysozyme (Ir-Lys) linked to the major histocompatability complex SL-A in the pig. *Immunogenetics* 7, 231–238.

Vaiman, M., Hauptmann, G. and Mayer, S. (1978b) Influence of the major histocompatiblity complex in the pig (SLA) on serum haemolytic complement levels. *Journal of Immunogenetics* 5, 59–65.

Vaske, D.A., Ruohonen-Lehto, M.K., Larson, R.G., Warner, C.M. and Rothschild, M.F. (1994a) Restriction fragment length polymorphisms at the porcine transporter associated with antigen processing 1 (TAP1) locus. *Journal of Animal Science* 72, 255.

Vaske, D.A., Liu, H.C., Larson, R.G., Warner, C.M. and Rothschild, M.F. (1994b) TaqI restriction fragment length polymorphisms at the porcine transporter associated with antigen processing 2 (TAP2) locus. *Journal of Animal Science* 72, 255.

Vogeli, P., Bertschinger, H.U., Stamm, M., Stricker, C., Hagger, C., Fries, R., Rapacz, J. and Stanzinger, G. (1996) Genes specifying receptors for F18 fimbriated *Escherichia coli*, causing oedema disease and postweaning diarrhoea in pigs, map to chromosome 6. *Animal Genetics* 27, 321–328.

Wilkie, B.N. and Mallard, B.A. (1999) Genetic effects on vaccination. *Advances in Veterinary Medicine* 41, 39–51.

Wilkie, B.N., Mallard, B.A., Quinton, M. and Gibson, J. (1998) Multi-trait selection for immune response: a possible alternative strategy for enhanced livestock health and productivity. In: Wiseman, J. *et al.* (eds) *Progress in Pig Science.* Nottingham University Press, Nottingham, pp. 29–38.

Winkler, I., Schreiner, F. and Harmeyer, J. (1986) Absence of renal 25-hydroxycholecalciferol-1-hydroxylase activity in a pig strain with vitamin D-dependent rickets. *Calcified Tissue International* 38, 87–94.

19 Lameness

O. Distl
Department of Animal Breeding and Genetics, School of Veterinary Medicine, Hannover, Germany

Summary

The chapter summarizes requirements for claw and leg quality as objectives in breeding programmes to improve longevity and lifetime performance. Several genetic investigations into claw traits have confirmed that implementation of selection on claw and leg quality should improve lifetime performance, longevity and claw soundness in dairy cows. Evidence for an unfavourable genetic correlation between milk yield traits and locomotion or foot problems comes from studies analysing feet and leg problems or diseases. The estimates also consistently indicate a genetic antagonism of moderate to medium size (r_g = 0.26–0.48). Measurement traits for claw and leg quality in cattle, based on phenotypic records, have to be defined carefully to maximize genetic correlations to the target traits, and several studies show that these claw and leg traits exhibit a sufficiently high additive genetic variation to achieve genetic improvement.

Introduction

Bovine lameness is a multifactorial health problem. The term 'lameness' describes a disturbance in locomotion and indicates damage to the musculoskeletal system. Diseases such as foot-and-mouth disease, and bovine virus diarrhoea/mucosal disease (BVD/MD), also show manifestations in the feet accompanied by lameness, but these diseases are not included in this consideration on lameness. The frequency of lameness can be measured as prevalence or incidence. Prevalence is the proportion of cases in an examined population at that moment. Incidence describes the number of new cases in a specified time period. The annual incidence of lameness is the number of observed new cases of lame animals in one year as a proportion of the number of healthy animals at the start of that year.

© CAB *International* 2000. *Breeding for Disease Resistance in Farm Animals*
(eds R.F.E. Axford, S.C. Bishop, F.W. Nicholas and J.B. Owen)

Reports on annual incidence of lameness in dairy cows ranged between 25% and 60%. Whitaker *et al.* (1983) found an annual incidence of 6.3% lame cows being treated by a veterinarian, whereas the incidence of cows requiring treatment and foot-care by farmers was three times higher (18.7%). Similiar figures were shown in databases from Israel, Scandinavia and Finland. Records on diseases of legs and feet kept by veterinary surgeons reveal annual or lactational incidences of about 5%, or less, in dairy cows.

Surveys on the prevalence of lameness or diseases of the digits have suggested a rather high incidence of the lameness problem in dairy herds (Ward, 1994). A Dutch study including 1141 Holstein Friesian cows found 21% lame cows (Reurink and Van Arendonk, 1987). In another Dutch study with 2121 Holstein and Dutch Friesian cows, 75% were affected by claw lesions, but only 1.2% of these cows were clinically lame (Smits *et al.*, 1992). The most common diagnoses were interdigital dermatitis (83.1%), pododermatitis aseptica/laminitis (75%) and digital dermatitis (17.6%). A French study comprising 160 herds and 4896 cows showed that 8.2% of the cows suffered from lameness and at least 25% of the cows were affected by heelhorn erosion, haemorrhages of the sole or symptoms of laminitis (Philipot *et al.*, 1990). Similar figures were found in first-lactation German Fleckvieh cows (Baumgartner *et al.*, 1990). In Denmark the incidence of claw diseases was studied at claw trimming. Sole ulcers were diagnosed in one or more than one foot in 20% and 29.7% of first-lactating cows, and in 23.5% and 24.7% of second- or higher lactating cows (Enevoldsen *et al.*, 1991a, b). Observed lameness in prevalence studies differs significantly from the number of lesions of digits, as not all lesions cause lameness.

The economic importance of lameness results from production losses. The estimated costs per cow due to lameness or digital diseases have been reported as £47 in the UK (Whitaker *et al.*, 1983), A$42.90 in Australia (Harris *et al.*, 1988), £81–392 also in the UK (Esslemont, 1990) and NLG230 in The Netherlands (D. Kooij *et al.*, personal communication 1995). Economic losses are related to decreased milk production and discarded milk, decreased fertility (prolonged calving interval, missing heat symptoms, increased number of inseminations), increased involuntary culling rate with possible lowered carcass value at slaughter, increased veterinary treatment and prophylactic costs and higher demand for labour (Fig. 19.1). As lameness is often painful, cows lie down longer and reduce their feed intake, followed by less milk production, weight and body condition loss. The negative effects on fertility due to lameness may be mediated through the negative energy balance and the reluctance to show heat symptoms.

Requirements for Claw and Leg Quality in Breeding Programmes

Genetic improvement of claw and leg conformation should enable animals to better resist influences of the environment that commonly cause disease or reduced performance. Claw and leg traits for breeding objectives have to be carefully defined and the traits chosen need to be tested for their usefulness in practical breeding and according to the environment to which the animals

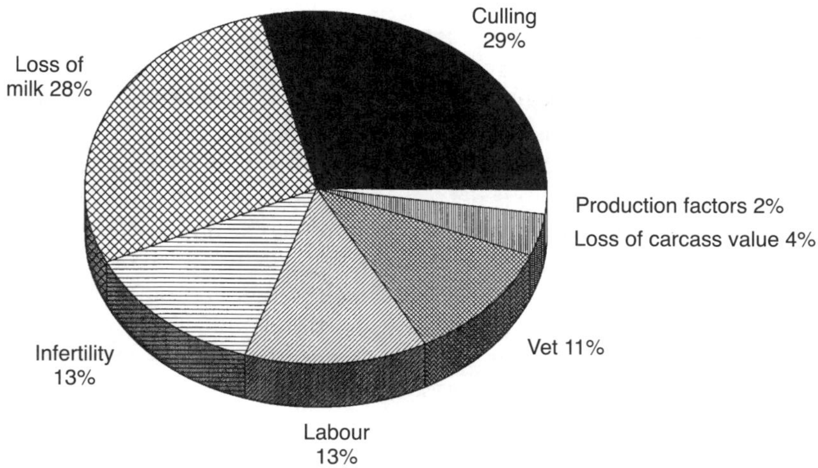

Fig. 19.1. Economic losses through diseases of feet in cattle.

are exposed. Possible negative side-effects on other characteristics of the musculoskeletal system should be avoided. Target traits to be used in breeding work may then be related to susceptibility to claw and leg diseases as well as to management requirements, such as, for example, low need for foot care and functional aspects of locomotion. High quality of the claw and leg improve longevity and lifetime performance. Minor disturbances of claw status, which may be subclinical and difficult to diagnose, will impair the animals' efficiency. Important parameters for claw and leg quality can only be identified when traits used in breeding are closely related to claw health, longevity, lifetime performance and functional efficiency of the animal. This definition implies that claw and leg quality cannot be recorded by just one trait. The traits necessarily seem to be complex and may be of differing importance, depending on the exposure to environmental effects. In particular, claw shape is a result of the interaction between individual factors and environment. Genetic components may respond differently to specific environments and in each specific environment other genetic components may play a predominant role.

Traits for Claw and Leg Quality in Cattle Based on Phenotypic Records

Selection procedures in animal breeding assume an additive genetic model including an infinite number of alleles, each with very small indistinguishable effects on the trait under consideration. The true genotypic value of an individual animal cannot be observed: however, procedures for prediction of breeding values are well developed to maximize genetic progress. The method of choice in this approach is to record phenotypic trait values and to use all additive genetic relationships among animals for ranking the animals by means of an animal model on a linear scale (BLUP, best linear unbiased prediction). Trait-based selection for claw and leg quality using records of males and females should be more effective than using only those traits in females.

Phenotypic traits to be used in genetic improvement of claw and leg characteristics should contribute to a decrease of the lameness problem in future generations. As the most common cause of lameness relates to the claws, more selection pressure should be put on claw traits. Genetic progress in claw and leg quality is determined by the correlated selection response of the traits chosen in the breeding programme. The approach to select on correlated traits aims to ensure that claw and leg quality are changed in such a direction as to be more efficient for a disease-free, highly productive and long-surviving cow. Therefore, effects genetically correlated with anatomical and physiological aspects of claw and leg function that have negative effects on production efficiency cannot have any advantage in selection.

Candidate traits for claw and leg quality have to meet certain requirements to be suitable for breeding purposes:

1. They must be objectively measurable or having a subjective score with high repeatability.
2. Feasible costs for recording: traits that have to be recorded in large numbers of animals require lower average recording costs per animal, whereas traits that need to be recorded only in a small number can justify higher average recording costs per animal.
3. Sufficient additive genetic variation of the target traits and the criteria used in practical breeding. The upper limit of genetic progress is given by the additive genetic variation of the target traits. The genetic and phenotypic correlations also influence the genetic progress. Claw and leg traits with low genetic correlations to the target traits are not useful in achieving genetic improvement in claw and leg quality. If measuring errors inflate the residual variation, heritability estimates are usually low and the genetic gain is substantially reduced.
4. Target traits for claw and leg quality have to be defined carefully. They should include the incidence of all relevant claw and leg diseases and of correlated diseases, as well as production traits that are related to claw and leg quality. Traits used for selection on claw and leg quality are often recorded in young animals, such as young bulls at an age of about 1 year and first-lactating cows, and therefore these traits should have predictive value for later life. In dual-purpose breeds with emphasis on meat production, the development of the size and burden of the claw in relation to body weight should be regarded when target traits for claw and leg quality are discussed.

The most common claw traits, summarized in Table 19.1, were discussed by the EAAP Working Group 'Claw Quality in Cattle' (Politiek *et al.*, 1986; Distl *et al.*, 1990) and Boelling and Pollot (1997). These traits were measures of claw shape, claw horn and inner structures of the claw. Several studies show that these claw and leg traits exhibit a sufficiently high additive genetic variation to achieve genetic improvement. Claw-shape measures can be recorded at moderate costs with high accuracy. Heritability estimates based on paternal half-sibs were mostly in the range from h^2 = 0.2 to h^2 = 0.4. The additive genetic variation of claw diseases is the limiting factor for the genetic progress

Table 19.1. Traits useful to record foot and leg quality in cattle.

Structural static traits
Subjective judgements or scores of feet and legs
Measures of claw-shape conformation
angle of dorsal border, length of dorsal border, heel length, diagonal length, heel height, area of ground surface
Claw horn properties
Physical traits
horn wear, resistance to abrasion, horn growth, water content, hardness, elasticity
Histological characteristics
number of tubules per mm^2, diameter of tubules and their cortex
Biochemical components
soluble proteins, types of cytokeratins, amino acid contents, minerals
Disease traits
Nationwide recording schemes for disease data
Recording at claw trimming one or two times a year
Specific recording for research projects
Functional traits
Subjective scoring of locomotion (scores from 1 to 9)
Pressure distribution under claws
Locomotion analysis by high-speed camera or ultrasonic measuring system

that could be achieved using claw and leg quality traits. Several studies have shown that additive genetic variance and heritabilities for claw and leg diseases in cows are of moderate size and mostly range between $h^2 = 0.15$ and $h^2 = 0.30$ (Petersen *et al.*, 1982; Junge, 1983; Nielsen and Smedegaard, 1984; Smit *et al.*, 1986; Reurink and Van Arendonk, 1987; Baumgartner *et al.*, 1990; Choi and McDaniel, 1993; Huang and Shanks, 1995). These studies provided evidence for polygenic inheritance of claw and leg diseases, although conclusions should be drawn carefully on genetic resistance against specific infectious agents. More detailed analyses are necessary to obtain further insight into the pathogenetics of infectious agents and their dynamics in housing systems. In sheep a possible involvement of genetic polymorphism of the class II region of the major histocompatibility complex was shown for footrot infection and antibody titres after vaccination against footrot (Litchfield *et al.*, 1993).

Heel erosion, sole ulcers, sole contusion and interdigital hyperplasia in cattle were found to be highly repeatable from one lactation to the next (Enevoldsen *et al.*, 1991a, b; Distl and Schmid, 1993). In a Swedish study comprising 169 cows from the first to the fifth lactation, the cow within-breed and environmental-group effect explained about 39% of the variation and was highest of all effects in the model (Ral *et al.*, 1993).

Locomotion scores proved to be heritable in two studies (Lawstuen *et al.*, 1987; Boelling and Pollot, 1998b). Heritability estimates were mostly around 0.10. Even if locomotion ability can be scored very easily and on a large scale, additive genetic variation is relatively small compared to other traits.

Claw measurements in young cows were significantly genetically and phenotypically correlated with the prevalence of claw diseases, longevity and lifetime performance of dairy cows. The most useful parameters were found to be angle of dorsal wall, length of dorsal border, heel depth, heel length and diagonal length (Nielsen and Smedegaard, 1984; Reurink and Van Arendonk, 1987; Rogers and McDaniel, 1989; Rogers *et al.*, 1989; Baumgartner and Distl, 1990; Choi and McDaniel, 1993). Baumgartner *et al.* (1990) developed a selection index for a progeny test based on claw-shape measurements in first-lactation cows. The traits to be genetically improved were all claw disorders in front and rear legs. They showed that the claw-shape measurements, length of the dorsal border, heel length and diagonal length gave a selection response of 63% relative to a direct selection against all claw diseases included in the index. Adding leg judgements to claw measures in the selection index increased expected genetic progress by about 20%.

A performance test of young bulls would provide the opportunity to record claw measurements in a standardized environment and seems to be economically more efficient than recording of about 40–50 daughters and their herdmates on farms (Fig. 19.2). The generation interval can be decreased and therefore the genetic improvement of claw and leg soundness can be more swiftly transmitted into the next cow population. However, the genetic correlations between claw-shape measures of young bulls and the frequency of claw diseases of their daughters were lower than the corresponding genetic correlations in daughters. Nevertheless, a substantial genetic gain can be achieved, amounting to 28–32% of the progress relative to the direct selection against diseases using the information of 40 daughters on farms (Table 19.2). Taking measurements and judgements of half-sibs and other relatives into account, the relative value of the selection pathway via the young bull increases to over

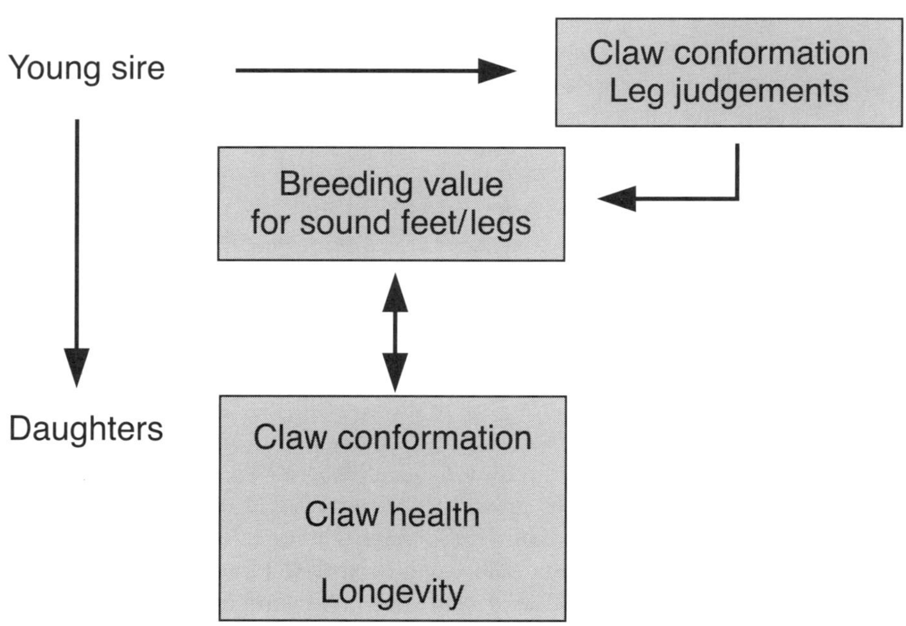

Fig. 19.2. Selection pathways for leg and claw quality in cattle.

Table 19.2. Relative genetic progress in soundness of feet and legs per generation using claw measures and leg judgements.

Information sources	Traits	Relative genetic progress (%)
Forty daughters on farms	All diseases of claws and legs	100 (reference value)
	All claw measures	67
	All claw measures and leg judgements	83
	Length of dorsal border, heel length, diagonal length	63
	Length of dorsal border	47
Young bull on test station	All claw measures	28
	All claw measures and leg judgements	32
Young bull and ten half-sibs on test station	All claw measures	39
	All claw measures and leg judgements	51
Young bull, ten half-sibs and mother (via further relatives)	All claw measures	51
	All claw measures and leg judgements	61

60%. Under a well-designed, half-sib structure of performance-tested young bulls, selection for claw and leg soundness in future daughter generations can be recommended. The accuracy of breeding values should be high enough to ensure a reliable selection decision and a sufficiently high variation of breeding values when measured on the same scale. The angle of the dorsal wall and judgements from the side view are the most important traits to be recorded in young bulls for selection decisions, as shown by the relative importance of single traits (Table 19.3).

McDaniel (1995) proposes a feet–leg selection index that includes claw diagonal length, leg score and a locomotion score recorded in female progeny groups on farms. Foot angle should be replaced by these better methods. Data from McDaniel's (1995) work show that daughters of bulls siring steeper foot

Table 19.3. Relative value of single selection traits.

	Relative value (%) as compared to all used traits	
Trait	Young bulls	Daughters
Angle of dorsal border	34	8
Area of ground surface	16	20
Length of dorsal border	14	47
Ratio of length of dorsal border to length of heel	13	–
Diagonal length	–	30
Hardness of claw horn	9	–

angle and straighter legs live longer than average. Until data on these traits are available, foot angle has to be used with some lesser emphasis on legs.

Chemical composition, determined by soluble proteins, and histological traits of claw horn characteristics exhibited medium to high additive genetic variance and heritabilities. Heritability estimates for mineral and amino acid contents in horn specimens from Swedish performance-tested bulls were low (Ral *et al.*, 1993). Microarchitecture of horn and contents, as well as types of soluble horn proteins, were not genetically correlated with claw shape measurements and wear rate of claw horn in paternal half-brothers (Distl *et al.*, 1982). These results seem contradictory to the opinion that a large number of horn tubules per mm^2 indicates good claw horn quality. Additional factors could be influencing wear of the claw horn, if both traits were compared at the same location of the horny capsule. Hardness and moisture content of claw horn displayed additive heritable components when half-sibs were kept in the same housing system (Distl *et al.*, 1982). In conclusion, measurements of microarchitecture, chemical and physical characteristics are difficult to standardize, rather expensive to record and, despite their rather high additive genetic component, most of them are lacking a clear genetic relationship to traits that express the breeding objectives.

Distribution of pressure underneath claws might be a valuable parameter for claw quality (Distl and Mair, 1993). Studies in a small sample of dairy cows indicate that differences exist between German Simmental and German Black and White cattle under the same housing system and management conditions. These breed differences in pressure distributions seem to be correlated with the incidence of heel erosion and sole contusions (Hubert, 1993; Hubert and Distl, 1994).

Selection Programmes for Claw and Leg Quality

Genetic investigations in claw traits showed that implementation of selection on claw and leg quality should improve lifetime performance, longevity and claw soundness in dairy cows (Fig. 19.3). Selection should be based on all available information of the individual's own performance, and that of progeny, half- and full-sibs and further relatives. Information should be combined in a BLUP-procedure. The claw and leg traits to be used in tested animals and in the different types of breeding programmes depend on costs and net returns of the claw and leg quality programme. As traits included in a selection index are combined on a monetary scale, the relative importance of claw and leg traits is given by their economic weight, put in the total merit index. Germany, Norway, Sweden and Denmark are examples of countries having included the trait 'legs and feet' in their total merit index. The relative weights compared to milk yield are 5:19 (0.26) in Norway, 3:29 (0.10, Swedish Red and White Cattle) and 3:26 (0.12, Swedish Friesian Cattle) in Sweden, 6:24 (0.25, Danish Red and White Cattle) and 7:22 (0.32, Danish Friesian Cattle) in Denmark. These indices were introduced in the Nordic countries at the end of the 1970s. The conclusions from this selection strategy were that the overall efficiency is

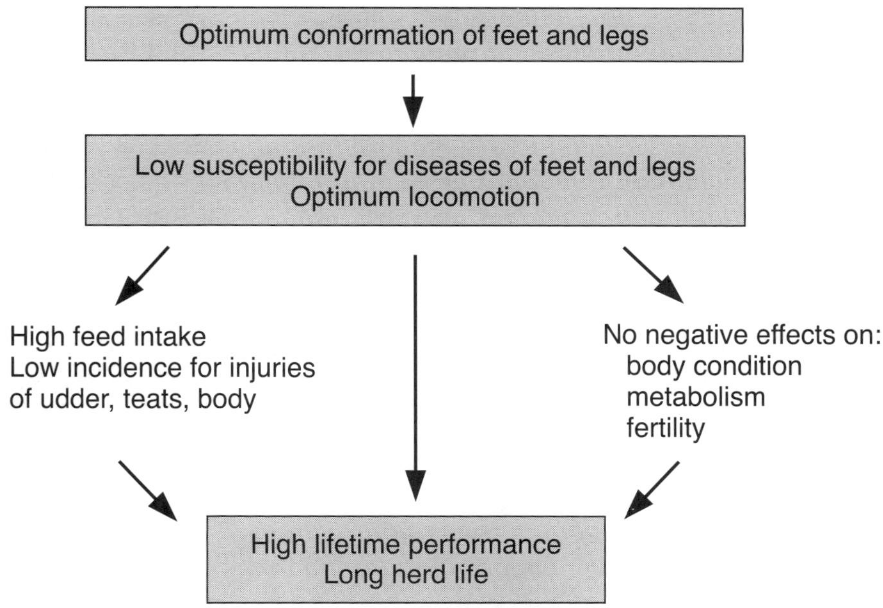

Fig. 19.3. Effects of optimum leg and claw quality on productivity in dairy cows.

superior to single-trait selection for milk yield because functional and constitutional traits are stabilized or improved (Andersen *et al.*, 1993).

Breeding programmes based on artificial insemination

Information on daughters in dairy farms

Subjective scores for claws and legs are widely used in progeny testing programmes as well as for selection of young bulls and bull dams. These visually recorded traits are based on a linear scoring system. Additive genetic variance and heritabilities mostly range between h^2 = 0.1 and h^2 = 0.3 (Distl, 1990; Brotherstone and Hill, 1991; Boldman *et al.*, 1992; Erf *et al.*, 1992; Klassen *et al.*, 1992; Misztal *et al.*, 1992; Smothers *et al.*, 1993; Mrode and Swanson, 1994). Traits mainly included are side view of the legs, foot angle, height of heel and spreading of claws. Data analysis from US Holsteins and Swiss Brown cattle showed that type traits are genetically correlated with productive life, and the reliability of breeding values for productive life could be increased by incorporation of breeding values for a number of type traits (Boldman *et al.*, 1992; Misztal *et al.*, 1992; Short and Lawlor, 1992; K.A. Weigel *et al.*, personal communication 1994; Vukasinovic, 1995). The estimated correlations indicated that continued selection for milk yield could result in undesirable effects on some conformational traits, but leg and feet traits would not be those most affected. Evidence for an unfavourable genetic correlation between milk yield traits and locomotion or foot problems comes from studies analysing feet and leg problems or diseases (Lyons *et al.*, 1991; Groen *et al.*,

1994; Uribe *et al.*, 1995). The estimates indicate consistently a genetic antagonism of moderate to medium size (r_g = 0.26–0.48). Permanent selection pressure on yields will lead to deterioration in the quality of feet and legs, making them more susceptible to disorders and exposing the cows to higher culling risk. Rogers (1993) emphasized that inclusion of non-yield traits, such as somatic-cell score, udder depth, teat placement and foot angle, with index weights 3 to 4 times less than milk yield would help to reduce negative correlations with selection for increased milk yield. Type classification records in Canadian Holsteins were phenotypically correlated with lifetime production around 0.07 for rear heel, bone and rear set, negative genetic correlations could be observed only for rear heel (–0.16 to –0.27) (Klassen *et al.*, 1992). Also, phenotypic and genetic associations have been established between herd life and feet and leg traits of the type classification system (Foster *et al.*, 1989; Boldman *et al.*, 1992; Short and Lawlor, 1992; Burke and Funk, 1993; Brotherstone, 1994; Vukasinovic, 1995).

As type classification takes place at the beginning of the first lactation, there should be no large combined effects between management and production records in later life that could lead to biased results. Genetic analyses of production records and type traits of legs and feet have to be done very carefully if selection for increased milk yield causes poorer claw and leg quality. In this case, cows that survive to a greater age express less desirable claw and leg traits.

Progeny testing of a sample of daughters in the field for measurable claw traits was recommended by the EAAP Working Group. Claw-shape measures should be taken on claws of one front and one rear leg, and traits selected were diagonal length, angle of dorsal wall, length of dorsal border and length of heel. Judgements of leg position proved valuable and should also be incorporated into the progeny test. Although costs at recording seem feasible, no breeding organization has decided to use claw measurements for progeny testing. Another interesting way forward could be to improve linear type classification systems. The Royal Dutch Cattle Syndicate (NRS) decided to introduce diagonal length as a claw trait in the Dutch conformation evaluation programme. Diagonal length is easy to judge visually if staff are uniformly trained. In general, claw and limb measures might be used in training programmes for type trait classifiers.

Information on young bulls and male progeny

Claw-shape measures as well as judgements of leg and feet can be obtained easily from sons that are progeny tested at stations for growth and carcass traits. As these programmes are widely used in dual-purpose and beef cattle breeds, recording of claw and leg traits should be possible in many countries. Heritability estimates of claw traits from male paternal half-sibs seem to be higher than those from field tests of female progeny groups (Distl *et al.*, 1982). It may be supposed that the error variance may be smaller under the more uniform environment of a test station. Genetic relationships between claw measures of sons and their paternal half-sisters may be expected to be at least as high as those estimated between performance-tested bulls and their female progeny. In Germany performance-test stations started in 1996–1997 to record claw measures in young bulls at the end of the test period for selection purposes. Claw measures included are angle of dorsal wall, length of dorsal border and diagonal, height of heel, hardness of claw horn at dorsal border and ground surface and, in some stations, also the area of ground surface. These measures are taken from the outer claws of one front and one rear leg.

Nucleus breeding programmes based on multiple ovulation and embryo transfer (MOET)

These breeding programmes were designed to make intensive use of individual performance, full- and half-sib information. The number of breeding animals is rather small (usually smaller than 100–400) and all animals are kept in a central test station. Nucleus breeding programmes provide good opportunities to include claw measurements and cinematographic or ultrasonic gait analyses. Also, the application of an electronic measurement system for recording pressure distributions underneath claws should be possible. These breeding programmes allow the possibility to develop and test new parameters for claw and leg quality traits. Parameters under test can be recorded at a test station with fairly similar conditions for all breeding animals and there are also many daughters and sons available from progeny tests. More effort in research should be put into these breeding programmes to estimate the necessary population parameters and to use the good conditions for further developments in claw and leg quality traits.

Marker-assisted selection (MAS)

The recent developments in mapping the bovine genome have made it possible to search for quantitative trait loci (QTL) (Womack, 1993; Bishop *et al.*, 1994). QTLs are quantitative traits marked by DNA polymorphisms, and the variance explained by these DNA markers can be handled in selection like single genes. QTLs are identified by exploiting the linkage between quantitative traits and the genetic markers. Heterozygous grandsires and their sons are genotyped and the quantitative traits are recorded in the daughters of these sons. Several projects are in progress to detect QTLs for milk production, type and health traits (Y. Da *et al.*, personal communication 1994; Georges *et al.*, 1995; Boichard, 1998; Gomez-Raya *et al.*, 1998; Reinsch *et al.*, 1998; Ron *et al.*, 1998). QTLs can only be found for feet and leg traits based on subjective records from many daughter groups. Studies in US Holsteins and French breeds (Holsteins, Normande, Montbéliard) have shown sire families segregating for QTLs in type traits and have confirmed a QTL for feet and leg judgements located on chromosome 9 in French breeds. This might be a very important area of further research if we wish to achieve genetic improvement of claw and leg quality by MAS. Another interesting aspect may be to detect candidate gene loci directly influencing claw and leg diseases. The study of laminitis suggests that the epidermal growth factor and its receptor could be tested as candidate genes. However, due to the multifactorial aetiology of many claw and leg diseases, the most relevant gene loci contributing to disease expression have to be identified before a molecular approach is feasible.

References

Andersen, B.B., Steine, T. and Pedersen, G.A. (1993) Breeding strategies for health and functional traits (with emphasis on cattle). *44th Annual Meeting of EAAP, Aarhus, Denmark, 16–19 August.*

Baumgartner, C. and Distl, O. (1990) Genetic and phenotypic relationships of claw disorders and claw measurements in first lactating German Simmental cows with stayability, milk production and fertility traits. *VIth International Symposium on Diseases of the Ruminant Digit, Liverpool, 16–20 July*, pp. 199–218.

Baumgartner, C., Distl, O. and Kräußlich, H. (1990) Eignung von Indikatormerkmalen für die Zucht auf Klauengesundheit beim Deutschen Fleckvieh. *Züchtungskunde* 62, 195–207, 208–221.

Bishop, M.D., Kappes, S.M., Keele, J.W., Stone, R.T., Sunden, S.L.F., Hawkins, G.A., Toldo, S.S., Fries, R., Grosz, M.D., Yoo, J. and Beattie, C.W. (1994) A genetic linkage map for cattle. *Genetics* 136, 619–639.

Boelling, D. and Pollot, G.E. (1997) The genetics of feet, legs and locomotion in cattle. *Animal Breeding Abstracts* 65, 1–11.

Boelling, D. and Pollot, G.E. (1998a) Locomotion, lameness, hoof and leg traits in cattle. I. Phenotypic influences and relationships. *Livestock Production Science* 54, 193–203.

Boelling, D. and Pollot, G.E. (1998b) Locomotion, lameness, hoof and leg traits in cattle. II. Genetic relationships and breeding values. *Livestock Production Science* 54, 205–215.

Boichard, D. (1998) QTL-detection with genetic markers in dairy cattle: an overview. *49th Annual Meeting of the EAAP, 24–27 August, Warsaw*, Paper CG3.1.

Boldman, K.G., Freeman, A.E., Harris, B.L. and Kuck, A.L. (1992) Prediction of sire transmitting abilities for herd life from transmitting abilities for linear type traits. *Journal of Dairy Science* 75, 552–563.

Brotherstone, S. (1994) Genetic and phenotypic correlations between linear type traits and production traits in Holstein–Friesian dairy cattle. *Animal Production* 59, 183–187.

Brotherstone, S. and Hill, W.G. (1991) Dairy herd life in relation to linear type traits and production. *Animal Production* 53, 279–287, 289–297.

Burke, B.P. and Funk, D.A. (1993) Relationship of linear type traits and herd life under different management systems. *Journal of Dairy Science* 76, 2773–2782.

Choi, Y.S. and Mc Daniel, B.T. (1993) Heritabilities of measures of hooves and their relation to other traits of Holsteins. *Journal of Dairy Science* 76, 1989–1993.

Distl, O. (1990) *Zucht auf Widerstandsfähigkeit gegen Krankheiten beim Rind.* Enke Verlag, Stuttgart.

Distl, O. and Mair, A. (1993) Computerized analysis of pedobarometric forces in cattle at the ground surface/floor interface. *Computers and Electronics in Agriculture* 8, 237–250.

Distl, O. and Schmid, D. (1993) Systematische Kontrolle der Klauengesundheit bei Kühen in ganzjähriger Laufstallhaltung. *Tierärztl. Prax.* 21, 27–35.

Distl, O., Graf, F. and Kräußlich, H. (1982) Genetische Variation von morphologischen, histologischen und elektrophoretischen Parametern bei Rinderklauen und deren phänotypischen und genetischen Beziehungen. *Züchtungskunde* 54, 106–123.

Distl, O., Koorn, D.S., McDaniel, B.T., Peterse, D., Politiek, R.D. and Reurink, A. (1990) Claw traits in cattle breeding programs: Report of the E.A.A.P. Working Group on 'Claw Quality in Cattle'. *Livestock Production Science* 25, 1–13.

Enevoldsen, C., Gröhn, Y.T. and Thysen I. (1991a) Sole ulcers in dairy cattle: associations with season, cow characteristics, disease, and production. *Journal of Dairy Science* 74, 1284–1298.

Enevoldsen, C., Gröhn, Y.T. and Thysen I. (1991b) Heel erosion and other interdigital disorders in dairy cows: associations with season, cow characteristics, disease, and production. *Journal of Dairy Science* 74, 1299–1309.

Erf, D.F., Hansen, L.B. and Lawstuen, D.A. (1992) Inheritance and relationships of workability traits and yield for Holsteins. *Journal of Dairy Science* 75, 1999–2007.

Esslemont, R. (1990) The costs of lameness in dairy herds. *VIth International Symposium on Diseases of the Ruminant Digit, Liverpool, 16–20 July*, pp. 237–251.

Foster, W.W., Freeman, A.E., Berger, P.J. and Kuck, A. (1989) Association of type traits scored linearly with production and herdlife of Holsteins. *Journal of Dairy Science* 72, 2651–2664.

Georges, M., Nielsen, D., Mackinnon, M., Mishra, A., Okimoto, R., Pasquino, A.T., Sargeant, L.S., Sorenson, A., Steele, M.R. and Zhao, X. (1995) Mapping quantitative trait loci controlling milk production in dairy cattle by exploiting progeny testing. *Genetics* 139, 907–920.

Gomez-Raya, L., Klungland, H., Vage, D.L., Olsaker, L., Fimland, E., Klemetsdal, G., Roenningen, K. and Lien, S. (1998) Mapping QTL for milk production traits in Norwegian cattle. *Proceedings of the 6th World Congress on Genetics Applied to Livestock Production, Armidale, Australia, 11–16 January*, 26, 429–432.

Groen, A.F., Hellinga, I. and Oldenbroek, J.K. (1994) Genetic correlations of clinical mastitis and feet and legs problems with milk yield and type traits in Dutch Black and White dairy cattle. *Netherlands Journal of Agricultural Science* 42, 371–378.

Harris, D.J., Hibburt, C.D., Anderson, G.A., Younis, P.J., Fitzpatrick, D.H., Dunn, A.C., Parsons, I.W. and McBeath, N.R. (1988) The incidence, cost and factors associated with foot lameness in dairy cattle in south-western Victoria. *Australian Veterinary Journal* 65, 171–176.

Huang, Y.C. and Shanks, R.D. (1995) Within herd estimates of heritabilities for six hoof characteristics and impact of dispersion of discrete severity scores on estimates. *Livestock Production Science* 44, 107–114.

Hubert, C. (1993) Einfluß von Rasse, Aufstallungssystem und Klauenpflege auf die Druckverteilung unter Rinderklauen sowie die Klauenhornhärte und -feuchte. Dissertation, University of Munich, Veterinary Faculty.

Hubert, C. and Distl, O. (1994) Untersuchung der Druckverteilung unter Rinderklauen bei den Rassen Deutsche Schwarzbunte und Deutsches Fleckvieh. *Züchtungskunde* 66, 327–338.

Junge, W. (1983) Klauenerkrankungen und ihre Ursachen bei Milchkühen. Dissertation, Christian-Albrecht University of Kiel, Agricultural Faculty.

Klassen, D.J., Monardes, H.G., Jairath, L., Cue, R.I. and Hayes, J.F. (1992) Genetic correlations between lifetime production and linearized type in Canadian Holsteins. *Journal of Dairy Science* 75, 2272–2282.

Lawstuen, D.A., Hansen, L.B. and Johnson, L.P. (1987) Genetic basis of secondary type traits for Holsteins. *Journal of Dairy Science* 70, 1633–1645.

Litchfield, A.M., Raadsma, H.W., Hulme, D.J., Brown, S.C., Nicholas, F.W. and Egerton,

J.R. (1993) Disease resistance in Merino sheep. *Journal of Animal Breeding and Genetics* 110, 321–334.

Lyons, D.T., Freeman, A.E. and Kuck, A.L. (1991) Genetics of Health traits in Holstein cattle. *Journal of Dairy Science* 74, 1092–1100.

McDaniel, B.T. (1995) Genetics and importance of feet and legs in dairy cattle. *46th Annual Meeting of the EAAP, Prague, Czech Republik, 3–7 September.*

Misztal, I., Lawlor, T.J., Short, T.H. and VanRaden, P.M. (1992) Multiple-trait estimation of variance components of yield and type traits using an animal model. *Journal of Dairy Science* 75, 544–551.

Mrode, R.A. and Swanson, G.J.T. (1994) Genetic and phenotypic relationships between conformation and production traits in Ayrshire Cattle. *Animal Production* 58, 335–338.

Nielsen, E. and Smedegaard, H.H. (1984) *Disease in Legs and Hooves with Black and White Dairy Cattle in Denmark.* Report No. 56, National Institute of Animal Science, Kopenhagen, Denmark.

Petersen, P.H., Nielsen, A.S., Buchwald, E. and Thysen, I. (1982) Genetic studies on hoof characters in dairy cows. *Journal of Animal Breeding and Genetics* 99, 286–291.

Philipot, J., Pluvinage, P., Cimarosti, I. and Luquet, F. (1990) On indicators of laminitis and heelhorn erosion in dairy cattle: a research based on the observation of digital lesions in the course of an ecopathological survey. *VIth International Symposium on Diseases of the Ruminant Digit, Liverpool, 16–20 July*, pp. 184–198.

Politiek, R.D., Distl, O., Fjeldaas, T., Heeres, J., McDaniel, B.T., Nielsen, E., Peterse, D.J., Reurink, A. and Strandberg, P. (1986) Importance of claw quality in cattle: review and recommendations to achieve genetic improvement. Report of the E.A.A.P. Working Group on 'Claw Quality in Cattle'. *Livestock Production Science* 15, 133–152.

Ral, G., Stalhammar, E.M. and Philipsson J. (1993) Studies on hoof disorders in Swedish dairy cattle breeds. *44th Annual Meeting of the EAAP, Aarhus, Denmark*, 16–19 August.

Reinsch, N., Xu, N., Thomsen, H., Looft, C., Kalm, E., Grupe, S., Kühn, C., Schwerin, M., Leyhe, B., Hiendleder, S., Erhard, G., Medjugorac, I., Russ, I., Förster, M., Brenig, B., Reents, R. and Averdunk, G. (1998) First results on somatic cell count loci from the ADR bovine mapping project. *Proceedings of the 6th World Congress on Genetics applied to Livestock Production, Armidale, Australia, 11–16 January*, 26, 426–428.

Reurink, A. and Van Arendonk, J.A.M. (1987) Relationships of claw disorders and claw measurements with efficiency of production in dairy cattle. *38th Annual Meeting of the EAAP, Lisboa, Portugal, 28 September–1 October.*

Rogers, G.W. (1993) Index selection using milk yield, somatic cell score, udder depth, teat placement, and foot angle. *Journal of Dairy Science* 76, 664–670.

Rogers, G.W. and McDaniel, B.T. (1989) The usefulness of selection for yield and functional type traits. *Journal of Dairy Science* 72, 187–193.

Rogers, G.W., McDaniel, B.T., Dentine, M.R. and Funk, D.A. (1989) Genetic correlations between survival and linear type traits measured in first lactation. *Journal of Dairy Science* 72, 523–527.

Ron, M., Heyen, D.W., Weller, I.J., Band, M., Feldmesser, E., Pasternak, H., Da, Y., Wiggans, G.R., Van Raden, P.M., Ezra, E. and Lewin, H.A. (1998) Detection and analysis of a locus affecting milk concentration in the US and Israeli dairy cattle populations. *Proceedings of the 6th World Congress on Genetics Applied to Livestock Production, Armidale, Australia, 11–16 January*, 26, 422–425.

Short, T.H. and Lawlor, T.J. (1992) Genetic parameters of conformation traits, milk yield, and herd life in Holsteins. *Journal of Dairy Science* 75, 1987–1998.

Smit, H., Verbeek, B., Peterse, D.J., Jansen, J., McDaniel, B.T. and Politiek, R.D. (1986) Genetic aspects of claw disorders, claw measurements and 'type' scores for feet in Friesian cattle. *Livestock Production Science* 15, 205–217.

Smits, M.C.J., Frankena, K., Metz, J.H.M. and Noordhuizen, J.P.T.M. (1992) Prevalence of digital disorders in zero-grazing dairy cows. *Livestock Production Science* 32, 231–244.

Smothers, C.D., Pearson, R.E., Hoeschele, I. and Funk, D.A. (1993) Herd final score and its relationship to genetic and environmental parameters of conformation traits of United States Holsteins. *Journal of Dairy Science* 76, 1671–1677.

Umbe, H.A., Kennedy, B.W., Martin, S.W. and Kelfon, D.F. (1995) Genetic parameters for common health disorders of Holstein cows. *Journal of Dairy Science* 78, 421–430.

Vukasinovic, N. (1995) Genetische Beziehungen zwischen Langlebigkeit, Milchleistung und Exterieur beim Schweizerischen Braunvieh. Dissertation, Technical University of Zurich.

Ward, W.R. (1994) Recent studies on the epidemiology of lameness. *VIIIth International Symposium on Disorders of the Ruminant Digit and International Conference on Bovine Lameness, Banff/Canada, 26–30 June*, pp. 197–203.

Whitaker, D.A., Kelly, J.M. and Smith, E.J. (1983) Incidence of lameness in dairy cows. *Veterinary Record* 113, 60–62.

Womack, J.E. (1993) The goals and status of the bovine gene map. *Journal of Dairy Science* 76, 1199–1203.

Index